MW01130295

Piping Systems Manual

About the Author

Brian Silowash, PE, CEM, LEED AP, is president of Innovative Design Engineering of America, LLC, a Pittsburgh-based engineering consulting firm specializing in facilities engineering, energy management, insurance consulting, and building inspections. He has worked as a maintenance foreman, a plant engineer, and an engineering consultant. Mr. Silowash has extensive experience in design, construction, and commissioning.

Piping Systems Manual

Brian Silowash

New York Chicago San Francisco
Lisbon London Madrid Mexico City
Milan New Delhi San Juan
Seoul Singapore Sydney Toronto

The McGraw-Hill Companies

Cataloging-in-Publication Data is on file with the Library of Congress.

McGraw-Hill books are available at special quantity discounts to use as premiums and sales promotions, or for use in corporate training programs. To contact a representative please e-mail us at bulksales@mcgraw-hill.com.

Piping Systems Manual

2 3 4 5 6 7 DOH 15 14 13 12 11

ISBN 978-0-07-159276-5
MHID 0-07-159276-8

Sponsoring Editor
Larry S. Hager

Acquisitions Coordinator
Alexis Richard

Editorial Supervisor
David E. Fogarty

Project Manager
Virginia Howe, Lone Wolf Enterprises, Ltd.

Copy Editor
Jacquie Wallace

Proofreader
Leona Woodson

Production Supervisor
Pamela A. Pelton

Composition
Lone Wolf Enterprises, Ltd.

Art Director, Cover
Jeff Weeks

In memory of my parents, George and Juliana Silowash

"Deep waters cannot quench love, nor floods sweep it away."
—Song of Songs 8:7

Contents

Chapter 6 Materials of Construction 83

Acknowledgments

The author wishes to thank R. Dodge Woodson of Lone Wolf Enterprises, Ltd., and Larry Hager of McGraw-Hill for providing the opportunity to prepare this work. Dodge's advice and knowledge of the publishing business were always welcome, and Larry was agreeable to the concept of this text. Without them, this work may never have been published. Thanks also to the entire team at McGraw-Hill who were always available to answer questions. Editorial Coordinator Alexis Richard was particularly helpful throughout the process. Copyeditor Jacquie Wallace's attention to detail was much appreciated. Production Manager Virginia Howe was always gracious, and I delighted in her ability to turn my raw manuscript into a book.

The support and encouragement of family and friends is not to be underestimated, especially in consideration of the need to execute engineering projects while simultaneously preparing a manuscript. Beyond that, tangible assistance was rendered by my brother George Silowash, who reviewed selected chapters for readability; my friend Mary McGrellis who prepared many of the illustrations; my nephew Ryan Silowash, who assisted with data entry; and my friend Dave Schwemmer, PE, who offered his expert insight into structural engineering.

The technical reviewers for this work included George Dorogy, PE, of Hatch; Chester Kos, PE of Hatch; Stephen N. Koslasky of Hatch; Norman Hunt, PE of Power Engineers, Inc.; and James S. McKinney, PE. I have had the special privilege of working closely with each of them, and aside from having enormous respect for their technical knowledge as mechanical engineers, I cherish their friendship.

Of the manufacturers and associations that offered the use of technical data, information, or artwork, those deserving special thanks are:

- American Concrete Pipe Association
- American Society of Mechanical Engineers
- American Welding Society
- Anvil International
- Association of Energy Engineers
- Bonney Forge Corporation
- Crane Energy Flow Solutions
- Dresser Piping Specialties
- Farris Engineering, division of Curtiss-Wright Flow Control Corporation
- Fiberglass Tank and Pipe Institute
- Fike Corporation
- Flexitallic
- Flowline Corporation
- Garlock
- Innovative Design Engineering of America, LLC
- ITT Goulds Pumps
- McGraw-Hill
- Pipeline Seal & Insulator, Inc.
- Ridgid Tool Company
- Victaulic Company
- Watts Regulator Company

Individuals who offered valuable technical expertise included:

- Ronald W. Haupt, P.E., Pressure Piping Engineering Associates, Inc., San Mateo, California
- David Diehl, COADE, Inc., Houston, Texas

My sincere apologies to anyone I may have overlooked.

Brian Silowash, PE, CEM, LEED AP
Innovative Design Engineering of America, LLC

CHAPTER 1

Introduction

I have for many years wanted to compile some thoughts about piping design. As a young engineer, I was often confronted with a problem that was new to me. Older engineers and superiors would often advise me to "check the Corinth job," or "see what we did five years ago on the XYZ project." I would dig through stacks of files and dozens of drawings, only to find that the problems were not the same, or what they had imagined as an existing solution existed only in their failing memories. Nothing was on paper that could be applied to the problem at hand. I suppose this sort of thing applies not just to piping design, but to every other aspect of engineering as well.

In any case, I would waste a lot of time looking for answers in the existing reference materials, only to discover that many texts were silent on the topic under investigation. I would then be forced to do a lot of research and draw my own conclusions.

An example of this was when I was responsible for the start-up of a hot oil calender system, circa 1984. The mill engineers and project managers were concerned over the cleanliness of the piping. My initial reaction was that someone should be watching what the contractors were doing as they fabricated and hung the pipe to ensure that the pipe remained clean. And although this seems to be a reasonable approach, it would not have assisted in this particular case. Nor is it common to bird-dog the fitters to ensure that hard hats, wrenches, 2 x 4's, etc. don't get left inside pipes.

Cleanliness of piping is not often addressed in the reference books. While there are standards for the cleanliness of hydraulic piping and piping found in the pharmaceutical and food and beverage industries, there was not a lot to choose from in the general arena of industrial service piping.

Many phone calls later, I was finally able to lay my hands on a copy of PFI Standard ES-5, *Cleaning of Fabricated Piping*. This was a three-page document published by the Pipe Fabrication Institute. At least now I had a starting point and was able to apply this standard to the system that was causing so much heartburn among my managers. Back in 1984, one had to rely on picking up a scent, persistence, and lots of phone calls and trips to the library. Now that we have the Internet, the playing field has been leveled, although a quick Internet search of "pipe cleanliness standards" proves that today the process is still no picnic.

There are many excellent reference materials available. Some of these are referenced in this manual, and no serious student of piping should be without the *Piping Handbook* by Nayyar, or earlier editions by Crocker and King.

This is not a scholarly manual. I have tried to organize it in a logical manner and make the information readable and easy to access. The reader will forgive me for stating certain opinions (which should be obvious in the text, and not to be confused with facts).

Further, this text is intended to be practical rather than comprehensive. I have tried to highlight the items a piping engineer will most likely encounter, rather than to attempt an encyclopedic volume. For example, while there is much wonderful information in ASME B31.1, I have touched only on the portions one might encounter in a "typical" piping job. Throughout the preparation of this manuscript, I was faced with trying to strike a balance between solving the tough problems we face every day, and overstating the obvious. A review of online discussion sites indicated to me that there really was no shortage of elementary questions out there, but in fairness to those who appear to be new to the profession, the more you delve into an issue, the less you seem to know[1]. And though I tried to remain practical, some subjects are irresistible, and so I couldn't resist footnoting that PTFE is the only known substance to which a gecko cannot stick.

The piping engineer for a project will encounter many issues outside of any strict definition of "piping." There will be process equipment such as tanks, heat exchangers, pumps, structures, and so on. Early in a project, the piping engineer is asked to determine the horsepower of the pumps, so that electrical equipment may be sized. This often occurs before complete process information is available. As the project continues, it is most often the piping engineer who becomes the focal point, the lightning rod, the bottle-neck. Operating and maintenance issues must always be considered, and are often left to the piping engineer to resolve. Broad knowledge of the other disciplines' needs, as well as the industry served, is often required. My task in writing this book was to concentrate on the piping side, though I have made some minor excursions into some of the areas described above. Perhaps if the publishers and the engineering community enjoy this book, they may permit me an opportunity to examine a broader scope at some later date.

Some Miscellaneous Thoughts on Piping

1. The trades should always be made aware that piping cleanliness is of the utmost importance. This certainly applies to the inside of the piping, valves, and fittings but also to sumps as well. Stressing this point will save a lot of time on startups.
2. Take advantage of "non-traditional" piping materials such as HDPE for underground applications. While these materials have been around for some time, "old-timers" may be reluctant to use them.
3. Determining the size of piping is usually a function of its velocity. Keep in mind that the installed cost of piping is primarily a function of labor costs and it really doesn't cost much more to increase one pipe size to reduce friction and also to allow for future capacity. On the other hand, one has to be aware of the application. Bigger is not always better, especially if you are dealing with slurries.
4. Be aware of the possibility of back flowing through Y-type strainers since these screens may be very flimsy and will collapse when the flow reverses through them.
5. Don't neglect startup considerations in the design of the piping system. Be sure that you have high point vents and low point drains, and have the spares and clearances to remove, clean, or replace strainer screens.
6. In some cases, you may have to consider the minimum and maximum flows through a line over its life. This is particularly important for slurries and gravity flow lines.
7. Nobody likes to pay for welders. This means that if you can minimize the number of welds, everyone (except the welders) will be happier.
8. Viton gaskets smell like cinnamon.

[1] Someone once defined an "expert" as "one who knows more and more about less and less, until he knows everything about nothing."

CHAPTER 2
Terminology

Actuator

A device mounted on a valve stem that is used to change the size of the valve aperture. Actuators may either be air-operated, motor-operated, or hydraulically-operated.

AHJ

Authority Having Jurisdiction (the code compliance officer).

Air Break

In a drainage system, an air break is a piping arrangement in which a drain discharges into another fixture or receptacle, without a direct connection, and at a point below the flood level rim and above the trap seal.

Air Gap

In a drainage system, an air gap is the unobstructed vertical distance that a liquid travels through the air between the outlet of a waste pipe and the flood level rim of the receptacle into which the waste pipe discharges.

In a water distribution system, the air gap is the unobstructed vertical distance that a liquid travels through the air between the lowest opening of any pipe (or faucet) supplying water to a tank, plumbing fixture, or other receptacle, and the flood level rim of that receptacle.

Angle of Repose

The angle with the horizontal at which a granular material remains stable.

ASHRAE

Association of Heating, Refrigeration, and Air Conditioning Engineers.
See www.ashrae.org.

ASME

American Society of Mechanical Engineers. An organization that has developed codes and standards for piping and many other items. The codes are in use throughout the world. See www.asme.org.

AWS

American Welding Society. An organization that has developed codes and standards for welding. See www.aws.org.

Ball Valve

A type of quarter-turn valve used for on/off and sometimes throttling applications. It consists of a ball mounted on the stem. The ball has a hole drilled through it, and is seated firmly against a seal inside the body of the valve. When the ball is rotated, the valve aperture is reduced because of the relative movement between the hole in the ball and the valve seat.

Baseplate

A flat plate machined to accept the mounting holes of a piece of equipment. It is anchored to the floor, and is usually, but not always, grouted. Grouting offers the best installation for preventing deflection and minimizing vibration.

BEP

1. Boiler External Piping. BEP is the piping that begins where the boiler proper ends, at the first circumferential weld, at the face of the first flange, or at the first threaded joint, and ends downstream of the stop valve. Refer to ASME B31.1, Paragraph 100.1.2
2. Best Efficiency Point. The Best Efficiency Point on a pump curve is that point at which the capacity and head intersect at the highest efficiency of the pump.

Bid Tab

Contraction of "Bid Tabulation", a table used to analyze the best choice of a set of competing proposals.

Blank

See "Blind."

Blind

A plate inserted between two flanges. There are three types: one with a hole in the plate, the same size as the inside pipe diameter; one with no hole, used to prevent the flow of fluid in the pipe; and a third which has both of the other types joined together by a short piece of steel, which pivots around a bolt hole in the flange (known as a "spectacle blind," due to its resemblance to a pair of eyeglasses). It may be swung into either position. Usually the open type first described has a handle through which a hole has been drilled. This enables the observer to determine whether the blind is open or closed. Open blinds are used to provide the spacing between the flanges for the occasions when a closed blind is to replace the open blind. These are most often used to provide safe access to a vessel. Sometimes also called a "blank."

Block-and-Bleed

A valve configuration in which the pressure upstream of a valve (the "block" valve) is able to be vented to atmosphere by another valve (the "bleed" valve). See also "double

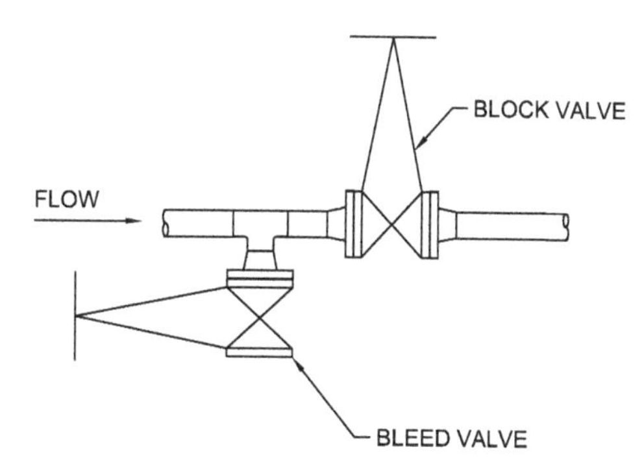

Figure 2.1 Block and bleed.

block and bleed." These configurations are most often used where safe access must be provided to a vessel, where blinds are not practical. See Figure 2.1.

BOP (Elevation)

Bottom-of-Pipe elevation.

Condensate

The liquid water that condenses out of a steam system as the steam loses its energy. It is usually recycled back to the boiler to provide feedwater, but it is sometimes wasted down the drain.

Contractor

The entity that installs the equipment or material, or that performs a specific service (e.g., pipe cleaning).

Cross Connection

An actual or potential connection between a potable water system and a potential source of contamination, which is not protected by an approved device designed to prevent flow from the potential contamination to the potable water system.

DBOO

Design/Build/Own/Operate. A plan in which an entity provides all four services, usually as a means of providing a service utility to a larger facility (e.g., a waste water treatment plant).

Dirt Leg

A vertical piece of pipe below a valve, with a horizontal tee that feeds a piece of equipment (often a furnace or water heater on a gas system, or a control valve or steam trap on a steam system). The bottom portion of the vertical leg is valved or capped so that when the upper valve is closed, dirt (particles, scale) may be emptied from the vertical leg. In natural gas service, the dirt leg may be used to trap condensation as well as dirt particles.

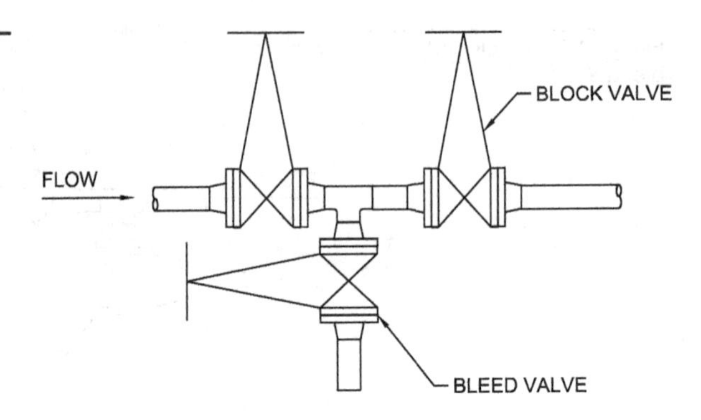

Figure 2.2 Double block and bleed.

Double Block and Bleed

A valve configuration in which the pressure upstream of a valve (the "block" valve) may be vented to atmosphere by another valve (the "bleed" valve) which is located upstream of yet another block valve. See Figure 2.2.

DWV

Drain, waste, vent piping, also known as soil pipe.

Elevation (drawing)

A drawing which shows a side view.

Engineer

The entity that provides the design services for a project.

Escutcheon

A circular cover plate with a hole in its center through which a pipe passes. It is used to cover the rough opening in a surface through which a pipe passes, to provide a finished appearance.

Expansion Joint

A device used to account for thermal expansion of pipes. It usually consists of a bellows, which compresses when the adjoining pipes expand.

Expansion Loop

A U-shaped offset designed in a piping system to provide flexibility for the thermal expansion of pipes.

Extra

An additional charge for work that is outside of the original scope of work.

Extrados

The outside radius of a bend.

Fabrication

A piece that is made in the shop or in the field, which may require welding, machining, and/or assembly.

Fire Loop

An underground water line outside the perimeter of a building. Part of the fire protection system, it provides water to the hydrants and may also supply the interior fire sprinkler system.

Fitting

A manufactured item that is used in a piping system to conveniently change the size or direction of the pipe. Examples are elbows, reducers, plugs, caps, and bushings.

Flood Level Rim

The level of the edge of a receptacle from which water overflows.

Flow Diagram

A drawing that shows the general flow characteristics of a piping system, including flowrate, temperature, and other parameters. Used to develop the P&IDs.

Fluid Dynamics

The study of how fluids behave in motion.

Gate Valve

A type of valve used for on/off applications. It consists of a plate that enters the fluid stream, seating against the valve body interior to provide a tight seal.

Globe Valve

A type of valve used for throttling and on/off applications. It consists of a disc mounted on the end of the valve stem. The disc rises and falls against a seat, effecting a seal or a variable opening which can be used to throttle the flow.

Greenfield

An undeveloped site for new construction.

HAZOP

HAZard and Operability analysis. A systematic means of identifying potential hazards so that they may be eliminated or mitigated.

Head

Pressure, converted into feet of water. Used for determining pressure requirements of pumps.

Header

A pipe that contains branch connections. Also known as the "run" pipe.

HVAC

Heating Ventilation and Air Conditioning.

Hydraulics

The study of how liquids behave at rest or in motion.

Hydronic

Piping that relates to hot and cold water in an HVAC system, and is used exclusively for heat transfer.

ID

Inside Diameter

Indirect Connection

A waste pipe that does not attach to the receptacle that it discharges into.

Intrados

The inside radius of a bend.

Invert Elevation

The elevation of the bottom of a formed (concrete or otherwise) trench or inside of a gravity drain pipe at the 6 o'clock position.

Lateral

A type of tee fitting in which the angle formed between the run pipe and the branch pipe is less than 90 degrees.

Lavatory

A sink with hot and cold water fixtures, and a drain to a sanitary sewer system.

Make-Up

The critical dimension of a threaded fitting.

MAWP

"Maximum Allowable Working Pressure." The MAWP is the maximum working pressure of the weakest component of a vessel.

Mfg

"Manufacturing." Not to be confused with "manufacturer."

Mfr

"Manufacturer."

MM

A designation for "million." Since M is the Roman numeral for 1000, MM is one-thousand thousands, or one million. For example a 2 MM tank has a capacity of 2 million gallons. Care should be exercised since some industries use M for million.

MOC

"Material of construction."

NBEP

Nonboiler external piping.

NDT

Non-destructive testing.

Nominal

A term used to indicate that something is "named" or called something. For instance, a standard 6 in diameter pipe has two pertinent diameters, neither of which is actually 6 in. The OD is 6.625 in, and the ID is 6.065 in. We say that the "nominal" diameter is 6 in. That is, we "call" it 6 in diameter as a matter of convenience, even though it is not exactly 6 in.

OD

Outside diameter.

OS&Y

Outside stem and yoke. A type of gate or globe valve in which the stem nut is held by two arms (the yoke), which rise out of the valve body.

Owner

The entity for whom the project is being developed. Usually also the operator of the facility.

P&ID

Piping and Instrumentation Diagram.

PFD

Process Flow Diagram. A drawing that shows the major equipment and general flow characteristics of a piping system. The PFD usually indicates flowrates and temperatures. Used to develop the P&IDs.

Plan

A drawing which depicts the overhead view.

Playpipe

A type of fire hose nozzle with a lever-actuated shut-off.

Pop-Off Valve

Same as a safety valve. An automatic pressure relieving device activated by the static pressure upstream of the valve, and characterized by the rapid, full-opening of the valve. May be used for steam, gas, or vapor service.

Pressure Relief Valve

A pressure relief device designed to reclose and prevent further venting of fluid after normal conditions have been restored. This is the generic term for any relief valve, whether it pops open or opens in proportion to the increase in pressure above the opening pressure.

Pup

A short length of pipe welded between two other lengths. Often used as a repair for a ruptured portion of piping.

Quad-Stenciled

A pipe labeled (stenciled) with four different specifications, meaning that it satisfies the requirements of any of the four specifications.

Reducer

A fitting that is used to change the size (diameter) of a pipe. Sometimes referred to as an increaser by the uninformed. The size is always stated as (larger diameter) × (smaller diameter).

Relief Valve

An automatic pressure-relieving device activated by the static pressure upstream of the valve, which opens in proportion to the increase in pressure above the opening pressure.

Re-Pad

A reinforcing pad. A ring placed around a branch connection in order to add strength to the connection between the run pipe and the branch pipe.

Rising Stem

A type of gate or globe valve in which the position of the orifice (open or closed) can be determined by examining the tip of the threaded stem. Compare with non-rising stem.

RTJ

Ring type joint. A type of flange with a grooved face to accept a ring gasket, common in the oil patch.

Rupture Disk

A pressure relief device that is non-reclosing. It consists of a thin disk held between two flanges. When the pressure reaches a set limit, the rupture disk is designed to rupture, relieving the pressure.

Safeguarding

Providing protective measures to minimize the risk of accidental damage to piping, or to mitigate the consequences of a possible pipe failure. See ASME B31.3, Appendix G.

Safety Relief Valve

A pressure relief valve activated by the static pressure upstream of the valve, which either opens rapidly (like a safety or pop-off valve), or opens in proportion to the increase in pressure above the opening pressure (like a relief valve).

Safety Valve

An automatic pressure relieving device activated by the static pressure upstream of the valve, and characterized by the rapid, full-opening of the valve. May be used for steam, gas, or vapor service. Also referred to as a pop-off valve.

Schedule

A nominal number which designates the pipe wall thickness.

Sleeper

A type of pipe support at grade upon which the pipe rests.

Slurry

A liquid that contains particles in suspension.

SMACNA

Sheet Metal and Air Conditioning Contractor's National Association.
See www.smacna.org.

Soil Stack

A stack that conveys fecal waste.

Soleplate

See "Baseplate."

Sovent®

A trade name for a proprietary sanitary sewage system.

Specialty

A manufactured item that performs a special function but is neither a valve nor a fitting. Examples include air vents, vacuum breakers, and strainers.

Spool

Originally a length of pipe with a flange on both ends, it has come to mean any prefabricated piece of pipe of any configuration, whether or not there are flanges on the ends.

Stack

A vertical pipe in a drainage system. It may be a vent, a soil pipe, a rainwater pipe, or other waste pipe.

Stanchion

An upright pipe support.

Steam Trap

An automatic valve used to remove condensate from a steam system.

STP

Standard Temperature and Pressure of air, defined by the International Union of Pure and Applied Chemistry as an absolute pressure of 1 bar (100 kPa or 14.5038 psia) and a temperature of 273.15K (0°C or 32°F). There is no universally accepted definition of STP in industry, so it is necessary to define this term in detail.

Swage

A forged fitting that reduces the size (diameter) of a pipe. It is longer than a reducer, and usually threaded on one end. Sometimes spelled swedge by the uninformed, due to the common mispronunciation.

Swedge

Common mispronunciation of "swage." See "Swage."

Take-out

The dimension of a valve or fitting that must be accounted for in the design and fabrication of a piping system. For instance, on a valve, it is the distance between the raised faces of the flanges. For threaded fittings it is also referred to as "make-up."

Throttle

To modulate the flow in a pipeline with a valve.

Trim

In reference to boilers, trim consists of valves, fittings, gages, or appurtenances installed on a boiler to provide control. Trim usually refers to safety valves, try cocks, low water cutoff switches, and water columns.

In reference to valves, trim consists of the seat rings, disk or facing of the disk, stem, and stem guide sleeves (the wearing surfaces).

Tri-Stenciled

A pipe labeled (stenciled) with three different specifications, meaning that it satisfies the requirements of any of the three specifications.

Valve

A device that is used to shut off or throttle the flow through a pipe.

Water-Distribution Piping

The piping inside a building that delivers both hot and cold water to plumbing fixtures.

Water-Service Piping

Piping which extends from a potable water source to the interior of a building.

Weight

Similar to "schedule," the "weight" is a nominal designation used to identify pipe wall thickness.

Weldment

A steel structure that is bonded by welding. Usually refers to the piece prior to any machining.

Weld-o-let®

A type of branch connection that welds onto the run pipe. Bonney Forge owns the trademark to the name "Weld-o-let®."

Wet Vent

A plumbing vent that also discharges waste water.

WOG

"Water-Oil-Gas." A designation that a fitting is suitable for these services, usually used with a specified pressure rating.

Workability

A characteristic that describes in relative terms the ability of a metal to be plastically deformed without fracturing.

Wrought

Past participle of "to work," hence, another name for "forged."

WSP

"Working steam pressure." A designation that a fitting is suitable for steam service at a specified pressure rating.

CHAPTER 3

Reference Materials

There are many excellent reference materials available to the piping engineer. Some are out of print, but still available in libraries or tucked away in the dusky recesses of a retired engineer's desk.

1. *Piping Handbook*, Mohinder L. Nayyar, McGraw-Hill, Inc., ISBN 0-07-046881-8. This handbook covers piping fundamentals, various types of piping systems, and does a good job of covering the various ASTM piping specs and ASME piping codes. This is a massive volume. The fifth and earlier editions were edited by Crocker and King, and are equally good references.
2. *Pipe Fitters Handbook*, Grinnell Supply Sales Company. This was a pocket reference book that used to be available from Grinnell. It contains comprehensive tables of pipe sizes, wall thicknesses, and take-out dimensions for fittings, but is now out-of-print.
3. *NAVCO Piping Datalog*, National Valve and Manufacturing Company, Pittsburgh, PA. A favorite among piping engineers and designers, this out-of-print booklet contains a wealth of data pertaining to pipes and fittings.
4. *Facility Piping Systems Handbook*, Michael Frankel, McGraw-Hill, Inc., ISBN 0-07-021891-9. Another excellent reference for various system types encountered in industrial, commercial, and institutional settings.
5. *National Plumbing Codes Handbook*, R. Dodge Woodson, McGraw-Hill, Inc., ISBN 0-07-071769-9. This is an excellent reference for plumbing projects. It provides a good summary of the various plumbing codes, as well as information useful to those studying to take the plumbing license exam.
6. *Piping and Pipe Support Systems*, Paul R. Smith and Thomas J. Van Laan, McGraw-Hill, Inc., ISBN 0-07-058931-3. This reference provides insight into the codes specific to performing stress analyses of piping systems.

7. *Marks' Standard Handbook for Mechanical Engineers,* Baumeister, Avallone, and Baumeister, McGraw-Hill, Inc., ISBN 0-07-004123-7. This encyclopedic reference contains a wealth of information concerning mechanical systems and materials.
8. Structural steel shapes catalogs published by any of the steel companies that roll structural steel. These catalogs are invaluable in the field for identifying the sizes of steel members and for detailing pipe supports.
9. Codes and Standards Training Institute (CASTI) publishes a variety of guide books including several that pertain to the ASME piping codes.
10. NIBCO Chemical Resistance Guide. Available from NIBCO, this reference provides excellent data regarding metals, plastics, and elastomers, and their ability to resist deterioration when exposed to a lengthy list of fluids.
11. Goulds Pump Manual. Aside from pump performance and selection data, this manual offers excellent design advice for centrifugal pump installations.

CHAPTER 4
Piping Codes

I once worked on a pulp and paper project in western Canada. The consulting firm I worked for was designing the piping for an evaporator set. Evaporators take the black liquor that is produced in the cooking of the wood fibers (the pulping process) and evaporate most of the water away, leaving a flammable liquid that can be burned in boilers to produce steam. The process also recovers most of the chemicals used in the pulping process.

We had about seven large stainless steel vessels that had nozzles protruding at all kinds of odd angles. Our piping was designed to attach to these nozzles. The piping spools had been prefabricated, and due to some problems with the way the nozzles were fabricated, the contractor was having trouble with the fit-up. The nozzles had been fabricated out of plate, and when they were rolled the ends were out of plane. It really was a poorly made fitting, and demanded a lot of grinding to mate with the adjoining pipe.

The contractor wanted a cost extra to cut off the nozzles and reweld a new longer length of pipe, which would cause a redesign of our piping.

I was the site engineer and suggested that the contractor cut the nozzle back a bit so that it would be in the same plane, and then insert a short pup to which the remainder of my piping could be welded.

"You can't do that! That's against code! No pup length can be less than half of the pipe diameter."

I didn't have a copy of the "code," and I didn't challenge the contractor to cite the code reference of which he spoke. Had I bothered to ask him the details of the particular code, I would have learned that he did not know. He *couldn't* have known, because there is no such code that prohibits welding a short pup.

Many contractors are code-savvy; many are not. To cite the codes is to speak with authority. It can be intimidating to challenge someone who appears to stand on firmer technical ground. But very often when you challenge someone to produce the code reference, you will find that it does not exist.

Many contractors and engineers have been trained to believe that something is part of a code when in fact it is not. They were told early in their career that something is code, or something violates a code. So without consulting the appropriate code to determine the validity of such a claim, it becomes part of the prejudices and superstitions that they bring to the job site. In the case of a pipe fitter, who could blame him for not consulting a code?

Well, why have codes in the first place?

Piping codes, like the Boiler and Pressure Vessel Code and National Electric Code have evolved over time in response to the need to continuously improve safety. Some of the ASME codes can trace their development back to historic disasters like the explosion of the steamship USS Sultana, which exploded on April 27, 1865 as it was transporting Union POW survivors from the Confederate South. Estimates range from 1300 to 1900 souls lost when one of the boilers exploded.

Another episode occurred in Boston on January 15, 1919, when a 2.3 MM gallon molasses tank ruptured, spewing its contents through the streets to a depth of 2 to 3 ft and killing 21 people. It is said that you can still smell molasses in the area to this day.

Incidents like these spawned public sentiment that "there ought to be a law," and so were born certain codes that regulate the design, manufacture, and operation of facilities that may pose a hazard to the public.

The codes, however, are not themselves "laws." Legislative bodies may adopt a code, or reference a code in a law, thereby requiring owners of equipment to abide by that particular code. Prudent engineers will contact the Authority Having Jurisdiction (AHJ) if they are uncertain about what codes must be applied.

But consider a manufacturing facility that may be staffed by unsophisticated personnel, who are nevertheless conscientious, at least in their desire to provide a profitable operation for the owners. Let's assume that the plant has undergone an evolution of updates, modifications, and expansions throughout the years. Perhaps there is no actual engineering staff. So there is a good chance that the plant personnel do not consult with the AHJ whenever there is a new modification. It doesn't take a lot of imagination to envision a scenario in which a piping failure might lead to property damage or worse.

There are many different piping codes in use throughout the world. These codes may be divided into the following basic groups:

1. Plumbing codes, intended to protect the public against unsanitary conditions.
2. Gas codes, intended to protect the public from hazardous fumes, fires, and explosions.
3. Industrial codes, intended to protect facilities and those working in them from catastrophic failures.

The location of the installation in question determines which plumbing and gas codes are to be applied. There is not a lot of difference between them, but there may be some differences, and local Authorities Having Jurisdiction (AHJ) should be consulted to determine which specific codes apply. Further, there may be amendments to the codes, which will be provided by the AHJ.

There has recently been a trend to standardize these codes within the U.S. In 1994, the International Code Council (ICC) was established as a nonprofit organization dedicated to developing a single set of comprehensive and coordinated national codes. Previously, there had been three code models in use throughout the U.S.:

- Building Officials and Code Administrators (BOCA)
- International Conference of Building Officials (ICBO)
- Southern Building Code Congress International (SBCCI).

These organizations created the ICC to resolve the technical disparities among the three sets of model codes.

Some of these codes may still be in force in various locations, so you should always check with the AHJ to determine what code needs to be applied to your project.

The newer, comprehensive code models developed by ICC are:

- International Building Code
- International Fire Code
- International Plumbing Code
- International Fuel Gas Code
- International Mechanical Code
- International Private Sewage Disposal Code
- International Residential Code – One and Two-Family Dwellings
- International Electrical Code
- International Energy Conservation Code
- International Property Maintenance Code
- International Zoning Code
- International Existing Building Code
- International Performance Code
- International Urban-Wildlife Interface Code.

Obviously, not all of these are piping codes. The list is provided for the sake of thoroughness.

The beauty of this system is that they are all fully compatible with each other, and represent an effort to unify the various other codes that have existed in the past. There are, however, some cumbersome features:

- Like all codes, you have to know where to look for the answer to the problem under investigation. For instance, you might reasonably expect that natural gas and hydronic piping would fall under the Mechanical Code. Not so. You will find hydronic piping and fuel oil in the Mechanical Code, but natural gas, being a fuel gas, will be found under the Fuel Gas Code.
- You really need access to most of these codes for them to be of any real value. The first six or seven are an absolute necessity, depending on what sort of projects you are involved in.
- Like all codes, they are written in *codespeak,* with lots of exceptions noted, and references to other parts of the code.
- As noted earlier, not all jurisdictions have adopted the ICC. You will still have to rely on the AHJ to tell you what codes are applicable.

When working on industrial piping projects, you may still have to rely on the plumbing and fuel gas codes mentioned above. However, any process piping or high pressure steam lines will lie outside the scope of the ICC codes. It is always a good idea to consult the "Scope" section of any code, to make sure that you are using the proper code. The Scope section will tell you what the purpose of the code is, and may provide hints of where else you may search. If you are looking for high pressure steam lines in an ICC code book, you will quickly find that they are out of the ICC's scope. Another set of codes is required.

Once again, there are many different piping codes for process and high-pressure lines in use throughout the world. Some countries, like Canada, take advantage of the considerable body of knowledge contained in the U.S. codes.

In the U.S. we follow the ASME Code for Pressure Piping, B31. The code was first published in 1935 by the ASA (American Standards Association, now known as ANSI, the American National Standards Institute). The responsibility for developing the code was assigned to the ASME (American Society of Mechanical Engineers).

The ASME Code is so extensive that it was more convenient to break it up into several separate documents, which represent various industries. The code now consists of:

- B31.1 Power Piping
- B31.3 Process Piping
- B31.4 Pipeline Transportation Systems for Liquid Hydrocarbons and Other Liquids
- B31.5 Refrigeration Piping
- B31.8 Gas Transportation and Distribution Piping
- B31.9 Building Services Piping
- B31.11 Slurry Transportation Piping Systems.

Usually, by the time you get involved in a project, most of the piping specifications are written, and the codes to be used have been laid out in the specifications. But how did the engineer who wrote the specs know which codes to apply? Much of the time the answer lies in the codes themselves. The codes will explain what their intended scope is. But the codes are often applied to piping systems that are outside their scope.

This sounds like it might be a big problem, but the intelligent application of a piping code outside of its scope is not necessarily bad.

Let's say that you are the engineer in charge of setting up the piping specs for a plant that is going to produce turbo-widgets for the up-and-coming e-widget industry. The heat is on to get into production right away, and the Chief Engineer is visiting you every 15 minutes to see if you have issued the specs for bid yet.

The first thing you notice after going to the code is that the e-widget industry is not represented. And you can't find any mention of piping systems used in the production of turbo-widgets.

But you know that the plant uses air for the turbo-stamping lines, it uses water to cool the ovens that bake the widgets, it uses hydraulics for the presses that assemble the turbo-widgets, and there is also high-pressure steam for heating the reactor vessels that produce the proprietary widget compounds.

Like many projects of this type, a review of the available codes indicates that your choices are probably going to be limited to B31.1, B31.3, and B31.9. If you have some HVAC refrigerant lines, you might have to rely on B31.5 as well.

Some of these codes are more stringent than others. For instance, the allowable stresses for a given piping material is less in B31.1 and B31.9 than it is in B31.3. It's the same material, made the same way. But the allowable stresses are different. In this respect, B31.1 and B31.9 are more stringent that B31.3.

You can always apply a more stringent code than is required for an application. Doing that does not compromise safety.

So you *could* specify the most stringent code for all of these services, but that might mean that your 2 in diameter low-pressure cooling water lines would be constructed out of Schedule 160 pipe. And by the way, since 2 in Schedule 160 pipe has a wall thickness of 0.344 in, the ID is reduced, and maybe you better use 3 in pipe so your head losses aren't so high.

But the Chief Engineer isn't crazy about the idea, since the piping costs are going to go way up. And a big part of engineering is doing the safest things in the most economical manner.

So you decide to use B31.1 for the steam and condensate systems, and B31.3 for everything else. You might want to use B31.1 for high-pressure hydraulics lines, especially if they are in an area where corrosion is a factor.

In the discussion that follows, the order in which the material is presented follows the same order as the Codes. The sections are numbered in accordance with the paragraphs as labeled in the Code. Referenced formulas and tables are identified as they are in the respective codes.

Note that this text is not a substitute for any of the codes. This text is provided to help the designer study and apply the codes. It is assumed that the code is available to the designer, and that this text is supplemental to it. We will not cover the codes completely, but will only hit the high points; those areas that are most likely to be encountered in a piping project, and those that can cause some of the most trouble.

Any code you use is going to be organized into numbered paragraphs. You will find that in applying a code, you will often be referred to other paragraphs elsewhere, especially where exceptions are involved. This can be frustrating, and it requires quite a bit of patience and perseverance. Working within the codes is a complicated process, and sometimes it may seem as though it is needlessly complicated. Perhaps the code writers may someday include the intent of what they want you to achieve in the codes. The Fine Print Notes in the National Electric Codes approach this philosophy, but it seems that all of the piping codes remain prescriptive. For example, the International Fuel Gas Code prohibits the use of cast iron bushings in Paragraph 403.10.4.5.2. If they explained that the reason they object to these fittings is because they are prone to undetected cracking during installation, then perhaps other installers would become aware of this problem as it applies to other applications or materials.

Of the seven B31 codes, most design engineers find that they spend most of their time dealing with B31.1, B31.3, and B31.9. The remainder of this chapter is confined to an overview of these codes.

ASME B31.1 Power Piping

"Power piping" in this case means the piping that is used around boilers. It is called power piping because often a boiler is used to make steam for power generation. This is done by converting pressure and temperature energy into kinetic energy in a turbine, which then produces electrical energy.

Mechanical engineering is the technical field associated with transforming energy into work. A steam generator is a perfect example of that. This code relates particularly to piping that would be found in electrical power plants, commercial and institutional plants, geothermal plants, and central heating and cooling plants.

This code is primarily concerned with the effects of temperature and pressure on the piping components.

100.1.2 Scope

Around a boiler, the scope of B31.1 begins where the boiler proper ends, at either:

1. The first circumferential weld joint
2. The face of the first flange
3. The first threaded joint

This piping is collectively referred to as "boiler external piping," since it is not considered part of the boiler.

Power piping may include steam, water, oil, gas, and air services. But it is not limited to these, and as mentioned before, there is nothing that says you cannot apply B31.1 to other piping systems unrelated to boilers or power generation, as long as they would not be better classified as within other piping codes, which may be more stringent.

100.1.3 Not in Scope

B31.1 does NOT apply to:

1. Components covered by the ASME Boiler and Pressure Vessel Code
2. Building heating and distribution steam and condensate piping if it designed for 15 psig or less
3. Building heating and distribution hot water piping if it is designed for 30 psig or less
4. Piping for hydraulic or pneumatic tools (and all of their components downstream of the first block valve off of the system header)
5. Piping for marine or other installations under federal control
6. Structural components
7. Tanks
8. Mechanical equipment
9. Instrumentation

102.3.2 Limits for Sustained and Displacement Stresses

This section addresses cyclic stresses among other things. Note that the Stress Range Reduction Factors apply only to thermal cycling and NOT to pressure cycles.

104.1 Pressure Design of Straight Pipe

This section contains several extremely useful formulas for determining either the design pressure of a particular pipe or the required wall thickness of a pipe operating at a certain pressure. These formulas are:

$t_m = [(PD_o) / 2(SE + Py)] + A$ Formula (3) in the code

$t_m = [(Pd + 2SEA + 2yPA)]/[2(SE + Py - P)]$ Formula (3A)

$P = [2SE(t_m - A)]/[D_o - 2y(t_m - A)]$ Formula (4)

$P = [2SE(t_m - A)]/[d - 2y(t_m - A) + 2t_m]$ Formula (4A)

where

t_m = Minimum Required Wall Thickness [in. or mm][1]

Piping is generally purchased based on commercially available schedules or wall thicknesses (unless specially ordered, which is usually prohibitively expensive). These thicknesses must take into account the mill tolerance, which may be as much as 12.5 percent less than the nominal thickness.

P = Internal design gage pressure [psig or kPa (gage)]

The pressure is either given or solved for in the equations.

D_o = Outside diameter of pipe [in or mm]

The outside diameter will be the OD of a commercially available pipe. An example of these data for carbon steel pipe can be found in Appendix A.1 of this text.

d = Inside diameter of pipe [in or mm]

The inside diameter will be the ID of a commercially available pipe.

S = Maximum Allowable Stress Values in Tension
for the material at the design temperature [psi or kPa]

These values are tabulated in ASME B31.1, Appendix A. Note that they are dependent on the temperature to which the material will be exposed. This temperature is the metal temperature. This would normally be the temperature of the fluid in the pipe, but if a pipe was to be exposed to a high temperature externally, it would be the fluid temperature outside the pipe.

Note that the values tabulated in Appendix A include the Weld Joint Efficiencies and the Casting Factors. Therefore, the tabulated values are the values of S, SE, or SF. See General Note (f) at the end of each table.

E = Weld Joint Efficiency, shown in ASME B31.1, Appendix A. These values depend on the material used and the method of manufacture. Naturally, if the material is a cast product, there is no weld. In that case the Casting Factor F is used.

F = Casting Factor shown in ASME B31.1, Appendix A. Where a cast material is used, the casting factor F takes the place of E in the equations above. Like the weld joint efficiency, these values depend on the material used and the method of manufacture.

A = Additional thickness [in or mm]

This value is used to compensate for:

1. Material that is removed in fabricating mechanical joints (threading or grooving)
2. Increased mechanical strength of the piping
3. Erosion or corrosion allowance
4. Cast pipe tolerances.

---

[1] Note that throughout this book we will make an effort to include both Imperial and metric units. Regardless of which system is utilized, the designer must always remain consistent in the use of the units chosen. Switching back and forth between systems of measure is not recommended. It is always recommended to include the units in any calculation so that they may be algebraically cancelled. This ensures that any appropriate conversion factors are applied within the calculation. See Example 4.1.

In the case of cast pipe, the code specifies the value of A to be:

0.14 in (3.56 mm) for centrifugally cast pipe

0.18 in (4.57 mm) for statically cast pipe

y = A coefficient used to account for material creep.

For ferritic[2] and austenitic steels, and some nickel alloys, the value ranges from 0.4 to 0.7, depending on temperature. This variation of y with temperature allows the wall thickness equation (Formula 3) to behave in accordance to the "Modified Lamé Equation" at low temperatures (with y = 0.4), and in accordance with a creep-rupture equation at high temperatures (with y = 0.7).[3]

For carbon steels (ferritic steels), y = 0.4 at temperatures less than 900°F (482°C)

For cast iron and nonferrous materials, y = 0.0

This reduces equation (3) to

$$t_m = (PD_o/2SE) + A$$

which is recognizable as Barlow's Formula.

There may be a different value of y for carbon and ferritic and austenitic stainless steels depending on the pipe geometry. If the ratio of the OD to the wall thickness is less than 6, then

$$y = d/(d + D_o)$$

In the commercially available pipe schedules, these ratios only occur in the heavier schedules in diameters 3½ in and smaller.

A typical value for corrosion allowance is 1/16 in for carbon steel pipe. As noted in 102.4.1, this is entirely up to the discretion of the designer. Sometimes the corrosion allowance may be as high as 0.1 in.

If the piping system contains bends (not elbows), then you also must compensate for thinning of the bends. These values are given in Table 102.4.5 of the code, and are reproduced here:

Radius of Bends[4]	Min. Thickness Recommended Prior to Bending
≥ 6 pipe diameters	1.06 t_m
5 pipe diameters	1.08 t_m
4 pipe diameters	1.14 t_m
3 pipe diameters	1.25 t_m

Note also that bends are not frequently used in industrial or commercial piping installations. They are probably most frequently used in tight quarters, such as on-board ships, but are also used in power house piping. The interested reader is referred to ASME B31.1 paragraph 102.4.5 (B) for the formulas for computing the required after-bend thickness.[4]

[2] Ferritic steels have a Body-Centered-Cubic structure and are more brittle at lower temperatures than austenitic steels. Carbon Steel and 405, 430, and 446 stainless steels are ferritic. Austenitic steels have a Face-Centered-Cubic structure, contain more nickel, and are more ductile at lower temperatures. The 300 series of stainless steels are austenitic.

[3] Burrows, W. R. (1954). A Wall-Thickness Formula for High-Pressure, High-Temperature Piping. *Transactions of the ASME*, 427–444.

[4] The pipe diameters referenced are the nominal diameters (see ASME B36.10M, Tables 2 and 4, and ASME B36.19M, Table 1).

104.3 Branch Connections

Branch connections are often made using fittings designed for the application. Such fittings are manufactured according to the standards listed in ASME B31.1, Table 126.1.

But branch connections are often made in other ways. Especially where large bore piping is concerned, some branch connections are made without the use of a manufactured fitting. It might be helpful now to make a distinction between "manufactured" and "fabricated." A **manufactured** fitting is one that could be purchased from a supplier. The supplier would sell it to you just as he received it from a factory. It would be (or *should* be) made to a specification; perhaps one of the specifications listed in ASME B31.1's Table 126.1. A **fabricated** fitting would be made in a shop, or in the field, with pieces of pipe and/or plate.

There are a few reasons why it may be advantageous to use a fabricated fitting instead of a manufactured fitting:

- Lack of availability of the specific fitting. For instance, a 20 in 304SS lateral may not be available.
- The cost of the manufactured fitting is very high compared to the ability to fabricate it.
- The cost of making the welds on a manufactured fitting is higher than on a fabricated fitting

Consider an 18" diameter run of piping, and a 14 in diameter branch connection. There are several ways that this could be accomplished.

One method might be to insert a full-size butt-welding tee, with an 18 in × 14 in reducer exiting the branch, as shown in Figure 4.1.

Another method might be to use a reducing tee as shown in Figure 4.2.

Figure 4.3 shows still a third method, in which the 14 in branch pipe is stubbed directly into the 18 in run pipe. The type of cuts required to join pipes like this is called a "fishmouth" by pipefitters.

Obviously, the types of fabrications one could concoct are limited only by one's imagination. In order to achieve safe designs that can withstand the same pressures as

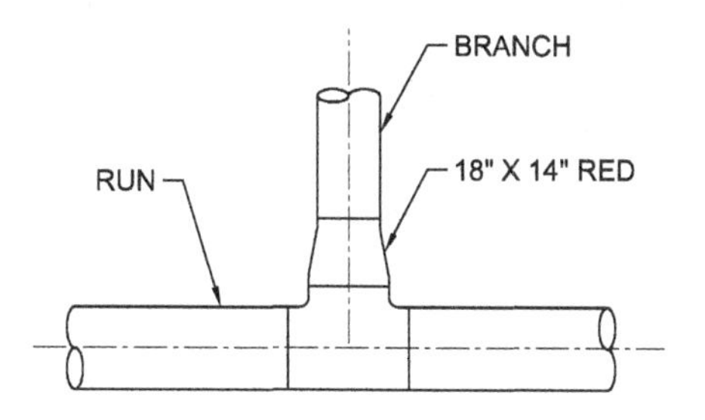

FIGURE 4.1 A branch connection using a full-size tee.

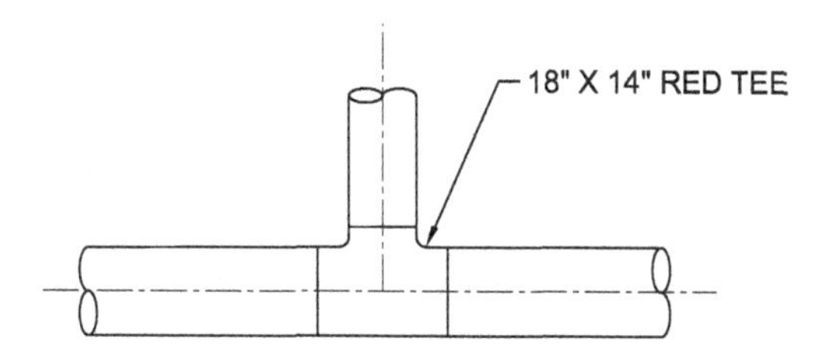

FIGURE 4.2 A branch connection with a reducing tee.

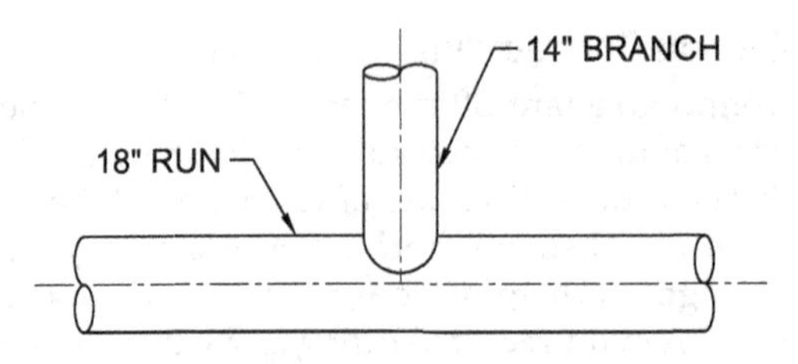

FIGURE 4.3 A branch connection using a nozzle weld without a fitting.

other pipe configurations, these fabrications are subject to the requirements of this code section.

Whenever a run pipe as shown in Figure 4.3 is cut to accommodate a branch connection, the strength of the run pipe is compromised due to the material that is removed. The larger the branch the more material that is removed, and the worse the situation.

If a manufactured fitting is chosen (for example in Figures 4.1 or 4.2 above) to provide the branch connection, and if the manufacturer's specification is listed in Table 126.1, then no further analysis is required. A situation similar to Figure 4.3 above might require additional engineering analysis, and Section 104.3 provides the guidance to perform that analysis.

The first thing to note is that fittings manufactured in accordance with the ASTM standards listed in ASME's Table 126.1 are satisfactory in meeting the code (Refer to Section 104.3.1 [B.1]). For example, if your specification designated that "all tees had to meet ASTM A234 Piping Fittings of Wrought Carbon Steel and Alloy Steel for Moderate and Elevated Temperatures," then you would be covered. No further qualifications would be necessary for those tees (other than to establish that they were installed in the piping system correctly[5]).

If, on the other hand, you decided that you really didn't want to spend the money on an 18 in diameter tee manufactured fitting, but would rather fabricate the fitting, then in order to satisfy the code, you would have to follow the requirement set forth in Section 104.3.1(D). These requirements specify the extent to which branch connections must be reinforced.

In practice, it is unusual to use a fabricated fitting when the branch diameter is the same size as the run diameter. The reason for this is the cut and weld for the branch would have to extend to the centerline of the pipe (halfway around the circumference). This would require extensive cutting, welding, and reinforcement, and would not be an economical choice. Most pipe specs indicate that full-size branch connections be made with a manufactured fitting (a tee).

For a single size reduction of the branch pipe, a reducing tee is often used. Below one size reduction, the branches are most often fabricated until the branch size is small enough that a manufactured welding fitting such as a Weld-o-let® may be used. These fittings are described in 104.3.1 (B.2). We will discuss these connections later in the chapter covering fittings.

Extruded outlets are described in Sections 104.3.1 (B.3) and 104.3.1 (G). They are manufactured by pulling a die through the wall of a pipe. Due to the custom nature of these fittings, they are unusual in general industrial applications, but can be found in the power industry.

This section pertains to branch connections where the axes of the main run and the branch intersect, and the angle formed by the main run of the pipe and the branch is between 45° and 90°.

[5] By "correctly," we mean installed in the correct location, correct orientation, with the axis of the run of the tee aligned with the axis of the adjoining pipe, and using the correct welding procedure.

If both of these conditions do not exist, additional tests or analysis must be performed to ensure that adequate strength is provided.

Section 104.3.1 (D) relates to branch connections subject to internal pressure. The code designates a region surrounding the intersection known as the "reinforcement zone." The reinforcement zone bounds the region of concern at the branch with a parallelogram. All of the analysis is confined to this zone, and any required reinforcement must fall within this zone.

Imagine a plane passing through the intersecting axes of the branch connection. See Figure 4.4. The discussion of "areas" in this code section refers to the cross-sectional areas that appear in this imaginary section.

The area of the material that is removed when the hole is cut in the run pipe must be offset by material that is present in other components within the reinforcement zone.

Credit is given for what is referred to as "excess pipe wall." Consider that piping systems are rarely operated at the maximum design pressure calculated by Formula (4) in Section 104.1.2. For one thing, the piping used is the commercially available pipe wall. This means that there is usually some inherent excess of pipe wall available in either the run pipe or the branch pipe or both. The code allows you to take this excess pipe wall into account when determining the need for additional reinforcement. This excess pipe wall would be the difference between the nominal pipe wall minus the mill

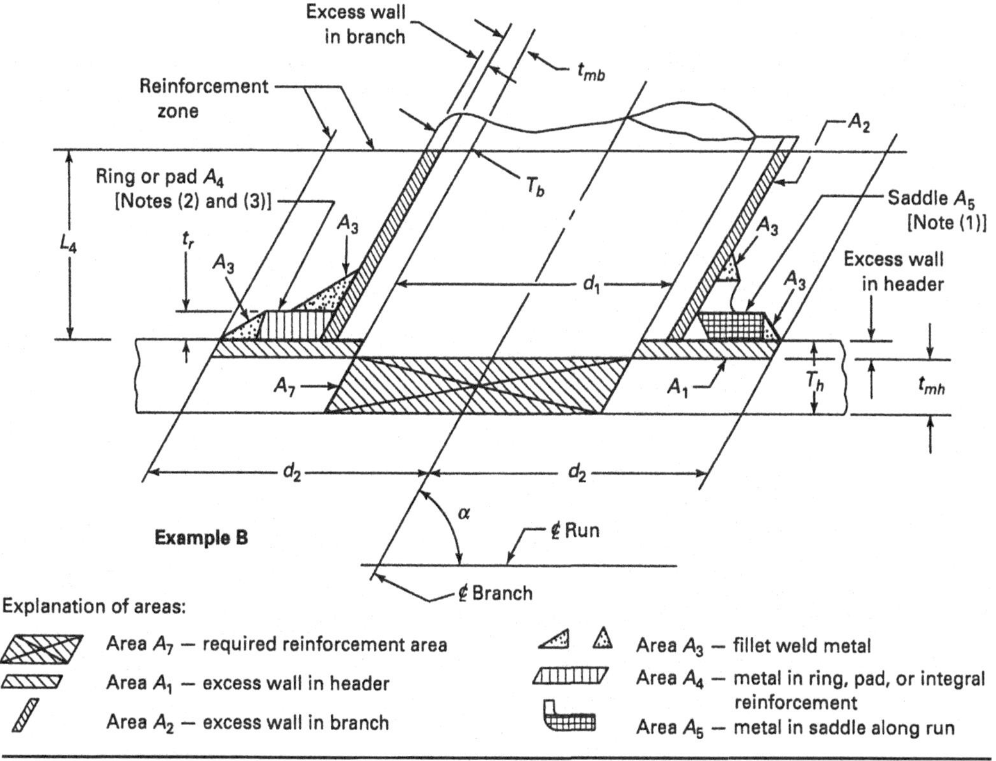

FIGURE 4.4 ASME B31.1-2001 Figure 104.3.1 (D) Example B, showing the various reinforcement areas for a branch connection. *Reprinted from B31.1-2001, by permission of The American Society of Mechanical Engineers. All rights reserved.*

tolerance, minus any additional thickness allowance, minus the wall thickness required by Formula (3) or (3A) in paragraph 104.1.2(A). In other words,

Excess pipe wall = (nominal wall thickness × 0.875) − (corrosion, erosion, threading, or grooving allowance) − (t_m, as calculated by Formula (3) or (3A))

The 0.875 term accounts for the mill tolerance of 12.5 percent.

Therefore, one place that can make up the amount of material removed by the hole in the run pipe is any excess pipe wall present in the reinforcement zone. This is designated A_1 for the run pipe (also known as the "header") and A_2 for the branch pipe.

Another source of excess material is the area of the fillet weld, designated A_3.

Sometimes a reinforcing pad (or "re-pad") is placed around the branch connection to add strength to the joint. The ratio of the width of the re-pad to its height should be as close as possible to 4:1 (within the limits of the reinforcing zone). It should never be less than 1:1. This material's area is designated A_4.

Still another method of reinforcing a branch connection is to weld on a saddle. These are limited to use on 90° branches. Their use in general industry is not as common as other methods of preparing branch connections, but they remain a viable alternative. The metal contained in the saddle along the run pipe in the reinforcement zone constitutes the additional metal that may be used to offset the material lost in cutting the hole in the run pipe. The area of the metal in the saddle along the run pipe is designated A_5.

So there are a total of five areas that can be added together to offset the loss of material created by the hole in the run pipe, which is designated as A_7. Since the pipe is expected to retain its integrity throughout its design life, the wall thickness expected at the end of the pipe's design life is the thickness that must be used in the calculations. The newer versions of the code call this pressure design area at the end of the service life A_6.

$$A_6 = (t_{mh} - A)\, d_1$$

where

t_{mh} = The required minimum wall thickness in the header (run) pipe.

A = The additional thickness to compensate for corrosion, erosion, grooving or threading

d_1 = The inside centerline longitudinal dimension of the branch opening in the run pipe, which would be equivalent to the ID of the branch pipe if the angle between the axes is 90°. The general form for d_1 is

$$d_1 = [D_{ob} - 2(T_b - A)]/\sin\alpha$$

where

α = the angle between the axes of the run and branch pipes.

The subscripts b and h designate the branch and header (or run) pipes respectively. If the angle between the branch and header pipes is other than 90°, then

$$A_7 = A_6\,(2 - \sin\alpha) = (t_{mh} - A)\, d_1\,(2 - \sin\alpha)$$

Note that if the angle between the axes is 90°, the value for A_7 reduces to

$$A_7 = A_6 = (t_{mh} - A)\, d_1$$

In order to satisfy the requirements of Paragraph 104.3.1, the following must be true:

$$A_7 \le A_1 + A_2 + A_3 + A_4 + A_5$$

Another way of stating this is that the required reinforcement area must be less than any combination of:

1. A_1 = Area of any excess pipe wall contained in the run
   $= (2d_2 - d_1)(T_h - t_{mh})$
2. A_2 = Area of any excess pipe wall contained in the branch.
   $= 2L_4(T_b - t_{mb})/\sin \alpha$
3. A_3 = Area of any welds beyond the outside diameters of either the run or branch, or of weld attachments of pads, rings, or saddles.
4. A_4 = Area provided by any rings, pads, or integral reinforcement.
5. A_5 = Area provided by a saddle on a right angle connection.
   $= (\text{OD of saddle} - D_{ob})\, t_r$

In practice, if you were using a saddle, you would not also have a re-pad, and vice-versa. Therefore, A_4 and A_5 can be considered mutually exclusive.

The code lists specific requirements for closely spaced branch connections, branch connections subject to external forces and moments.

The Pipe Fabrication Institute publishes worksheets (designated ES36) that aid in these calculations.

One of the piping projects I worked on was a high pressure descale system for a steel mill. These systems blast the scale from rolling operations off of the strip of steel as it passes through the rolling mills. The scale is removed through a combination of thermal shock and high velocity water jets. The water pressures may be in the 3000 to 4500 psi range. Obviously, great care must be taken to ensure a safe piping system.

We were having trouble finding available fittings for the descale piping, due to the very heavy wall thicknesses required by the piping code. The design pressure was 3900 psi, and we were using ASME B31.3. As usual, the design schedule was incredibly tight, and no one was in the mood to hear that the project might be delayed because the engineer was specifying hard-to-find materials. The equipment vendor took me aside and told me that my calculations were too conservative, since I had allowed for a mill tolerance of 87.5 percent of the published wall thicknesses, as required by the code.

"I worked in a pipe mill," he told me, "and we got paid by how many tons of pipe we produced! I can tell you that the mills don't skimp on wall thickness. It's to their advantage to make the walls thicker since they sell by weight!"

An argument like that doesn't carry much weight. You should avoid trouble by meeting or exceeding the code requirements.

Example 4.1
Given: Design Pressure: 165 psig
Temperature: 675°F
Service: Superheated Steam inside a powerhouse
Code: ASME B31.1
Design: A fabricated 4 in branch connection at 90°, coming off an 8 in diameter header.

Solution:

Let's assume that the material of construction is garden-variety carbon steel: A53 Grade A, ERW (electric resistance welded) pipe[6].

We must determine the minimum wall thicknesses of each of the pipes. From Formula (3) in Section 104.1, we have

$$t_m = [(PD_o) / 2\,(SE + Py)] + A$$

P and D_o are given. From Table A-1 of B31.1, we find that the maximum allowable stress value for A53, Grade A, ERW pipe is 10.2 ksi from minus 20 to 650°F. However, our example has a temperature that is higher than 650°F. We are permitted to interpolate to arrive at the design stress. The value at 700°F is 9.9 ksi, so we will use SE = 10.05 ksi. The value of y is taken to be 0.4, since we are using carbon steel pipe.

Because we are working with carbon steel, we assume a corrosion allowance of 0.0625 in. This will ensure that the pipe will remain serviceable at the design pressure over a period of time that it will take for the pipe to corrode 1/16 in from the inside diameter. If the line is located in a harsh environment without any maintenance, we might choose to provide an additional corrosion allowance to the outside of the pipe. For our example, the line is inside a building and protected from external corrosion. "Fitness for service" calculations may be made during the life of the piping using ultrasonic thickness measurements taken at various points to monitor material lost due to corrosion or erosion.

Therefore, for the header

$$\begin{aligned} t_{mh} &= [(PD_o) / 2\,(SE + Py)] + A \\ &= (165\text{ psig})(8.625\text{ in}) / 2[10{,}050\text{ psi} + (165\text{ psig})(0.4)] + 0.0625\text{ in} \\ &= 0.070\text{ in} + 0.0625\text{ in} \\ &= 0.133\text{ in} \end{aligned}$$

The minimum wall thickness that is commercially available for 8 in diameter CS pipe is Schedule 20, with a thickness of 0.250 in. This would be suitable, but Schedule 40 is the standard thickness (sometimes also called the "weight" of the pipe), so we choose Schedule 40 with a thickness of 0.322 in. We note that Schedule 40 could have a wall thickness as low as

0.322 in × 0.875 = 0.282 in due to the 12.5 percent mill tolerance on wall thickness.

This minimum wall thickness of the run pipe is known as T_h.

Therefore, by selecting Schedule 40, we see that we have an excess pipe wall of

$$0.282\text{ in} - 0.133\text{ in} = 0.149\text{ in}$$

We perform the same exercise for the branch:

$$\begin{aligned} t_{mb} &= [(PD_o) / 2(SE + Py)] + A \\ &= (165\text{ psig psi})(4.500\text{ in}) / 2[10{,}050\text{ psi} + (165\text{ psig psi})(0.4)] + 0.0625\text{ in} \\ &= 0.037\text{ in} + 0.0625\text{ in} \\ &= 0.099\text{ in} \end{aligned}$$

[6] ASTM A53 Grade A is not suitable for prolonged exposure to temperatures above 800°F (427°C), since the carbon phase may be converted to graphite.

The minimum wall thickness that is commercially available for 4 in diameter pipe is Schedule 40 (standard weight), with a thickness of 0.237 in. But we note that with the mill tolerance, it could be as thin as

$$0.237 \text{ in} \times 0.875 = 0.207 \text{ in}$$

This value is known as T_b.

We select Schedule 40, and note that the branch pipe has an excess pipe wall of

$$0.207 \text{ in} - 0.099 \text{ in} = 0.108 \text{ in}$$

We may now calculate A_7.

$$\begin{aligned} d_1 &= \text{Inside diameter of the branch pipe} = 4.026 \text{ in} \\ A_7 &= (t_{mh} - A)\, d_1 \\ &= (0.133 \text{ in} - 0.0625 \text{ in})\,(4.026 \text{ in}) \\ &= 0.284 \text{ in}^2 \end{aligned}$$

Now we examine the excess areas, and determine the need for reinforcements. We must first determine the bounds of the reinforcement zone.

d_2 = The half-width of the reinforcing zone
= The greater of d_1 or $(T_b - A) + (T_h - A) + d_1/2$, but in no case more than the outside diameter of the header pipe.

The OD of the header is 8.625 in and $d_1 = 4.026$ in. The other term to check is:

$$\begin{aligned} (T_b - A) + (T_h - A) + d_1/2 &= (0.207 \text{ in} - 0.0625 \text{ in}) + \\ &\quad (0.282 \text{ in} - 0.0625 \text{ in}) + 4.026 \text{ in}/2 \\ &= 0.145 \text{ in} + 0.220 \text{ in} + 2.013 \text{ in} \\ &= 2.377 \text{ in} \end{aligned}$$

Therefore, $d_2 = 4.026$ in. This establishes the width of the reinforcement zone. We must now find the height of the reinforcement zone, L_4.

L_4 is the smaller of $2.5\,(T_b - A) + t_r$ or $2.5\,(T_h - A)$, where

t_r is the thickness of a reinforcing pad, if used. If the branch attachment is a manufactured connection, it may contain additional metal for reinforcing the connection. This is termed "integrally reinforced," and if this is the case, then the value of t_r is the height of a 60° right triangle whose perpendicular legs align with the ODs of the run and branch, and which lies completely within the area of the extra metal reinforcement.

In our example, we will presume that we will not have a re-pad. We already know that this is a fabricated piece, so we are not using an integrally reinforced connection. Therefore $t_r = 0$ and

$$L_4 = \text{the smaller of } 2.5\,(0.207 \text{ in} - 0.0625 \text{ in}) + 0 = 0.361 \text{ in}$$

OR

$$2.5\,(0.282 \text{ in} - 0.0625 \text{ in}) = 0.549 \text{ in}$$

Therefore, $L_4 = 0.361$ in

$$A_1 = (2d_2 - d_1)(T_h - t_{mh}) = [2(4.026 \text{ in}) - 4.026 \text{ in}](0.282 \text{ in} - 0.133 \text{ in}) = 0.600 \text{ in}^2$$

We see that this is already larger than A_7, so we can stop here and be satisfied that the branch connection meets the code requirements.

But what if the problem had been at a much higher pressure?

Example 4.2

Given: The same design and MOC as Example 4.1, but with a design pressure of 1000 psig. Use ASME B31.1. (The saturation temperature at P = 1000 psig is 546°F, so the steam is still superheated if it is at 675°F.)

Solution:

We start as before.

$$\begin{aligned} t_{mh} &= [(PD_o) / 2\,(SE + Py)] + A \\ &= (1000 \text{ psig})(8.625 \text{ in}) / 2[10{,}050 \text{ psi} + (1000(0.4)] + 0.0625 \text{ in} \\ &= 0.413 \text{ in} + 0.0625 \text{ in} \\ &= 0.475 \text{ in} \end{aligned}$$

Schedule 100 has a wall thickness of 0.594 in. This is a very uncommon schedule, but for the purpose of this exercise we will use it. Under ordinary circumstances, an engineer would choose the next thicker common schedule, which in this case would be Schedule XXS (double-extra-strong) with a wall thickness of 0.875 in. But in this exercise, let's see how much excess metal lies inside the reinforcement zone, and what must be done if it is not enough to satisfy the code.

$$T_h = 0.594 \text{ in} \times 0.875 = 0.520 \text{ in}$$

Therefore, by selecting Schedule 100, the excess pipe wall is

$$T_h - t_{mh} = 0.520 \text{ in} - 0.475 \text{ in} = 0.045 \text{ in}$$

Again, we perform the same exercise for the branch:

$$\begin{aligned} t_{mb} &= [(PD_o) / 2(SE + Py)] + A \\ &= (1000 \text{ psig})(4.500 \text{ in}) / 2[10{,}050 \text{ psi} + (1000 \text{ psig})(0.4)] + 0.0625 \text{ in} \\ &= 0.215 \text{ in} + 0.0625 \text{ in} \\ &= 0.278 \text{ in} \end{aligned}$$

4 in diameter Schedule 80 (extra-strong) has a thickness of 0.337 in, so we choose it.

$$T_b = 0.337 \text{ in} \times 0.875 = 0.295 \text{ in}$$

The branch pipe has an excess pipe wall of

$$T_b - t_{mb} = 0.295 \text{ in} - 0.278 \text{ in} = 0.017 \text{ in}$$

We now calculate the required reinforcement area A_7.

$$\begin{aligned} d_1 &= \text{inside diameter of the branch pipe} = 3.826 \text{ in} \\ A_7 &= (t_{mh} - A)\, d_1 \\ &= (0.475 \text{ in} - 0.0625 \text{ in})\,(3.826 \text{ in}) \\ &= 1.578 \text{ in}^2 \end{aligned}$$

Now we examine the excess areas, and determine the need for reinforcements. We first determine the bounds of the reinforcement zone.

d_2 = The half-width of the reinforcing zone
= The greater of d_1 or $(T_b - A) + (T_h - A) + d_1/2$, but in no case more than the outside diameter of the header pipe

The OD of the header remains 8.625 in, and d_1 = 3.826 in

$$(T_b - A) + (T_h - A) + d_1/2 = (0.295 \text{ in} - 0.0625 \text{ in}) + (0.520 \text{ in} - 0.0625 \text{ in}) + 3.826 \text{ in} / 2$$
$$= 0.233 \text{ in} + 0.458 \text{ in} + 1.913 \text{ in}$$
$$= 2.604 \text{ in}$$

Therefore, d_2 = 3.826 in. This establishes the width of the reinforcement zone. We must now find the height of the reinforcement zone, L_4.

L_4 is the smaller of 2.5 $(T_b - A) + t_r$ or 2.5 $(T_h - A)$, where

Again, let's first assume that we will not have a re-pad, and that $t_r = 0$

$$L_4 = 2.5\ (0.295 \text{ in} - 0.0625 \text{ in}) + 0 = 0.581 \text{ in}$$

OR

$$L4 = 2.5\ (0.520 \text{ in} - 0.0625 \text{ in}) = 1.144 \text{ in}$$

Therefore, L_4 = 0.581 in

$A_1 = (2d_2 - d_1)(T_h - t_{mh})$ = 0.045 in × 3.826 in = 0.172 in^2. Clearly, we must continue to look elsewhere for the extra material.

$A_2 = 2L_4(T_b - t_{mb})$ = 2 × 0.581 in × 0.017 in = 0.020 in^2. It appears that we will need a lot more metal. We now have accounted for 12 percent of what we need to satisfy the code requirement.

A_3 is the fillet weld metal. Typically, the fillet weld thickness to the root of the weld is never more than the thinnest of the two pieces to be welded. In this case, the thinnest piece is the 4 in Schedule 80 wall at 0.337 in. We will specify a 5⁄16 in weld. See Figure 4.5. This means that the cross sectional area of the weld is ½ × 5⁄16 in × 5⁄16 in, or 0.156 in^2. See Figure 4.6 and Figure 4.7. But the weld exists on both sides of the branch, so we double it.

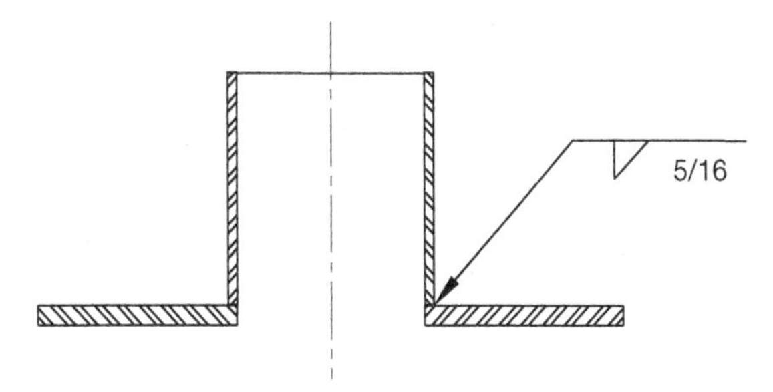

FIGURE 4.5 A fillet weld between branch and run.

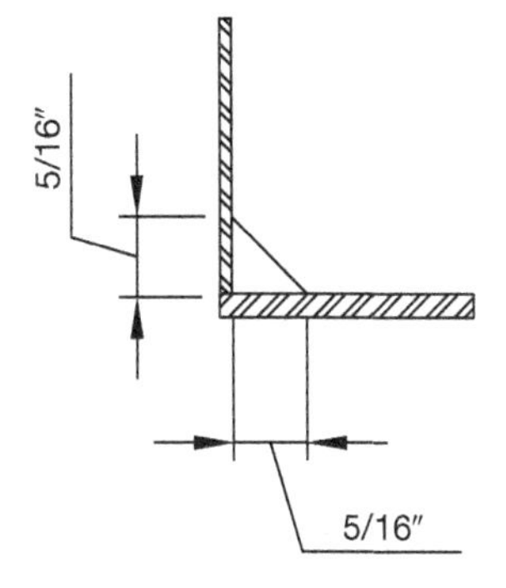

FIGURE 4.6 A fillet weld area.

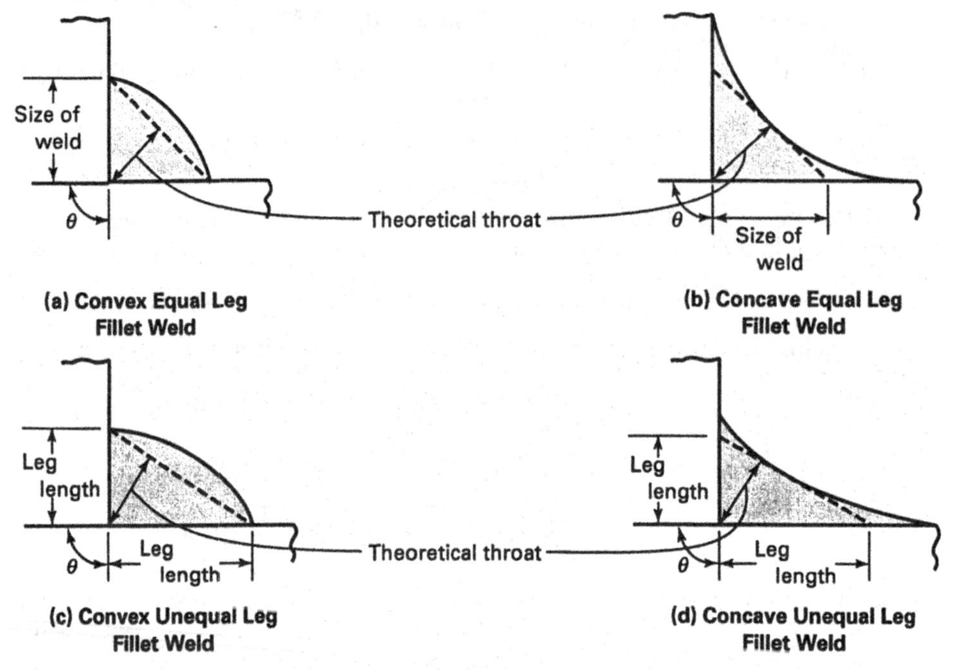

GENERAL NOTES:

(a) The "size" of an equal leg fillet weld shall be described by the leg length of the largest inscribed isoceles triangle.

(b) The "size" of an unequal leg fillet weld shall be described using both leg lengths and their location on the members to be joined.

(c) Angle θ, as noted in the above figures, may vary from the 90 deg angle as shown based on the angle between the surfaces to be welded.

(d) For an equal leg fillet weld where the angle θ between the members being joined is 90 deg, the theoretical throat shall be 0.7 × leg length. For other fillet welds, the theoretical throat shall be based on the leg lengths and the angle θ between the members to be joined.

(e) For all fillet welds, particularly unequal leg fillet welds with angle θ less than 90 deg, the theoretical throat shall lie within the cross section of the deposited weld metal and shall not be less than the minimum distance through the weld.

Figure 4.7 Fillet weld size from ASME B31.1-2001, Figure 127.4.4 (A). *Reprinted from B31.1-2001, by permission of The American Society of Mechanical Engineers. All rights reserved.*

Our total excess metal at this point is now

$$0.172 \text{ in}^2 + 0.020 \text{ in}^2 + 2(0.156 \text{ in}^2) = 0.504 \text{ in}^2$$

We still need $1.578 \text{ in}^2 - 0.504 \text{ in}^2 = 1.074 \text{ in}^2$

We need to add a reinforcing pad, preferably one with an aspect ratio of 4:1. We note that the half-width of the reinforcing zone is 3.826, and that the OD of the branch pipe is 4.5 in, leaving us with

$$3.826 \text{ in} - 4.5 \text{ in}/2 = 1.576 \text{ in}$$

as a limit on the width of a re-pad. If we aim for a 4:1 aspect ratio on this (the aspect ratio recommended by the code), we can consider a plate with a thickness of $\frac{3}{8}$ in and a width of 1.5 in. We recognize that from a practical standpoint, the ID of the re-pad has to be at least 5/8 in larger than the OD of the branch pipe to allow room for the fillet weld that connects the branch to the run pipe. This joint must be welded prior to the addition of the re-pad. This reduces the ID of the repad, which will leave us with a repad width of 1.188 in. The annulus between the ID of the repad and the OD of the branch pipe will be filled with weld, so as a convenience we can consider this weld metal to be part of the repad metal, even though technically it belongs to the A_3 weld area.

This yields an additional area of

$$A_4 = 2 \times 0.375 \text{ in} \times 1.5 \text{ in} = 1.125 \text{ in}^2$$

We now have

$$A_7 \leq A_1 + A_2 + A_3 + A_4 + A_5$$

$$1.578 \text{ in}^2 \leq 0.172 \text{ in}^2 + 0.020 \text{ in}^2 + 2(0.156) \text{ in}^2 + 1.125 \text{ in}^2 + 0 = 1.629 \text{ in}^2$$

This satisfies the code.

104.3.3 Miters

Miters are perfectly acceptable fabricated fittings for pressure piping, if constructed in accordance with the requirements of Paragraph 104.3.3. Note however that they are usually only used in large bore piping where manufactured elbows are either unavailable or very expensive. Miters require much fit-up and welding. It is easier to simply purchase an elbow that conforms to one of the standards listed in Table 126.1 if these are available.

119 Expansion and Flexibility

A review of this section of the code quickly reveals that the calculations for performing a flexibility and stress analysis are quite complex and cumbersome. These calculations are best performed with computer programs specifically designed to model and analyze piping systems.

There is, however, a formula which may be used to determine whether the more complex analysis is required. The conditions for applying this formula are:

- The piping system must be of uniform size
- The formula pertains only to ferrous systems
- It must be restrained only by two anchors (one at each end)
- It must be less than 7000 cycles during its life (which is fairly common)
- It should not be used with unequal leg U-bends ($L/U > 2.5$)
- It should not be used for large diameter, thin-wall pipe
- There is no assurance that the terminal reactions will be acceptably low.

If the above caveats are met, then the piping system requires no additional stress analysis if it satisfies the following equation:

$$DY/(L\text{-}U)^2 \leq 30\, S_A/E_C \text{ for Imperial units, or}$$
$$DY/(L\text{-}U)^2 \leq 208{,}000\, S_A/E_C \text{ for metric units}$$

where

D = nominal pipe size [in or mm]

Y = Resultant of movements to be absorbed by the pipe lines [in or mm]

L = Developed length of the piping [ft or m]

U = Length of a straight line joining the anchors [ft or m]

E_C = Modulus of Elasticity at room temperature [psi or kPa]

S_A = Allowable displacement stress range per Formula (1A)
$= f\,(1.25Sc + 0.25Sh)$

f = cyclic stress range factor for noncorroded pipe[7]
$= 6/N^{0.2} \leq 1.0$

N = total number of equivalent reference displacement stress range cycles expected during the service life of the piping

Sc = basic material allowable stress at the minimum metal temperature expected [psi or kPa][8]

Sh = basic material allowable stress at the maximum metal temperature expected [psi or kPa]

Example 4.3

Given: ASTM A53, Grade B pipe configuration as shown in Figure 4.8. The system is for daily heat-up of a process, with a design life of 15 years.

Diameter: 6 in

Operating Temperature: 250°F

Determine whether a stress analysis is required.

Solution:

$$D = 6 \text{ in}$$

$$\Delta T = 250°F - 70°F = 180°F$$

Coefficient of Linear Expansion $\alpha = 6.8 \times 10^{-6}$ in/in/°F from ASME B31.1 Table B-1
Modulus of Elasticity $E = 29.5 \times 10^6$ psi from ASME B31.1 Table C-1

$$S_c = 15.0 \text{ ksi}, S_h = 15.0 \text{ ksi from ASME B31.1 Table A-1}$$

$$N = (1 \text{ cycle/day})(365 \text{ days/year})(15 \text{ year}) = 5475 \text{ cycles}$$

$$f = 6/(5475)^{0.2} = 1.07, \text{ but } f \leq 1.0, \text{ so } f = 1.0$$

$$S_A = f(1.25Sc + 0.25Sh) = (1.0)[(1.25)(15.0 \text{ ksi}) + (0.25)(15.0 \text{ ksi})]$$

$$= 22.5 \text{ ksi}$$

$$30\, S_A/E_C = 22{,}500 \text{ psi}/(29.5 \times 10^6 \text{ psi}) = 0.022$$

Let δx, δy, δz be the displacements due to thermal expansion

$$\delta x = \alpha L_x(\Delta T) = (6.8 \times 10^{-6} \text{ in/in/°F})(8 \text{ ft} \times 12\text{in/ft})(180°F) = 0.12 \text{ in}$$

$$\delta y = \alpha L_y(\Delta T) = (6.8 \times 10^{-6} \text{ in/in/°F})(4 \text{ ft} \times 12\text{in/ft})(180°F) = 0.06 \text{ in}$$

$$\delta z = \alpha L_y(\Delta T) = (6.8 \times 10^{-6} \text{ in/in/°F})(6 \text{ ft} \times 12\text{in/ft})(180°F) = 0.09 \text{ in}$$

---

[7] A minimum value for f is 0.15, which results in an allowable displacement stress range for a total number of equivalent reference displacement stress range cycles greater than 10^8 cycles.

[8] Paragraph 102.4.3 states that the joint efficiency factor E does not need to be applied to the basic material allowable stresses, Sc and Sh. Also, for materials that have a minimum tensile strength higher than 70 ksi (480 MPa), the values of S_c and S_h must be no greater than 20 ksi (140 MPa).

Figure 4.8 Example 4.3.

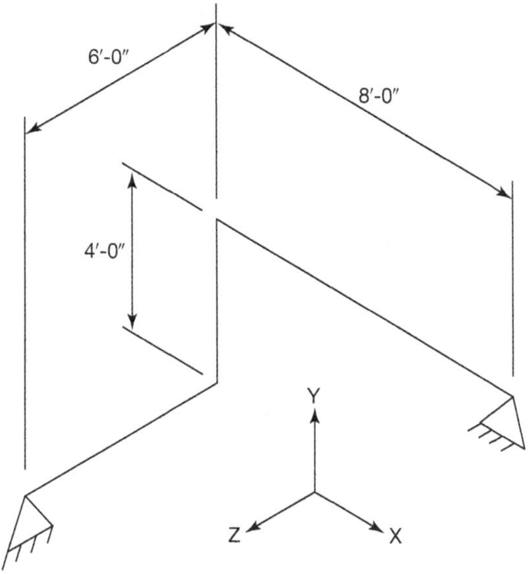

$$Y = (\delta x^2 + \delta y^2 + \delta z^2)^{1/2} = [(0.12\text{ in})^2 + (0.06\text{ in})^2 + (0.09\text{ in})^2]^{1/2}$$
$$= 0.16\text{ in}$$
$$L = 8\text{ ft} + 4\text{ ft} + 6\text{ ft} = 18\text{ ft}$$
$$U = (X^2 + Y^2 + Z^2)^{1/2} = [(8\text{ ft})^2 + (4\text{ ft})^2 + (6\text{ ft})^2]^{1/2}$$
$$= 10.8\text{ ft}$$
$$DY/(L - U)^2 = 6 \times 0.16/(18 - 10.8)^2 = 0.019 \leq 0.022$$

Therefore, a more detailed stress analysis is not required. The code takes care to note that this is an empirical relationship, with no proof of accuracy or consistently conservative results. Note that we did not apply units to this equation, since it is empirical, and the units do not cancel. Further, just because the calculation showed that a more detailed stress analysis was not required, it gives us no indication of the forces applied at the anchors.

Earlier versions of ASME B31.1 gave a simpler equation:

$$DY/(L\text{-}U)^2 \leq 0.03 \text{ for Imperial units, or}$$
$$DY/(L\text{-}U)^2 \leq 208.3 \text{ for metric units}$$

These formulas had the same conditions applied as the newer code version, but were much easier to use. Had we applied this equation, we would have determined that the system did not require a more formal analysis.

Because of the many limitations on the application of these formulas, it is probably better to just perform a more thorough stress analysis and disregard this limited flexibility analysis altogether.

137 Pressure Tests

After a pipe system is installed in the field, it is usually pressure tested to ensure that there are no leaks. Once a system is in operation, it is difficult, if not impossible, to repair leaks.

ASME B31.1 has established procedures for applying pressure tests to piping systems. There are generally two types of pressure tests applied to a piping system. One is a hydrotest and the other is a pneumatic test. The hydrotest is greatly preferred for the following reasons:

- Leaks are easier to locate.
- A hydrotest will lose pressure more quickly than a pneumatic test if leaks are present.
- Pneumatic tests are more dangerous, due to the stored pressure energy and possibility of rapid expansion should a failure occur.

On the other hand, if a piping system cannot tolerate trace levels of the testing medium (for instance, a medical oxygen system) then a pneumatic test is preferred.

137.4 Hydrostatic Testing

It is important to provide high point vents and low point drains in all piping systems to be hydrotested. The high point vents are to permit the venting of air, which if trapped during the hydrotest may result in fluctuating pressure levels during the test period. The drains are to allow the piping to be emptied of the test medium prior to filling with the operating fluid. (Low point drains are always a good idea though since they facilitate cleaning and maintenance.)

A hydrotest is to be held at a test pressure not less than 1.5 times the design pressure. The system should be able to hold the test pressure for at least 10 minutes, after which the pressure may be reduced to the design pressure while the system is examined for leaks. A test gauge should be sensitive enough to measure any loss of pressure due to leaks, especially if portions of the system are not visible for inspection.

The test medium for a hydrotest is usually clean water, unless another fluid is specified by the Owner. Care must be taken to select a medium that minimizes corrosion.

137.5 Pneumatic Testing

The test medium must be nonflammable and nontoxic. It is most often compressed air, but may also be nitrogen, especially for fuel gases or oxygen service. Note that compressed air often contains both oil and water, so care must be exercised in specifying an appropriate test medium.

A preliminary pneumatic test is often applied, holding the test pressure at 25 psig to locate leaks prior to testing at the test pressure. The test pressure for pneumatic tests is to be at least 1.2 but not more than 1.5 times the design pressure. The pneumatic test must be held at least 10 minutes, after which time it must be reduced to the lower of the design pressure or 100 psig (700 kPa gage) until an inspection for leaks is conducted.

If a high degree of sensitivity is required, other tests are available such as mass-spectrometer or halide tests.

Other portions of ASME B31.1 discuss various fittings, load cases, pipe hangers, systems specific to boiler piping, and welding requirements. We will follow-up on some of these areas in later portions of this book.

ASME B31.3 Process Piping

The term "process piping" may be thought of as any piping that does not fall under the other B31 codes. It is generally considered to be the piping that one may find in chemical plants, refineries, paper mills, and other manufacturing plants.

This code is structured similar to B31.1 in that it is organized into chapters, parts, and paragraphs. Note that while the paragraphs of B31.1 are numbered in the 100s, those in B31.3 are numbered in the 300s. This convention follows throughout the B31 codes.

There are several very important concepts in this code that should be identified before we delve too far into the particulars. Because we have entered into the realm of process piping, it is necessary to recognize some of the inherent hazards associated with handling dangerous chemicals.

One concept is "damaging to human tissue." This describes a fluid that could cause irreversible damage to the skin, eyes, or mucous membranes unless the fluid is promptly flushed away with water, or medication or antidotes are administered promptly.

In many manufacturing facilities, you will see eyewash stations or emergency showers, or a device which is a combination of both. These have evolved over the years into fairly sophisticated units, which may have remote alarms if activated to alert plant personnel to the fact that someone is in trouble. The Occupational Safety and Health Administration 29 CFR 1910.151 states that "...where the eyes or body of any person may be exposed to injurious corrosive materials, suitable facilities for quick drenching or flushing of the eyes and body shall be provided within the work area for immediate emergency use."

But what constitutes "suitable facilities?" The American National Standards Institute (ANSI) has developed a *voluntary* standard ANSI Z358.1 which recommends, among other things, that the shower be no more than 10 seconds walking time away from the location of the hazard without requiring the use of steps or ramps, that the water should be tepid, unless this would create an additional hazard due to acceleration of chemical reactions, and that the water be disposed of in a manner that will not create additional hazards. Tepid water can be mixed at the device using an ASSE 1071 device.

In practice, you may sometimes observe eyewash/shower units that have not been tested in years and that are covered with dirt and grime. This is an indication of a poorly maintained facility. The drains of eyewash and safety showers are often not piped to a sewer and many are not located near a floor drain. This may create some reluctance to test these devices. This is unfortunate, since the supply lines should be periodically flushed.

300.1 Scope

The scope of this code includes all fluids. This scope specifically excludes the following:

1. Piping with an internal design pressure between 0 and 15 psi (105 kPa)
2. Power boilers and BEP which is required to be in accordance with B31.1
3. Tubes inside fired heaters
4. Pressure vessels, heat exchangers, pumps, or compressors.

300.2 Definitions

There are several very important definitions included in this paragraph under the term "fluid service":

a) Category D fluids are those in which all of the following apply:
   1. The fluid is nonflammable, nontoxic, and not damaging to human tissue.
   2. The design pressure does not exceed 150 psig (1035 kPa).
   3. The design temperature is between -20°F and 366°F (-29°C and 186°C).

b) Category M fluids are those in which a single exposure to a very small quantity could lead to serious irreversible harm, even if prompt restorative measures are taken.

c) High pressure fluids are those in which the Owner has specified that the pressures will be in excess of that allowed by the ASME B16.5 PN420 (Class 2500) rating for the specified design temperature and material group.

d) Normal fluids are everything else that does not fit into the above categories. These are the fluids most often used with this code.

301 Design Conditions

This section requires the designer to consider the various temperatures, pressures, and loads that the piping system may be subject to. While it is a good checklist, most of the items contained are common sense.

301.3.2 Uninsulated Components

This paragraph describes how to determine the design temperature of uninsulated piping and components. Of particular interest is the description of how to determine component temperatures using the fluid temperature. The instructions indicate that for fluid temperatures above 150°F (65°C) the temperature for uninsulated components shall be no less than a certain percentage of the fluid temperature. For example, the temperature used for lap joint flanges shall be 85 percent of the fluid temperature.

Note that unless you use the absolute temperature in degrees Rankine or Kelvin, such a calculation has no meaning, since a percentage cannot be applied to the Fahrenheit or Celsius scales.

302 Design Criteria

Note that B31.3 also has a Table 326.1 that corresponds to B31.1's Table 126.1. A comparison between the two tables shows that Table 126.1 is focused more on steel pipe and fittings, while Table 326.1 pertains more to nonmetallic pipe and fittings. The obvious reason is that process piping deals with more fluids that are corrosive to steel. In many cases, thermoplastics, thermosetting plastics, and resins will be more appropriate materials for the fluids handled in the purview of the process piping code.

This set of paragraphs states that if the components listed in Table 326.1 are rated for a specific temperature/pressure condition, then they are suitable for the design pressures and temperatures allowed by this code. If they have no specific temperature/pressure rating, but are instead based on the ratings of straight seamless pipe, then the component must be de-rated by 12.5 percent, less any mechanical and corrosion allowances.

In other words, you have to determine the minimum wall thickness of the straight pipe based on the design temperature and pressure, as well as the mechanical and corrosion allowances. Once you apply the mill tolerance of 12.5 percent, you will be safe in selecting a fitting that satisfies the same requirements as the straight pipe to which it is connected.

302.2.4 Allowances for Pressure and Temperature Variations

There are paragraphs in both B31.1 and B31.3 that describe allowable deviations from operating conditions. These are called "allowances for pressure and temperature variations." The rules for such operating excursions are not complicated, but in practice industrial users do not chart how often the operating pressures exceed the allowable pressures.

Most often, any pressure excursions are prevented through the use of pressure relief devices, such as pressure relief valves, pressure safety valves, or rupture disks.

Also, it is important to note that the allowable stresses are temperature dependent. So if there are temperature excursions (as allowed for in both B31.1 and B31.3) the allowable stress may vary. Unless someone has taken the trouble to build a database of the relationships between operating temperature and pressure, and allowable temperature and pressure, then the designer will be well-advised to base the design pressure on the MAXIMUM temperature that the system will ever see, and not to rely on the allowance for temperature variations.

Therefore, from a practical standpoint, it is best to not rely upon any allowances for temperature or pressure excursions above the design conditions. Choose your design conditions so that the temperature and pressure will not be exceeded.

304.1.1 Pressure Design of Straight Pipe

As we noted above, paragraph 302 requires us to calculate the required wall thickness to satisfy the design temperature and pressure conditions. We did the same thing for B31.1. But B31.3 handles things a little differently.

$$t_m = t + c \qquad \text{Formula (2) in the code}$$

$$t = PD/2(SEW + PY) \qquad \text{Formula (3a)}$$

$$t = P(d+2c)/2[SEW - P(1\text{-}Y)] \qquad \text{Formula (3b)}$$

where

t_m = Minimum required wall thickness [in or mm]. This minimum wall thickness includes any mechanical, corrosion, or erosion allowances.

If the piping system contains bends (not elbows), then you also must compensate for thinning of the bends, as in ASME B31.1. Because this is not common, the interested reader is referred to ASME B31.3 Paragraph 304.2.1 for the formulas required to determine after-bend thicknesses

t = Pressure design thickness, as determined by any of the Formulas (3a) through (3b) [in or mm].

c = Mechanical, corrosion, or erosion allowances [in or mm]. Note that for unspecified tolerances on thread or groove depth, the code specifies that an additional 0.02 in (0.5 mm) shall be added to the depth of the cut to take the unspecified tolerance into account.

T = Pipe wall thickness, either measured or minimum per purchase specification [in or mm].

Unless specially ordered (which is usually prohibitively expensive) piping is generally purchased based on commercially available schedules (or wall thicknesses). These thicknesses must take into account the mill tolerance which may be as much as 12.5 percent less than the nominal thickness.

Therefore, under ordinary circumstances, the pipe wall thickness (T) will be 87.5 percent of the thickness of the listed schedule.

d = Inside diameter of pipe [in or mm].

P = Internal design gage pressure [psig or kPa (gage)]

The pressure is either given, or solved for in the equations.

D = Outside diameter of pipe [in or mm]

The outside diameter will be the OD of a commercially available pipe. Carbon steel pipe dimensions are shown in Appendix 1 of this text.

E = Quality Factor from ASME Table A-1A or A-1B. Table A-1A relates exclusively to castings. Table A-1B relates to longitudinal weld joints. The quality factor is a means of de-rating the pressure based on the material and method of manufacture. Thus, for A106 seamless pipe, the quality factor E = 1.00.

Casting quality factors may be increased if the procedures and inspections listed in ASME B31.3 Table 302.3.3C are utilized.

The quality factors are in place to account for imperfections in castings, such as inclusions and voids. Machining all of the surfaces of a casting to a finish of 250 micro inches (6.3 μm) improves the effectiveness of surface examinations such as magnetic particle, liquid penetrant, or ultrasonic examinations.

The Quality Factor E is analogous to the Weld Joint Efficiency E or Casting Factor F in B31.1. **But note that the Quality Factor E in B31.3 is NOT included in the stress values provided in B31.3 Tables A-1 and A-2. See Paragraph 302.3.1(a).**

S = Stress in material at the design temperature [psi or kPa].

These values are tabulated in ASME B31.3, Appendix A. Note that they are dependent on the temperature to which the material will be exposed. This temperature is the metal temperature. This would normally be the temperature of the fluid in the pipe, but if a pipe was to be exposed to a high temperature externally, it would be the fluid temperature outside the pipe. See also Paragrah 301.

Once again, note that the values tabulated in ASME B31.3 Appendix A DO NOT include the Quality Factors. Therefore, the tabulated values are only the values of S.

W = Weld Joint Strength Factor. This factor accounts for the long term strength of weld joints at elevated temperatures. In the absence of specific data such as creep testing, W is taken as 1.0 at temperatures of 950°F (510°C) and below. W falls linearly to 0.5 at 1500°F (815°C).

Y = A coefficient used to account for material creep, as in B31.1. The table of Y coefficients in B31.3 is virtually identical to the table given in B31.1. As previously noted, the variation of Y with temperature allows the wall thickness equation to behave in accordance to the "Modified Lamé Equation" at low temperatures (with Y = 0.4), and in accordance with a creep-rupture equation at high temperatures (with Y = 0.7).

The values are taken from Table 304.1.1 for $t < D/6$. For ductile metals (including steel), the value is 0.4 across the range of temperatures. For $t \geq D/6$,

$$Y = (d + 2c)/(D + d + 2c)$$

Note that the difference between Formulas (3a) and (3b) is that (3a) begins with the OD of the pipe, and (3b) begins with the ID of the pipe.

B31.3 Chapter VII deals with nonmetallic piping and piping lined with nonmetals. Paragraph A304 explains how to calculate minimum wall thicknesses in such cases. The method and equations closely parallel Paragraph 304.1.1. The Allowable Stresses are replaced with Hydrostatic Design Stresses for nonmetals in Table B-1.

Pressure design of high-pressure piping (pressures in excess of the Class 2500 rating for the design temperature and material group), is covered in Paragraph K304. The equations given use Table K-1 for the basic allowable stresses. These stresses are higher than those listed in Table A-1 for the same materials.

Until recent editions of the code introduced the Weld Joint Strength Reduction Factor W, the code permitted the use of a simpler Formula (3b) for calculating pressure design thickness:

$$t = PD/2SE$$

The computation was simpler and because the denominator did not contain the addition of the PY term, this led to consistently more conservative results. With the possibility of $W < 1.0$ in the equation, this formula could lead to less conservative pressure design thicknesses.

However, if the metal temperatures are less than 950°F (510°C) or in the absence of specific creep testing data, then $W = 1.0$ and this formula could still be applied with confidence.

304.3 Branch Connections

Similar to B31.1, B31.3 permits fabrication of branch connections. We are faced with the following prerequisites:

1. The run pipe diameter-to-thickness ratio $(D_h/T_h) < 100$ and the branch-to-run diameter ratio (D_b/D_h) is not greater than 1.0. If we examine the tables of commercially-available pipe data, we see that the first condition in which we might see a diameter-to-thickness ratio in excess of 100 would be 24 in diameter Schedule 5S, which has a wall thickness of 0.218 in. For thicknesses above the standard wall thickness, there is little chance that the ratio will exceed 100. So it is clear that while this is an important consideration for large bore, thin wall pipes, it is a situation that most of us are not likely to encounter. Looking next at the requirement that the branch-to-run ratio is not greater than 1.0, we see that this means only that it is impossible to stub a larger branch onto a smaller run. And if we tried to do that, we might be inclined to reverse the names of the branch and run. In other words, the branch pipe is always the smaller diameter, unless they are both the same diameter. Whichever pipe is designated as the run pipe, must still satisfy $(D_h/T_h) < 100$.
2. If $D_h/T_h \geq 100$, the branch diameter D_b has to be less than one-half the run diameter D_h.
3. The angle between the branch and run is at least 45°. See Figure 4.8. This is analogous to B31.1's angle α. We will examine the similarities and differences between B31.1 and B31.3 regarding the branch connection calculations.
4. The axes of the branch and run pipe must intersect each other. This was also a requirement of B31.1.

Once the prerequisites are established, we can examine any requirements for reinforcing the branch connection.

Here is an example of how the codes complicate matters. The names assigned to the various reinforcing areas in B31.3 are different than those we examined in B31.1. See Table 4.1 for a comparison of the terminology as they pertain to branch connections.

For B31.3:

A_1 = Required reinforcement area
$= t_h d_1 (2 - \sin \beta)$ Formula (6)

A_2 = Area due to excess thickness in the run pipe wall
$= (2d_2 - d_1)(T_h - t_h - c)$ Formula (7)

A_3 = Area due to excess thickness in the branch pipe wall
$= 2L_4 (T_b - t_b - c) / \sin \beta$ Formula (8)

A_4 = Area provided by welds and attached reinforcements

d_1 = Effective length of run pipe removed at the branch
$= [D_b - 2(T_b - c)] / \sin \beta$

d_2 = Half-width of the reinforcement zone
= The greater of d_1 or $(T_b - c) + (T_h - c) + d_1/2$, but not more than D_h

L_4 = Height of the reinforcement zone
= The lesser of $2.5(T_b - c) + T_r$ or $2.5(T_h - c)$

T_r = Minimum thickness of re-pad or saddle made from pipe

β = Smaller of the angles between the intersecting axes of the Branch and Run pipes

The subscripts b and h refer to the branch and run pipe (or header pipe) respectively.

In order to satisfy the reinforcement requirements of B31.3,

$$A_2 + A_3 + A_4 \geq A_1 \quad \text{Formula (6a)}$$

As in B31.1, we have a reinforcement zone bounded by the parallelogram shown in Figure 4.9. Also, as in B31.1, there are also specific requirements for closely spaced nozzles (overlapping reinforcement zones) and branch connections subject to external pressure, forces, or moments.

Let's turn our attention to some examples so that we can compare B31.1 with B31.3.

Example 4.4

We will use the same design parameters as Example 4.1, but will instead analyze it using the rules of B31.3.

Given: Design Pressure: 165 psig

Temperature: 675°F

Service: Superheated Steam inside a powerhouse

Code: ASME B31.3

Design: A fabricated 4 in branch connection at 90°, coming off an 8 in diameter header.

Solution:

Again we assume that the material of construction is A53 Grade A, ERW (electric resistance welded) pipe.

We have to determine the required wall thickness of each pipe, by applying one of the formulas in Section 304.1.2.

By Formula (3a):

$$t = PD/2(SEW + PY)$$

P and D are given, and we know that W = 1.0 since the temperature is less than 950°F. We need to obtain the values of S, E, and Y.

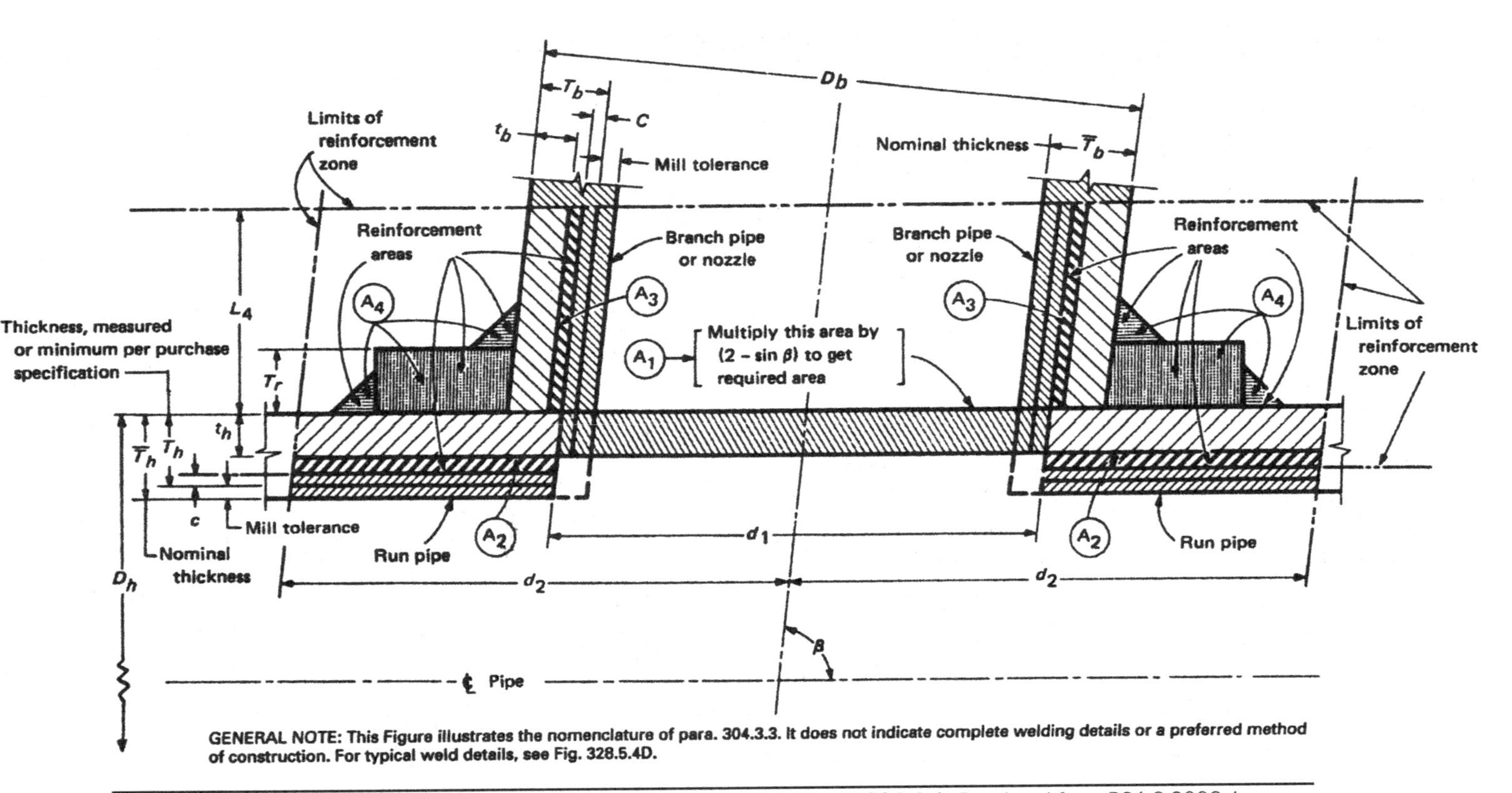

GENERAL NOTE: This Figure illustrates the nomenclature of para. 304.3.3. It does not indicate complete welding details or a preferred method of construction. For typical weld details, see Fig. 328.5.4D.

FIGURE 4.9 Branch connection nomenclature from ASME B31.3-2006, Figure 304.4.4. *Reprinted from B31.3-2006, by permission of The American Society of Mechanical Engineers. All rights reserved.*

Area	B31.1 and B31.9	B31.3
Required Reinforcement Area	A_7	A_1
Area due to Excess Thickness in Run Pipe Wall	A_1	A_2
Area due to Excess Thickness in Branch Pipe Wall	A_2	A_3
Weld Area	A_3	A_4
Re-pad	A_4	A_4
Saddle	A_5	A_4
Half-Width of Reinforcement Zone along Run Pipe Axis	d_2	d_2
Height of Reinforcement Zone	L_4	L_4
Smaller Angle between the intersecting axes of the Branch and Run Pipes	α	β

TABLE 4.1 Comparison of Branch Connection Terminology

We find S in Table A-1. For A-53 Grade A, we find that the basic allowable stress (S) at 675°F lies between 14.4 and 14.5 ksi. We interpolate and use 14,450 psi, and note a significant increase in the allowable stresses assigned to the same material we used in Example 4.1. Under the tables in B31.1, we found that at this temperature, A-53 Grade A had an allowable stress of 10,050 psi. This is a difference of nearly 44 percent.

In comparing the two codes you will find that, in general, at least at temperatures up to about 200°F, the stresses in B31.1 are approximately 75 percent of those in B31.3.

We find E in Table A1-B. For A-53, there are three classes or types of pipe listed. Our example uses ERW pipe, so we see that the Basic Quality Factor for Longitudinal Weld Joints, E_j is 0.85.

Carbon steel is a ferritic steel, so in Table 304.1.1 we find Y = 0.4.

Applying Formula (3a):

$$t_h = (165 \text{ psi})(8.625 \text{ in}) \,/\, 2[(14{,}450 \text{ psi})(0.85)(1.0) + (165 \text{ psi})(0.4)] = 0.0576 \text{ in}$$

Note that we have not yet applied any corrosion allowance.

Applying Formula (3a) for the branch connection.

$$t_b = (165 \text{ psi})(4.500 \text{ in}) \,/\, 2[(14{,}450 \text{ psi})(0.85)(1.0) + (165 \text{ psi})(0.4)] = 0.0301 \text{ in}$$

Next we apply the corrosion allowance and find a commercial schedule that will satisfy the pressure requirements of B31.3.

Let c = 0.0625 in. This is a typical value.

$$t_{mh} = t_h + c = 0.0576 \text{ in} + 0.0625 \text{ in} = 0.1201 \text{ in}$$

We need to select a commercially-available wall thickness that is at least 0.120 inch, minus the 12.5 percent mill tolerance on wall thickness. Schedule 20 has a wall thickness of 0.250 in, which would satisfy the required thickness. But let's use the Schedule 40 pipe that we also used in Example 4.1, so that we can draw a more meaningful comparison between B31.1 and B31.3.

$$T_h = 0.322 \text{ in} \times 0.875 = 0.282 \text{ in}$$

For the 4 in branch the minimum wall thickness will be

$$t_{mb} = 0.0301 \text{ in} + 0.0625 \text{ in} = 0.0926 \text{ in}$$

For the 4 in branch, we select Schedule 40. Schedule 10 is also available in 4 in carbon steel, but light schedules were introduced to save money on low-hazard services such as fire-protection and compressed air. We will want to use Schedule 40, which has a wall thickness of 0.237 in. We also used Schedule 40 for the branch pipe in Example 4.1.

$$T_b = 0.237 \text{ in} \times 0.875 = 0.207 \text{ in}$$

We now calculate the reinforcement zone dimensions.

$$\begin{aligned} d_1 &= [D_b - 2\,(T_b - c)] \,/\, \sin\beta \\ &= [4.500 \text{ in} - 2\,(0.207 \text{ in} - 0.0625 \text{ in})] \,/\, \sin 90^\circ \\ &= 4.500 \text{ in} - 0.289 \text{ in} \\ &= 4.211 \text{ in} \end{aligned}$$

d_2 = the greater of d_1 or $(T_b - c) + (T_h - c) + d_1/2$, but not more than D_h

$$(T_b - c) + (T_h - c) + d_1/2 = (0.237 \text{ in} - 0.0625 \text{ in}) + (0.282 \text{ in} - 0.0625 \text{ in}) + 4.211 \text{ in}/2 = 2.500 \text{ in}$$

$$D_h = 8.625 \text{ in}$$

Therefore, $d_2 = d_1 = 4.211$ in

L_4 = the lesser of $2.5\,(T_b - c) + T_r$ or $2.5\,(T_h - c)$

($T_r = 0$ since we don't yet know if we have a re-pad.)

$$\begin{aligned} 2.5\,(T_b - c) + T_r &= 2.5\,(0.207 \text{ in} - 0.0625 \text{ in}) + 0 \\ &= 0.361 \text{ in} \end{aligned}$$

$$\begin{aligned} 2.5\,(T_h - c) &= 2.5(0.282 \text{ in} - 0.0625 \text{ in}) \\ &= 0.549 \text{ in} \end{aligned}$$

Therefore, $L_4 = 0.361$ in

We may now compute the required reinforcement area A_1 using Formula (6).

$$\begin{aligned} A_1 &= t_h d_1\,(2 - \sin\beta) \\ &= (0.0576 \text{ in})(4.211 \text{ in})(2 - \sin 90^\circ) \\ &= 0.243 \text{ in}^2 \end{aligned}$$

A_2 = Area due to excess thickness in the run pipe wall

$$= (2d_2 - d_1)\,(T_h - t_h - c) \qquad \text{Formula (7)}$$

$$= [2\,(4.211 \text{ in}) - 4.211 \text{ in}]\,(0.282 \text{ in} - 0.0576 \text{ in} - 0.0625 \text{ in})$$

$$= 0.682 \text{ in}^2$$

We note that we can stop here, since the area due to excess pipe wall in the run exceeds the required reinforcement area. Comparing the results of B31.1 with B31.3, we see that:

Description	B31.1	B31.3
Required Reinforcement Area	0.284 in^2	0.243 in^2
Area due to Excess Thickness in Run Pipe Wall	0.600 in^2	0.682 in^2
Allowable Stress	SE = 10,050 psi	SE = 12,300 psi

These were the identical pipe configurations and materials. We eventually arrived at the same result, that is, no additional material such as a re-pad was required. But notice the difference in the values calculated between the two codes.[9]

Let's re-work the same problem with the design parameters given in Example 4.2.

Example 4.5
Given: Design Pressure: 1000 psig
Temperature: 675°F
Service: Superheated Steam inside a powerhouse
Material: A53 Grade A, ERW (electric resistance welded) pipe.
Code: ASME B31.3
Design: A fabricated 4 in branch connection at 90°, coming off an 8 in diameter header.

Solution:

By Formula (3a):

$$t_h = (1000\text{ psi})(8.625\text{ in}) / 2[(14{,}450\text{ psi})(0.85)(1.0) + (1000\text{ psi})(0.4)] = 0.340\text{ in}$$

$$t_b = (1000\text{ psi})(4.500\text{ in}) / 2[(14{,}450\text{ psi})(0.85)(1.0) + (1000\text{ psi})(0.4)] = 0.177\text{ in}$$

Let c = 0.0625 in

$$t_{mh} = t_h + c = 0.340\text{ in} + 0.0625\text{ in} = 0.403\text{in}$$

We select Schedule 80, with a wall thickness of 0.500 in. The actual minimum wall thickness of 8 in diameter Schedule 80 pipe, less the mill tolerance is

$$T_h = 0.500\text{ in} \times 0.875 = 0.438\text{ in}$$

For the 4 in branch the minimum wall thickness will be

$$t_{mb} = 0.177\text{ in} + 0.0625\text{ in} = 0.240\text{ in}$$

For the 4 in branch, we select Schedule 80, with a wall thickness of 0.337 in.

$$T_b = 0.337\text{ in} \times 0.875 = 0.295\text{ in}$$

We now calculate the reinforcement zone dimensions.

[9] Note also that for a superheated steam line as described, the appropriate code application would be ASME B31.1. The B31.3 example is provided for instruction only.

$$
\begin{aligned}
d_1 &= [D_b - 2(T_b - c)] / \sin\beta \\
&= [4.500 \text{ in} - 2(0.295 \text{ in} - 0.0625 \text{ in})] / \sin 90^\circ \\
&= 4.500 \text{ in} - 0.465 \text{ in} \\
&= 4.035 \text{ in}
\end{aligned}
$$

d_2 = the greater of d_1 or $(T_b - c) + (T_h - c) + d_1/2$, but not more than D_h

$$(T_b - c) + (T_h - c) + d_1/2 = (0.295 \text{ in} - 0.0625 \text{ in}) + (0.438 \text{ in} - 0.0625 \text{ in}) + 4.035 \text{ in}/2 = 2.626 \text{ in}$$

$D_h = 8.625$ in

Therefore, $d_2 = d_1 = 4.035$ in

L_4 = the lesser of $2.5(T_b - c) + T_r$ or $2.5(T_h - c)$

($T_r = 0$ since we don't yet know if we have a re-pad)

$$
\begin{aligned}
2.5(T_b - c) + T_r &= 2.5(0.295 \text{ in} - 0.0625 \text{ in}) + 0 \\
&= 0.581 \text{ in}
\end{aligned}
$$

OR

$$
\begin{aligned}
2.5(T_h - c) &= 2.5(0.438 \text{ in} - 0.0625 \text{ in}) \\
&= 0.939 \text{ in}
\end{aligned}
$$

Therefore, $L_4 = 0.581$ in

We may now compute the required reinforcement area A_1 using Formula (6).

$$
\begin{aligned}
A_1 &= t_h d_1 (2 - \sin\beta) \\
&= (0.340 \text{ in})(4.035 \text{ in})(2 - \sin 90^\circ) \\
&= 1.372 \text{ in}^2
\end{aligned}
$$

A_2 = Area due to excess thickness in the run pipe wall

$$
\begin{aligned}
&= (2d_2 - d_1)(T_h - t_h - c) \qquad \text{Formula (7)} \\
&= [2(4.035 \text{ in}) - 4.035 \text{ in}](0.438 \text{ in} - 0.340 \text{ in} - 0.0625 \text{ in}) \\
&= 0.143 \text{ in}^2
\end{aligned}
$$

A_3 = Area due to excess thickness in the branch pipe wall

$$
\begin{aligned}
&= 2L_4(T_b - t_b - c) / \sin\beta \qquad \text{Formula (8)} \\
&= 2(0.581 \text{ in})(0.295 \text{ in} - 0.177 \text{ in} - 0.0625 \text{ in}) / \sin 90^\circ \\
&= 0.065 \text{ in}^2
\end{aligned}
$$

A_4 = Area provided by welds and attached reinforcements

Let's assume that the welds are 5/16 in per leg. The weld area will be:

$$A_{4\,(\text{weld})} = 2(1/2)(0.3125 \text{ in})(0.3125 \text{ in}) = 0.098 \text{ in}^2$$

The factor "2" is because we have the same fillet weld triangle on both sides of the branch. We still have a deficit of reinforcing material of:

$$1.372 \text{ in}^2 - 0.143 \text{ in}^2 - 0.065 \text{ in}^2 - 0.098 \text{ in}^2 = 1.066 \text{ in}^2$$

We need a re-pad. B31.3 does not mention the aspect ratio of 4:1 that B31.1 recommends. But we can still use it, provided it does not lay outside the reinforcement zone. The half-width of the reinforcement zone is $d_2 = 4.035$ in. The OD of the branch pipe is 4.500 in. Again we want the branch fillet welded to the run before the repad is applied, so the minimum ID of the repad is

$$4.5 \text{ in} + 2(5/16 \text{ in}) = 5.125 \text{ in}$$

This leaves us with

$$4.035 \text{ in} - 5.125 \text{ in}/2 = 1.473 \text{ in as the limit for the width of a re-pad.}$$

We select a thickness approximately one-fourth of the width. If we use a 3/8 in plate by 1¼ in wide, then

$$A_{4(\text{re-pad})} = 2 \times 0.375 \text{ in} \times 1.25 \text{ in} = 0.9375 \text{ in}^2$$

In practice, the 5/16 in annulus between the ID of the repad and the OD of the branch pipe will be filled with weld to the height of the repad, so this adds to the available metal within the reinforcement zone

$$A_{4(\text{weld})} = 2(3/8 \text{ in})(5/16 \text{ in}) + .098 \text{ in}^2 = 0.332 \text{ in}^2$$

The additional metal added by the repad and weld is

$$A_4 = 0.9375 \text{ in}^2 + 0.332 \text{ in}^2 = 1.26 \text{ in}^2$$

and the required reinforcement area is satisfied.

Note that we must choose a re-pad material that is compatible with the metallurgy of the adjoining pipes. Also, if the allowable stress of the re-pad is less than that of the run pipe, its area must be increased in proportion to the ratio of the allowable stress of the run pipe over the re-pad. If the re-pad allowable stress is more than the allowable stress of the run pipe, no credit may be taken.

Comparing the results of this example to Example 4.2, we see that we were able to use a Schedule 80 header for B31.3 instead of a Schedule 100 header for B31.1.

305.2 Specific Requirements

This section describes what piping may be used for certain services. A review of the four fluid services reveals that the most benign service is Category D, followed by normal fluids, and then perhaps high pressure fluids with Category M fluids constituting the most hazardous service. If you were dealing with water in a plant it would most likely be a Category D fluid service, regardless of whether it is potable or cooling water or general service water. Most water service in a manufacturing facility does not exceed 150 psig or 366°F.

Paragraph 305.2.1 specifies that there are three pipe specifications that are suitable only for these benign Category D services. They are:

- API 5L, Furnace Butt-Welded
- ASTM A53, Type F (also Furnace Butt-Welded)
- ASTM A 134 if made from other than ASTM A285 plate

These pipes are not suitable for the more hazardous services of B31.3. The furnace butt-weld pipes have a Weld Joint Quality Factor E_j of only 0.6. These pipes are made with a continuous longitudinal butt weld. The furnace butt weld process is a continuous forge weld that is made through the application of mechanical pressure. It is not as strong as electric resistance welded pipe, electric fusion welded pipe, or seamless pipe.

Paragraph 305.2.2 discusses pipe that requires safeguarding. The two pipe specifications that require safeguarding for services other than Category D are:

- ASTM A 134 if made from ASTM A285 plate
- ASTM A139.

Because safeguarding is an added expense in terms of both design and installation, the designer would be better off using a more suitable pipe material. But this avoids the concept of safeguarding altogether, which may be required in other circumstances.

For example, suppose you are faced with designing a piping system that handles dilute hydrochloric acid. You realize that it is corrosive to carbon steel, and stainless steel is not suitable due to the possibility of stress corrosion cracking. You select a PVC piping system that is impervious to the HCl, but might not hold up so well against fork truck traffic. The pipe system must be "safeguarded."

ASME B31.3 Appendix G addresses the concept of safeguarding. Safeguarding is the provision of protective measures to minimize the risk of accidental damage to a piping system, or to mitigate the consequences of a possible pipe failure. Such provisions include, but are not limited to:

- Physical barriers
- Guards around pipe flanges to prevent spraying of fluids if a gasket fails
- Isolation of hazardous areas
- Installation of fire protection systems
- Process controls to shut down systems in the event of a failure
- Grounding of static charges to prevent ignition of flammable vapors
- Implementation of special operating or maintenance procedures.

These are considerations that are often addressed in a HAZOP analysis.

Paragraph 305.2.3 lists piping that may be used under severe cyclic conditions. Severe cyclic conditions are defined in Paragraph 300.2 as those in which the number of cycles exceeds 7000, and the Displacement Stress Range (S_E) exceeds the 80 percent of the Allowable Displacement Stress Range (S_A).

As in ASME B31.1, the cycles referred to in this section of B31.3 are temperature cycles and not pressure cycles.

The allowable displacement stress (S_A) is dependent only on the material, the temperature fluctuations during the operating cycles, and the number of cycles.

The Displacement Stress Range (S_E) is dependent only on the loading conditions and the pipe geometry.

The calculation of the resultant stresses is best left to a computer program, as it becomes cumbersome. For the moment, it is enough to realize that severe cyclic conditions are not common, since thermal cycles above 7000 over the expected life of the system are not common.

319.4 Flexibility Analysis

Like B31.1, there is a simple test to determine if a more complex stress analysis is required. The prerequisites are the same as for B31.1, but the equation is slightly different, using the outside diameter rather than the nominal diameter. The threshold values are different as well.

$$Dy/(L - U)^2 \leq K_1 \qquad \text{(Formula 16) in the code}$$

where

$K_1 = 30\, S_A/E_a\ (in/ft)^2$

$= 208{,}000\, S_A/E_a\ (mm/m)^2$

S_A = Allowable Displacement Stress [ksi or MPa]

E_a = Modulus of Elasticity at 70°F [ksi or MPa]

f = may be as high as 1.2 for some ferrous materials.

D = Outside diameter of the pipe[in or mm]

Example 4.6

Using the same system as Example 4.3, determine if a more complex analysis is required using Formula 16.

Solution: The left hand side of the equation would appear to be identical to the left hand side of the equation we used for B31.1 in Example 4.3. Interestingly, or perhaps oddly, ASME B31.3 uses a different Coefficient of Thermal Expansion than we found in B31.1. This may affect the thermal displacements.

$\alpha = 6.49 \times 10^{-6}$ in/in/°F from ASME B31.3 Table C-3

$\delta x = \alpha L_x(\Delta T) = (6.49 \times 10^{-6}$ in/in/°F)(8 ft × 12 in/ft)(180°F) = 0.11 in

$\delta y = \alpha L_y(\Delta T) = (6.49 \times 10^{-6}$ in/in/°F)(4 ft × 12 in/ft)(180°F) = 0.06 in

$\delta z = \alpha L_y(\Delta T) = (6.49 \times 10^{-6}$ in/in/°F)(6 ft × 12 in/ft)(180°F) = 0.08 in

$y = (\delta x^2 + \delta y^2 + \delta z^2)^{1/2} = [(0.11\ in)^2 + (0.06\ in)^2 + (0.08\ in)^2]^{1/2} = 0.15$ in

U = 10.8 ft and L = 18 ft as before. $Dy/(L - U)^2 = (6.625)(0.15)/(18 - 10.8)^2 = 0.019$

$S_A = f\,(1.25\, S_c + 0.25\, S_h)$

f = 1.07 this time, since the upper limit is 1.2.

$S_c = S_h$ = 20.0 ksi from ASME B31.3 Table A-1

S_A = (1.07)[(1.25)(20.0 ksi) + (0.25)(20.0 ksi)] = 32,100 psi

$E_a = 29.5 \times 10^6$ psi from ASME Table C-6

$K_1 = 30(32{,}100\ psi/29.5 \times 10^6\ psi) = 0.032$

Then $Dy/(L - U)^2 = 0.019 \leq 0.032$ and the system again does not require a more complex analysis, other than to determine if the loads applied at the anchors are excessive.

345 Testing

B31.3 requires that all piping designed in accordance with B31.3 be leak tested. For the benign Category D fluids, a service test may be conducted using the service fluid as the test medium, and setting the test pressure at the operating pressure. This is in lieu of conducting a hydrostatic test. Of course, a hydrostatic test may be applied at the Owner's discretion. It is not required however.

The B31.3 hydrostatic test is similar to that described in B31.1. It is most often conducted with clean water, unless that would pose a problem such as contamination or corrosion, and it is held for 10 minutes at 1.5 times the design pressure.

Due to the possibility of brittle fracture of nonmetallic piping which may be found in systems under the scope of B31.3, a pneumatic leak test requires a pressure relief device having a set pressure of the test pressure plus the smaller of 50 psi or 10 percent of the test pressure.

Because chemical piping can involve core complicated equipment and piping designs, there may be additional factors to be considered in a pressure test.

Internal piping of a jacketed line should be tested at the more critical of either the internal or jacket design pressure (Paragraph 345.2.5).

Because there may be elevated temperatures, Paragraph 345.4.2 includes a provision for establishing a more appropriate test pressure:

$$P_T = 1.5\,P\,(S_T/S) \qquad \text{Equation (24) in the code}$$

and $S_T/S < 6.5$

where

P_T = Minimum test gage pressure

P = Internal design gage pressure

S_T = Stress value at test temperature

S = Stress value at design temperature as listed in Table A-1.

For high pressure piping (pressures in excess of the Class 2500 rating for the design temperature and material group), the limit of $S_T/S < 6.5$ does not apply. Further, the allowable stresses are taken from Table K-1 rather than from Table A-1. (Paragraph K345.4.2)

ASME B31.9 Building Services Piping

The scope of this code envelopes industrial, institutional, commercial, public buildings, and multi-unit residences. Because the most demanding service that one might encounter in such a facility would be steam and condensate, one might expect this code to rely on ASME B31.1. In fact, there are many similarities between B31.9 and B31.1.

Both codes cover boiler external piping. However, B31.9 includes steam boilers up to 15 psig maximum, while B31.1 uses 15 psig as a lower limit of its scope. Similarly, B31.9 includes water heating units up to 160 psig maximum, while B31.1 uses 160 psig as its lower limit for hot water.

904 Pressure Design of Components

B31.9 permits pressure and wall thickness calculations to be performed in accordance with B31.1. Alternately, it permits these to be calculated using the following formulas:

$$t_m = (PD/2SE) + A \qquad \text{(Formula 1)}$$

$$P = 2SE\,(t_m - A)/D \qquad \text{(Formula 2)}$$

where the variables are defined as in ASME B31.1. As with B31.1, the Maximum Allowable Stress Values tabulated in Table A-1 already include the weld joint efficiency factor E. That is, the values tabulated are equal to SE.

Branch connection strength follows ASME B31.1.

Minimum flexibility is similar to the older version of the ASME B31.1 flexibility analysis, i.e.

$DY/(L\text{-}U)^2 \leq 0.03$ for Imperial units, or
$DY/(L\text{-}U)^2 \leq 208.3$ for metric units

Under B31.9:

- No cast iron fittings are permitted
- No more than two anchors and no intermediate restraints are present (same as B31.1)
- Two different pipe sizes may be used, but they may differ by only one size
- The least wall thickness is at least 75 percent of the greatest wall thickness.

These flexibility criteria are always to be applied with great care, and are invalid for unequal leg U-bends in which $L/U > 2.5$. Further, just because the system appears to satisfy the flexibility criterion, there is no assurance that the reactions at the anchors will be acceptable.

Summary of Code Comparisons

Allowable stress values for the most common metal piping at temperatures up to 100°F:

B31.1: 1/4 of the specified minimum tensile strength
B31.3: 1/3 of the specified minimum tensile strength
B31.9: 1/4 of the specified minimum tensile strength

Tabulated stress values shown in the "A" tables:

B31.1: Includes Quality Factor E or F
B31.3: Does NOT include Quality Factor. See B31.3 Tables A-1A or A-1B
B31.9: Includes Quality Factor E or F

Minimum required wall thickness:

B31.1: $t_m = [(PD_o) / 2(SE + Py)] + A$ Formula (3)
B31.3: $t_m = PD/2(SEW + PY) + c$ Formula (2) and Formula (3a)
B31.9: $t_m = PD/2SE + A$ Formula (1)

Internal design pressure:

B31.1: $P = [2SE(t_m - A)] / [D_o - 2y(t_m - A)]$ Formula (4)
B31.3: $P = 2(SEW + PY)(t_m - c) / D_o$
B31.9: $P = 2SE(t_m - A) / D_o$ Formula (2)

Minimum flexibility analysis:

The system is identical to a successfully operating installation or replaces one with a satisfactory service record, or satisfies the following:

B31.1: $DY/(L\text{-}U)^2 \leq 30\ S_A/E_C$ for Imperial units or
$DY/(L\text{-}U)^2 \leq 208{,}000\ S_A/E_C$ for metric units
B31.3: Virtually identical to B31.1, but there may be some slight variations in material values.

B31.9: $DY/(L\text{-}U)^2 \leq 0.03$ for Imperial units (Formula 9)

CHAPTER 5

Specifications and Standards

Starting with a blank sheet of paper can be either intimidating or liberating. Very often in a project the piping specifications are available in some form or another. They may be specific to the current project or left over from an earlier project but deemed to be suitable for the one you are working on. Most engineers collect specifications as they move from project to project. These specs tend to evolve over time, with appendages growing out of them to handle a situation that stung some poor engineer once upon a time. One danger in recycling old specifications is that they often refer to out-of-date or obsolete specifications and standards.

Whereas codes which are adopted by regulating authorities are required to be met or exceeded, specifications and standards are not necessarily mandatory requirements of regulating authorities, although they are often cited within the codes.

Standards are prepared by trade associations to provide guidance on how materials, equipment, or systems should be installed or operated. An example of this is the Compressed Gas Association, which offers technical advice such as CGA G-4 *Oxygen* . This standard provides design considerations that assist in the safe operation of oxygen piping systems. Some standards set dimensions for fittings, so that every time you need to connect one 150 lb 6 in flange to another 150 lb 6 in flange, the bolt holes line up and the raised faces of the flanges mirror each other on the opposite sides of the gasket. This should be true even if the flanges are made by different manufacturers.

Specifications can mean either those that are project-specific, or those that are developed to establish the requirements of a material like carbon steel or a type of pipe like A-53 Grade B, or a valve.

Project-specific specifications are those that are prepared by the engineer to describe the quality of the material and workmanship for a project. They may take the form of an outlined narrative or a table. Project-specific specifications will be discussed in a later chapter.

The other type of specification is most often developed by a standards organization, government agency, or trade association. See Table 5.1. When we speak of "specifications" in this chapter, this is the type of specification that we mean.

The codes, standards, and specifications are usually identified using the following convention:

ACRO SPEC-YR

Where

ACRO = the acronym of the organization which has developed the code, standard, or specification

SPEC = an alphanumeric identifier

YR = the year of the latest revision

The codes, standards, or specifications are updated in regular intervals so that the latest technology is available, and so it is important to know which revision is the latest. In this book, we will omit the revision year since it will be sufficient to identify the document.

A list of some of the common codes, standards, and specifications appears in Table 5.2. While this list is not meant to be exhaustive, it represents a practical collection of applicable references in one place. The engineer may search for available references using this table.

Note that some of the standards share joint responsibility between two organizations. This occurs frequently between ASME and ANSI, and a good example is one of the codes for flanges from NPS ½ to NPS 24. The standard is ASME/ANSI B16.5. This is often referred to as ANSI B16.5, and the convention in this book will be to use the organization that is more commonly identified with the standard. If the reader were to look for the standard, it would be easy to locate using only the one organization. But when the standard is located in a library, or on-line, it will be identified as ASME/ANSI B16.5.

For piping and fittings, it is often necessary to identify both the material specification and a specification that addresses the dimensional requirements of the component.

The presence of an asterisk next to a specification listed in Table 5.2 indicates that the specification is also listed in ASME B31.1 Table 126.1. Table 126.1 is important because it identifies "standard" piping components, which are suitable for use at temperature-pressure ratings specified by the manufacturing standard.

Table 5.3 lists the various ASME stamps required by the ASME codes.

Acronym	Name	Contact Data
ANSI	American National Standards Institute (formerly American Standards Association, ASA)	ANSI 1819 L Street, NW 6th floor Washington, DC 20036 USA (202) 293-8020 www.ansi.org
API	American Petroleum Institute	API 1220 L Street, NW Washington, DC 20005-4070 USA (202)682-8000 www.api.org
ASHRAE	American Society of Heating, Refrigeration, and Air-Conditioning Engineers	ASHRAE 1791 Tullie Circle, N.E. Atlanta, GA 30329 USA (800) 527-4723 (U.S. and Canada only) (404) 636-8400 www.ashrae.org
ASME	American Society of Mechanical Engineers	ASME International Three Park Avenue New York, NY 10016-5990 USA (800) 843-2763 (U.S/Canada) 001-800-843-2763 (Mexico) 973-882-1167 (outside North America) www.asme.org
ASTM	American Society for Testing and Materials	ASTM International 100 Barr Harbor Drive PO Box C700 West Conshohocken, PA 19428-2959 USA (610) 832-9500 www.astm.org
AWWA	American Water Works Association	American Water Works Association 6666 W. Quincy Ave Denver, CO 80235 USA (303) 794.7711 (800) 926.7337 www.awwa.org

Table 5.1 Some Important Standards Organizations and Trade Associations Pertaining to Piping

(continued on next page)

Acronym	Name	Contact Data
CGA	Compressed Gas Association	Compressed Gas Association 4221 Walney Road, 5th Floor Chantilly, VA 20151 USA (703) 788-2700 www.cganet.com
CSA	Canadian Standards Association	CSA 5060 Spectrum Way Mississauga, Ontario L4W 5N6 CANADA (416) 747-4000 (800) 463-6727 www.csa.ca
DIN	Deutches Institut für Normung	Deutsches Institut für Normung e. V. Burggrafenstraße 6 10787 Berlin Germany Phone: +49 30 2601-0 www.din.de
FM	FM Global	FM Global 1301 Atwood Avenue P.O. Box 7500 Johnston, RI 02919 USA (401) 275 3000 www.fmglobal.com
ICC	International Code Council	International Code Council 500 New Jersey Avenue, NW 6th Floor Washington, DC 20001-2070 USA (888) 422-7233 www.iccsafe.org
ISO	International Organization for Standardization	International Organization for Standardization (ISO) 1, ch. de la Voie-Creuse, Case postale 56 CH-1211 Geneva 20, Switzerland Phone +41 22 749 01 11 www.iso.org

Table 5.1 *(continued)*

Acronym	Name	Contact Data
JIS	Japanese Industrial Standards	Japanese Standards Association 4-1-24 Akasaka Minato-ku Tokyo 107-8440 Japan Tel: +81-3-3583-8005 www.jsa.or.jp
NFPA	National Fire Protection Association	NFPA 1 Batterymarch Park Quincy, Massachusetts 02169-7471 USA (617) 770-3000 (800) 344-3555 www.nfpa.org
NSF	NSF International (formerly National Sanitation Foundation)	NSF International P.O. Box 130140 789 N. Dixboro Road Ann Arbor, MI 48113-0140 USA (734) 769-8010 (800) NSF-MARK www.nsf.org
PFI	Pipe Fabrication Institute	PFI 511 Avenue of America's, # 601 New York, NY 10011 USA (866) 913-3434 www.pfi-institute.org
PPI	Plastics Pipe Institute	PPI 105 Decker Court, Suite 825 Irving TX, 75062 USA (469) 499-1044 www.plasticpipe.org
SAE	Society of Automotive Engineers	SAE World Headquarters 400 Commonwealth Drive Warrendale, PA 15096-0001 USA (877) 606-7323 (U.S. and Canada only) (724) 776-4970 (outside U.S. and Canada) www.sae.org

Table 5.1 *(continued)*

ASME B31.1 Table 126.1	Organization	ID Number	Title
	ANSI	B16.20	Metallic Gaskets for Pipe Flanges; Ring-Joint, Spiral-Wound, and Jacketed
	ANSI	B16.5	Pipe Flanges and Flanged Fittings: NPS 1/2 through 24
	ANSI	D5421	Standard Specification for Contact Molded "Fiberglass" (Glass-Fiber-Reinforced Thermosetting Resin) Flanges
	ANSI	F2015	Standard Specification for Lap Joint Flange Pipe End Applications
	ANSI	F704	Standard Practice for Selecting Bolting Lengths for Piping System Flanged Joints
*	ANSI	Z223.1	National Fuel Gas Code (NFPA 54)
*	API	5L	Line Pipe
	ASME	A112.1.2	Air Gaps in Plumbing Systems
	ASME	A112.14.1	Backwater Valves
	ASME	A112.18.1M	Plumbing Fixture Fittings
	ASME	A112.21.3M	Hydrants for Utility and Maintenance Use
	ASME	A112.3.1M	Supports for Off-the -Floor Plumbing Fixtures for Public Use
	ASME	A112.4.1	Water Heater Relief Valve Drain Tubes
	ASME	A13.1	Scheme for Identification of Piping Systems
*	ASME	B1.20.1	Pipe Threads, General Purpose (Inch)
*	ASME	B1.20.3	Dryseal Pipe Threads (Inch)
*	ASME	B16.1	Cast Iron Pipe Flanges and Flanged Fittings
*	ASME	B16.3	Malleable Iron Threaded Fittings
*	ASME	B16.4	Cast Iron Threaded Fittings
*	ASME	B16.5	Pipe Flanges and Flanged Fittings
*	ASME	B16.9	Factory-Made Wrought Steel Buttwelding Fittings
*	ASME	B16.10	Face-to-Face and End-to-End Dimensions of Valves
*	ASME	B16.11	Forged Steel Fittings, Socket Welded and Threaded
	ASME	B16.12	Cast Iron Threaded Drainage Fittings
*	ASME	B16.14	Ferrous Pipe Plugs, Bushings and Locknuts with Pipe Threads
*	ASME	B16.15	Cast Bronze Threaded Fittings
*	ASME	B16.18	Cast Copper Alloy Solder Joint Pressure Fittings

Table 5.2 Practical List of Applicable Codes, Standards, and Specifications. Not all specifications listed in ASME Table 126.1 are listed within this table.

ASME B31.1 Table 126.1	Organization	ID Number	Title
*	ASME	B16.20	Metallic Gaskets for Pipe Flanges - Ring Joint, Spiral-Wound, and Jacketed
*	ASME	B16.21	Nonmetallic Flat Gaskets for Pipe Flanges
*	ASME	B16.22	Wrought Copper and Copper Alloy Solder Joint Pressure Fittings
	ASME	B16.23	Cast Copper Solder Joint Drainage Fittings (DWV)
*	ASME	B16.24	Cast Copper Alloy Pipe Flanges and Flanged Fittings: Class 150, 300, 400, 600, 900, 1500 and 2500
*	ASME	B16.25	Buttwelding Ends
	ASME	B16.26	Cast Copper Alloy Fittings for Flared Copper Tubes
	ASME	B16.28	Wrought Steel Buttwelding Short Radius Elbows and Returns
	ASME	B16.29	Wrought Copper and Copper Alloy Solder Joint Drainage Fittings - DWV
	ASME	B16.32	Cast Copper Alloy Solder Joint Fittings for Sovent® Drainage Systems
	ASME	B16.33	Manually Operated Metallic Gas Valves for Use in Gas Piping Systems up to 125 psig (Sizes 1/2" through 2")
*	ASME	B16.34	Valves - Flanged, Threaded, and Welding End
	ASME	B16.36	Orifice Flanges
	ASME	B16.38	Large Metallic Valves for Gas Distribution (Manually Operated 2 1/2" to 12", 125 psig Maximum)
	ASME	B16.39	Malleable Iron Threaded Pipe Unions
	ASME	B16.40	Manually Operated Thermoplastic Gas Shutoffs and Valves in Gas Distribution Systems
	ASME	B16.41	Functional Qualification Requirements for Power Operated Active Valve Assemblies for Nuclear Power Plants
*	ASME	B16.42	Ductile Iron Pipe Flanges and Flanged Fittings, Classes 150 and 300
	ASME	B16.45	Cast Iron Fittings for Sovent ® Drainage Systems
*	ASME	B16.47	Large Diameter Steel Flanges: NPS 26 through NPD 60
*	ASME	B16.48	Steel Line Blanks

Table 5.2 *(continued)*

ASME B31.1 Table 126.1	Organization	ID Number	Title
	ASME	B31.1	Power Piping
	ASME	B31.2	Fuel Gas Piping
*	ASME	B31.3	Process Piping
*	ASME	B31.4	Liquid Transportation Systems for Hydrocarbons, LPG, Anhydrous Ammonia and Alcohols
	ASME	B31.5	Refrigeration Piping
*	ASME	B31.8	Gas Transmission and Distribution Piping
	ASME	B31.9	Building Services Piping
	ASME	B31.11	Slurry Transportation Piping Systems
	ASME	B31G	Manual for Determining the Remaining Strength of Corroded Pipelines
	ASME	B73.1M	Specification for Horizontal End-Suction Centrifugal Pumps for Chemical Process
	ASME	B73.2M	Specification for Vertical In-Line Centrifugal Pumps for Chemical Process
*	ASME	Section I	Power Boilers
*	ASME	Section II	Materials Part A - Ferrous Material Specifications
*	ASME	Section II	Materials Part B - Nonferrous Material Specifications
*	ASME	Section II	Materials Part C - Specifications for Welding Rods, Electrodes, and Filler Metals
*	ASME	Section II	Materials Part D - Properties
*	ASME	Section III	Rules for Construction of Nuclear Power Plant Components
*	ASME	Section III	Subsection NCA General Requirements for Division 1 and Division 2
*	ASME	Section III	Division 1 Subsection NB Class 1 Components
*	ASME	Section III	Division 1 Subsection NC Class 2 Components
*	ASME	Section III	Division 1 Subsection ND Class 3 Components
*	ASME	Section III	Division 1 Subsection NE Class MC Components
*	ASME	Section III	Division 1 Subsection NF Supports
*	ASME	Section III	Division 1 Subsection NG Core Support Structures
*	ASME	Section III	Division 1 Subsection NH Components in Elevated Temperature Service
*	ASME	Section III	Division 1 Appendices

Table 5.2 *(continued)*

ASME B31.1 Table 126.1	Organization	ID Number	Title
*	ASME	Section III	Division 2 Code for Concrete Reactor Vessels and Containments
*	ASME	Section IV	Heating Boilers
*	ASME	Section IX	Welding and Brazing Qualifications
*	ASME	Section V	Nondestructive Examination
*	ASME	Section VI	Recommended Rules for the Care and Operation of Heating Boilers
*	ASME	Section VII	Recommended Guidelines for the Care of Power Boilers
*	ASME	Section VIII	Pressure Vessels Division 1
*	ASME	Section VIII	Pressure Vessels Division 2 Alternative Rules
*	ASME	Section X	Fiber-Reinforced Plastic Pressure Vessels
*	ASME	Section XI	Rules for Inservice Inspection of Nuclear Power Plant Components
*	ASTM	A47	Ferritic Malleable Iron Castings
*	ASTM	A48	Gray Iron Castings
*	ASTM	A53	Pipe, Steel, Black and Hot-Dipped, Zinc-Coated, Welded and Seamless
	ASTM	A74	Cast Iron Soil Pipe and Fittings
*	ASTM	A105	Forgings, Carbon Steel, for Piping Components
*	ASTM	A106	Seamless Carbon Steel Pipe for High-Temperature Service
*	ASTM	A108	Aluminum-Alloy Permanent Mold Castings
*	ASTM	A126	Gray Iron Castings for Valves, Flanges, and Pipe Fittings
*	ASTM	A134	Pipe, Steel, Electric-Fusion (Arc)-Welded (Sizes NPS 16 and Over)
*	ASTM	A135	Electric-Resistance-Welded Steel Pipe
*	ASTM	A139	Electric-Fusion (Arc)-Welded Steel Pipe (NPS 4 and Over)
*	ASTM	A181	Forgings, Carbon Steel for General Purpose Piping
*	ASTM	A182	Forged or Rolled Alloy-Steel Pipe Flanges, Forged Fittings, and Valves and Parts for High-Temperature Service
*	ASTM	A193	Alloy-Steel and Stainless Steel Bolting Materials for High-Temperature Service
*	ASTM	A194	Carbon and Alloy Steel Nuts for Bolts for High-Pressure and High-Temperature Service

(continued on next page)

ASME B31.1 Table 126.1	Organization	ID Number	Title
*	ASTM	A197	Cupola Malleable Iron
	ASTM	A211	Spiral-Welded Pipe – Standard withdrawn in 1993
*	ASTM	A216	Steel Castings, Carbon Suitable for Fusion Welding for High Temperature Service
*	ASTM	A217	Steel Castings, Martensitic Stainless and Alloy, for Pressure-Containing Parts Suitable for High-Temperature Service
*	ASTM	A234	Piping Fittings of Wrought Carbon Steel and Alloy Steel for Moderate and Elevated Temperature Services
	ASTM	A252	Welded and Seamless Steel Pipe Piles
*	ASTM	A268	Seamless and Welded Ferritic And Martensitic Stainless Steel Tubing for General Service
*	ASTM	A278	Gray Iron Castings for Pressure-Containing Parts for Temperatures Up to 650°F (350°C)
*	ASTM	A307	Carbon Steel Bolts and Studs, 60,000 psi Tensile Strength
*	ASTM	A312	Seamless and Welded Austenitic Stainless Steel Pipes
*	ASTM	A320	Alloy-Steel Bolting Materials for Low-Temperature Service
*	ASTM	A333	Seamless and Welded Steel Pipe for Low-Temperature Service
*	ASTM	A335	Seamless Ferritic Alloy-Steel Pipe for High-Temperature Service
*	ASTM	A336	Alloy Steel Forgings for Pressure and High-Temperature Parts
	ASTM	A338	Malleable Iron Flanges, Pipe Fittings, and Valve Parts for Railroad, Marine, and Other Heavy Duty Service at Temperatures Up to 650°F (345°C)
*	ASTM	A350	Forgings, Carbon and Low-Alloy Steel, Requiring Notch Toughness Testing for Piping Components
*	ASTM	A351	Steel Castings, Austenitic, for High-Temperature Service
*	ASTM	A354	Quenched and Tempered Alloy Steel Bolts, Studs and Other Externally-Threaded Fasteners
*	ASTM	A358	Electric-Fusion-Welded Austenitic Chromium-Nickel Alloy Steel Pipe for High-Temperature Service
*	ASTM	A369	Carbon and Ferritic Alloy Steel Forged and Bored Pipe for High-Temperature Service

Table 5.2 *(continued)*

ASME B31.1 Table 126.1	Organization	ID Number	Title
*	ASTM	A376	Seamless Austenitic Steel Pipe for High-Temperature Central-Station Service
*	ASTM	A377	Index of Specifications for Ductile-Iron Pressure Pipe
	ASTM	A381	Standard Specification for Metal-Arc-Welded Steel Pipe for Use With High-Pressure Transmission Systems
*	ASTM	A389	Steel Castings, Alloy, Specially Heat-Treated for Pressure-Containing Parts Suitable for High-Temperature Service
*	ASTM	A395	Ferritic Ductile Iron Pressure-Retaining Castings for Use at Elevated Temperatures
*	ASTM	A403	Wrought Austenitic Stainless Steel Piping Fittings
*	ASTM	A409	Welded Large Diameter Austenitic Steel Pipe for Corrosive or High-Temperature Service
*	ASTM	A420	Piping Fittings of Wrought Carbon Steel and Alloy Steel for Low-Temperature Service
*	ASTM	A426	Centrifugally Cast Ferritic Alloy Steel Pipe for High-Temperature Service
*	ASTM	A437	Alloy-Steel Turbine-Type Bolting Material Specially Heat Treated for High Temperature Service
*	ASTM	A449	Quenched and Tempered Steel Bolts and Studs
*	ASTM	A450	General Requirements for Carbon, Ferritic Alloy, and Austenitic Alloy Steel Tubes
*	ASTM	A451	Centrifugally Cast Austenitic Steel Pipe for High-Temperature Service
*	ASTM	A453	High-Temperature Bolting Materials, With Expansion Coefficients Comparable to Austenitic Steels
	ASTM	A523	Plain End Seamless and Electric-Resistance-Welded Steel Pipe for High-Pressure Pipe-Type Cable Circuits
	ASTM	A524	Seamless Carbon Steel Pipe for Atmospheric and Lower Temperatures
*	ASTM	A530	General Requirements for Specialized Carbon and Alloy Steel Pipe
*	ASTM	A536	Ductile Iron Castings
*	ASTM	A587	Electric-Resistance-Welded Low-Carbon Steel Pipe for the Chemical Industry

(continued on next page)

ASME B31.1 Table 126.1	Organization	ID Number	Title
	ASTM	A648	Steel Wire, Hard Drawn for Prestressing Concrete Pipe
*	ASTM	A671	Electric-Fusion-Welded Steel Pipe for Atmospheric and Lower Temperatures
*	ASTM	A672	Electric-Fusion-Welded Steel Pipe for High-Pressure Service at Moderate Temperatures
	ASTM	A674	Polyethylene Encasement for Ductile-Iron Pipe for Water or Other Liquids
*	ASTM	A691	Carbon and Alloy Steel Pipe, Electric-Fusion-Welded for High-Pressure Service at High Temperatures
	ASTM	A694	Carbon and Alloy Steel Forgings for Pipe Flanges, Fittings, Valves, and Parts for High-Pressure Transmission Service
*	ASTM	A714	High-Strength Low-Alloy Welded and Seamless Steel Pipe
	ASTM	A716	Ductile Iron Culvert Pipe
	ASTM	A733	Welded and Seamless Carbon Steel and Austenitic Stainless Steel Pipe Nipples
	ASTM	A742	Steel Sheet, Metallic Coated and Polymer Precoated for Corrugated Steel Pipe
	ASTM	A746	Ductile Iron Gravity Sewer Pipe
	ASTM	A760	Corrugated Steel Pipe, Metallic-Coated for Sewers and Drains
	ASTM	A761	Corrugated Steel Structural Plate, Zinc-Coated, for Field-Bolted Pipe, Pipe-Arches, and Arches
	ASTM	A762	Corrugated Steel Pipe, Polymer Precoated for Sewers and Drains
*	ASTM	A789	Standard Specification for Seamless and Welded Ferritic/Austenitic Stainless Steel Tubing for General Service
*	ASTM	A790	Seamless and Welded Ferritic/Austenitic Stainless Steel Pipe
	ASTM	A796	Structural Design of Corrugated Steel Pipe, Pipe-Arches, and Arches for Storm and Sanitary Sewers and Other Buried Applications
	ASTM	A798	Installing Factory-Made Corrugated Steel Pipe for Sewers and Other Applications
	ASTM	A807	Installing Corrugated Steel Structural Plate Pipe for Sewers and Other Applications

Table 5.2 *(continued)*

ASME B31.1 Table 126.1	Organization	ID Number	Title
	ASTM	A813	Single- or Double-Welded Austenitic Stainless Steel Pipe
	ASTM	A814	Cold-Worked Welded Austenitic Stainless Steel Pipe
*	ASTM	A815	Wrought Ferritic, Ferritic/Austenitic, and Martensitic Stainless Steel Piping Fittings
	ASTM	A849	Post-Applied Coatings, Pavings, and Linings for Corrugated Steel Sewer and Drainage Pipe
	ASTM	A861	High-Silicon Iron Pipe and Fittings
	ASTM	A862	Application of Asphalt Coatings to Corrugated Steel Sewer and Drainage Pipe
	ASTM	A865	Threaded Couplings, Steel, Black or Zinc-Coated (Galvanized) Welded or Seamless, for Use in Steel Pipe Joints
	ASTM	A872	Centrifugally Cast Ferritic/Austenitic Stainless Steel Pipe for Corrosive Environments
	ASTM	A885	Steel Sheet, Zinc and Aramid Fiber Composite Coated for Corrugated Steel Sewer, Culvert, and Underdrain Pipe
	ASTM	A888	Hubless Cast Iron Soil Pipe and Fittings for Sanitary and Storm Drain, Waste, and Vent Piping Applications
	ASTM	A926	Test Method for Comparing the Abrasion Resistance of Coating Materials for Corrugated Metal Pipe
*	ASTM	A928	Ferritic/Austenitic (Duplex) Stainless Steel Pipe Electric Fusion Welded with Addition of Filler Metal
	ASTM	A929	Steel Sheet, Metallic-Coated by the Hot-Dip Process for Corrugated Steel Pipe
	ASTM	A930	Life-Cycle Cost Analysis of Corrugated Metal Pipe Used for Culverts, Storm Sewers, and Other Buried Conduits
	ASTM	A943	Spray-Formed Seamless Austenitic Stainless Steel Pipes
	ASTM	A949	Spray-Formed Seamless Ferritic/Austenitic Stainless Steel Pipe
	ASTM	A954	Austenitic Chromium-Nickel-Silicon Alloy Steel Seamless and Welded Pipe
	ASTM	A978	Composite Ribbed Steel Pipe, Precoated and Polyethylene Lined for Gravity Flow Sanitary Sewers, Storm Sewers, and Other Special Applications

(continued on next page)

ASME B31.1 Table 126.1	Organization	ID Number	Title
	ASTM	A984	Steel Line Pipe, Black, Plain-End, Electric-Resistance-Welded
	ASTM	A998	Structural Design of Reinforcements for Fittings in Factory-Made Corrugated Steel Pipe for Sewers and Other Applications
	ASTM	A999	General Requirements for Alloy and Stainless Steel Pipe
	ASTM	A1005	Steel Line Pipe, Black, Plain End, Longitudinal and Helical Seam, Double Submerged-Arc Welded
	ASTM	A1006	Steel Line Pipe, Black, Plain End, Laser Beam Welded
*	ASTM	B111	Copper and Copper-Alloy Seamless Condenser Tubes and Ferrule Stock
*	ASTM	B148	Aluminum-Bronze Sand Castings
*	ASTM	B26	Aluminum-Alloy Sand Castings
*	ASTM	B42	Seamless Copper Pipe, Standard Sizes
*	ASTM	B43	Seamless Red Brass Pipe, Standard Sizes
*	ASTM	B61	Steam or Valve Bronze Castings
*	ASTM	B62	Composition Bronze or Ounce Metal Castings
*	ASTM	B68	Seamless Copper Tube, Bright Annealed
*	ASTM	B75	Seamless Copper Tube
*	ASTM	B88	Standard Specification for Seamless Copper Water Tube
*	ASTM	B161	Nickel Seamless Pipe and Tube
*	ASTM	B165	Nickel-Copper Alloy (UNS N04400) Seamless Pipe and Tube
*	ASTM	B167	Nickel-Chromium-Iron Alloy (UNS N06600, N06601, N06603, N06690, N06693, N06025, and N06645) and Nickel-Chromium-Cobalt-Molybdenum Alloy (UNS N06617) Seamless Pipe and Tube
*	ASTM	B210	Aluminum Alloy Drawn Seamless Tubes
*	ASTM	B241	Aluminum-Alloy Seamless Pipe and Seamless Extruded Tube
*	ASTM	B247	Aluminum and Aluminum-Alloy Die, Hand, and Rolled Ring Forgings
*	ASTM	B251	General Requirements for Wrought Seamless Copper and Copper-Alloy Tube
*	ASTM	B280	Standard Specification for Seamless Copper Tube for Air Conditioning and Refrigeration Field Service

Table 5.2 *(continued)*

ASME B31.1 Table 126.1	Organization	ID Number	Title
*	ASTM	B283	Copper and Copper-Alloy Die Forgings (Hot Pressed)
*	ASTM	B302	Threadless Copper Pipe, Standard Sizes
	ASTM	B306	Standard Specification for Seamless Copper Drainage Tube (DWV)
*	ASTM	B315	Seamless Copper Alloy Pipe and Tube
*	ASTM	B361	Factory-Made Wrought Aluminum and Aluminum-Alloy Welding Fittings
*	ASTM	B366	Factory-Made Wrought Nickel and Nickel Alloy Fittings
*	ASTM	B367	Titanium and Titanium Alloy Castings
*	ASTM	B381	Titanium and Titanium Alloy Forgings
*	ASTM	B407	Nickel-Iron-Chromium Alloy Seamless Pipe and Tube
*	ASTM	B423	Nickel-Iron-Chromium-Molybdenum-Copper Alloy (UNS N08825 and N08821) Seamless Pipe and Tube
*	ASTM	B462	UNS N06030, UNS N06022, UNS N06200, UNS N08020, UNS N08024, UNS N08026, UNS N08367, UNS N10276, UNS N10665, UNS N10675, UNS R20033 Alloy Pipe Flanges, Forged Fittings and Valves and Parts for Corrosive High-Temperature Service
*	ASTM	B464	Welded (UNS N08020, N08024, N08026 Alloy) Pipe
*	ASTM	B466	Seamless Copper-Nickel Pipe and Tube
*	ASTM	B467	Welded Copper-Nickel Pipe
*	ASTM	B468	Welded (UNS N08020, N08024, N08026) Alloy Tubes
*	ASTM	B546	Electric Fusion-Welded Ni-Cr-Co-Mo Alloy (UNS N06617), Ni-Fe-Cr-Si Alloys (UNS N08330 and UNS N08332), Ni-Cr-Fe-Al Alloy (UNS N06603), Ni-Cr-Fe Alloy (UNS N06025), and Ni-Cr-Fe-Si Alloy (UNS N06045) Pipe
*	ASTM	B547	Aluminum and Aluminum-Alloy Formed and Arc-Welded Round Tube
*	ASTM	B564	Nickel and Alloy Forgings
*	ASTM	B584	Copper Alloy Sand Castings for General Applications
*	ASTM	B603	Welded Copper-Alloy Pipe
*	ASTM	B619	Welded Nickel and Nickel-Cobalt Alloy Pipe
*	ASTM	B622	Seamless Nickel and Nickel-Cobalt Alloy Pipe and Tube
*	ASTM	B626	Welded Nickel and Nickel-Cobalt Alloy Tube
*	ASTM	B673	UNS N08904, UNS N08925, and UNS N08926 Welded Pipe

(continued on next page)

ASME B31.1 Table 126.1	Organization	ID Number	Title
*	ASTM	B674	UNS N08904, UNS N08925, and UNS N08926 Welded Tube
*	ASTM	B677	UNS N08904, UNS N08925, and UNS N08926 Seamless Pipe and Tube
*	ASTM	B704	Welded UNS N06625 and N08825 Alloy Tubes
*	ASTM	B705	Nickel-Alloy (UNS N06625 and N08825) Welded Pipe
*	ASTM	B729	Seamless UNS N08020, UNS N08026, and UNS N08024 Nickel-Alloy Pipe and Tube
	ASTM	B819	Standard Specification for Seamless Copper Tube for Medical Gas Systems
*	ASTM	B861	Titanium and Titanium Alloy Seamless Pipe
*	ASTM	B862	Titanium and Titanium Alloy Welded Pipe
	ASTM	D2241	Standard Specification for Polyvinyl Chloride Pressure-Rated Pipe (SDR Series)
	ASTM	E84	Standard Test Method for Surface Burning Characteristics of Building Materials
	AWWA	C104/ A21.4	Cement-Mortar Lining for Ductile-Iron Pipe and Fittings for Water
	AWWA	C105/ A21.5	Polyethylene Encasement for Ductile-Iron Pipe Systems
*	AWWA	C110/ A21.10	Ductile-Iron and Gray-Iron Fittings for Water
*	AWWA	C111/ A21.11	Rubber-Gasket Joints for Ductile-Iron Pressure Pipe and Fittings
*	AWWA	C115/ A21.15	Flanged Ductile-Iron Pipe with Ductile-Iron or Gray-Iron Threaded Flanges
	AWWA	C116/ A21.16	Protective Fusion-Bonded Epoxy Coatings Int. & Ext. Surf. Ductile-Iron/Gray-Iron Fittings
*	AWWA	C150/ A21.50	Thickness Design of Ductile-Iron Pipe
*	AWWA	C151/ A21.51	Ductile-Iron Pipe, Centrifugally Cast, for Water or Other Liquids
*	AWWA	C153/ A21.53	Ductile-Iron Compact Fittings for Water Service
*	AWWA	C200	Steel Water Pipe 6 in (150 mm) and Larger
	AWWA	C203	Coal-Tar Protective Coatings & Linings for Steel Water Pipelines, Enamel & Tape, Hot-Applied

Table 5.2 *(continued)*

ASME B31.1 Table 126.1	Organization	ID Number	Title
	AWWA	C205	Cement-Mortar Protective Lining and Coating for Steel Water Pipe, 4 in (100 mm) and Larger, Shop Applied
	AWWA	C206	Field Welding of Steel Water Pipe
*	AWWA	C207	Steel Pipe Flanges for Waterworks Service, Sizes 4 in Through 144 in (100 mm Through 3,600 mm)
*	AWWA	C208	Dimensions for Fabricated Steel Water Pipe Fittings
	AWWA	C209	Cold-Applied Tape Coatings for the Exterior of Special Sections, Connections, and Fittings for Steel Water Pipelines
	AWWA	C210	Liquid-Epoxy Coating Systems for the Interior and Exterior of Steel Water Pipelines
	AWWA	C213	Fusion-Bonded Epoxy Coating for the Interior and Exterior of Steel Water Pipelines
	AWWA	C214	Tape Coating Systems for the Exterior of Steel Water Pipelines
	AWWA	C215	Extruded Polyolefin Coatings for the Exterior of Steel Water Pipelines
	AWWA	C216	Heat-Shrinkable Cross-Linked Polyolefin Coatings for the Exterior of Special Sections, Connections, and Fittings for Steel Water Pipelines
	AWWA	C217	Petrolatum and Petroleum Wax Tape Coatings for the Exterior of Connections and Fittings for Steel Water Pipelines
	AWWA	C218	Coating the Exterior of Aboveground Steel Water Pipelines and Fittings
	AWWA	C219	Bolted, Sleeve-Type Couplings for Plain-End Pipe
	AWWA	C220	Stainless-Steel Pipe, 1/2 in (13 mm) and Larger
	AWWA	C221	Fabricated Steel Mechanical Slip-Type Expansion Joints
	AWWA	C222	Polyurethane Coatings for the Interior and Exterior of Steel Water Pipe and Fittings
	AWWA	C223	Fabricated Steel and Stainless Steel Tapping Sleeves
	AWWA	C224	Nylon-11 Based Polyamide Coating System for the Interior and Exterior of Steel Water Pipe and Fittings
	AWWA	C225	Fused Polyolefin Coating Systems for the Exterior of Steel Water Pipelines

(continued on next page)

ASME B31.1 Table 126.1	Organization	ID Number	Title
	AWWA	C226	Stainless Steel Fittings for Waterworks Service, Sizes 1/2 in Through 72 in (13 mm Through 1,800 mm)
*	AWWA	C300	Reinforced Concrete Pressure Pipe, Steel-Cylinder Type
*	AWWA	C301	Prestressed Concrete Pressure Pipe, Steel-Cylinder Type
*	AWWA	C302	Reinforced Concrete Pressure Pipe, Noncylinder Type
	AWWA	C303	Concrete Pressure Pipe, Bar-Wrapped, Steel-Cylinder Type
*	AWWA	C304	Design of Prestressed Concrete Cylinder Pipe
	AWWA	C400	Asbestos–Cement Pressure Pipe, 4 in–16 in (100 mm–400 mm), for Water Dist. & Trans.
	AWWA	C401	Selection of Asbestos–Cement Pressure Pipe, 4 in–16 in (100 mm-400 mm), for Water Dist. Sys.
	AWWA	C402	Asbestos-Cement Transmission Pipe, 18 In Through 42 in (450 mm Through 1,050 mm) for Water Supply Service
	AWWA	C403	The Selection of Asbestos–Cement Transmission Pipe, Sizes 18 in Through 42 in (450 mm Through 1,050 mm),
*	AWWA	C500	Metal-Seated Gate Valves for Water Supply Service
	AWWA	C502	Dry-Barrel Fire Hydrants
	AWWA	C503	Wet-Barrel Fire Hydrants
*	AWWA	C504	Rubber-Seated Butterfly Valves
	AWWA	C507	Ball Valves, 6 in Through 48 in (150 mm Through 1,200 mm)
	AWWA	C508	Swing-Check Valves for Waterworks Service, 2 in (50 mm) Through 24 in (600 mm) NPS
*	AWWA	C509	Resilient-Seated Gate Valves for Water Supply Service
	AWWA	C510	Double Check Valve Backflow Prevention Assembly
	AWWA	C511	Reduced-Pressure Principle Backflow Prevention Assembly
	AWWA	C512	Air Release, Air/Vacuum, and Combination Air Valves for Waterworks Service
	AWWA	C513	Open-Channel, Fabricated-Metal, Slide Gates and Open-Channel, Fabricated-Metal Weir Gates
	AWWA	C515	Reduced-Wall, Resilient-Seated Gate Valves for Water Supply Service

Table 5.2 *(continued)*

ASME B31.1 Table 126.1	Organization	ID Number	Title
	AWWA	C517	Resilient-Seated Cast-Iron Eccentric Plug Valves
	AWWA	C540	Power-Actuating Devices for Valves and Slide Gates
	AWWA	C550	Protective Epoxy Interior Coatings for Valves and Hydrants
	AWWA	C560	Cast-Iron Slide Gates
	AWWA	C561	Fabricated Stainless Steel Slide Gates
	AWWA	C563	Fabricated Composite Slide Gates
*	AWWA	C600	Installation of Ductile Iron Water Mains and their Appurtenances
	AWWA	C605	Underground Installation of Polyvinyl Chloride (PVC) Pressure Pipe and Fittings for Water
*	AWWA	C606	Grooved and Shouldered Joints
	AWWA	C651	Disinfecting Water Mains
	AWWA	C652	Disinfection of Water-Storage Facilities
	AWWA	C653	Disinfection of Water Treatment Plants
	AWWA	C654	Disinfection of Wells
	AWWA	C900	Polyvinyl Chloride (PVC) Pressure Pipe and Fabricated Fittings, 4 in–12 in (100 mm-300 mm), for Water Transmission and Distribution
	AWWA	C901	Polyethylene (PE) Pressure Pipe and Tubing, ½ in (13 mm) Through 3 in (76 mm), for Water Service
	AWWA	C903	Polyethylene-Aluminum-Polyethylene Composite Pressure Pipes
	AWWA	C904	Cross-Linked Polyethylene (PEX) Pressure Pipe, ½ in (12mm) Through 3 in (76mm), for Water Service
	AWWA	C905	Polyvinyl Chloride (PVC) Pressure Pipe and Fabricated Fittings, 14 in–48 in (350 mm–1,200 mm)
	AWWA	C906	Polyethylene (PE) Pressure Pipe and Fittings, 4 in (100 mm) Th. 63 in (1,600 mm), for Water Dist. and Trans.
	AWWA	C907	Injection-Molded Polyvinyl Chloride (PVC) Pressure Fittings, 4 in Through 12 in (100 mm Through 300 mm)
	AWWA	C909	Molecularly Oriented Polyvinyl Chloride (PVCO) Pressure Pipe, 4 in–24 in (100 mm–600 mm), for Water Distribution
	AWWA	C950	Fiberglass Pressure Pipe

(continued on next page)

ASME B31.1 Table 126.1	Organization	ID Number	Title
	CGA	E-1	Standard Connections for Regulator Outlets, Torches, and Fitted Hose for Welding and Cutting Equipment
	CGA	E-2	Hose Line Check Valve Standards for Welding and Cutting
	CGA	E-3	Pipeline Regulator Inlet Connection Standards
	CGA	E-4	Standard for Gas Pressure Regulators
	CGA	E-5	Torch Standard for Welding and Cutting
	CGA	E-6	Standard for Hydraulic Type Pipeline Protective Devices
	CGA	E-7	American National and CGA Standard for Medical Gas regulators and Flowmeters
	CGA	G-1	Acetylene
	CGA	G-1.3	Acetylene Transmission for Chemical Synthesis
	CGA	G-2	Anhydrous Ammonia
	CGA	G-3	Sulfur Dioxide
	CGA	G-4	Oxygen
	CGA	G-4.1	Cleaning Equipment for Oxygen Service
	CGA	G-4.4	Industrial Practices for Gaseous Oxygen Transmission and Distribution Piping
	CGA	G-5	Hydrogen
	CGA	G-5.4	Standard for Hydrogen Piping at Consumer Locations
	CGA	G-5.5	Hydrogen Vent Systems
	CGA	G-6	Carbon Dioxide
	CGA	G-6.1	Standard for Low Pressure Carbon Dioxide Systems at Consumer Sites
	CGA	G-6.5	Standard for Small Stationary Low Pressure Carbon Dioxide Systems
	CGA	G-7	Compressed Air for Human Respiration
	CGA	G-8.1	Standard for Nitrous Oxide Systems at Consumer Sites
	CGA	G-12	Hydrogen Sulfide
	CGA	P-8	Safe Practices Guide for Air Separation Plants
	CGA	P-11	Metric Practice Guide for Compressed Gas Industry
	CGA	P-19	Recommended Procedures for Nitrogen Purging of Tank Cars
*	MSS	SP-9	Spot Facing for Bronze, Iron and Steel Flanges

Table 5.2 *(continued)*

ASME B31.1 Table 126.1	Organization	ID Number	Title
*	MSS	SP-25	Standard Marking System for Valves, Fittings, Flanges and Unions
*	MSS	SP-42	Class 150 Corrosion Resistant Gate, Globe, Angle and Check Valves with Flanged and Butt Weld Ends
*	MSS	SP-43	Wrought Stainless Steel Butt-Welding Fittings
	MSS	SP-44	Steel Pipeline Flanges
*	MSS	SP-45	Bypass and Drain Connections
*	MSS	SP-51	Class 150LW Corrosion Resistant Cast Flanges and Flanged Fittings
*	MSS	SP-53	Quality Standard for Steel Castings and Forgings for Valves, Flanges and Fittings and Other Piping Components - Magnetic Particle Exam Method
*	MSS	SP-54	Quality Standard for Steel Castings for Valves, Flanges, and Fittings and Other Piping Components - Radiographic Examination Method
*	MSS	SP-55	Quality Standard for Steel Castings for Valves, Flanges, Fittings, and Other Piping Components - Visual Method for Evaluation of Surface Irregularities
*	MSS	SP-58	Pipe Hangers and Supports - Materials, Design, and Manufacture
*	MSS	SP-6	Standard Finishes for Contact Faces of Pipe Flanges and Connecting-End Flanges of Valves and Fittings
	MSS	SP-60	Connecting Flange Joint Between Tapping Sleeves and Tapping Valves
*	MSS	SP-61	Pressure Testing of Steel Valves
	MSS	SP-65	High Pressure Chemical Industry Flanges and Threaded Stubs for Use with Lens Gaskets
*	MSS	SP-67	Butterfly Valves
*	MSS	SP-68	High Pressure Butterfly Valves with Offset Design
*	MSS	SP-69	ANSI/MSS Edition Pipe Hangers and Supports - Selection and Application
	MSS	SP-70	Cast Iron Gate Valves, Flanged and Threaded Ends
	MSS	SP-71	Gray Iron Swing Check Valves, Flanged and Threaded Ends
	MSS	SP-72	Ball Valves with Flanged or Butt-Welding Ends for General Service

(continued on next page)

ASME B31.1 Table 126.1	Organization	ID Number	Title
	MSS	SP-73	Brazing Joints for Copper and Copper Alloy Pressure Fittings
*	MSS	SP-75	Specification for High Test Wrought Butt Welding Fittings
	MSS	SP-77	Guidelines for Pipe Support Contractual Relationships
	MSS	SP-78	Cast Iron Plug Valves, Flanged and Threaded Ends
*	MSS	SP-79	Socket-Welding Reducer Inserts
*	MSS	SP-80	Bronze Gate, Globe, Angle and Check Valves
	MSS	SP-81	Stainless Steel, Bonnetless, Flanged Knife Gate Valves
*	MSS	SP-83	Class 3000 Steel Pipe Unions, Socket Welding and Threaded
	MSS	SP-85	Cast Iron Globe & Angle Valves, Flanged and Threaded Ends
	MSS	SP-86	Guidelines for Metric Data in Standards for Valves, Flanges, Fittings and Actuators
	MSS	SP-88	Diaphragm Valves
*	MSS	SP-89	Pipe Hangers and Supports - Fabrication and Installation Practices
	MSS	SP-90	Guidelines on Terminology for Pipe Hangers and Supports
	MSS	SP-91	Guidelines for Manual Operation of Valves
	MSS	SP-92	MSS Valve User Guide
*	MSS	SP-93	Quality Standard for Steel Castings and Forgings for Valves, Flanges, and Fittings and Other Piping Components - Liquid Penetrant Exam Method
*	MSS	SP-94	Quality Standard for Ferritic and Martensitic Steel Castings for Valves, Flanges, and Fittings and Other Piping Components - Ultrasonic Exam Method
*	MSS	SP-95	Swaged Nipples and Bull Plugs
*	MSS	SP-96	Guidelines on Terminology for Valves and Fittings
	MSS	SP-97	Integrally Reinforced Forged Branch Outlet Fittings - Socket Welding, Threaded and Buttwelding Ends
	MSS	SP-98	Protective Coatings for the Interior of Valves, Hydrants, and Fittings
	MSS	SP-99	Instrument Valves

Table 5.2 *(continued)*

ASME B31.1 Table 126.1	Organization	ID Number	Title
	MSS	SP-100	Qualification Requirements for Elastomer Diaphragms for Nuclear Service Diaphragm Type Valves
	MSS	SP-101	Part-Turn Valve Actuator Attachment - Flange and Driving Component Dimensions and Performance Characteristics
	MSS	SP-102	Multi-Turn Valve Actuator Attachment - Flange and Driving Component Dimensions and Performance Characteristics
	MSS	SP-103	Wrought Copper and Copper Alloy Insert Fittings for Polybutylene Systems
	MSS	SP-104	Wrought Copper Solder Joint Pressure Fittings
*	MSS	SP-105	Instrument Valves for Code Applications
*	MSS	SP-106	Cast Copper Alloy Flanges and Flanged Fittings, Class 125, 150 and 300
	MSS	SP-108	Resilient-Seated Cast Iron-Eccentric Plug Valves
	MSS	SP-109	Welded Fabricated Copper Solder Joint Pressure Fittings
	MSS	SP-110	Ball Valves Threaded, Socket-Welding, Solder Joint, Grooved and Flared Ends
	MSS	SP-111	Gray-Iron and Ductile-Iron Tapping Sleeves
	MSS	SP-112	Quality Standard for Evaluation of Cast Surface Finishes - Visual and Tactile Method.
	MSS	SP-113	Connecting Joint between Tapping Machines and Tapping Valves
	MSS	SP-114	Corrosion Resistant Pipe Fittings Threaded and Socket Welding, Class 150 and 1000
	MSS	SP-115	Excess Flow Valves for Natural Gas Service
	MSS	SP-116	Service Line Valves and Fittings for Drinking Water Systems
	MSS	SP-117	Bellows Seals for Globe and Gate Valves
	MSS	SP-118	Compact Steel Globe & Check Valves - Flanged, Flangeless, Threaded & Welding Ends (Chemical & Petroleum Refinery Service)
	MSS	SP-119	Factory-Made Wrought Belled End Socket-Welding Fittings
	MSS	SP-120	Flexible Graphite Packing System for Rising Stem Steel Valves (Design Requirements)
	MSS	SP-121	Qualification Testing Methods for Stem Packing for Rising Stem Steel Valves

(continued on next page)

ASME B31.1 Table 126.1	Organization	ID Number	Title
	MSS	SP-122	Plastic Industrial Ball Valves
	MSS	SP-123	Non-Ferrous Threaded and Solder-Joint Unions for Use With Copper Water Tube
	MSS	SP-124	Fabricated Tapping Sleeves
	MSS	SP-125	Gray Iron and Ductile Iron In-Line, Spring-Loaded, Center-Guided Check Valves
	MSS	SP-126	Steel In-Line Spring-Assisted Center Guided Check Valves
	MSS	SP-127	Bracing for Piping Systems Seismic-Wind-Dynamic Design, Selection, Application
	MSS	SP-129	Copper-Nickel Socket-Welding Fittings and Unions
	MSS	SP-130	Bellows Seals for Instrument Valves
	MSS	SP-131	Metallic Manually Operated Gas Distribution Valves
	MSS	SP-132	Compression Packing Systems for Instrument Valves
	MSS	SP-133	Excess Flow Valves for Low Pressure Fuel Gas Appliances
	NFPA	NFPA 13	Installation of Sprinkler Systems
	NFPA	NFPA 13E	Fire Department Operations in Properties Protected by Sprinkler and Standpipe Systems
	NFPA	NFPA 13R	Installation of Sprinkler Systems in Residential Occupancies up to and Including Four Stories in Height
	NFPA	NFPA 14	Installation of Standpipe and Hose Systems
	NFPA	NFPA 15	Water Spray Fixed Systems for Fire Protection
	NFPA	NFPA 16	Installation of Deluge Foam-Water Sprinkler and Foam-Water Spray Systems
	NFPA	NFPA 16A	Installation of Closed-Head Foam-Water Sprinkler Systems
	NFPA	NFPA 17	Dry Chemical Extinguishing Systems
	NFPA	NFPA 17A	Wet Chemical Extinguishing Systems
	NFPA	NFPA 18	Wetting Agents
*	NFPA	NFPA 1963	Fire Hose Connections
	NFPA	NFPA 20	Installation of Centrifugal Fire Pumps
	NFPA	NFPA 214	Water-Cooling Towers
	NFPA	NFPA 22	Water Tanks for Private Fire Protection
	NFPA	NFPA 24	Installation of Private Fire Service Mains and their Appurtenances

Table 5.2 *(continued)*

ASME B31.1 Table 126.1	Organization	ID Number	Title
	NFPA	NFPA 25	Inspection, Testing and Maintenance of Water-Based Fire Protection Systems
	NFPA	NFPA 30	Flammable and Combustible Liquids Code
	NFPA	NFPA 31	Installation of Oil-Burning Equipment
	NFPA	NFPA 325	Fire Hazard Properties of Flammable Liquids, Gases, and Volatile Solids
	NFPA	NFPA 328	Control of Flammable and Combustible Liquids and Gases in Manholes, Sewers, and Similar Underground Structures
	NFPA	NFPA 37	Installation of Stationary Combustion Engines and Gas Turbines
	NFPA	NFPA 45	Fire Protection for Laboratories Using Chemicals
	NFPA	NFPA 49	Hazardous Chemicals Data
	NFPA	NFPA 491	Hazardous Chemical reactions
	NFPA	NFPA 497	Classification of Flammable Liquids, Gases or Vapors and of Hazardous (classified) Locations for Electrical Installations in Chemical Process Areas
	NFPA	NFPA 50	Bulk Oxygen Systems at Consumer Sites
	NFPA	NFPA 50A	Gaseous Hydrogen Systems at Consumer Sites
	NFPA	NFPA 50B	Liquified Hydrogen Systems at Consumer Sites
	NFPA	NFPA 51	Design and Installation of Oxygen-Fuel Gas Systems for Welding, Cutting, and Allied Processes
	NFPA	NFPA 54	National Fuel Gas Code
	NFPA	NFPA 57	Liquified Natural Gas (LNG) Vehicular Fuel Systems
	NFPA	NFPA 58	Storage and Handling of Liquified Petroleum Gases
	NFPA	NFPA 59	Storage and Handling of Liquified Petroleum Gases at Utility Gas Plants
	NFPA	NFPA 600	Industrial Fire Brigades
	NFPA	NFPA 70	National Electric Code
	NFPA	NFPA 99C	Gas and Vacuum Systems
	NSF	14	Plastics Piping System Components and Related Materials
	NSF	61	Drinking water system components - Health effects
	PFI	ES1	Internal Machining and Solid Machined Backing Rings For Circumferential Butt Welds

(continued on next page)

ASME B31.1 Table 126.1	Organization	ID Number	Title
	PFI	ES2	Method of Dimensioning Piping Assemblies
	PFI	ES3	Fabricating Tolerances
	PFI	ES4	Hydrostatic Testing of Fabricated Piping
	PFI	ES5	Cleaning of Fabricated Piping
	PFI	ES7	Minimum Length and Spacing for Welded Nozzles
	PFI	ES11	Permanent Marking on Piping Materials
*	PFI	ES16	Access Holes, Bosses, and Plugs for Radiographic Inspection of Pipe Welds
	PFI	ES20	Wall Thickness Measurement by Ultrasonic Examination
	PFI	ES21	Internal Machining and Fit-up of GTAW Root Pass Circumferential Butt Welds
	PFI	ES22	Recommended Practice for Color Coding of Piping Materials
*	PFI	ES24	Pipe Bending Methods, Tolerances, Process and Material Requirements
	PFI	ES25	Random Radiography of Pressure Retaining Girth Butt Welds
	PFI	ES26	Welded Load Bearing Attachments to Pressure Retaining Piping Materials
	PFI	ES27	"Visual Examination" The Purpose, Meaning and Limitation of the Term
	PFI	ES29	Internal Abrasive Blast Cleaning of Ferritic Piping Materials
	PFI	ES30	Random Ultrasonic Examination of Butt Welds
	PFI	ES31	Standard for Protection of Ends of Fabricated Piping Assemblies
	PFI	ES32	Tool Calibration
	PFI	ES34	Temporary Painting/Coating of Fabricated Piping
	PFI	ES35	Nonsymmetrical Bevels and Joint Configurations for Butt Welds
	PFI	ES36	Branch Reinforcement Work Sheets.
	PFI	ES36	Branch Reinforcement Work Sheets
	PFI	ES37	Standard for Loading and Shipping of Piping Assemblies
	PFI	ES39	Fabricating Tolerances for Grooved Piping Systems

Table 5.2 *(continued)*

ASME B31.1 Table 126.1	Organization	ID Number	Title
	PFI	ES40	Method of Dimensioning Grooved Piping Assemblies
	PFI	ES41	Standard for Material Control and Traceability of Piping Components
	PFI	ES42	Standard for Positive Material Identification of Piping Components using Portable X-Ray Emission Type Equipment
	PFI	ES43	Standard for Protection of Austenitic Stainless Steel and Nickel Alloy Materials
	PFI	ES44	Drafting Practices Standard
	PFI	ES45	Recommended Practice for Local Post-Weld Heat Treatment
	PFI	ES46	Bar Coding
	PFI	TB1	Pressure - Temperature Ratings of Seamless Pipe Used in Power Plant Piping Systems
	PFI	TB3	Guidelines Clarifying Relationships and Design Engineering Responsibilities Between Purchasers' Engineers and Pipe Fabricator or Pipe Fabricator Erector
	PFI	TB5	Information Required for the Bidding of Pipe Fabrication
	PFI	TB7	Guidelines for Fabrication and Installation of Stainless Steel High Purity Distribution Systems
	PFI	TB8	Recommended Practice for the Fabrication of Polyvinylidene Fluoride (PVDF) Piping
	PFI	TB9	Customary Fitting, Forging, Plate and Bar Materials used with Pipe
	SAE	J518	Hydraulic Flanged Tube, Pipe, and Hose Connections, Four-Bolt Split Flange Type

ASME Code Section	Name	Description
Power Boilers – Section I	S	Power Boilers
	A	Power Boiler Assemblies
	E	Electric Boilers
	M	Miniature Boilers
	PP	Pressure Piping
	V	Power Boiler Safety Valves
Heating Boilers – Section IV	H	Cast Iron Heating Boilers or Other Heating Boilers
	HLW	Lined Potable Water Heaters
	HV	Heating Boilers Safety Valves
Pressure Vessels – Section VIII Division 1	U	Pressure Vessels
	UM	Miniature Vessels
	UV	Pressure Vessels Safety Valves
	UD	Pressure Vessels Rupture Disks
Pressure Vessels – Section VIII Division 2	U2	Alternative Rules for Pressure Vessels
Pressure Vessels – Section VIII Division 3	U3	High Pressure Vessels
	UV3	Safety Valves for High Pressure Vessels
Reinforced Plastic Vessels – Section X	RP	Fiber-Reinforced Plastic Pressure Vessels
Transport Tanks – Section XII	T	Transport Tanks
	TV	Transport Tanks Safety Valves
	TD	Transport Tanks Pressure Relief Devices
Nuclear Stamps	N	Nuclear Components
	NPT	Nuclear Partials
	NA	Nuclear Installation and Shop Assembly
	N3	Storage and Transport of Nuclear Fuel
Nuclear Certificates of Accreditation	NS	Nuclear Supports
	QSC	Material Organization
National Board Inspection Code	R	Repair and Alteration
	VR	Repair of Safety Valves

Table 5.3 ASME Stamps, Symbols, and Certifications

CHAPTER 6

Materials of Construction

At a Fortune 500 company, a young engineer had designed a hot water system for a pilot plant using PVC. He may have thought he was doing a good job by saving money over the cost of a copper or steel system, but when one of the lines ruptured, it showered two workers with scalding water. One was hospitalized. At 140°F, PVC is only capable of withstanding 22 percent of its working pressure at 73°F.

At a plant I visited in Chicago, I observed a PVC line extending out of an air receiver on a compressed air system. Not only was this time bomb a poor application of material, but the compressor was an unanchored reciprocating compressor, which had "walked" on the floor, further stressing the PVC.

PVC is a perfectly acceptable material for the applications for which it is designed. If you expose it to high temperature, high pressure, shock loads, or the sun, you can expect failure in a short time. The failure may be accompanied by property damage, shrieking, or worse.

There are really only a few piping materials in common use. You will often see reference books provide data on materials like borosilicate glass piping, but applications for this are uncommon outside of laboratory settings. One obsolete use of borosilicate piping is for coolant piping in electronics equipment that used vacuum tubes.

Likewise, data for wood stave piping and terra cotta are common, even though such applications are out-of-date. One can still observe wood stave construction in old cooling tower installations[1] and terra cotta house drains, but no one would ever design a new system using these materials[2].

So the remaining common materials break down into the following categories:

Ferrous Metal Piping

- Cast iron
- Ductile iron
- Carbon steel
- Stainless steel

Non-ferrous metal piping

- Copper
- Brass
- Titanium
- Aluminum

[1] At a paper mill in North Carolina, wooden water tanks were still in use as recently as 2004. The maintenance superintendent loved them because they required no painting and did not corrode. Small leaks were sealed by tossing a few handfuls of sawdust into the open top of the tank.

[2] Although some in the "green" building movement advocate clay pipe as an environmentally friendly MOC.

Plastics

- PVC
- CPVC
- PB
- HDPE
- ABS
- PEX

Composites

- FRP
- Concrete
- Lined pipe

Many of these materials are represented by technical organizations that publish manuals to aid in the design of piping systems constructed of such materials. These manuals contain useful data for applications that are specific to the material under consideration. It's also worthy to note that these organizations are very competitive with one another. Aside from providing technical data, they often cast the products they represent in a favorable light compared to other products. It is therefore the engineer's responsibility to determine which is the best material for the application, based on the available data, economics, and prior experience.

Upon examining any of the various tables that describe what sizes and wall thicknesses of pipe are available, the reader will note that the critical dimension in every case is the outside diameter. In the world of piping, the outside diameter only occasionally matches exactly with the nominal diameter. This is essentially the difference between piping and tubing.

Tubing is specified in two ways:

1. It may be specified by OD and wall thickness
2. It may be specified by ID and wall thickness.

Piping is specified by nominal diameter and wall thickness. The wall thickness may be referred to as "pressure class," "thickness class," "schedule," or "weight," depending on the material.

The selection of materials is of paramount importance to the piping engineer. Some of the criteria used to select a material include chemical compatibility, system cleanliness requirements, service life, allowable stress, availability, ease of repair, and economy.

Casting versus Forging

In addition to specifying the correct material for a pipe or fitting, it is important to understand the way in which the item was manufactured. In general, metals that are cast are more brittle, and therefore more susceptible to fracture, than are forged metals. However, there are varying degrees of strength among cast metals as well. Hence, cast iron is weaker than cast steel. This is why cast iron reducing bushings are not permitted for use in fuel gas piping, but cast steel bushings are.

Given a choice, the forged materials will be the better selection due to their ability to withstand abuse. Over-tightening a cast threaded fitting will often result in a stress fracture. Castings are not easily welded. Castings may have surface imperfections due to improper cleaning of the mold.

Forging work-hardens the metal and imparts greater durability. Forged metals withstand higher stresses. Forged steel fittings are easily welded. The term "wrought," as in wrought iron or wrought copper indicates a forging process.

However, engineering is the science of making choices, or compromises, and those choices are often based as much on economics as technical merit. Castings often represent the least expensive installed cost.

Cast Iron Pipe

Commercially manufactured cast iron contains between 2 and 6.67 percent carbon. These metals are exceptionally strong in compression, but are very brittle. They have very low ductility and malleability, and cannot be drawn, rolled, or worked at room temperature. A sharp blow with a sledge hammer can crack a cast iron pipe.

Cast irons melt readily however, and can be cast into complicated shapes and machined. This property suits them for some valve bodies.

A distinction is drawn between "cast" iron and "ductile" iron. While ductile irons are also cast, the terms distinguish the metallographic structures of the materials. Cast iron describes a metallographic structure in which the carbon exists in the form of graphite flakes. The graphite is essentially carbon, chemically uncombined with any other elements. This structure is known as "gray cast iron" or simply "cast iron." The larger the graphite flakes are, the weaker the metal is. The flakes form stress concentrations at the microscopic level.

Applications

Cast iron was probably the first metal used for piping. The first recorded use of it was for a fountain in Langensalza, Germany circa 1562. A water distribution system was installed in France in 1664 for the palace of Versailles, and is allegedly still in use.

Because it is so brittle, it is not often used for pressure piping applications, although ASME B31.3 contains Basic Allowable Stress data for gray cast iron pipe. Cast iron pipe is now used primarily for drain, waste, and vent (DWV) applications, which is also known as "soil pipe."

Once used for water distribution piping, cast iron is now relegated to drain, waste, and vent service, due to the improved metallurgy provided by other materials, especially ductile iron. Cast iron water mains are still in service throughout the world however, in aging infrastructure.

Applicable Specifications

ANSI/AWWA	C105/A 21.5	American National Standard for Polyethylene Encasement for Ductile-Iron Pipe Systems
ANSI/AWWA	C110/A21.10	Ductile-Iron and Gray-Iron Fittings for Water
ANSI/AWWA	C115/A21.15	Flanged Ductile-Iron Pipe with Ductile-Iron or Gray-Iron Threaded Flanges
ANSI/AWWA	C116/A21.16	Protective Fusion-Bonded Epoxy Coatings for the Interior and Exterior Surfaces of Ductile-Iron/ Gray-Iron Fittings for Water Supply Service
ASME/ANSI	B 16.45	Cast Iron Fittings for Sovent® Drainage Systems.
ASSE	1043	Cast Iron Fittings for Sovent® Drainage Systems
ASTM	A-74	Standard Specification for Cast Iron Soil Pipe and Fittings
ASTM	A-674	Standard Practice for Polyethylene Encasement for Ductile Iron Pipe for Water or Other Liquids.
ASTM	A-888	Standard Specification for Hubless Cast Iron Soil Pipe and Fittings for Sanitary and Storm Drain, Waste, and Vent Piping Applications

ASTM	C-1277	Standard Specification for Shielded Couplings Joining Hubless Cast Iron Soil Pipe and Fittings
ASTM	C-564	Standard Specification for Rubber Gaskets for Joining Cast Iron Soil Pipe and Fittings
CISPI	301	Standard Specification for Hubless Cast Iron Soil Pipe and Fittings for Sanitary and Storm Drain, Waste, and Vent Piping Applications
CISPI	310	Specification for Couplings for Use In Connection With Hubless Cast Iron Soil Pipe and Fittings for Sanitary and Storm Drain, Waste, and Vent Piping Applications
CSA	B 602	Mechanical Couplings for Drain, Waste, and Vent Pipe and Sewer Pipe (Canadian Standards Association)
CSA	B 70	Cast Iron Soil Pipe, Fittings, and Means of Joining (Canadian Standards Association)

Manufacture of Cast Iron Pipe

Centrifugal casting is used to produce cast iron pipe. In this process, a water-cooled metal or sand-lined mold is spun about its longitudinal axis, while the premeasured molten iron alloy is poured into the mold. The centrifugal force flings the iron against the walls of the mold where it remains until solidified. Cast iron fittings are produced using static molds.

There are two types of cast iron (CI) pipe. The "hubless" variety is a straight cylinder manufactured in accordance with either ASTM A-888 or CISPI 301 whose ends are joined with mechanical couplings. Small diameters are also sometimes threaded. The hubless variety, also known as "no hub," is available in sizes from 1½ in to 10 in diameter. This variety has its own set of wall thicknesses. See Table 6.1.

The other variety, called "bell and spigot," is supplied with a single enlarged end (the "bell") and manufactured in accordance with ASTM A-74. See Figure 6.1. The narrower spigot of one piece is inserted into the bell of another. The joints are sealed with an integral elastomeric gasket, or with a lead and oakum joint. The elastomer gasket in an integral gasket bell and spigot joint is usually neoprene. This type of joint can deflect up to 5° without compromising the integrity of the joint.

Oakum is a hemp or jute rope that is coated with pine tar. It is packed into the annulus between the OD of the plain end spigot and the ID of the bell. Molten lead is

Figure 6.1 A cast iron bell.

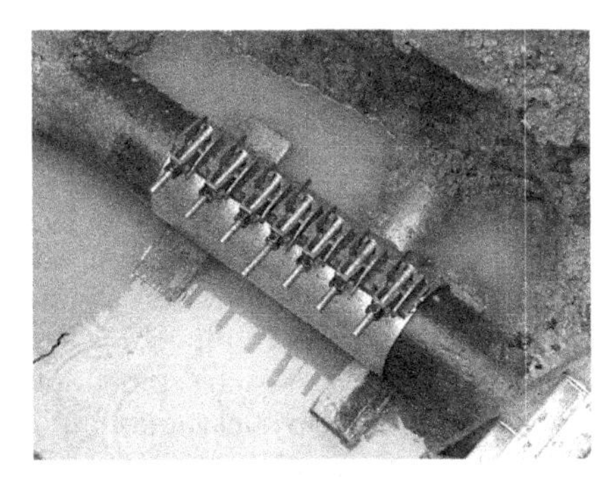

Figure 6.2 A hubless coupling used to repair a failed bell-and-spigot joint.

then poured into the remaining space and once cooled sufficiently, it is rammed with a caulking tool. This forms a strong seal that is impervious to water and root penetration. Of course, this is a labor-intensive joint, but it is still used in modern applications; mostly for repair work.

Lead wool is yet another method of sealing bell and spigot joints. This material looks like steel wool, but is made entirely of lead, and is caulked into the joints with a hammer and chisel.

Bell and spigot cast iron pipe is available in diameters from 2 in to 15 in. It is also available in two weight classes, designated "service weight" also called "standard weight," and "extra heavy weight." See Table 6.1.

From the early 1900s through approximately 1970 a lead substitute was used for bell and spigot joints. Called "leadite," this nonmetallic sulfur compound was melted like lead and poured into the joints, forming a brittle, vitreous seal. This seal was popular due to a favorable cost versus lead. It was also supposed to be easier to work than lead, with less material required to seal the joints. Differences in coefficients of thermal expansion between leadite and cast iron, as well as the brittle nature of the material, have caused line breaks, and many municipalities that have leadite joints are currently undergoing repair and replacement campaigns to eliminate these joints.

The mechanical joint in a hubless coupling consists of a neoprene sleeve which is compressed against the ODs of the two abutting pipes (See Figure 6.2). Another type of mechanical joint utilizes a stainless steel shield with an elastomeric sleeve and two or more hose clamps.

Cast iron pipe can be cut to length in a variety of ways. The most common tool in the field is probably the soil pipe cutter (Figure 6.3) which has a series of hardened steel

Figure 6.3 A soil pipe cutter. *Credit: RIDGID®*

wheels mounted on a chain. The chain is clamped to the circumference of the pipe and slowly tightened, resulting in a circumferential stress concentration that rises until the pipe is cracked along the plane formed by the wheels. The pipe or cutter may also be rotated to provide additional stress points along the circumference. Abrasive saws may also be used, as well as power hack saws and even cold chisels.

Cast iron soil pipe can be used above ground or below ground. It can support the weight of soil in underground applications. As with any aboveground installation, care must be taken to properly support CI pipe. This is especially true due to its weight and the method of joining. Each end of each segment must be supported.

CI pipe can withstand some physical abuse but is not able to withstand sharp blows. This is precisely why it can be cut using soil pipe cutters or chisels. It can resist soil pressures and is a good choice for underground drains. See Table 6.1.

Although CI pipe has a long service life, it can also be expected to display some tuberculation over time, and may not provide a smooth flow surface due to the accumulation of solids that may form on localized corrosion. CI pipe which was used in potable water service was often cement lined to prevent iron dissolution into the water, and to protect the interior of the pipe from corrosion.

Low noise transmission is one benefit of CI drains. A low coefficient of thermal expansion relative to other materials means that noise resulting from thermal expansion may also be minimized.

> In a company apartment in Savannah, Georgia, I was sometimes awakened by creaking under the bathroom sink. I traced the source of the noise to the expansion of a shared PVC drain line between my sink and the one in the apartment opposite mine. It seems that the early risers in that apartment were using hot water in the sink which caused the drain line to expand. The holes cut through the walls were neither "gasketed" to cushion the relative motion between the pipe and the wall, nor firestopped, and the expansion of the PVC drain against the wooden cabinet created a very annoying creaking. PVC has a coefficient of linear expansion five times higher than cast iron.

CI pipe is a noncombustible material that does not require firestopping through walls. This may provide an economic incentive to use CI pipe in applications that require firestopping.

Note that cast iron is used as a material for valve bodies and equipment, and in those cases sometimes requires a flange connection. In these cases, a full-face gasket is used for the 125/150 lb pressure class in order to prevent the cast iron flange from breaking. The gasket would act as a fulcrum and might crack the cast iron flange when the flange bolts are torqued. For this same reason, a flat faced flange should not be mated to a raised face flange.

Because a bell and spigot joint can be pulled apart under internal pressure loads (including water hammer) CI pipe must be restrained. A common restraint method is to use concrete thrust blocks.

Example 6.1

Given: Underground 12 in diameter, service weight water main operating at 100 psi. The flow rate is 2800 GPM. Soil bearing pressure is 2500 PSF. See Figure 6.4. The density of water is 62.4 lb/ft^3.

NO HUB

Pipe Size (in)	OD (in)	Nominal Thickness (in)	P (lb/ft2)	Condition 1 Trench Width 18 in	24 in	36 in	P (lb/ft2)	Condition 2 Trench Width 18 in	24 in	36 in	P (lb/ft2)	Condition 3 Trench Width 18 in or Greater
1½	1.90	0.16	57434	299	299	299	95720	499	499	499	191400	1000
2	2.35	0.16	36300	189	189	189	60400	315	315	315	120900	1000
3	3.35	0.16	17100	89	89	89	28500	148	148	148	57000	475
4	4.38	0.19	14000	73	73	73	23300	121	121	121	46600	388
5	5.30	0.19	9400	49	49	49	15700	81	81	81	31300	261
6	6.30	0.19	6600	42	34	34	10900	70	57	57	21900	182
8	8.38	0.23	5400	45	35	28	9000	75	58	47	18000	150
10	10.56	0.28	5000	42	42	26	8200	68	68	43	16800	140

SERVICE WEIGHT

Service Pipe Size (in)	OD (in)	Nominal Thickness (in)	P (lb/ft2)	Condition 1 Trench Width 18 in	24 in	36 in	P (lb/ft2)	Condition 2 Trench Width 18 in	24 in	36 in	P (lb/ft2)	Condition 3 Trench Width 18 in or Greater
2	2.30	0.17	43300	225	225	225	72100	376	376	376	144200	1200
3	3.30	0.17	20000	104	104	104	33400	174	174	174	66800	557
4	4.30	0.18	13000	68	68	68	21600	112	112	112	43200	360
5	5.30	0.18	8400	44	44	44	14000	73	73	73	28000	233
6	6.30	0.18	5900	38	31	31	9800	63	51	51	19600	163
8	8.38	0.23	5400	45	35	28	9000	75	58	47	18000	150
10	10.50	0.28	5100	43	43	27	8500	71	71	44	17000	142
12	12.50	0.28	3600	30	30	23	5900	49	49	38	11900	99
15	15.88	0.36	3700	31	31	31	6100	51	51	51	12200	102

P = Design Soil Pressure = Sum of dead and live load pressures at the level of the top of the pipe.
Condition 1 = No pipe bedding; hard trench bottom; continuous line support.
Condition 2 = Bedding placed for uniform support; soil under haunches of pipe should be compacted.
Condition 3 = Select, loose soil envelope placed about the pipe as packing, with a dense soil arch compacted up over the envelope.

TABLE 6.1 Maximum Allowable Trench Depths for Cast Iron Soil Pipe. Maximum depths are height of soil above top of pipe in feet. CISPI

(continued on next page)

EXTRA HEAVY WEIGHT

Extra Heavy Pipe Size (in)	OD (in)	Nominal Thickness (in)	P (lb/ft2)	Condition 1 Trench Width			P (lb/ft2)	Condition 2 Trench Width			P (lb/ft2)	Condition 3 Trench Width
				18 in	24 in	36 in		18 in	24 in	36 in		18 in or Greater
2	2.38	0.19	51100	266	266	266	85200	444	444	444	144200	1420
3	3.50	0.25	40200	209	209	209	67000	349	349	349	66800	1116
4	4.50	0.25	23500	122	122	122	39200	204	204	204	43200	653
5	5.50	0.25	15400	99	80	80	25700	165	34	134	28000	428
6	6.50	0.25	10900	70	57	57	18100	116	94	94	19600	302
8	8.62	0.31	9500	79	61	49	15800	131	101	82	18000	263
10	10.75	0.37	8600	72	72	55	14400	120	120	92	17000	240
12	12.75	0.37	6100	51	51	39	10100	84	84	65	11900	169
15	15.88	0.44	5500	50	50	50	9200	77	77	77	12200	153

P = Design Soil Pressure = Sum of dead and live load pressures at the level of the top of the pipe.
Condition 1 = No pipe bedding; hard trench bottom; continuous line support.
Condition 2 = Bedding placed for uniform support; soil under haunches of pipe should be compacted.
Condition 3 = Select, loose soil envelope placed about the pipe as packing, with a dense soil arch compacted up over the envelope.

Table 6.1 *(continued)*

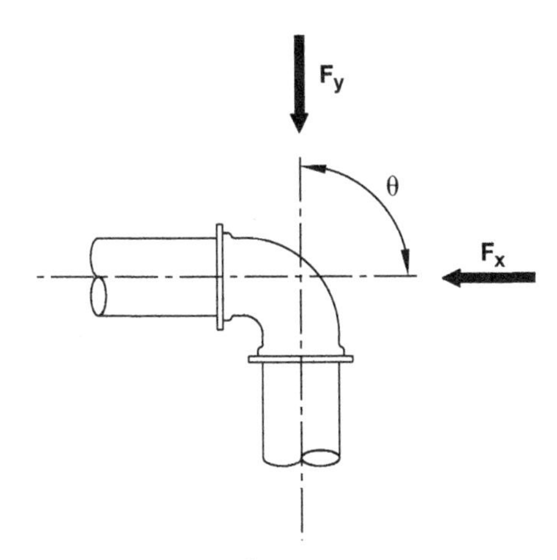

FIGURE 6.4 Example 6.1 thrust block.

Find: Size of thrust block required at a 90° elbow.

Solution: The force to be restrained is the sum of the forces created by the internal pressure in the pipe and the velocity of the fluid impacting the elbow.

A 12 in diameter service weight pipe has an OD of 12.5 in and a wall thickness of 0.28 in. The area over which the pressure force is applied is therefore

$$A = (3.14/4) \times [12.5 \text{ in} - 2(0.28 \text{ in})]^2 = 112 \text{ in}^2$$

The velocity is

$$v = \frac{Q}{A}$$

Where $Q = (2800 \text{ gal/min})\ (0.1337 \text{ ft}^3/\text{gal}) = 374 \text{ ft}^3/\text{min} = 6.2 \text{ ft}^3/\text{sec}$

$$v = \frac{Q}{A} = \frac{(374 \text{ ft}^3/\text{min})}{(112 \text{ in}^2)\left(\dfrac{\text{ft}^2}{144 \text{ in}^2}\right)} = 481 \text{ ft/min}$$

$$= 58.01 \text{ ft/sec}$$

We solve for the force created by the pressure:

$$F = PA = 100 \text{ lb/in}^2 \times 112 \text{ in}^2 = 11{,}200 \text{ lb}$$

The momentum equation is given by:

$$F = m \circ v = \frac{(6.2 \text{ ft}^3/\text{sec})\ (62.4 \text{lb/ft}^3)\ (8.01 \text{ft/sec})}{32.2 \text{ ft/sec}^2} = 96 \text{ lb}$$

Therefore the total force that must be absorbed by the soil is

$$F = 11{,}200 \text{ lb} + 96 \text{ lb} = 11{,}296 \text{ lb}$$

The bearing area over which the thrust block must act is therefore

$$(11{,}296 \text{ lb})/(2500 \text{ lb/ft}^2) = 4.5 \text{ ft}^2$$

A 2 ft by 2.25 ft thrust block would be appropriate. For an elbow, two blocks are required, placed where the arrows are shown in the figure. Note that the thickness of the thrust block must also be accounted for. The block must be sufficiently thick to withstand the compressive force applied without shearing (reinforcing may be required). The calculation currently makes no allowance for water hammer loads, which may increase the force substantially. Finally, there is often very little care paid to thrust blocks during the construction phase. Forms are rarely used. Instead, a hole is dug, a pile of concrete is poured, the hole is covered, and the pipe very often leaks. The surface of the thrust block that bears against the soil should be flat.

Any time you specify a thrust block, great care must be taken in the field to ensure that the thrust block will satisfy the requirements. Field inspections are advisable. In lieu of this, a better choice will be restrained joints designed to accommodate the thrust loads.

Ductile Iron Pipe

Ductile iron (DI) pipe was developed in 1948, and soon replaced cast iron pipe in pressure applications. The metallographic structure of ductile iron is such that the graphite exists in the form of nodules. These compact nodules do not interrupt the metallurgical matrix like the graphite flakes in cast iron. The result is a material that is stronger and tougher than cast iron.

Applications

Like cast iron, ductile iron is used for sewage service. But it finds additional use in liquid service; especially for water, and especially potable water.

Applicable Specifications

ASME	B16.42	Ductile Iron Pipe Flanges and Flanged Fittings, Classes 150 and 300
AWWA	C104/A21.4	Cement-Mortar Lining for Ductile-Iron Pipe and Fittings for Water
ASTM	A716	Ductile Iron Culvert Pipe
ASTM	A746	Ductile Iron Gravity Sewer Pipe
AWWA	C105/A21.5	Polyethylene Encasement for Ductile-Iron Pipe Systems
AWWA	C110/A21.10	Ductile-Iron and Gray-Iron Fittings for Water
AWWA	C111/A21.11	Rubber-Gasket Joints for Ductile-Iron Pressure Pipe and Fittings
AWWA	C115/A21.15	Flanged Ductile-Iron Pipe with Ductile-Iron or Gray-Iron Threaded Flanges
AWWA	C116/A21.16	Protective Fusion-Bonded Epoxy Coatings Int. & Ext. Surf. Ductile-Iron/Gray-Iron Fittings
AWWA	C150/A21.50	Thickness Design of Ductile-Iron Pipe
AWWA	C151/A21.51	Ductile-Iron Pipe, Centrifugally Cast, for Water or Other Liquids
AWWA	C153/A21.53	Ductile-Iron Compact Fittings for Water Service

Manufacture of DI Pipe

Ductile iron pipe is manufactured in the same way as cast iron pipe; that is, using a centrifugal casting process to form the lengths of pipe.

Ductile iron pipe is available in five pressure classes, defined as the rated working pressure of the pipe, based on a minimum yield strength of 42,000 psi and a 2.0 safety factor which is applied to the working pressure plus a surge pressure of 100 psi.

When DI pipe was introduced, the outside diameters of the sizes between 4 and 48 in inclusive were selected to be identical to that of cast iron pipe to make the transition to the new material easier. Accessories and fittings were then compatible.

Example 6.2

Given: 12 in diameter DI pipe at 350 psi working pressure
100 psi surge pressure
42,000 psi yield stress

Find: Required wall thickness using a factor of safety of 2

By Barlow's formula,

$$T = PD/2S$$

where

T = Wall thickness
P = Pressure
D = Outside diameter
S = Yield stress

Then

$$T = (350 \text{ psi} + 100 \text{ psi})(13.20 \text{ in}) / 2\,(42{,}000 \text{ psi})$$
$$= 0.071 \text{ in}$$

Next we apply the safety factor

$$T = 2 \times 0.071 \text{ in} = 0.14 \text{ in}$$

DI pipe manufacturers add both a "service allowance" and a casting tolerance to this thickness to arrive at the commercial wall thickness. The service allowance is 0.08 in. The casting tolerances are given in Table 6.2 as 0.06 in for this diameter of pipe.

Adding the calculated wall thickness to the service allowance and the casting tolerance yields a commercial wall thickness of

$$t = 0.14 \text{ in} + 0.08 \text{ in} + 0.06 \text{ in} = 0.28 \text{ in}$$

which is the value shown in Table 6.3 for the nominal wall thickness of a 12 in diameter Pressure Class 350 DI pipe.

There are 12 standard wall thickness classes for DI pipe. These classes are analogous to "schedules" or "weights" that are used for steel pipe. The DI pipe classes are divided into two categories:

1. The Pressure Class is named after the working pressure of the pipe, as shown in the example above. It allows for a pressure surge of 100 psi above the working pressure, and includes a yield stress of 42,000 psi for the DI material, a factor of safety of 2, and also the service allowance and casting tolerances described in the above example.

Nominal Diameter (in)	Minus Tolerance (in)
4 – 8	0.05
10 – 12	0.06
14 – 42	0.07
48	0.08
54 – 64	0.09

Table 6.2 Casting tolerances for ductile iron pipe.

2. The other set of classes are called "Special Thickness Classes" and they are identified with the numbers 50 through 56. These numbers are nominal identifiers and have no physical meaning (unlike the Pressure Class names which identify the working pressure values). These Special Thickness Classes are often specified since they permit a larger variety of wall thicknesses for an application than would otherwise be available from the Pressure Class thicknesses.

The combination of the two classes provides a more extensive menu from which to select an economical wall thickness. The minimum wall thickness manufactured is 0.25 in.

The outer surface of DI pipe is normally coated with a 1 mil thick asphaltic coating in accordance with AWWA C151. The inner surfaces are normally furnished with the standard cement lining as specified in AWWA C104. The inside surface may also be furnished uncoated, with asphaltic coating, coal tar epoxy (a resin and tar combination), or various proprietary coatings, depending on the application.

The cement inner lining is by far the most common, in order to limit corrosion and improve flow characteristics. Cement linings are adequate for water temperatures up to 212°F (100°C). If handled roughly or stored for long periods, the cement lining may be subject to cracks, and sometimes even looseness. Vendor literature states that this does not inhibit the effectiveness of the lining. It is thought that exposure of the lining to water causes the cracks to close due to swelling of the cement as the water is absorbed into the microstructure of the lining. Even so, AWWA C104 provides a procedure for repairing damaged cement lining. A three part Portland cement and two part clean sand mixture is prepared with enough water to provide a slump of 5 to 8 in. This mixture is applied with a paintbrush and allowed to cure slowly before use. Alternatively, an asphaltic coating may be applied over the damaged area. See Table 6.4 for thicknesses of linings.

Fabrication and Assembly of Ductile Iron Pipe

DI pipe may be cut using:

- Abrasive saws
- Torches
- Milling cutters
- Portable guillotine saws.

There are several common methods in use to join DI pipe. One is a bell-and-spigot joint, which, like some cast iron pipe, uses a flexible gasket to provide the seal. Another method is the use of flanges. Still another is the mechanical joint.

Size in	Outside Diameter in	Pressure Class (Working Pressure in psi)					Special Thickness Classes						
		150	200	250	300	350	50	51	52	53	54	55	56
4	4.80	—	—	—	—	0.25	—	0.26	0.29	0.32	0.35	0.38	0.41
6	6.90	—	—	—	—	0.25	0.25	0.28	0.31	0.34	0.37	0.40	0.43
8	9.05	—	—	—	—	0.25	0.27	0.30	0.33	.036	0.39	0.42	0.45
10	11.10	—	—	—	—	0.26	0.29	0.32	0.35	0.38	0.41	0.44	0.47
12	13.20	—	—	—	—	0.28	0.31	0.34	0.37	0.40	0.43	0.46	0.49
14	15.30	—	—	0.28	0.30	0.31	0.33	0.36	0.39	0.42	0.45	0.48	0.51
16	17.40	—	—	0.30	0.32	0.34	0.34	0.37	0.40	0.43	0.46	0.49	0.52
18	19.50	—	—	0.31	0.34	0.36	0.35	0.38	0.41	0.44	0.47	0.50	0.53
20	21.60	—	—	0.33	0.36	0.38	0.36	0.39	0.42	0.45	0.48	0.51	0.54
24	25.80	—	0.33	0.37	0.40	0.43	0.38	0.41	0.44	0.47	0.50	0.53	0.56
30	32.00	0.34	0.38	0.42	0.45	0.49	0.39	0.43	0.47	0.51	0.55	0.59	0.63
36	38.30	0.38	0.42	0.47	0.51	0.56	0.43	0.48	0.53	0.58	0.63	0.68	0.73
42	44.50	0.41	0.47	0.52	0.57	0.63	0.47	0.53	0.59	0.65	0.71	0.77	0.83
48	50.80	0.46	0.52	0.58	0.64	0.70	0.51	0.58	0.65	0.72	0.79	0.86	0.93
54	57.56	0.51	0.58	0.65	0.72	0.79	0.57	0.65	0.73	0.81	0.89	0.97	1.05
60	61.61	0.54	0.61	0.68	0.76	0.83	—	—	—	—	—	—	—
64	65.67	0.56	0.64	0.72	0.80	0.87	—	—	—	—	—	—	—

TABLE 6.3 Ductile Iron Sizes and Wall Thicknesses. The thicknesses shown are adequate for the rated water working pressure plus a surge allowance of 100 psi. Values are based on a minimum yield strength in tension of 42,000 psi and 2.0 safety factor times the sum of working pressure and 100 psi surge allowance.

Size in	Nominal Pipe Length ft	Standard Thickness Minimum Thickness in	Standard Thickness Weight Per Foot lb	Standard Thickness Weight Per Length lb	Double Thickness Minimum Thickness in	Double Thickness Weight Per Foot lb	Double Thickness Weight Per Length lb
4	18	1/16	0.87	17	1/8	1.71	31
6	20	1/16	1.3	26	1/8	2.57	51
8	20	1/16	1.74	35	1/8	3.45	69
10	20	1/16	2.15	43	1/8	4.28	86
12	20	1/16	2.57	51	1/8	5.12	102
14	20	3/32	4.49	90	3/16	8.93	179
16	20	3/32	5.13	103	3/16	10.19	204
18	20	3/32	5.76	115	3/16	11.47	229
20	20	3/32	6.4	128	3/16	12.73	255
24	20	3/32	7.68	154	3/16	15.31	306
30	20	1/8	12.76	255	1/4	25.42	508
36	20	1/8	15.31	306	1/4	30.51	610
42	20	1/8	17.82	356	1/4	35.53	711
48	20	1/8	20.35	407	1/4	40.6	812
54	20	1/8	22.89	458	1/4	45.68	914
60	20	1/8	24.71	494	1/4	49.32	986
64	20	1/8	26.35	527	1/4	52.61	1052

TABLE 6.4 Cement Linings per ANSI/AWWA C104/A21.4

"Mechanical joint" is a generic term used to describe the joining of two pipe ends by any means other than welding, brazing, soldering, or caulking. In describing cast or ductile iron pipes, the term "mechanical joint" most often means a joint other than a flanged joint that is made by bolting.

Mechanical joints are designed with a gland that compresses a gasket. In some designs, the joint is also restrained from pulling apart.

It is worth noting that whenever a cut pipe is to be inserted into a gasketed bell and spigot joint, the cut end must be beveled in order to prevent damaging the gasket.

In underground systems using bell and spigot joints, the same problems with thrust blocks apply to ductile iron pipe as to cast iron pipe. Leakage through flexible gasket bell-and-spigot joints is estimated to be approximately 1 GPH/1000 ft of pipe at 150 psi.

Carbon Steel

Carbon steel piping is the type that is most often used in industrial applications. It has the advantage of wide availability, high strength, and myriad connection systems and fittings.

Many grades of carbon steel pipe are available. These grades vary due to metallurgy and manufacture of the pipe itself. To the pipefitter, there is essentially no difference between the various grades. He will ply his trade in the same manner irrespective of the ASTM number. In critical applications, a welder may choose a different electrode or current, depending on the grade of steel used. And once the material is specified, the piping engineer or designer will also pay no particular attention to the grade of steel used.

Applications

Carbon steel piping is used for many liquid and gas services, both above and below ground. It is also widely used for steam systems. It is inappropriate for corrosive services but is used for caustic services. It may be used for potable water if appropriate linings are applied to prevent iron dissolution.

Applicable Standards

ASME	B16.11	Forged Steel Fittings, Socket Welded, and Threaded
ASME	B16.25	Buttwelding Ends
ASME	B16.28	Wrought Steel Buttwelding Short Radius Elbows and Returns
ASTM	A53	Pipe, Steel, Black and Hot-Dipped, Zinc-Coated, Welded, and Seamless
ASTM	A106	Seamless Carbon Steel Pipe for High-Temperature Service
ASTM	A134	Pipe, Steel, Electric-Fusion (Arc)-Welded (Sizes NPS 16 and Over)
ASTM	A135	Electric-Resistance-Welded Steel Pipe
ASTM	A139	Electric-Fusion (Arc)-Welded Steel Pipe (NPS 4 and Over)
ASTM	A182	Forged or Rolled Alloy-Steel Pipe Flanges, Forged Fittings, and Valves and Parts for High-Temperature Service
ASTM	A211	Spiral-Welded Pipe – Standard Withdrawn in 1993
ASTM	A252	Welded and Seamless Steel Pipe Piles
ASTM	A333	Seamless and Welded Steel Pipe for Low-Temperature Service
ASTM	A335	Seamless Ferritic Alloy-Steel Pipe for High-Temperature Service
ASTM	A369	Carbon and Ferritic Alloy Steel Forged and Bored Pipe for High-Temperature Service
ASTM	A381	Standard Specification for Metal-Arc-Welded Steel Pipe for Use With High-Pressure Transmission Systems
ASTM	A426	Centrifugally Cast Ferritic Alloy Steel Pipe for High-Temperature Service
ASTM	A523	Plain End Seamless and Electric-Resistance-Welded Steel Pipe for High-Pressure Pipe-Type Cable Circuits
ASTM	A524	Seamless Carbon Steel Pipe for Atmospheric and Lower Temperatures
ASTM	A530	General Requirements for Specialized Carbon and Alloy Steel Pipe
ASTM	A691	Carbon and Alloy Steel Pipe, Electric-Fusion-Welded for High-Pressure Service at High Temperatures
ASTM	A694	Carbon and Alloy Steel Forgings for Pipe Flanges, Fittings, Valves, and Parts for High-Pressure Transmission Service
ASTM	A714	High-Strength Low-Alloy Welded and Seamless Steel Pipe
ASTM	A733	Welded and Seamless Carbon Steel and Austenitic Stainless Steel Pipe Nipples
ASTM	A865	Threaded Couplings, Steel, Black or Zinc-Coated (Galvanized) Welded or Seamless, for Use in Steel Pipe Joints

ASTM	A984	Steel Line Pipe, Black, Plain-End, Electric-Resistance-Welded
ASTM	A1005	Steel Line Pipe, Black, Plain End, Longitudinal and Helical Seam, Double Submerged-Arc Welded
ASTM	A1006	Steel Line Pipe, Black, Plain End, Laser Beam Welded
AWWA	C200	Steel Water Pipe 6 in. (150 mm) and Larger
AWWA	C203	Coal-Tar Protective Coatings & Linings for Steel Water Pipelines, Enamel and Tape, Hot-Applied
AWWA	C205	Cement-Mortar Protective Lining and Coating for Steel Water Pipe, 4 in. (100 mm) and Larger, Shop Applied
AWWA	C206	Field Welding of Steel Water Pipe
AWWA	C207	Steel Pipe Flanges for Waterworks Service, Sizes 4 in. through 144 in. (100 mm through 3600 mm)
AWWA	C208	Dimensions for Fabricated Steel Water Pipe Fittings
AWWA	C209	Cold-Applied Tape Coatings for the Exterior of Special Sections, Connections, and Fittings for Steel Water Pipelines
AWWA	C210	Liquid-Epoxy Coating Systems for the Interior and Exterior of Steel Water Pipelines
AWWA	C213	Fusion Bonded Epoxy Coating for the Interior and Exterior of Steel Water Pipelines
AWWA	C214	Tape Coating Systems for the Exterior of Steel Water Pipelines
AWWA	C215	Extruded Polyolefin Coatings for the Exterior of Steel Water Pipelines
AWWA	C216	Heat-Shrinkable Cross-Linked Polyolefin Coatings for the Exterior of Special Sections, Connections, and Fittings for Steel Water Pipelines
AWWA	C217	Petrolatum and Petroleum Wax Tape Coatings for the Exterior of Connections and Fittings for Steel Water Pipelines
AWWA	C218	Coating the Exterior of Aboveground Steel Water Pipelines and Fittings
AWWA	C219	Bolted, Sleeve-Type Couplings for Plain-End Pipe
AWWA	C221	Fabricated Steel Mechanical Slip-Type Expansion Joints
AWWA	C222	Polyurethane Coatings for the Interior and Exterior of Steel Water Pipe and Fittings
AWWA	C223	Fabricated Steel and Stainless Steel Tapping Sleeves
AWWA	C224	Nylon-11 Based Polyamide Coating System for the Interior and Exterior of Steel Water Pipe and Fittings
AWWA	C225	Fused Polyolefin Coating Systems for the Exterior of Steel Water Pipelines

There is wide variety in carbon steel piping materials, as may be seen by examining the list of materials contained in the piping codes. We have seen that the allowable stress is used to determine what wall thickness is required. The allowable stress is a function of both the metallurgy of the material and the method of manufacture.

The various piping specifications provided by organizations such as ASTM and API provide guidelines for both the metallurgy and the method of manufacture. And a given specification may prescribe multiple chemistries and multiple methods of manufacture.

For instance, ASTM A-53 describes a piping material that may be manufactured from steel made in the open hearth, basic oxygen furnace, or electric furnace, and then formed into pipe by either furnace welding, electric welding, or being pierced with a mandrel and extruded. Each of these variables produces a final product with a different yield stress. See Table 6.5.

It is therefore imperative that the engineer spells out exactly the specification he needs for an application. It is not sufficient to merely specify ASTM A-53 for example.

Sometimes a material will satisfy more than one specification, and in order to reduce inventory, a piping supplier will offer "tri-stenciled" or "quad-stenciled" piping. This means that the piping satisfies three or four piping specifications, with all three or four specs stenciled on the exterior of the pipe. These designations often mean:

"Tri-stenciled"

- ASTM A-53 Grade B
- ASTM A-106 Grade B
- API 5L – X42.

"Quad-Stenciled"

- ASTM A-53 Grade B
- ASTM A-106 Grade B
- API-5L Grade B
- API 5L – X42.

		Grade A	Grade B
Type F Furnace Butt Weld			
C	max %	0.30	-
Mn	max %	1.20	-
Tensile Strength per A-53	ksi	48.00	-
Yield Strength per A-53	ksi	30.00	-
B 31.1 Maximum Allowable Stress (SE) at 100 deg F	ksi	7.20	-
B 31.3 Maximum Allowable Stress (SE) at 100 deg F	ksi	9.60	-
Type E Electric Resistance Weld			
C	max %	0.25	0.30
Mn	max %	0.95	1.20
Tensile Strength per A-53	ksi	48.00	60.00
Yield Strength per A-53	ksi	30.00	35.00
B 31.1 Maximum Allowable Stress (SE) at 100 deg F	ksi	10.20	12.80
B 31.3 Maximum Allowable Stress (SE) at 100 deg F	ksi	13.60	20.00
Type S Seamless			
C	max %	0.25	0.30
Mn	max %	0.95	1.20
Tensile Strength per A-53	ksi	48.00	60.00
Yield Strength per A-53	ksi	30.00	35.00
B 31.1 Maximum Allowable Stress (SE) at 100 deg F	ksi	12.00	15.00
B 31.3 Maximum Allowable Stress (SE) at 100 deg F	ksi	13.60	20.00

Table 6.5 Example of Different Yield Strengths within One Specification (ASTM A-53)

But it is important to note that designations such as tri-stenciled or quad-stenciled may include different specs than those noted above. The best way to proceed is to specify the exact specification that is required.

Manufacture of Carbon Steel Pipe

Seamless Pipe

Seamless pipe is produced by heating a round billet or square bloom of steel, and then piercing it with a bullet-shaped piercer, over which the steel is stretched. This is followed by another piercer that opens the hole even more and further elongates the hollow cylinder. A series of straightening rollers, sizing rollers, and heating, cooling, and inspection processes results in a seamless pipe that may now be cut to length and end finished. The final product is hydrostatically tested, inspected, coated if required, and stenciled with the specification.

Because it is a homogeneous substance with no weld stresses, seamless pipe is the strongest variety. For example, the maximum allowable stresses are higher for ASTM Type A-53 Type S (Seamless) than for any of the other varieties of A-53.

The piercing of a round billet is a process that incredibly was first performed in 1845, but proved too technically challenging to produce long lengths of pipe until about 1895.

Electric Resistance Weld (ERW) Pipe

ERW pipe is made from coils that are cupped longitudinally by forming rolls and a fin pass section of rolls that brings the ends of the coil together to form a cylinder. These ends are passed through a high frequency welder which heats the steel to 2600°F and squeezes the ends together to form a fusion weld. The weld is heat treated to remove welding stresses and the pipe is cooled, sized to the proper OD, straightened, and cut to length.

There is an optional process that can also be employed to size the pipe while at the same time increasing the transverse yield strength. Incredibly, this process involves hydraulically expanding the pipe. The ends are sealed, and water is forced into the pipe at a pressure high enough to plastically deform the steel to the desired OD and wall thickness. This obviously results in a hydrotest of the weld as well, but each length of pipe is also subjected to a separate hydrotest and is then straightened. The welds are ultrasonically tested, the ends are prepared, and the pipe is visually inspected, coated, and stenciled.

Double Submerged Arc Weld (DSAW) Pipe

DSAW pipe starts with a plate that is edge milled to ensure that the edges to be joined will be parallel. The edges are then crimped upward and a ram forms the plate into a U-shape. Next the U is placed in an "O"-ing press, which completes the cylinder and prepares it for tack welding. The pipe is welded on the inside, then again on the outside. An inspection of the weld follows, and then the pipe is placed into a mechanical expander that plastically deforms the wall in short longitudinal increments to achieve the appropriate diameter. Additional hydrostatic and NDT is next followed by end beveling and an X-ray inspection of each end of the pipe weld. The finished pipe is visually inspected, coated, and stenciled.

Furnace Weld, Butt Weld or Continuous Weld (CW) Pipe

These all refer to the same process, and in fact, the term "Furnace Butt Weld" is also synonymous. Steel strip coil (called a "skelp") is uncoiled and fed into a roller leveler which flattens the steel. An endless strip is created prior to further processing, so the ends of the coil are cut square and welded to the trailing end of the preceding coil. The continuous strip enters a furnace where it is heated to approximately 2450°F (1343°C). The edge of the strip is raised to approximately 2600°F (1427°C) by an oxygen lance (nozzle) as it exits the furnace. Forming rolls gradually bend the strip into a circle where a set of rollers welds the seam using the heat contained in the strip and the pressure exerted by the welding rollers. The pipe OD is reduced, and the wall thickness is achieved in a stretching mill. A saw cuts the pipe to length, and the pipe enters a sizing mill that reduces the pipe to the final OD. The pipe is straightened, end finished, hydrotested, coated, stenciled, and inspected.

Spiral-Welded Pipe

ASTM A211 was a specification for spiral-welded pipe. This specification was withdrawn in 1993. Spiral-welded pipe is currently manufactured to meet the specifications of ASTM A139 or AWWA C200. It is used primarily by utilities for water distribution service.

Spiral-welded pipe is produced from coils of steel that are unwound and flattened. The flattened strip is formed by angled rollers into a cylinder of the desired diameter. Interior and exterior submerged arc welds seal the spiral seam. At the end of the coil, a new coil is butt welded to the trailing edge of the pipe, forming a cross seam. The pipe is cut to length and the ends are beveled if required.

Available joints for this type of pipe include butt weld, lap weld, and rubber gasketed joints. The lap weld and rubber gasketed joints require a bell to be formed on one end of the pipe. This is created by the manufacturer with a hydraulic expander. After forming the ends, the pipes are hydrostatically tested and then lined with cement. The pipe is next heated to eliminate moisture, and the exterior is blasted to prepare it for the application of an exterior coating.The coating consists of a primer or adhesive, followed by a dielectric tape, and finally a polyethylene wrap.

Sizes of spiral-welded pipe are available from 24 to 144 in, and wall thicknesses up to 1 in.

Wall Thicknesses of Carbon Steel Pipe

Carbon steel piping is described by the nominal diameter and the "weight" or "schedule." The weight or schedule is merely a description of the wall thickness.

As with the DI pressure classes, the carbon steel pipe schedules have evolved over the years into what amounts to a menu of diameters and wall thicknesses that have little to do with either the actual diameters or the wall thicknesses. Everything about these schedules must be considered "nominal." And the engineer who must perform field measurements of existing installations would be well-advised to either memorize the outside diameters of some of the more common pipes, or to carry with him a copy of the schedules. See Appendix 1 for carbon steel pipe data.

The wall thicknesses known as Standard, Extra Strong (XS), and Double Extra Strong (XXS) are from an older system of describing wall thickness known as Iron Pipe Size (IPS). This system has carried over into the relatively newer system, called the Nominal Pipe Size (NPS), and many (but not all) of the old IPS sizes are duplicated within the NPS schedules:

- Standard is identical to Schedule 40 up to and including 10 in. Starting at 12 in, the wall thickness of Standard pipe remains at 0.375 in.

- Extra Strong is identical to Schedule 80 up to and including 8 in. Starting at 8 in, the wall thickness of XS pipe remains at 0.500 in.
- Double Extra Strong has no related schedule. It is thicker than Schedule 160 up to 6 in diameter, and starting at 8 in diameter it is thinner than Schedule 160.

The NPS system is used throughout North America and the United Kingdom. Europe also uses the same schedules, however they are referred to as the Diameter Nominal (DN) system, and they are identified by a metric approximation of the nominal diameter. Thus, a 1 in pipe specified for use in Canada may be specified in the European Union as DN 25. DN pipe is sometimes also referred to as DIN or ISO pipe.

Not all of the schedules are commercially available in every pipe diameter. Some schedules are not produced, and some are produced only if a minimum tonnage is ordered.

This is seldom a problem though, since most of the time the engineer will specify or work in the common schedules:

- 10 or 10S
- Standard
- 40
- XS
- 80
- 160
- XXS

Note that the published wall thicknesses may be 12.5 percent thicker than the actual thickness.

Sizes of Carbon Steel Pipe

Many specifications for industrial piping limit the sizes to be used on a project. Rarely will any size below 1 in be specified, for instance, since the smaller diameters are more difficult to support, and the difference in cost is negligible.

There is also a prejudice against 1¼ in and 2½ in pipe in industrial projects due to the costs of stocking pipe and fittings that are so close in size to other economical selections. HVAC projects are more apt to use these sizes for hydronic piping.

The availability of fittings and valves is another concern when specifying piping, since any pipe must necessarily mate with fittings. For this reason, 3½ in and 5 in diameter is hardly ever specified. As might be expected, 5 in would fill the gap between 4 in and 6 in nicely in terms of flow velocities. But it is never specified for industrial projects.

The suffix "S" on some of the schedules indicates that these are available in corrosion resistant (stainless) steel. Carbon steel is available in Schedule 10 for all sizes except 42 in. This is sometimes useful for low-pressure applications, and finds use in fire protection piping.

Fabrication and Assembly of Carbon Steel Pipe

Carbon steel piping is cut to length with oxy-fuel torches or abrasive saws. It is joined in several ways:

- Welded
- Threaded and Coupled (T&C)

- Flanged
- Mechanical couplings.

No doubt the bulk of industrial piping above 2 in diameter is welded construction. Welded construction provides leak-proof, rigid joints, and although welding is a relatively expensive method of joining pipes, it is the method most employed.

T&C piping is generally not used in sizes above 2 in, even though the ANSI standard for NPT threads goes all the way to 24 in diameter. Above 2 in diameter it is difficult to torque the fittings sufficiently to seal a joint under pressure. The most common thread is the tapered National Pipe Thread (NPT), but others are also available.

The NPT is an interference fit between mating threads. The joints are lubricated and sealed with a joint compound or with PTFE (Teflon®) tape. Care must obviously be taken to choose a sealant that is compatible with the fluid conveyed and the operating temperature.

Flanges offer a means of providing an easy connection to equipment and to other pipes and valves. Flanges for carbon steel piping have a raised face, unlike the DI flanges. This raised face allows the bolt torque to be transferred over a smaller area, resulting in sufficient gasket compression for the application. Flanges will also be discussed in detail in a later chapter.

Manufacturers have developed clever methods for joining pipes mechanically. Some of the incentives for avoiding welds may be to speed field assembly, reduce costs, or eliminate the need for hot work in hazardous environments. Some of these mechanical couplings are shown in Figure 6.5. These fittings are prepared by bolting or crimping with a gasket. Flexibility or rigidity may be designed into the joint, depending on the type of coupling selected.

FIGURE 6.5 Mechanical joints, left-to-right, top-to-bottom: Dresser Style 38 coupling, Dresser mechanical coupling manufactured in 1890, Victaulic Style 77, Victaulic Style 597, which crimps to the adjoining pipe with a special rolling tool. *Credits: Style 38 and 1890 Couplings, Dresser Piping Specialties, Dresser Inc., Victaulic images reprinted with permission of Victaulic Company.*

Stainless Steel Piping

Stainless steel piping is used whenever corrosion resistance is desired. The addition of chromium is primarily responsible for the corrosion resistant properties of stainless steels. Even though stainless steel exhibits excellent corrosion protection, it would be very unusual to use it in an underground application. A more economical solution would be to use carbon steel with cathodic protection.

Austenitic stainless steels are identified by the 300 series. These alloys contain a maximum of 0.15 percent carbon and a minimum of 16 percent chromium, along with nickel and/or manganese. Austenitic stainless steels are nonmagnetic and have the best high-temperature strengths of all of the stainless steels.

Ferritic stainless steels belong to the 400 series. These steels contain 14 to 27 percent chromium and are magnetic.

Martensitic stainless steels contain 11.5 to 18 percent chromium, and are also magnetic. They are sometimes used in valve components.

Most stainless piping is of the austenitic variety. The most common grades for piping are 304 and 316. There are also special subsets of these which contain lower carbon, making them less susceptible to carbide precipitation. These are designated by the suffix "L."

Stainless is susceptible to stress corrosion cracking and so exposure to chlorine compounds must be avoided.

Applications

Stainless steel piping is used wherever iron dissolution cannot be tolerated, as in the production of foods, beverages, and pharmaceuticals. It is often used in uninsulated industrial services to avoid the need to paint the exterior of pipes. The added cost of stainless steel piping can often be offset by the cost of painting and repainting the exterior of carbon steel pipes over the expected lifetime of the installation. The cost of using stainless piping can be reduced further if one is able to take advantage of the 5S or 10S lightweight schedules.

Applicable Specifications

ASTM	A312	Seamless, Welded, and Heavily Cold Worked Austenitic Stainless Steel Pipes
ASTM	A358	Electric-Fusion-Welded Austenitic Chromium-Nickel Alloy Steel Pipe for High-Temperature Service
ASTM	A376	Seamless Austenitic Steel Pipe for High-Temperature Central-Station Service
ASTM	A409	Welded Large Diameter Austenitic Steel Pipe for Corrosive or High-Temperature Service
ASTM	A451	Centrifugally Cast Austenitic Steel Pipe for High-Temperature Service
ASTM	A790	Seamless and Welded Ferritic/Austenitic Stainless Steel Pipe
ASTM	A813	Single- or Double-Welded Austenitic Stainless Steel Pipe

ASTM	A814	Cold-Worked Welded Austenitic Stainless Steel Pipe
ASTM	A872	Centrifugally Cast Ferritic/Austenitic Stainless Steel Pipe for Corrosive Environments
ASTM	A928	Ferritic/Austenitic (Duplex) Stainless Steel Pipe Electric Fusion Welded with Addition of Filler Metal
ASTM	A943	Spray-Formed Seamless Austenitic Stainless Steel Pipe
ASTM	A949	Spray-Formed Seamless Ferritic/Austenitic Stainless Steel Pipe
ASTM	A954	Austenitic Chromium-Nickel-Silicon Alloy Steel Seamless and Welded Pipe
ASTM	A999	General Requirements for Alloy and Stainless Steel Pipe
AWWA	C220	Stainless-Steel Pipe, 1/2 in (13 mm) and Larger
AWWA	C226	Stainless Steel Fittings for Waterworks Service, Sizes 1/2 in through 72 in (13 mm through 1800 mm)

Note that in ASME B31.1, Table A-3 references ASTM A213 seamless stainless pipe. This is a specification for boiler and heat exchanger tubes. The corresponding piping specification is ASTM A312.

Manufacture of Stainless Steel Pipe

Aside from differences in the welding procedures, stainless pipe is manufactured similarly to carbon steel pipe.

As noted above, there are special wall thicknesses available in stainless steel piping to reduce the material costs. (A crude rule-of-thumb is that the material cost of stainless steel is three times more than carbon steel.) Manufacturers therefore developed lighter wall thicknesses to make stainless steel more attractive[3].

The mill tolerance of 12.5 percent must be applied to stainless steel piping as well as to carbon. Usually, any corrosion allowance can be eliminated in the thickness calculation though, since stainless resists corrosion under most applications.

Fabrication and Assembly of Stainless Steel Piping

One technique that is applied to cutting stainless steel is "air carbon arc gouging." This technique can be applied to a wide range of metals, but it is often used to quickly cut through stainless steel. An electric arc is generated between a copper-coated carbon electrode and the metal to be cut. This is done using an electric arc welder. When the metal melts at the arc, a jet of air blows the molten metal away. This method removes metal quickly, but it is very noisy due to the high electric current and high-pressure air. The resulting cut is typically very clean and does not result in carbon absorption by the surrounding metal since the molten metal is quickly removed by the air jet. Low air flow can lead to carbon hardening and cracking of the metal in the area of the cut.

Except for different welding procedures, stainless steel piping is fabricated and assembled in much the same way as carbon steel piping. Refer to Table 6.6 for a summary of the various welding procedures.

---

[3] Carbon steel is also available in Schedule 10S.

Abbreviation	Method	Description	Applications
GMAW	Gas-Metal Arc Welding	Electrode is wire filler metal which is fed through a gun. A shielding gas prevents oxidation of the puddle.	Carbon Steel, Stainless Steel
GTAW	Gas-Tungsten Arc Welding	Non-consumable tungsten electrode produces arc. A shielding gas (He, Ar) prevents oxidation of the weld puddle. Separate filler metal is added.	Stainless Steel, Aluminum, Titanium, Magnesium
Heli-Arc		Same as GTAW	
MIG	Metal Inert Gas	Same as GMAW	
SMAW	Shielded Metal Arc Welding ("Stick" Welding)	Consumable electrode (rod) is flux-covered filler metal.	Carbon Steel
TIG	Tungsten Inert Gas	Same as GTAW	

Table 6.6 Summary of Electric Welding Methods

Carbide precipitation is a phenomenon that occurs when austenitic (300 series) stainless steels are heated, as in welding. When heated, the carbides attract chromium atoms. This appears as a dark band along the weld. The surrounding region of the stainless contains less chromium since it is bound with the carbides. This results in a region in which the corrosion resistant properties are deficient.

There are two ways to prevent carbide precipitation during welding. One is to use a low-carbon grade of stainless, identified by the suffix "L" (for "low-carbon"). The other option is to stabilize the stainless steel with the addition of columbium or titanium. These elements have an affinity for carbon, and their carbides are also corrosion resistant. In practice, it is more common to use the low-carbon grades of stainless.

Attachments are often welded to piping when installing pipe supports. It is impractical for these attachments to be stainless due to the cost. In order to protect the piping from carbon precipitation, a stainless steel "poison pad" is usually welded to the pipe as shown in Figure 6.6. Any carbon precipitation occurs in the pad.

Copper Tubing

Copper tubing is essentially unalloyed pure copper.

Copper is a ductile material that can easily be drawn into tubing. It resists corrosion under many conditions, and is therefore a suitable material for potable water service[4]. Copper pipes have been found in ancient Egypt for conveying bath water. Like many metals, copper is a germicide.

[4] On June 7, 1991, the U.S. Environmental Protection Agency has published 40 CFR Part 141, known as the Lead and Copper Rule. The intent of this regulation is to require water utilities to limit lead concentrations in drinking water at customers' taps. The limits are 15 ppb for lead and 1.3 ppm for copper. Note that copper is also an essential mineral for plant and animal life.

Type	Color	Applications	Drawn	Annealed
K	Green	Hot and Cold Water (Service and Distribution), Fire Protection, Solar, Fuel Oil, HVAC, Snow Melting, Compressed Air, Natural Gas, LPG, Vacuum	Yes	Yes
L	Blue	Hot and Cold Water (Service and Distribution), Fire Protection, Solar, Fuel Oil, HVAC, Snow Melting, Compressed Air, Natural Gas, LPG, Vacuum	Yes	Yes
M	Red	Hot and Cold Water (Service and Distribution), Fire Protection, Solar, Fuel Oil, HVAC, Snow Melting, Vacuum	Yes	Not Available
DWV	Yellow	Drain/Waste/Vent HVAC, Solar	Yes	Not Available
ACR	Blue	Air Conditioning & Refrigeration Natural Gas, LPG, Compressed Air	Yes	Yes
OXY/MED	Type K – Green Type L – Blue	Medical Gases, Vacuum	Yes	Not Available

Table 6.7 Summary of Copper Tubing

Figure 6.6 A poison pad.

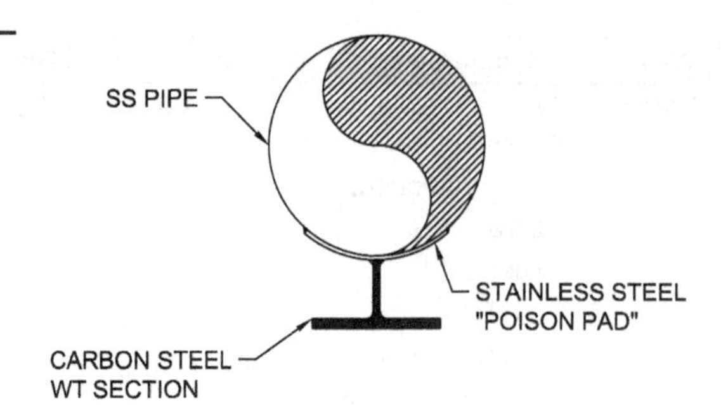

Applications

Copper tubing is an excellent material for conveying both hot and cold water, so it is used primarily for plumbing and hydronics. It frequently finds use in refrigerant piping. Due to its ease of fabrication and corrosion resistance, it is also used for high-purity applications such as medical gases.

In industrial settings copper is sometimes used for instrument air, vacuum, fuel oil, or fuel gas systems.

Copper resists corrosions due to the formation of an oxide layer against the pipe. If the oxide layer is attacked, pitting can occur. This pitting usually occurs where there is localized turbulence. The turbulence may be created due to high velocities in combination with perturbances such as fittings. Once the pitting begins, the turbulence increases and the pitting process accelerates. When used in hot water recirculation (to provide hot water at the tap more quickly), most industry guidelines recommend velocities in the 3 to 5 ft per second range. Some engineers recommend the low end of this range to further reduce the turbulence that has been implicated in this pitting action.

Applicable Specifications

ANSI	B16.15	Cast Copper Alloy Threaded Fittings
ANSI	B16.18	Cast Copper Alloy Solder Joint Pressure Fittings
ANSI	B16.22	Wrought Copper and Copper Alloy Solder Joint Pressure Fittings
ANSI	B16.23	Cast Copper Alloy Solder Joint Drainage Fittings DWV
ANSI	B16.24	Bronze Pipe Flanges and Flanged Fittings
ANSI	B16.26	Cast Copper Alloy Fittings for Flared Copper Tubes
ANSI	B16.29	Wrought Copper and Wrought Copper-Alloy Solder Joint Drainage Fittings DWV
ANSI	B16.50	Wrought Copper and Copper Alloy Braze-Joint Pressure Fittings
ASTM	B88	Standard Specification for Seamless Copper Water Tube
ASTM	B280	Standard Specification for Seamless Copper Tube for Air Conditioning and Refrigeration Field Service
ASTM	B306	Standard Specification for Seamless Copper Drainage Tube (DWV)
ASTM	B819	Standard Specification for Seamless Copper Tube for Medical Gas Systems
MSS	SP104	Wrought Copper LW Solder Joint Pressure Fittings
MSS	SP109	Welded Fabricated Copper Solder Joint Pressure Fittings

Manufacture of Copper Tubing

Copper is melted in a furnace where the temperature is brought several hundred degrees Fahrenheit above the melting point. Impurities are removed through various slagging techniques, with the result being a pool of 99.9 percent copper. Copper cast into solid billets is pierced with a mandrel in much the same way that seamless carbon steel pipe is produced. Hollow billets and pierced billets are extruded, and then drawn through dies to reduce the outside diameter.

Drawing results in work-hardening. The tubes become stiff, and this as-drawn condition is one of the two tempers available for copper tubing. Alternately, copper tubing is available in a soft annealed condition, usually sold in coils, but also available in straight lengths.

There are six different types of copper tubing, as shown in Table 6.7. Under ASTM B88, the wall thicknesses are described by "type" rather than "schedule."

- Type K has the heaviest wall thickness for a given size. The next heaviest is Type L, with Type M being the lightest.[5]
- Types K, L, M and DWV and Medical Gas (OXY/MED) tubes are specified by nominal diameters, with their actual ODs always 1/8 in larger than the nominal size.
- Type ACR is designated by its actual OD.
- OXY/MED tubing is available in Types K and L, but is manufactured in accordance with ASTM B819.

See Table 6.8 for available copper tubing sizes and wall thicknesses.

Fabrication and Assembly of Copper Tubing

Copper tubing may be cut to length with a hacksaw or abrasive saw, but is usually cut with a tubing cutter. These repeatedly score the tube with hardened wheels, displacing the soft copper until the tube is cut all the way through. Cutters are available with small swing radii to permit field cuts in tight quarters.

Copper tubing is joined by soldering, brazing, compression fittings, or grooved-end connections.

Fittings for soldering or brazing are available as either wrought or cast, but wrought fittings are preferred for brazing. The term "wrought" (often misspelled as "wrot" in the trades and catalogs) means "worked," and these fittings are forged as opposed to being cast. Cast fittings contain copper, tin, lead, and zinc.

Soldered Connections

Also called sweat connections, solder joints are made with a variety of fittings that slip over the OD of the adjoining tube. The small clearance provides a means of filling the annulus with solder using capillary action. Solder can be made to flow up inside a joint.

While lead[6] solder is still available, its use in potable water was banned in the United States in 1988. Solder for potable water systems now contains tin and antimony.

---

[5] A useful mnemonic device is "thicK." The only letter in the word "thick" among K, L, and M is "K," and this has the thickest wall.

[6] The chemical symbol for lead is Pb, from the Latin plumbum. This is the word from which "plumber" is derived. Plumber used to mean "lead-worker," since lead pipes were used extensively in plumbing systems.

Nominal Size in	Outside Diameter in	Schedule	Wall Thickness in	Inside Diameter in	Flow Area in²	Flow Area ft²
1/8	0.125	ACR - Annealed	0.030	0.065	0.003	0.00002
3/16	0.187	ACR - Annealed	0.030	0.127	0.013	0.00009
1/4	0.250	ACR - Annealed	0.030	0.190	0.028	0.00020
	0.375	K	0.035	0.305	0.073	0.00051
	0.375	L	0.030	0.315	0.078	0.00054
5/16	0.312	ACR - Annealed	0.032	0.248	0.048	0.00034
3/8	0.375	ACR - Annealed	0.032	0.311	0.076	0.00053
	0.375	ACR - Drawn	0.030	0.315	0.078	0.00054
	0.500	K	0.049	0.402	0.127	0.00088
	0.500	L	0.035	0.430	0.145	0.00101
	0.500	M	0.025	0.450	0.159	0.00110
1/2	0.500	ACR - Annealed	0.032	0.436	0.149	0.00104
	0.500	ACR - Drawn	0.035	0.430	0.145	0.00101
	0.625	K	0.049	0.527	0.218	0.00151
	0.625	L	0.040	0.545	0.233	0.00162
	0.625	M	0.028	0.569	0.254	0.00177
5/8	0.625	ACR - Annealed	0.035	0.555	0.242	0.00168
	0.625	ACR - Drawn	0.040	0.545	0.233	0.00162
	0.750	K	0.049	0.652	0.334	0.00232
	0.750	L	0.042	0.666	0.348	0.00242
3/4	0.750	ACR - Annealed	0.035	0.680	0.363	0.00252
	0.750	ACR - A or D	0.042	0.666	0.348	0.00242
	0.875	K	0.065	0.745	0.436	0.00303
	0.875	L	0.045	0.785	0.484	0.00336
	0.875	M	0.032	0.811	0.517	0.00359
7/8	0.875	ACR - A or D	0.045	0.785	0.484	0.00336
1	1.125	K	0.065	0.995	0.778	0.00540

Table 6.8 Copper Tube Sizes

Nominal Size in	Outside Diameter in	Schedule	Wall Thickness in	Inside Diameter in	Flow Area in^2	Flow Area ft^2
	1.125	L	0.050	1.025	0.825	0.00573
	1.125	M	0.035	1.055	0.874	0.00607
1 1/8	1.125	ACR - A or D	0.050	1.025	0.825	0.00573
1 1/4	1.375	K	0.065	1.245	1.217	0.00845
	1.375	L	0.055	1.265	1.257	0.00873
	1.375	M	0.042	1.291	1.309	0.00909
	1.375	DWV	0.040	1.295	1.317	0.00915
1 3/8	1.375	ACR - A or D	0.055	1.265	1.257	0.00873
1 1/2	1.625	K	0.072	1.481	1.723	0.01196
	1.625	L	0.060	1.505	1.779	0.01235
	1.625	M	0.049	1.527	1.831	0.01272
	1.625	DWV	0.042	1.541	1.865	0.01295
1 5/8	1.625	ACR - A or D	0.060	1.505	1.779	0.01235
2	2.125	K	0.083	1.959	3.014	0.02093
	2.125	L	0.070	1.985	3.095	0.02149
	2.125	M	0.058	2.009	3.170	0.02201
	2.125	DWV	0.042	2.041	3.272	0.02272
2 1/8	2.125	ACR - Drawn	0.070	1.985	3.095	0.02149
2 1/2	2.625	K	0.095	2.435	4.657	0.03234
	2.625	L	0.080	2.465	4.772	0.03314
	2.625	M	0.065	2.495	4.889	0.03395
2 5/8	2.625	ACR - Drawn	0.080	2.465	4.772	0.03314
3	3.125	K	0.109	2.907	6.637	0.04609
	3.125	L	0.090	2.945	6.812	0.04730
	3.125	M	0.072	2.981	6.979	0.04847
	3.125	DWV	0.045	3.035	7.234	0.05024
3 1/8	3.125	ACR - Drawn	0.090	2.945	6.812	0.04730
3 1/2	3.625	K	0.120	3.385	8.999	0.06249
	3.625	L	0.100	3.425	9.213	0.06398
	3.625	M	0.083	3.459	9.397	0.06526

Table 6.8 Copper Tube Sizes *(continued)*

(continued on next page)

Nominal Size in	Outside Diameter in	Schedule	Wall Thickness in	Inside Diameter in	Flow Area in²	Flow Area ft²
3 5/8	3.625	ACR - Drawn	0.100	3.425	9.213	0.06398
4	4.125	K	0.134	3.857	11.684	0.08114
	4.125	L	0.110	3.905	11.977	0.08317
	4.125	M	0.095	3.935	12.161	0.08445
	4.125	DWV	0.058	4.009	12.623	0.08766
4 1/8	4.125	ACR - Drawn	0.110	3.905	11.977	0.08317
5	5.125	K	0.160	4.805	18.133	0.12593
	5.125	L	0.125	4.875	18.665	0.12962
	5.125	M	0.109	4.907	18.911	0.13133
	5.125	DWV	0.072	4.981	19.486	0.13532
6	6.125	K	0.192	5.741	25.886	0.17976
	6.125	L	0.140	5.845	26.832	0.18634
	6.125	M	0.122	5.881	27.164	0.18864
	6.125	DWV	0.083	5.959	27.889	0.19368
8	8.125	K	0.271	7.583	45.162	0.31362
	8.125	L	0.200	7.725	46.869	0.32548
	8.125	M	0.170	7.785	47.600	0.33056
	8.125	DWV	0.109	7.907	49.104	0.34100
10	10.125	K	0.338	9.449	70.123	0.48697
	10.125	L	0.250	9.625	72.760	0.50528
	10.125	M	0.212	9.701	73.913	0.51329
12	12.125	K	0.405	11.315	100.554	0.69829
	12.125	L	0.280	11.565	105.046	0.72949
	12.125	M	0.254	11.617	105.993	0.73606

TABLE 6.8 Copper Tube Sizes *(continued)*

Cleanliness of the joining surfaces is of paramount importance to forming a leak-free joint. After cleaning (with emery cloth, if required), the joint is coated with flux. While the flux promotes a clean pool of solder, the application of excessive flux has been implicated in cold water pitting of copper tubing. The flux also acts as a wetting agent for the solder.

Solder should not be used where temperatures will exceed 250°F (121°C).

Brazed Connections

Brazing is defined as the joining of two metals with a third dissimilar metal at a temperature higher than soldering. Brazing is sometimes referred to as "silver soldering." Brazed connections are often required when working with ACR tubing,

The National Institutes of Health recommends that if copper tubing domestic water lines contain lead solder, they be flushed for a minute or two before using the water to cook with or drink. This is only necessary if they have not been operated for awhile (for instance, overnight). You can tell if the water has been flushed enough by checking its temperature. Stagnant water in cold water pipes will warm up inside the structure. When the water gets colder, you know that you are drawing fresh water from the distribution piping out in the street. This is another reason why you should only use cold water for cooking. Hot water leaches lead faster than cold water does.

since they form stronger joints than soldered connections. Brazing can accommodate operating temperatures as high as 350°F (17°C). The same fittings are used for brazing as for soldering, although wrought fittings are preferred for brazing.

Compression Fittings

Flared connections are a type of compression connection most easily made with annealed Type K or L tubing. Types K, L, or M hard temper may also be flared after annealing the end of the tube to be flared.

Because a flared connection is essentially a union with a metal-to-metal seal between the ends of the tubing, it is essential for the end to be squarely cut and free of burrs. A flaring tool shapes a cup on the tube end, and a nut behind the cup tightens into a cone-shape that engages the interior of the flare.

Another type of compression fitting is made with a ferrule that digs into the OD of the tube whenever a compression nut is tightened around it.

No-Solder Push-On Joints

Recently introduced as an alternative to soldering are a variety of fittings that join copper tubing by inserting it into the joint. See Figure 6.7. The end of the joint contains a plastic sleeve that fits inside the copper tubing. The fitting itself is a cast copper material that contains a ring of sharp barbs or teeth around its ID. These teeth bite into the OD of the copper tube when inserted. Once inserted, the tubing cannot be removed. If a trial fit-up is required for measuring, a plastic removal sleeve is used to prevent the fitting from permanently attaching to the tubing.

Figure 6.7 Push-on copper tubing couplings require no soldering skills.

The assembly of copper tubing using these fittings requires no special skills such as soldering. This results in a potential labor savings over construction using sweat fittings. However, the cost of these fittings is considerably more than sweat fittings.

Note that the plastic sleeve inside these fittings reduces the ID of the tubing significantly, so if many of these connections are used in a single run, some restriction of flow may result.

Brass Pipe

Brass pipe is manufactured from an alloy of 85 percent copper and 15 percent zinc. It is an uncommon piping material, but is sometimes used for potable water pressure pipe where hot work is impractical. It is also sometimes used for drainage due to its ability to resist specific corrosives. Brass pipe may be threaded, flanged, or soldered.

Applicable Specifications

ANSI	B16.15	Cast Copper Alloy Threaded Fittings
ANSI	B16.18	Cast Copper Alloy Solder Joint Pressure Fittings
ANSI	B16.24	Bronze Pipe Flanges and Flanged Fittings
ANSI	B16.26	Cast Copper Alloy Fittings for Flared Copper Tubes
ANSI	B16.29	Wrought Copper and Wrought Copper-Alloy Solder Joint Drainage Fittings DWV

Titanium Piping

Titanium is often thought to be an exotic material for piping systems, yet it is the ninth most common element in the earth's crust. Among common piping materials, only iron and aluminum are more abundant. There is approximately six times more titanium present than copper. Titanium resists corrosion, has high strength, and low weight.

Applications

Titanium piping is used in the petrochemical, pulp and paper, food processing, and power generation industries. It is compatible with:

- Chlorides
- Hydrogen sulfide
- Dilute hydrochloric acid
- Ammonia
- Hydrogen

Applicable Specifications

ASTM	B861	Standard Specification for Titanium and Titanium Alloy Seamless Pipe
ASTM	B862	Standard Specification for Titanium and Titanium Alloy Welded Pipe
ASTM	B363	Standard Specification for Seamless and Welded Unalloyed Titanium and Titanium Alloy Welding Fittings

Manufacture

Titanium pipe is available as either seamless or welded. Seamless pipe is available in sizes from ⅛ to 6 in NPS, and in schedules 10S, 40S, and 80S to match the same wall thicknesses as steel pipe. Welded titanium pipe is available in sizes from ¾ to 24 in NPS, in schedules 10S and 40S.

There are many grades of titanium piping, but the most common is Grade 2. This is an unalloyed grade with a minimum yield strength of 40 ksi and a minimum tensile strength of 50 ksi. Other grades of titanium pipe are Grades 1, 3, 7, and 12.

Fabrication and Assembly of Titanium Pipe

Titanium piping can be cut with an abrasive wheel. It is welded using the TIG or MIG processes, but the welding process for titanium is sensitive to air currents and contamination more so than for other metals.

Aluminum Piping and Tubing

Aluminum is a lightweight metal that is approximately one-third the density of steel. It resists corrosion, and may be alloyed with magnesium, manganese, or silicon.

Applications

Aluminum piping systems are not very common, but they are sometimes used for pneumatic conveying and compressed gas applications, including enrichment of uranium through centrifugal cascades. It resists corrosion, but is not compatible with acids, mercury, or strong alkalis.

At least one manufacturer has designed a clever aluminum piping system for use in plant compressed air systems. The pipes are joined with a proprietary mechanical coupling which seals the ends together with a sleeve. The advantages are alleged to be low installation cost due to the lightweight of the pipe, ease of completing the joint, and flexibility for changing the configuration. Because aluminum will not corrode, a further advantage over steel is that the airstream remains clean.

One caution regarding compressed air systems is that they are notorious for developing leaks, and even small leaks rob the system of efficiency. The cost to compress air is so high that every leak results in lost energy dollars in a very short time. Therefore, any improvements that can be made to seal a compressed air system will likely offset installation costs over the life of the system. The emphasis should be on providing a leak-free system.

Applicable Specifications

ASTM B210 Standard Specification for Aluminum and Aluminum-Alloy Drawn Seamless Tubes

ASTM B241 Standard Specification for Aluminum and Aluminum-Alloy Drawn Seamless Pipe and Seamless Extruded Tube

Manufacture

Seamless aluminum pipe and tubing is formed by extruding a hollow billet over a mandrel. Seamless tubes may be drawn through a die to achieve closer tolerances on diameter and wall thicknesses.

Non-Heat Treatable Alloys

Drawn tube in Alloy 3003 has excellent resistance to chemical corrosion. It has excellent workability and weldability. The 3000 series of alloys contain manganese.

Drawn tube in Alloy 5085 provides high strength after welding. It too has excellent resistance to chemical corrosion. The 5000 series of alloys contain magnesium.

Heat-Treatable Alloys

Alloy 6061 is available in both tube and pipe. It is almost twice as strong as Alloy 3003, and has good workability, weldability, and corrosion resistance. The 600 series of alloys contain magnesium and silicon.

Alloy 6063 is also available in both tube and pipe. It is more resistant to corrosion than 6061. It can also be anodized.

The anodization process is usually applied to aluminum, magnesium, zinc, and their alloys. It consists of applying an oxide coating on the surface of the metal in order to impart qualities such as wear resistance or improve the appearance of the metal. Anodizing builds a thicker oxide layer than would otherwise form from simple exposure to air.

Aluminum pipe is available as Schedule 40, in sizes from ⅛ to 12 in, and as Schedule 80 in sizes from 1 to 5 in. Tubing is available in sizes ranging from 3⁄16 to 10½ in.

Fabrication and Assembly

Being a soft metal, aluminum is easily cut with hand or power hacksaws. Abrasive saws are not used on the soft alloys, since they tend to smear the metal rather than cut it.

Joints are brazed or welded using alloy fillers, and although it is weldable, skill is required to produce acceptable joints. Mechanical couplings are also used.

PVC (Polyvinyl Chloride) Piping

PVC is a thermoplastic polymer. PVC piping is used in pressure piping as well as in plumbing drainage. It is widely available, relatively inexpensive, and the subject of some controversy over its environmental impact. It is the most widely used material for plastic piping.

The chlorine component of PVC constitutes 57 percent of the weight of the molecule, so less petroleum is required in the production of PVC than in other plastics.

PVC pipe is manufactured from compounds that contain no plasticizers. Plasticizers such as phthalates are known to leach out of the plastic, and have been a health concern due to a link with hormone activity. Because phthalates are not present in PVC pipe, it may be considered inert when exposed to potable water. The controversy surrounding its use is due to the environmental impact of dioxins released during its manufacture and disposal by incineration. Post-consumer recycling of PVC on an industrial scale has been difficult, but recent technologies offer hope for economical recycling with reduced environmental impact. PVC pipe manufacturers claim to recycle nearly 100 percent of waste product.

Although there is not a uniform standard for color of PVC piping, the following color scheme generally applies:

- White – Cold water, DWV
- Blue – Cold water
- Dark Gray – Cold water, industrial
- Green – Sewer.

Applications

PVC is widely used for cold water pressure piping and DWV piping. It is also used in many industrial applications.

Pipe materials that have a low modulus of elasticity will produce lower surge pressures than materials with higher moduli of elasticity. Therefore, a PVC piping system will generate a lower pressure surge (water hammer) than a carbon steel system (all other things being equal). Joint restraints under pressure surge conditions are always important to consider.

PVC pipe used for potable water distribution must be certified for potable water service.

Though PVC contains stabilizers to help prevent ultraviolet degradation, PVC should be painted with latex paint if it is to be exposed to sunlight, and pipe stored outdoors should be covered with an opaque tarp. Better applications are underground or inside structures. Above ground support of PVC pipe can be costly in order to avoid excessive sagging in between supports.

PVC is not suitable for hot water, since it has a maximum temperature limit of only 140°F (60°C). Being a thermoplastic, the strength drops off quickly as the temperature rises above ambient, and the working pressure must be de-rated for temperatures above 73°F (23°C). See Table 6.9.

PVC is never to be used for compressed gases, nor is it to be tested with gases under pressure.

Applicable Specifications

ASTM	D1784	Standard Specification for Rigid Poly(Vinyl Chloride) (PVC) Compounds and Chlorinated Poly(Vinyl Chloride) (CPVC) Compounds
ASTM	D1785	Standard Specification for Poly(Vinyl Chloride) (PVC) Plastic Pipe, Schedules 40, 80, and 120
ASTM	D2241	Standard Specification for Polyvinyl Chloride Pressure-Rated Pipe (SDR Series)
ASTM	D2466	Standard Specification for Poly(Vinyl Chloride) (PVC) Plastic Pipe Fittings, Schedule 40
ASTM	D2467	Standard Specification for Poly(Vinyl Chloride) (PVC) Plastic Pipe Fittings, Schedule 80
ASTM	D2564	Standard Specification for Solvent Cements for Poly(Vinyl Chloride) (PVC) Plastic Piping Systems
ASTM	D2665	Standard Specification for Poly(Vinyl Chloride) (PVC) Plastic Drain, Waste, and Vent Pipe and Fittings

ASTM	D2672	Standard Specification for Joints for IPS PVC Pipe Using Solvent Cement
ASTM	D2680	Standard Specification for Acrylonitrile-Butadiene-Styrene (ABS) and Polyvinyl Chloride (PVC) Composite Sewer Piping
ASTM	D2729	Standard Specification for Poly(Vinyl Chloride) (PVC) Sewer Pipe and Fittings
ASTM	D2855	Standard Practice for Making Solvent-Cemented Joints with Poly(Vinyl Chloride) (PVC) Pipe and Fittings
ASTM	D2949	Standard Specification for 3.25-in Outside Diameter Poly(Vinyl Chloride) (PVC) Plastic Drain, Waste, and Vent Pipe and Fittings
ASTM	D3034	Standard Specification for Type PSM Poly(Vinyl Chloride) (PVC) Sewer Pipe and Fittings
ASTM	F402	Standard Practice for Safe Handling of Solvent Cements, Primers, and Cleaners Used for Joining Thermoplastic Pipe and Fittings
ASTM	F679	Standard Specification for Poly(Vinyl Chloride) (PVC) Large-Diameter Plastic Gravity Sewer Pipe and Fittings
ASTM	F758	Standard Specification for Smooth-Wall Poly(Vinyl Chloride) (PVC) Plastic Underdrain Systems for Highway, Airport, and Similar Drainage
ASTM	F949	Standard Specification for Poly(Vinyl Chloride) (PVC) Corrugated Sewer Pipe With a Smooth Interior and Fittings
AWWA	C605	Underground Installation of Polyvinyl Chloride (PVC) Pressure Pipe and Fittings for Water
AWWA	C900	Polyvinyl Chloride (PVC) Pressure Pipe and Fabricated Fittings, 4-12 in (100-300 mm), for Water Transmission and Distribution
AWWA	C905	Polyvinyl Chloride (PVC) Pressure Pipe and Fabricated Fittings, 14-48 in (350-1200 mm)
AWWA	C907	Injection-Molded Polyvinyl Chloride (PVC) Pressure Fittings, 4-12 in (100-300 mm)
AWWA	C909	Molecularly Oriented Polyvinyl Chloride (PVCO) Pressure Pipe, 4-24 in (100-600 mm), for Water Distribution
CSA	B137.0	Definitions, general requirements, and methods of testing for thermoplastic pressure piping
CSA	B137.2	Polyvinylchloride (PVC) injection-moulded gasketed fittings for pressure applications
CSA	B137.3	Rigid polyvinylchloride (PVC) pipe and fittings for pressure applications
CSA	B181.0	Definitions, general requirements, and methods of testing for thermoplastic nonpressure pipe
CSA	B182.1	Plastic drain and sewer pipe and pipe fittings
CSA	B182.11	Standard practice for the installation of thermoplastic drain, storm, and sewer pipe and fittings
CSA	B182.2	PSM type polyvinylchloride (PVC) sewer pipe and fittings

CSA	B182.4	Profile polyvinylchloride (PVC) sewer pipe and fittings
CSA	B182.7	PSM type multilayer polyvinylchloride (PVC) sewer pipe having reprocessed-recycled content
NSF	14	Plastics Piping System Components and Related Materials
NSF	61	Drinking water system components - Health effects

Manufacture

PVC pipe is produced from vinyl chloride monomer which is reacted to form polymer chains. The plastic is extruded through dies into pipe of the required diameter and wall thickness.

Depending on the method used to join the pipe in the field, the pipe may be made with bell-and-spigot ends or with plain ends suitable for solvent welding, mechanical joints, or threading (in Schedule 80). Threaded connections are not recommended for temperatures above 110°F (43°C).

PVC pipe is available in both solid wall and a cellular core construction. The cellular core includes solid inner and outer layers that are simultaneously extruded around a cellular core.

A visit to the plumbing aisle of any home center store reveals that PVC drain pipes for under-sink P-traps have extremely thin walls. These drain pipes are commonly connected with molded threads and slip joint compression unions. Fortunately these lines are under little more pressure than a few inches of water, and so it is possible to connect them without leaks. But it can be a struggle.

PVC wall thicknesses are available in Schedules 40, 80, and 120, as well as in "Dimension Ratios". Dimension Ratios (DR) are sometimes also referred to as Standard Dimension Ratios (SDR). These dimensionless values are the ratio of the OD of the pipe to the wall thickness, and are also inversely proportional to working pressures. Thus, the lower the DR, the higher the working pressure. See Table 6.10 and Appendix 2.

Operating Temp		De-Rating Factor
(°F)	(C)	
73	23	1.00
80	27	0.88
90	32	0.75
100	38	0.62
110	43	0.51
120	49	0.40
130	54	0.31
140	60	0.22

Table 6.9 Factors for De-Rating PVC Due to Temperature

Figure 6.8 Plumbing code violation. Solvent welded joint (PVC welded to ABS test plug fitting).

Fabrication and Assembly of PVC Pipe

Small diameter PVC pipe may be cut with a hacksaw. Larger diameter PVC pipe may be cut with a circular saw or similar equipment. In either case, the cut should be square, deburred, and beveled if it is to be inserted into a bell-and-spigot joint. Further, an insertion line should be marked to indicate full-seating of the joint.

Solvent welded pipe must be cleaned and primed prior to application of the solvent. The primers are colored purple so that visual inspection can quickly determine if the pipe was properly primed. The solvent is applied to both the male and female portions to be joined, and the male end is inserted and twisted one-quarter turn to assure adequate adhesion. The joint should be restrained for 30 seconds to prevent push-out. The solvents actually dissolve the PVC and re-harden, forming very strong joints. Excess solvent should be removed, and puddling of the solvent is to be avoided.

Note that solvent cement joints are not permitted between different types of plastic pipe. See Figure 6.8. Such joints are to be made with mechanical couplings.

Schedule PVC pipe is made to the same outside dimensions as carbon steel pipe. Sometimes PVC (especially underground fire protection systems) is used in conjunction with ductile iron fittings, rather than with PVC fittings. These joints are restrained with metal clamps.

Threaded connections should be sealed using paste sealants rather than Teflon® tape. The reason is that tape products can cause deformation of the female fitting, leading to cracking.

ASTM E84, "Standard Test Method for Surface Burning Characteristics of Building Materials," is a test for quantifying the ability of a material to catch fire when exposed to a source of ignition, and to determine how much smoke is generated when the material is exposed to a flame. The smoke and noxious gases given off when a material burns are often more deadly than the flames of a fire. Considerable care needs to be given to the selection of materials that are placed in areas where smoke could be delivered into occupied spaces during a fire. Hence, electrical engineers are concerned with "plenum-rated" insulation on cables that are installed in plenum returns to the HVAC system. Any smoke generated there could be forced into occupied spaces during a fire.

Similarly, piping engineers need to be aware of what the local building code requires of plastic piping materials that are installed in a plenum. These codes generally require that combustibles located in ducts or plenums have a Flame Spread Index of not more than 25 and a Smoke Developed Rating of not more than 50 when tested in accordance with ASTM E84.

SDR	Working Pressure psi	Nominal Pipe Size in	OD in	Average ID in	Min. Wall in	Nominal Wt./Ft. lb/ft
13.5	315	1/2	0.840	0.696	0.062	0.110
21	200	3/4	1.050	0.910	0.060	0.136
21	200	1	1.315	1.169	0.063	0.180
21	200	1-1/4	1.660	1.482	0.079	0.278
21	200	1-1/2	1.900	1.700	0.090	0.358
21	200	2	2.375	2.129	0.113	0.550
21	200	2-1/2	2.875	2.581	0.137	0.797
21	200	3	3.500	3.146	0.167	1.168
21	200	3-1/2	4.000	3.597	0.190	1.520
21	200	4	4.500	4.046	0.214	1.927
21	200	5	5.563	5.001	0.265	2.948
21	200	6	6.625	5.955	0.316	4.185
21	200	8	8.625	7.756	0.410	7.069
26	160	1	1.315	1.175	0.060	0.173
26	160	1-1/4	1.660	1.512	0.064	0.233
26	160	1-1/2	1.900	1.734	0.073	0.300
26	160	2	2.375	2.173	0.091	0.456
26	160	2-1/2	2.875	2.635	0.110	0.657
26	160	3	3.500	3.210	0.135	0.966
26	160	3-1/2	4.000	3.672	0.154	1.250
26	160	4	4.500	4.134	0.173	1.569
26	160	5	5.563	5.108	0.214	2.411
26	160	6	6.625	6.084	0.255	3.414
26	160	8	8.625	7.921	0.332	5.784
26	160	10	10.750	9.874	0.413	8.971
26	160	12	12.750	11.711	0.490	12.620
26	160	14	14.000	12.860	0.538	15.205
26	160	16	16.000	14.696	0.615	19.877
26	160	18	18.000	16.533	0.692	25.156
26	160	20	20.000	18.370	0.769	31.057
26	160	24	24.000	22.043	0.923	44.744
41	100	18	18.000	17.061	0.439	16.348
41	100	20	20.000	18.956	0.488	20.196
41	100	24	24.000	22.748	0.585	29.064

Table 6.10 PVC Dimension Ratios

CPVC (Chlorinated PolyVinyl Chloride) Piping

CPVC is also a thermoplastic. It is manufactured from the same resin as PVC, but undergoes an additional reaction in which chlorine replaces some of the hydrogen in the monomer. This yields a chlorine content of between 63 and 69 weight percent, whereas PVC contains 57 percent chlorine by weight. The result is a higher temperature at which CPVC softens, and this allows CPVC piping to be used in some domestic hot water applications.

Applications

CPVC is used in water-distribution, less commonly for water-service. Water distribution implies both hot water and potable cold water service, as inside a building. CPVC can support temperatures up to 180°F (82°C)[7]. See Table 6.11.

CPVC is also used to handle corrosive fluids, and is sometimes also used in fire suppression. Because it offers higher strength at elevated temperatures than PVC, it may be applied in more applications in industrial services.

Because it is a plastic material, it must never be used for compressed gas services.

Like PVC, CPVC should not be installed outdoors without protection from ultraviolet exposure. While it contains UV stabilizers, manufacturers recommend painting it with latex paint to protect it from UV degradation.

Applicable Specifications

ASTM	D1784	Standard Specification for Rigid Poly(Vinyl Chloride) (PVC) Compounds and Chlorinated Poly(Vinyl Chloride) (CPVC) Compounds
ASTM	D2846	Standard Specification for Chlorinated Poly(Vinyl Chloride) (CPVC) Plastic Hot- and Cold-Water Distribution Systems
ASTM	F437	Standard Specification for Threaded Chlorinated Poly(Vinyl Chloride) (CPVC) Plastic Pipe Fittings, Schedule 80
ASTM	F438	Standard Specification for Socket-Type Chlorinated Poly(Vinyl Chloride) (CPVC) Plastic Pipe Fittings, Schedule 40
ASTM	F439	Standard Specification for Chlorinated Poly (Vinyl Chloride) (CPVC) Plastic Pipe Fittings, Schedule 80
ASTM	F441	Standard Specification for Chlorinated Poly(Vinyl Chloride) (CPVC) Plastic Pipe, Schedules 40 and 80
ASTM	F442	Standard Specification for Chlorinated Poly(Vinyl Chloride) (CPVC) Plastic Pipe (SDR-PR)
ASTM	F493	Standard Specification for Solvent Cements for Chlorinated Poly(Vinyl Chloride) (CPVC) Plastic Pipe and Fittings
CSA	B137.0	Definitions, general requirements, and methods of testing for thermoplastic pressure piping
CSA	B137.6	Chlorinated polyvinylchloride (CPVC) pipe, tubing, and fittings for hot and cold-water distribution systems
CSA	B181.0	Definitions, general requirements, and methods of testing for thermoplastic nonpressure pipe

[7] The International Plumbing Code requires the use of piping having a minimum pressure rating of 100 psi (6.89 bar) at 180°F (82°C) for the hot water portion of a water distribution system.

CSA	B181.2	Polyvinylchloride (PVC) and chlorinated polyvinylchloride (CPVC) drain, waste, and vent pipe and pipe fittings
CSA	B182.1	Plastic drain and sewer pipe and pipe fittings
CSA	B182.11	Standard practice for the installation of thermoplastic drain, storm, and sewer pipe and fittings
NSF	14	Plastics Piping System Components and Related Materials
NSF	61	Drinking water system components - Health effects

Manufacture of CPVC Pipe

CPVC pipe is extruded from resins in the same manner as PVC pipe. It is available in Schedules 40 and 80, as well as in SDR dimensions in sizes from ¼ through 12 in. It is available in copper tube sizes for plumbing applications in sizes from ½ through 2 in.

CPVC is usually cream colored, but may also be gray or white.

Fabrication and Assembly of CPVC Pipe

CPVC is cut to length with hacksaws, circular saws, abrasive saws, or with tubing cutters that contain special plastic cutting wheels. Cutting wheels designed for metal are not suitable to cut CPVC. Ratchet cutters are also sometimes used, although their use should be limited to temperatures above 50°F (10°C) to prevent the CPVC from cracking.

The ends should be beveled and deburred.

CPVC is joined with either a one-step cement that does not require a primer, or a two-step process that does require a primer. The one-step cement is yellow; the cement for the two-step process is orange. The pipes are assembled in the same manner as for PVC, that is, with a quarter-turn that should be restrained for 30 seconds to prevent push-out. Excess cement should be avoided to prevent puddling, and should be wiped from the joint after insertion.

Supports should be located every 3 ft for 1 in diameter and smaller, and every 4 ft for sizes exceeding 1 in diameter.

Operating Temp		
(°F)	(°C)	De-Rating Factor
73	23	1.00
80	27	1.00
90	32	0.91
100	38	0.82
110	43	0.77
120	49	0.65
130	54	0.62
140	60	0.50
150	66	0.47
160	71	0.40
170	77	0.32
180	82	0.25
200	93	0.20

Table 6.11 Factors for De-Rating CPVC Due to Temperature

Polybutylene (PB) Piping

Polybutylene piping was never used for industrial applications, although it has been used for both water distribution and water service piping. At least two class action lawsuits were filed in the United States alleging defects in the manufacture or installation of PB piping. The PB was favored by installers because it is a flexible system which is "forgiving" in terms of routing and avoiding interferences. The installer did not have to be dimensionally precise.

The problem with the material is thought to be that chlorine in public water supplies attacks the PB tubing and some of the fittings. This results in scaling, flaking, and brittleness, which culminates in failure of the system

PB piping was joined using acetal, brass, or copper fittings over which the tubing was placed and then secured with a metal band. PB tubing was colored blue, gray, or black.

Polyethylene (PE) and High-Density Polyethylene (HDPE) Piping

Polyethylene is a polymer thermoplastic that excels in certain underground applications. It belongs to a class of polymers known as polyolefins. It tolerates abuse and may be deformed without compromising its strength. It is made by the polymerization of ethylene with propylene, butene, or hexene. The result is a long polymer chain which can be modified to adjust the desired properties of the bulk material which will be extruded into pipe.

The three properties of density, molecular weight, and molecular weight distribution influence the physical properties of the material. In that sense, it should be noted that the composition of HDPE is analogous to steel. Various chemicals can be used to modify the properties of the final product in a manner similar to alloying materials in steels.

Early PE compositions had many side-branches coming off of the main polymer chain. These side branches prevented the polymer chains from packing together very

During a large steel mill project several years ago, the piping contractor suggested using HDPE for much of the underground water piping on the grounds that it would save time and money. We engineers exchanged skeptical glances, thinking they were just trying to pull a fast one. They explained how they felt it would be a poor application to use HDPE underground inside the melt shop, just in case a ladle of steel got loose, and we certainly agreed.

The contractor submitted some vendor information including a sample of the fusion weld. We immediately took turns trying to break the weld by hand or by propping it on a chair rung and kicking it. It wouldn't budge. One of the designers suggested that this material might become brittle in the cold Ohio winters. (It would be exposed to the elements prior to burial). We left it in a freezer overnight, and went through the same exercise with no difference in the results. We decided that it would be satisfactory after reviewing some additional data with the vendor.

I became a supporter of HDPE for underground applications. It installs quickly, is less expensive than steel, and is nearly impervious. Sometimes the contractors are ahead of the curve.

tightly, and so were called "Low Density Polyethylene" (LDPE). These are not often used for piping, since the higher density varieties exhibit more desirable qualities, such as improved tensile strength, improved low-temperature brittleness, higher softening point, and increased chemical resistance.

Applications

HDPE is often used to convey water or low-pressure natural gas. It does not corrode, and its flow characteristics are very good, since the surfaces are so smooth. The resistance to corrosion can make it a more economical choice for underground piping than coated and wrapped carbon steel, provided that the fluids are compatible with the material, and the pressures can be accommodated.

HDPE is only used above ground in installations that are regarded as temporary. It is snaked over grade in such applications. Due to its extreme flexibility (low Young's Modulus) it would require extensive support for above ground installation.

Applicable Specifications

AASHTO	F2136	Standard Test Method for Notched, Constant Ligament-Stress (NCLS) Test to Determine Slow-Crack Growth Resistance of HDPE Resins or HDPE Corrugated Pipe
AASHTO	M294	Corrugated Polyethylene Pipe, 12 to 24 in Diameter
API	15LE	Specification for Polyethylene Line Pipe (American Petroleum Institute)
ASTM	D2104	Polyethylene (PE) Plastic Pipe, Schedule 40
ASTM	D2239	Polyethylene (PE) Plastic Pipe (SIDR-PR) Based on Controlled Inside Diameter
ASTM	D2447	Polyethylene (PE) Plastic Pipe, Schedules 40 to 80, Based on Outside Diameter
ASTM	D2609	Plastic Insert Fittings for Polyethylene (PE) Plastic Pipe
ASTM	D2683	Socket Type Polyethylene Fittings for Outside Diameter Controlled Polyethylene Pipe and Tubing
ASTM	D2683	Socket-Type Polyethylene Fittings for Outside Diameter-Controlled Polyethylene Pipe and Tubing
ASTM	D2737	Polyethylene (PE) Plastic Tubing
ASTM	D3035	Polyethylene (PE) Plastic Pipe (SDR-PR) Based on Controlled Outside Diameter
ASTM	D3261	Butt Heat Fusion Polyethylene (PE) Plastic Fittings for Polyethylene (PE) Plastic Pipe and Tubing
ASTM	D3350	Polyethylene Plastics Pipe and Fittings Materials
ASTM	F1025	Selection and Use of Full-Encirclement-Type Band Clamps for Reinforcement or Repair of Punctures or Holes in Polyethylene Gas Pressure Pipe
ASTM	F1055	Electrofusion Type Polyethylene Fittings for Outside Diameter Controlled Polyethylene Pipe and Tubing
ASTM	F1056	Socket Fusion Tools for Use in Socket Fusion Joining Polyethylene Pipe or Tubing and Fittings

ASTM	F1248	Test Method for Determination of Environmental Stress Crack Resistance (ESCR) of Polyethylene Pipe
ASTM	F1282	Standard Specification Polyethylene/Aluminum/ Polyethylene (PE-AL -PE) Composite Pressure Pipe
ASTM	F1473	Notch Tensile Test to Measure the Resistance to Slow Crack Growth of Polyethylene Pipes and Resins
ASTM	F1533	Deformed Polyethylene (PE) Liner
ASTM	F1563	Tools to Squeeze Off Polyethylene (PE) Gas Pipe or Tubing
ASTM	F1606	Standard Practice for Rehabilitation of Existing Sewers and Conduits with Deformed Polyethylene (PE) Liner
ASTM	F1734	Practice for Qualification of a Combination of Squeeze Tool, Pipe, and Squeeze-Off Procedure to Avoid Long-Term Damage in Polyethylene (PE) Gas Pipe
ASTM	F1759	Standard Practice for Design of High-Density Polyethylene (HDPE) Manholes for Subsurface Applications
ASTM	F1804	Determine Allowable Tensile Load For Polyethylene (PE) Gas Pipe During Pull-in Installation
ASTM	F1901	Polyethylene (PE) Pipe and Fittings for Roof Drain Systems
ASTM	F1924	Plastic Mechanical Fittings for Use on Outside Diameter Controlled Polyethylene Gas Distribution Pipe and Tubing
ASTM	F1962	Standard Guide for Use of Maxi-Horizontal Directional Drilling for Placement of Polyethylene Pipe or Conduit Under Obstacles, Including River Crossing
ASTM	F1973	Factory Assembled Anodeless Risers and Transition Fittings in Polyethylene (PE) Fuel Gas Distribution Systems
ASTM	F2164	Standard Practice for Field Leak Testing of Polyethylene (PE) Pressure Piping Systems Using Hydrostatic Pressure
ASTM	F2206	Standard Specification for Fabricated Fittings of Butt-Fused Polyethylene (PE) Plastic Pipe, Fittings, Sheet Stock, Plate Stock, or Block Stock
ASTM	F2231	Standard Test Method for Charpy Impact Test on Thin Specimens of Polyethylene Used in Pressurized Pipes
ASTM	F2263	Standard Test Method for Evaluating the Oxidative Resistance of Polyethylene (PE) Pipe to Chlorinated Water
ASTM	F405	Corrugated Polyethylene (PE) Tubing and Fittings
ASTM	F667	Large Diameter Corrugated Polyethylene (PE) Tubing and Fittings
ASTM	F714	Polyethylene (PE) Plastic Pipe (SIDR-PR) Based on Controlled Outside Diameter
ASTM	F771	Polyethylene (PE) Thermoplastic High-Pressure Irrigation Pipeline Systems
ASTM	F810	Smooth Wall Polyethylene (PE Pipe for Use in Drainage and Waste Disposal Absorption Fields
ASTM	F858	Insertion of Flexible Polyethylene Pipe into Existing Sewers
ASTM	F894	Polyethylene (PE) Large Diameter Profile Wall Sewer and Drain Pipe

ASTM	F905	Qualification of Polyethylene Saddle Fusion Joints
ASTM	F982	Polyethylene (PE) Corrugated Pipe with a Smooth Interior and Fittings
AWWA	C901	Polyethylene (PE) Pressure Pipe and Tubing, ½ in (13 mm) through 3 in (76 mm), for Water Service
AWWA	C906	Polyethylene (PE) Pressure Pipe and Fittings, 4 in (100 mm) through 63 in (1600 mm), for Water Distribution
AWWA	M55	PE Pipe - Design and Installation
CSA	B137.1	Polyethylene (PE) Pipe, Tubing, and Fittings for Cold-Water Pressure Services
CSA	B137.4	Polyethylene Piping Systems for Gas Services (Canadian Standards Association)
CSA	B137.4.1	Electrofusion-type Polyethylene (PE) Fittings for Gas Services
CSA	B137.9	Polyethylene/Aluminum/Polyethylene (PE-AL-PE) Composite Pressure-Pipe Systems
CSA	B182.6	Profile polyethylene (PE) Sewer Pipe and Fittings for Leak-Proof Sewer Applications
CSA	B182.8	Profile Polyethylene (PE) Storm Sewer and Drainage Pipe and Fittings
FM	1613	Polyethylene (PE) Pipe and Fittings for Underground Fire Protection Service
ISO	4427	Polyethylene (PE) Pipes for Water Supply
ISO	4437	Buried polyethylene (PE) Pipes for the Supply of Gaseous Fuels - Metric Series - Specifications
NSF	14	Plastics Piping System Components and Related Materials
NSF	61	Drinking Water System Components - Health Effects
PPI	MS-2	Model Specification for Polyethylene Plastic Pipe, Tubing and Fittings for Natural Gas Distribution
PPI	MS-3	Model Specification for Polyethylene Plastic Pipe, Tubing and Fittings for Water Mains and Distribution
PPI	TN-13	General Guidelines for Butt, Saddle, and Socket Fusion of Unlike Polyethylene Pipes and Fittings
PPI	TN-15	Resistance of Solid Wall Polyethylene Pipe to a Sanitary Sewage Environment
PPI	TN-16	Rate Process Method for Projecting Performance of Polyethylene Piping Components
PPI	TN-4	Odorants in Plastic Fuel Gas Distribution Systems
PPI	TN-6	Polyethylene (PE) Coil Dimensions
PPI	TR-22	Polyethylene Plastic Piping Distribution Systems for Components of Liquid Petroleum Gas
PPI	TR-33	Generic Butt Fusion Joining Procedure for Polyethylene Gas Pipe
PPI	TR-34	Disinfection of Newly Constructed Polyethylene Water Mains
PPI	TR-35	Chemical and Abrasion Resistance of Corrugated Polyethylene Pipe

PPI	TR-36	Hydraulic Considerations for Corrugated Polyethylene Pipe
PPI	TR-37	CPPA Standard Specification (100-99) for Corrugated Polyethylene (PE) Pipe for Storm Sewer Applications
PPI	TR-38	Structural Design Method for Corrugated Polyethylene Pipe
PPI	TR-39	Structural Integrity of Non-Pressure Corrugated Polyethylene Pipe
PPI	TR-40	Evaluation of Fire Risk Related to Corrugated Polyethylene Pipe

Manufacture of PE pipe

PE resin is extruded into pipe diameters ranging from ½ to 63 in diameter. It is often pigmented solid black, but may be coextruded with other colors as stripes to denote different services. There are three different dimensioning systems to describe the wall thickness of PE piping:

1. Dimension Ratio – The wall thickness is determined using the diameter divided by the wall thickness. This provides a ratio that is based on an allowable working pressure for all of the diameters in a given Dimension Ratio (DR).
2. Iron Pipe Size Inside Diameter (SIDR) – The ID of the PE pipe matches the ID of Schedule 40 steel pipe, and the wall thicknesses vary in accordance with the pressure rating of the pipe. This designation is available in smaller sizes (approximately 2 in and less).
3. Copper Tube Size Outside Diameter (CTS) – The pipe ODs are the same as for copper tubing, i.e., the OD is 1/8 in larger than the nominal size. These are also available only in smaller sizes.

The dimension ratio system is the one used most often for industrial piping, but the pipe manufacturers specify the dimension ratio in two ways:

$$IDR = ID/t \text{ for pipes manufactured according to a controlled ID}$$

and

$$DR = OD/t \text{ for pipes manufactured according to a controlled OD}$$

These two terms are important in determining the allowable working pressure of the pipe. The plastics industry has developed equations (called ISO equations) for calculating the allowable stresses that look similar to the familiar Barlow's formula, with modifications for the environment, the operating temperature, and the dimension ratio.

Thus, for pipe manufactured to a controlled OD,

$$P = [2\,(HDB) \times F_E \times F_T] \,/\, (DR - 1)$$

where

HDB = Hydrostatic Design Basis (see Table 6.12)

F_E = Environmental Design Factor based on the service fluid (see Table 6.13)

F_T = Service Temperature Design Factor (see Table 6.14)

and for pipe manufactured to a controlled ID,

$$P = [2\,(HDB) \times F_E \times F_T] / (IDR + 1)$$

Certain dimension ratios that meet ASTM-specified number series are known as "Standard Dimension Ratios" (SDR) or SIDR, and this terminology is also found in vendor data. Standard Dimension Ratios are: 41, 32.5, 26, 21, 17, 13.5, 11, 9, and 7.3. There is approximately a 25 percent difference in minimum wall thickness from one SDR to the next.

Care must be taken to use the HDB values and not the HDS values. The HDS is the Hydrostatic Design Stress, and

$$HDS = HDB \times F_E$$

Molded PE fittings require a larger body to provide additional wall thickness for a given pressure rating. Similarly, fabricated fittings made from straight pipe segments are usually derated at least one SDR.

Property	PE 4710	PE 3408	PE 2406
Density Cell Class per ASTM D3350	4	3	2
Slow Crack Growth (SCG) Cell Class per ASTM D3350	7	4	4
Hydrostatic Design Stress (HDS) at 73°F (23°C)	1000 psi (6.89 MPa)	800 psi (5.52 Mpa)	600 psi (4.14 MPa)
Hydrostatic Design Basis at 73°F (23°C)	2000 psi (13.79 MPa)	1600 psi (11.04 MPa)	1250 psi (8.62 MPa)
Maximum recommended temperature for Pressure Service	Check with manufacturer	140°F (60°C)	140°F (60°C)
Maximum Recommended Temperature for Non-Pressure Service	Check with manufacturer	180°F (82°C)	180°F (82°C)

TABLE 6.12 Properties of PE Plastics for Piping

Fluid	F_E
Fluids such as potable and process water, benign chemicals, dry natural gas (non-federally regulated), brine, CO_2, H_2S, wastewater, sewage, glycol/anti-freeze solutions	0.50
Dry natural gas (Federally regulated under CFR Title 49, Part 192),	0.32
Fluids such as solvating/permeating chemicals in pipe or soil (typically hydrocarbons) in 2% or greater concentrations, natural or other fuel-gas liquid condensates, crude oil, fuel oil, gasoline, diesel, kerosene, hydrocarbon fuels	0.25

TABLE 6.13 Environmental Design Factors, F_E, for HDPE Pipe

	F_T	
Service Temp	PE 3408	PE 2406
40°F (4 C)	1.20	1.10
60°F (16 C)	1.08	1.04
73°F (23 C)	1.00	1.00
100°F (38 C)	0.78	0.92
120°F (49 C)	0.63	0.86
140°F (60 C)	0.50	0.80

Table 6.14 PE Service Temperature Design Factors, F_T

Fabrication and Assembly of PE Pipe

Because it is nonmagnetic and used underground almost exclusively, during installation PE pipe is often buried with a metallic tape so that it may be more easily located after burial with a metal detector.

PE pipe is cut to length with a saw, knife, run-around cutter, or guillotine cutter. While mechanical couplings and compression fittings are available, their suitability for use with PE must be verified by the manufacturer. This is due to the extremely low coefficient of friction of polyethylene. Most often, PE pipe is fusion welded.

Fusion welding is conducted with a fusion machine that clamps both ends of the pipe to be joined. Cutters face the joint so that the ends are square and smooth. A heating plate is inserted between the ends and is electrically heated so that a portion of the ends of the pipe are at melt temperature. The machine brings the ends together under force, and holds them until the bond cools and solidifies. The result is a weld bead on the inside and outside surfaces of the pipe, which is stronger than the adjoining pipe. See Figure 6.9.

Another type of fusion weld can be made with an electrofusion sleeve, of the type shown in Figure 6.10. These sleeves fit over the ends of two adjoining pipes, and an electrical source is connected to two terminals. A heating coil built into the sleeve heats the sleeve and pipe to the melting point, and the electric source is withdrawn at a specified time. The process bonds the inside surface of the sleeve to the outside surfaces of the two pipes, and is satisfactory for making field repairs when a fusion machine is unavailable. The sleeves may also be used for regular assembly of the pipe, especially in remote locations, but the full butt weld provided by a fusion machine is regarded as a superior weld to the electrofusion method.

Neither threading nor solvent cementing is possible with PE pipe.

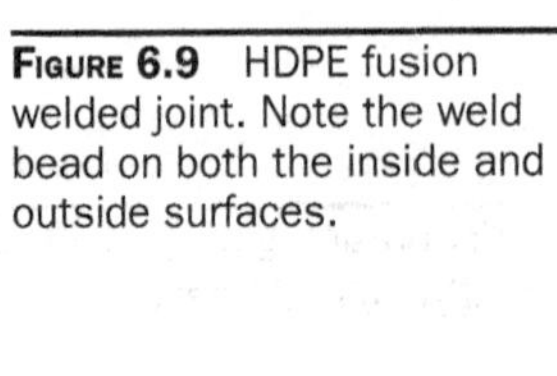

Figure 6.9 HDPE fusion welded joint. Note the weld bead on both the inside and outside surfaces.

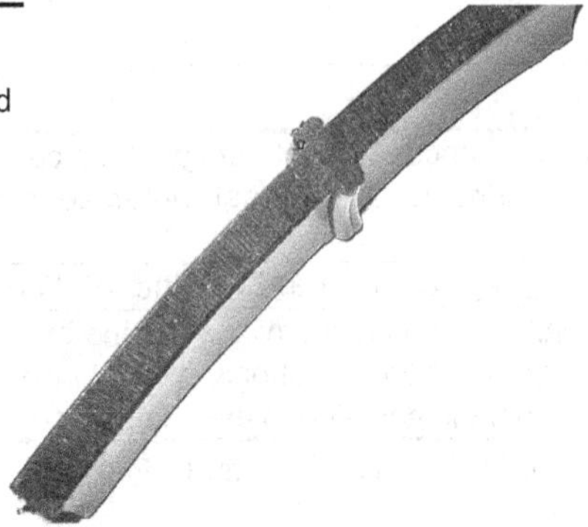

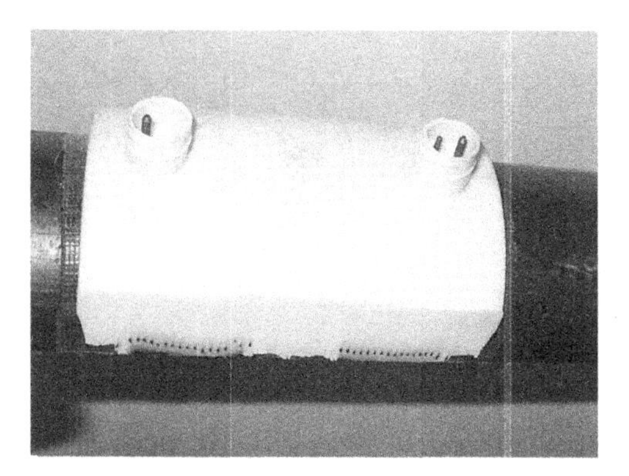

FIGURE 6.10 HDPE fusion weld sleeve. This coupling is used for repairs or for field joints where it may be impractical to use a fusion weld machine. Note the electrical windings which melt the coupling to the pipe upon application of the electric current across the terminals.

An underground acid waste line at a steel mill was installed with double-containment and a leak detector. Both the inner and outer pipes were HDPE, with a leak detector that consisted of an electric cable whose impedance would be monitored by a sophisticated controller (See Figure 6.11). The vendor promised that the cable and controller could identify the location of a leak within 3 ft.

Due to poor planning, an imposed sense of urgency from the owner, and the philosophy that "there is never enough time to do it right, but always enough time to do it over," the commissioning engineer permitted the acid line to be used before the controller was connected.

The line leaked.

The annular space became flooded with acid waste, and now there was no way to use the flooded cable to determine where the leak was. The commissioning engineer called me at the office, and left the horrifying news. Now we had to figure out a way to find where the leak was over a distance of about 1500 ft. The line was buried approximately 6 ft. There were pull ports for the cable located about every 200 ft.

The supplying vendor offered no advice to fix the problem. I contacted a competing vendor, who offered to send a technician out with sensitive sonic detectors. The thinking was if the break was large enough, the inner pipe could be purged with air at moderate pressure, and the escaping air would be heard with the device as we walked the surface of the line down. It didn't work.

Finally, the system was flooded with water under pressure, and all of the caps on the pull ports were removed. Geysers observed at the pull ports identified within 200 ft of where the leak was. The line was excavated between those ports, and the leak was repaired.

The cause of the leak was traced to a faulty weld on the interior pipe. It seems that during installation, the adjoining faces were cut with the fusion welding machine at the end of a shift, with the weld put off until the following morning. Falling temperatures overnight produced a temperature gradient that left the inner pipe contracted and the outer pipe expanded. When the weld was made, a cold joint resulted on the inner pipe. Of course, this could not be observed once the outer pipe was fused. PE pipe has a high Coefficient of Linear Expansion compared to other pipe materials.

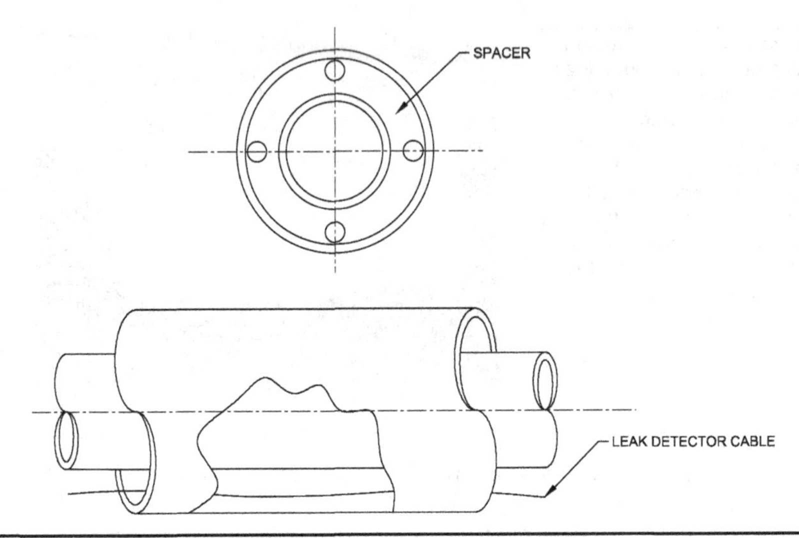

Figure 6.11 A double-containment pipe.

"Snaking" is a method of introducing slack in the pipe as it is laid in the trench. This accounts for contraction of the pipe prior to backfilling and also increases the grip of the soil on the pipe. The pipe should be cooled to a temperature close to the soil prior to cutting to length, since contraction can reduce the length by 1 in per 100 ft for every 10°F drop in temperature.

Jigs can be made to temporarily pinch off the flow in a PE pipe. Consult the pipe manufacturer for design details and allowable duration.

Acrylonitrile-Butadiene-Styrene (ABS) Piping

ABS is another thermoplastic used primarily for DWV service. It is a rigid black (sometimes dark gray) product that is easy to work with, is inexpensive, and is therefore a popular choice for residential and commercial construction.

Applications

ABS is acceptable for use as a water service pipe, but is most often used in DWV service. It is also sometimes specified for pressurized lines for crude oil, pumped waste, salt water, and irrigation applications.

Bare ABS pipes are not suitable for use in plenums, since their Flame Spread and Smoke Developed Indices exceed code requirements for this use. ABS pipes installed in plenums would have to be boxed in with gypsum board or wrapped with a suitable insulation whose properties are acceptable for use in plenums.

Applicable Specifications

ASTM D2751 Standard Specification for Acrylonitrile-Butadiene-Styrene (ABS) Sewer Pipe and Fittings

ASTM F628 Standard Specification for Acrylonitrile-Butadiene-Styrene (ABS) Schedule 40 Plastic Drain, Waste, and Vent Pipe (With a Cellular Core) Schedule 40 Plastic Drain, Waste, and Vent Pipe and Fittings

FIGURE 6.12 Some typical plastic pipe solvents.

ASTM D2661 Standard Specification for Acrylonitrile-Butadiene-Styrene (ABS)

ASTM D3965 Standard Specification for Rigid Acrylonitrile-Butadiene-Styrene (ABS) Materials for Pipe and Fittings

CSA B181.1 Acrylonitrile-Butadiene-Styrene (ABS) Drain, Waste, and Vent Pipe, and Pipe Fittings

CSA B181.5 Coextruded Acrylonitrile-Butadiene-Styrene/Polyvinylchloride (ABS/PVC) Drain, Waste, and Vent Pipe

Manufacture

Like the other thermoplastics, ABS is extruded into the desired size and wall thickness. It is available in Schedule 40 in diameters 1½ in, 2 in, 3 in, 4 in, and 6 in, with solid wall or cellular core construction. A full complement of fittings is also available.

Fabrication and Assembly of ABS Pipe

ABS is easily cut with a hand saw. The edges should be cut square and deburred before applying solvent cement. As noted earlier, solvent cements are not to be used to join different types of plastics.

ABS may be joined with mechanical joints, but the International Plumbing Code restricts the use of these to underground locations, unless otherwise approved.[8] It is difficult to imagine why a code official would object to using mechanical joints to connect an above ground component like the test plug fitting in Figure 6.8, but a better installation would be to use the same material throughout and avoid the issue altogether.

Solvent joints are made in the same manner as with other thermoplastic solvent joints. That is, the fittings are to be rotated one-quarter turn and held in place long enough to prevent push-out of the pipe. The solvent must be compatible with ABS. See Figure 6.12.

Cross-Linked Polyethylene (PEX) Piping

Cross-linked polymers are thermosetting polymers, as opposed to thermoplastic polymers. The difference is that the long polymer chains of thermosets are chemically bonded to other long polymer chains during the processing which occurs at high temperatures. Thermosets are still affected by heat, but not to the extent that thermoplastics

[8] For water supply and distribution piping, cf. IPC 605.10.1. For sanitary drainage, cf. IPC 705.2.1.

are. Thermoset plastics are not recyclable, in the sense that they cannot be ground up and remelted to form the same compound.

Applications

PEX is used in hot and cold water piping for water distribution systems and in radiant heating applications including snow melt systems. It is not intended for outdoor use.

PEX tubing for heating is available with an oxygen barrier to prevent oxygen from corroding cast iron system components. The oxygen barriers are available as a PEX-Aluminum-PEX composite system in which the aluminum prevents oxygen from permeating the tube. The other oxygen barrier is identifiable as a red coating on the PEX tube exterior.

Applicable Specifications

ASTM	F876	Cross-linked Polyethylene (PEX) Tubing
ASTM	F877	Cross-linked Polyethylene (PEX) Plastic Hot- and Cold-Water Distribution Systems
ASTM	F1281	Standard Specification for Cross-linked Polyethylene/Aluminum/ Cross-linked Polyethylene (PEX-AL-PEX) Pressure Pipe
ASTM	F1807	Metal Insert Fittings Utilizing a Copper Crimp Ring for SDR 9 Cross-linked Polyethylene (PEX) Tubing
ASTM	F1865	Mechanical Cold Expansion Insert Fitting With Compression Sleeve for Cross-linked Polyethylene (PEX) Tubing
ASTM	F1960	Mechanical Cold Expansion Insert Fittings with PEX Reinforcing Rings for Use with Cross-linked (PEX) Tubing
ASTM	F1961	Metal Mechanical Cold Flare Compression Fittings with Disc Spring for Cross-linked Polyethylene (PEX) Tubing
ASTM	F1974	Standard Specification for Metal Insert Fittings for Polyethylene/ Aluminum/Polyethylene and Cross-linked Polyethylene/ Aluminum/Cross-linked Polyethylene Composite Pressure Pipe
ASTM	F2023	Standard Test Method for Evaluating the Oxidative Resistance of Cross-linked Polyethylene (PEX) Tubing and Systems to Hot Chlorinated Water
ASTM	F2080	Cold Expansion Fittings with Metal Compression Sleeves for Cross-linked Polyethylene (PEX) Pipe
ASTM	F2098	Stainless Steel Clamps for Securing SDR 9 Cross-linked Polyethylene (PEX) Tubing to Metal Insert Fittings
ASTM	F2159	Standard Specification for Plastic Insert Fittings Utilizing a Copper Crimp Ring for SDR9 Cross-linked Polyethylene (PEX) Tubing
ASTM	F2262	Standard Specification for Cross-linked Polyethylene/Aluminum/ Cross-linked Polyethylene Tubing OD Controlled SDR9
AWWA	C903	Polyethylene-Aluminum-Polyethylene Composite Pressure Pipes

AWWA	C904	Cross-Linked Polyethylene (PEX) Pressure Pipe, 1/2 in (12 mm) through 3 in (76 mm), for Water Service
CSA	B137.10	Crosslinked Polyethylene/Aluminum/Cross-Linked Polyethylene (PEX-AL-PEX) Composite Pressure-Pipe Systems
CSA	B137.5	Cross-linked Polyethylene (PEX) Tubing Systems for Pressure Applications

Manufacture of PEX Tubing

There are three methods used to manufacture PEX tubing:

1. The "Engel" method (PEX-A) uses a special extruder with a plunger action in which peroxide is added to the resin. The cross-linking occurs through a combination of pressure and high temperature, and the tubing is extruded through a die.
2. The "Silane" method (PEX-B) attaches a silane molecule to the base polyethylene molecule with the aid of a catalyst. After extrusion through a die, the tubing is heated with steam or hot water to cross-link (thermoset) the polymer.
3. The electron beam method (PEX-C) begins with normal HDPE tubing which is exposed to high-energy electron beam radiation. This removes hydrogen atoms from the chain and causes them to bond at the location of the missing hydrogen atoms.

The designations PEX-A, PEX-B, and PEX-C are more common in Europe. The means of manufacture however does not significantly affect performance, and the three varieties may be treated as identical. More important is whether the tubing is certified to meet the appropriate specification.

PEX is available in Copper Tube Sizes ¼ through 2 in. It is usually available in coils but can also be purchased in 20 ft straight lengths. The wall thicknesses vary to achieve an SDR 9 rating. PEX is rated for long-term service at temperatures up to 180°F (82°C) and a working pressure of 100 psi.

Fabrication and Assembly of PEX Tubing

Because it is a flexible material, PEX finds favor among some plumbers because the need for dimensional accuracy during routing is eliminated. Still, there are good and bad practices in routing and workmanship, and the tube has limits on how much bend it can accept without compromising the strength of the product. Similarly, adequate support of any piping system is always important, whether or not the system is flexible. The maximum recommended support of horizontal tube runs is 32 in (800 mm) for tubing up to 1 in diameter. This would be a practical spacing across the range of sizes however.

The tubing may be cut with plastic tubing cutters. A variety of methods are used to join the tubes. Some require the tube to be expanded over a fitting, and some require a ring to be crimped over the end of the tube. Regardless, it is important to provide square cuts without burrs on the ends of the tube.

Fiberglass Reinforced Plastic (FRP) Piping

Fiberglass Reinforced Plastic (FRP) is a composite material with wide use due to its chemical resistance and strength. The resins are thermosetting epoxy or polyester and the glass fibers are embedded within these resins to impart mechanical strength. The orientation of the fibers and the resin blend can be manipulated to obtain the desired quality for the specific application.

Applications

FRP piping is used to convey chemicals in liquid or vapor phases. It often finds use in fume scrubbers, ductwork, stacks, and fuel, acids, caustics, and solvents piping. It therefore is frequently used in oil field applications. It is also used to convey water in industrial applications. FRP may be used in both above and below ground installations.

Because FRP does not conduct electricity, caution must be exercised in handling flammable liquids so that static electricity discharge does not create an ignition source[9]. Test data appears to indicate that the velocity of the fluid inside the pipe is responsible for static electricity build-up, and a practical maximum limit for fuels inside nonconductive piping such as FRP is thought to be 12 ft per second (3.6 m/sec). Additional precautions such as grounding metal valves and fittings are also recommended, as well as wrapping the FRP with a copper wire helix and grounding it at regular intervals of approximately 500 ft (150 m).

FRP has a maximum temperature rating of 250°F (121°C).

Static electricity can also become a problem in some metallic piping systems. At one plant where I worked, we had a pellet handling system that pneumatically conveyed polypropylene pellets from railcars to extruder hoppers. The system was made out of plain end aluminum tubing, with stainless steel mechanical clamps that sealed the tubes with a white elastomer gasket around the joint, securing it within the SS sleeve with a 3-bolt clamp.

The construction engineer instructed the contractor to remove the brass inserts on the clamps, reckoning that these were supplied to hold the clamps together during shipment. These brass strips were actually supplied to provide a continuous electrical path across the joints so that the entire system could be grounded. High velocity polypropylene can create some serious static charges.

Upon realizing that the system was now ungrounded, the plant engineer decided to provide a ground by welding aluminum lugs on both sides of each joint and connecting an external wire. This had the advantage of providing a visible ground, but it had the disadvantage of preventing the removal of the clamps, which could not be slid off of the joint due to the lugs.

Applicable Specifications

ANSI	D5421	Standard Specification for Contact Molded "Fiberglass" (Glass-Fiber-Reinforced Thermosetting Resin) Flanges
API	14LR	Specification for Low Pressure Fiberglass Line Pipe
API	15HR	Specification for High Pressure Fiberglass Line Pipe
API	RP 15TL4	Recommended Practice for Care and Use of Fiberglass Tubulars
ASTM	D2996	Filament Wound Fiberglass (Glass-Fiber-Reinforced Thermosetting Resin) Pipe

[9] Sullivan D. Curran, PE. (n.d.). *Static Electricity in Fuel Handling Facilities*. Retrieved November 22, 2008, from Fiberglass Tank and Pipe Institute: http://www.fiberglasstankandpipe.com/static.htm.

ASTM	D2997	Centrifugally Cast Fiberglass (Glass-Fiber-Reinforced Thermosetting Resin) Pipe
ASTM	D3262	Fiberglass (Glass-Fiber-Reinforced Thermosetting Resin) Sewer Pipe
ASTM	D3754	Fiberglass (Glass-Fiber-Reinforced Thermosetting Resin) Sewer and Industrial Pressure Pipe
AWWA	C950	Fiberglass Pressure Pipe
UL	971	Nonmetallic Underground Piping for Flammable Liquids

Manufacture of FRP

FRP pipe is made by winding glass filaments around a mandrel. These glass filaments are responsible for imparting the mechanical strength of the pipe, and ratings up to 450 psi can be achieved. The resins are responsible for the resistance to corrosion, and often an interior or exterior lining may be applied to offer increased corrosion resistance. These linings do not contain much, if any, reinforcement, and so they must not be included in strength calculations.

In a sense, FRP piping may be thought of as tubing since it is built up over a mandrel (the ID is set by the size of the mandrel) and the wall thickness may be varied to achieve the desired physical properties. In other words, FRP is specified by ID and wall thickness, just like tubing.

Special "oil field tubular" can be manufactured with pressure ratings up to 3000 psi, but these are for "down-hole" applications.

FRP pipe diameters range from 1 to 108 in.

Fabrication and Assembly of FRP Pipe

FRP piping is commonly cut to length with abrasive saws. The joints may be bell and spigot, threaded, butted, or mechanical joint. Except for mechanical joints, the others are generally secured with adhesive and wrapped with a glass cloth for additional strength.

Concrete Pipe

Concrete pipe may be either reinforced (RCP) or non-reinforced, and is used in pressure and gravity applications.

Applications

Concrete pipe is often used for storm and sanitary sewers. Pressurized services are usually nonpotable cooling or process water.

Applicable Specifications

ASTM	C14	Standard Specification for Concrete Sewer, Storm Drain, and Culvert Pipe
ASTM	C76	Reinforced Concrete Culvert, Storm Drain, and Sewer Pipe (AASHTO M170)
ASTM	C361	Reinforced Concrete Low-Head Pressure Pipe
ASTM	C443	Joints for Circular Concrete Sewer and Culvert Pipe, Using Rubber Gaskets (AASHTO M198)

ASTM	C478	Precast Reinforced Concrete Manhole Sections (AASHTO M199)
ASTM	C497	Standard Methods of Testing Concrete Pipe, Sections or Tile (AASHTO T33)
ASTM	C507	Reinforced Concrete Elliptical Culvert, Storm Drain, and Sewer Pipe (AASHTO M207)
ASTM	C655	Reinforced Concrete D-Load Culvert, Storm Drain, and Sewer Pipe (AASHTO M242)
ASTM	C822	Standard Definitions of Terms Relating to Concrete Pipe and Related Products
ASTM	C924	Testing Concrete Pipe Sewer Lines by Low-Pressure Air Test Method
ASTM	C969	Infiltration and Exfiltration Acceptance Testing of Installed Precast Concrete Pipe Sewer Lines
ASTM	C1103	Joint Acceptance Testing of Installed Concrete Pipe Sewer Lines
ASTM	C1131	Least Cost (Life Cycle) Analysis of Concrete Culvert, Storm Sewer, and Sanitary Sewer Systems
ASTM	C1214	Concrete Pipe Sewer Lines by Negative Air Pressure (Vacuum) Test Method
ASTM	C1244	Standard Test Method for Concrete Sewer Manholes by Negative Air Pressure (Vacuum) Test (Metric)
ASTM	C1433	Precast Reinforced Concrete Box Sections for Culverts, Storm Drains and Sewers (AASHTO M259)
AWWA	C300	Reinforced Concrete Pressure Pipe, Steel-Cylinder Type
AWWA	C301	Prestressed Concrete Pressure Pipe, Steel-Cylinder Type
AWWA	C302	Reinforced Concrete Pressure Pipe, Noncylinder Type
AWWA	C303	Concrete Pressure Pipe, Bar-Wrapped, Steel-Cylinder Type
AWWA	C304	Design of Prestressed Concrete Cylinder Pipe

Manufacture of Concrete Pipe

Nonreinforced concrete pipe is available in sizes from 4 to 36 in, in two strength classes: standard and extra strength. Reinforced concrete pipe is available in sizes from 12 to 144 in. RCP is available is five strength classes, according to the load requirements set forth in ASTM C 76 (AASHTO M170). See Table 6.15.

The 0.01 in crack loads were developed in an effort to quantify "visible" cracks. As used in ASTM C76, the 0.01 in crack is a test criterion for concrete pipe tested in a three-edge bearing test and is not intended as a criterion of failure.

Class	0.01 in Crack	Ultimate
I	800 lb	1,200 lb
II	1,000 lb	1,500 lb
III	1,350 lb	2,000 lb
IV	2,000 lb	3,000 lb
V	3,000 lb	3,750 lb

Table 6.15 RCP Strength Classes

Fabrication and Assembly of Concrete Pipe

Concrete pipe is manufactured with gasketed bell joint ends, so assembly involves driving the straight spigot end into the bell. For small diameters, the spigot may be driven into the bell with a pry bar. For larger diameters a come-along may be used, although this task is often done using the bucket of a backhoe. The backhoe straddles the trench, and the bucket presses against the bell, driving the opposite end (the spigot) into the bell of the adjoining pipe. Concrete pipe manufacturers warn against using this method, unless precautions are taken to prevent localized stresses from being placed on the bell. The direction of lay of bell-and-spigot concrete pipe should be with the bells upstream. Figure 6.13 provides additional information regarding the handling and installation of concrete pipe.

It is frequently necessary to join concrete pipe to a manhole. In the past, these were built with masonry. It is far more common for these to be prefabricated concrete structures now, with the penetrations already cast in. The annulus between the OD of the pipe and the ID of the penetration is often sealed with an elastomeric band that may be tightened to form a watertight seal. See Figure 6.14. Care must be exercised to ensure that the plane of the seal lies entirely within the plane cut by the penetrating pipe in the radius of curvature of the manhole wall. See Figure 6.15.

Asbestos Cement Pipe

Asbestos has been used in various forms since ancient times. The developed ancient cultures marveled at its properties, and wove tablecloths out of it that could be cleaned by immersing in fire. The ancient Romans noticed that slaves who worked with the mineral fibers sometimes developed lung disease.

Use of asbestos grew throughout the industrial revolution and it was used in many products for its resistance to heat and its ability to be woven into cloth. It has been used as a component of cement piping since 1913 in Italy, and production of asbestos cement pipe in the U.S. began in 1929.

Manufacture of asbestos cement pipe continues today in spite of the health hazards associated with asbestos. The use of asbestos in new construction is banned by the European Union, Australia, Japan, and New Zealand.

Some controversy exists over the lack of a direct link to health hazards through the use of asbestos in water piping. Several studies have been conducted, but the results to date do not appear to prove a link between various cancers and the use of asbestos cement pipe in water supplies. Usually, the route of exposure to asbestos that causes the greatest concern is inhalation. While this is a potential route of exposure to workers who demolish or install asbestos cement pipes, it is not a normal exposure route to consumers who utilize tap water, unless the fibers are freed from the cement-asbestos matrix. This can occur if the lines are exposed to acidic water that tends to dissolve the cement.

So there are two concerns for consumers: ingestion of asbestos fibers, and inhalation of asbestos fibers due to the evaporation of water, as in a clothes dryer, a humidifier, or a vaporizer.

Thousands of miles of asbestos cement water piping have been installed throughout North America alone, and tests in some cities have revealed millions of fibers present in a quart of tap water.

Though the jury is still out on the health risks associated with the ingestion of fibers from asbestos cement piping, the prudent engineer might give serious consideration to erring on the safe side, especially since many years sometimes pass between exposure

Concrete Pipe Installation Procedures

Concrete Pipe Installation Procedures briefly outline some important steps in concrete pipe installation. They are intended only as a guide and do not replace or supersede project specifications or contract documents.

American
Concrete Pipe
Association
(972) 506-7216
Fax (972) 506-7682
email: info@concrete-pipe.org
www. concrete-pipe.org

February 2007

Unloading

Handling

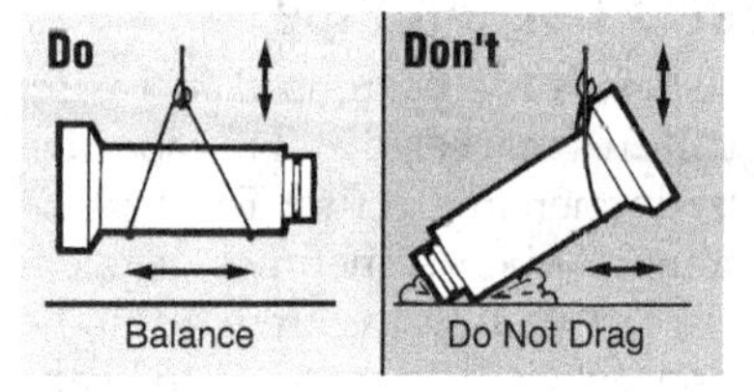

Stockpiling

Figure 6.13 Concrete pipe installation procedures. *American Concrete Pipe Association.*

Excavation & Foundation Preparation

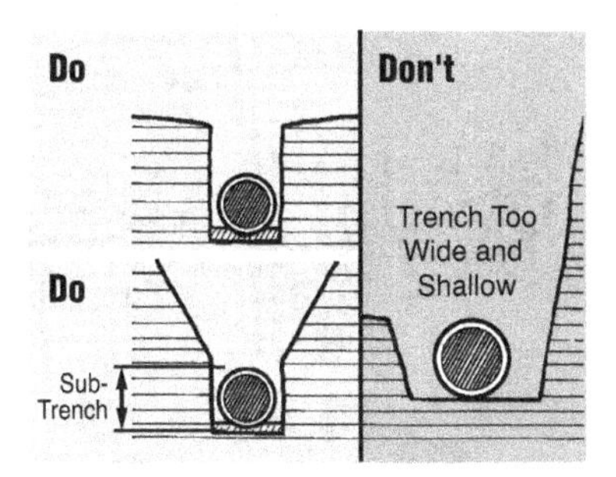

Pipe Bedding

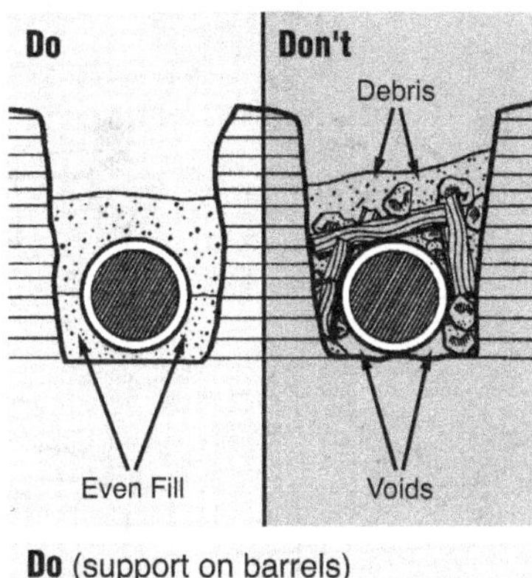

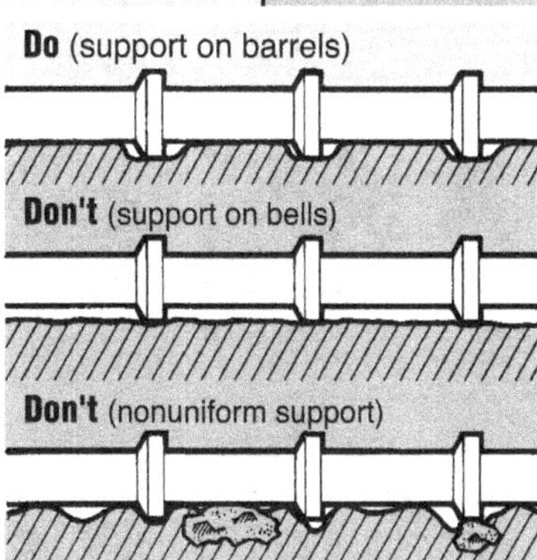

Alignment Line & Grade

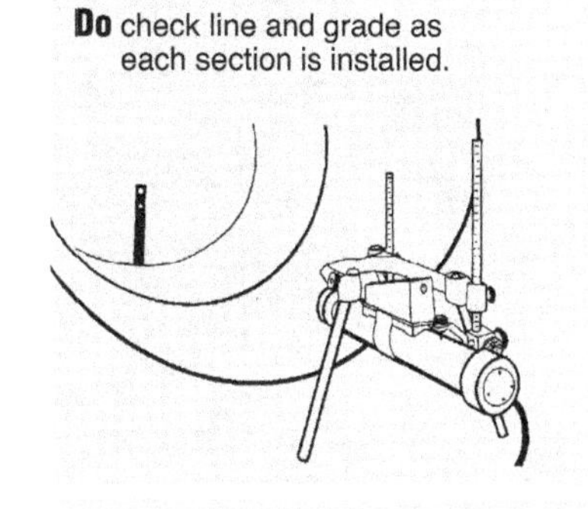

Do check line and grade as each section is installed.

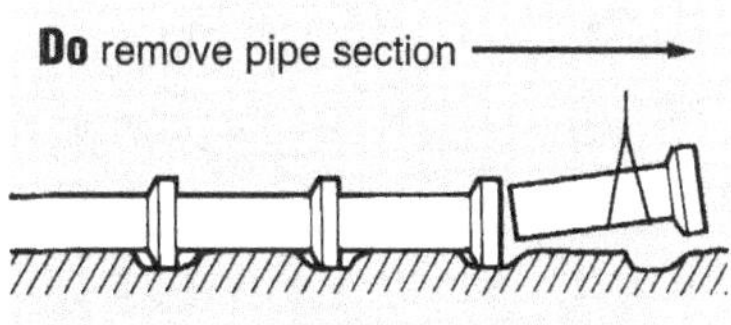

Don't adjust pipe alignment or grade with pipe in the home position.

Warning

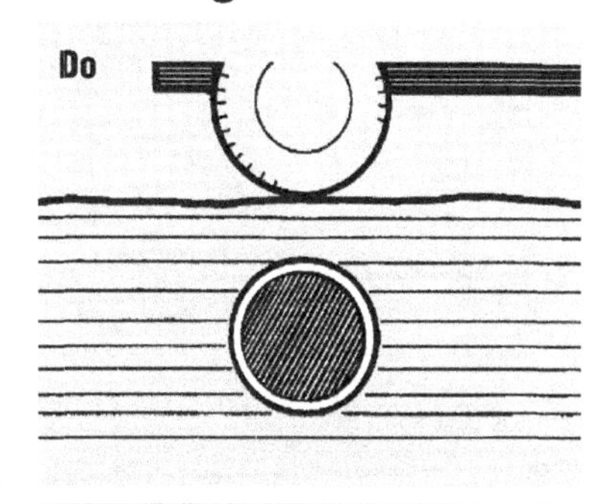

Don't operate heavy construction equiptment over the pipe until adequate cover is in place.

(continued on next page)

Preparation & Jointing

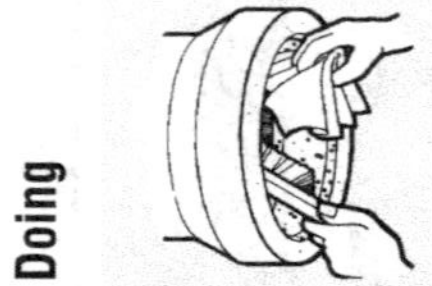
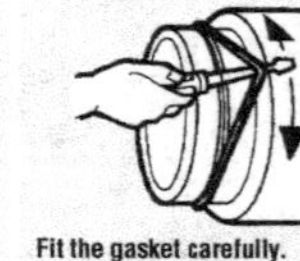
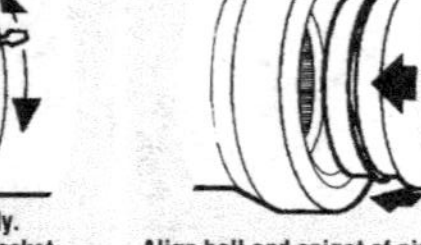
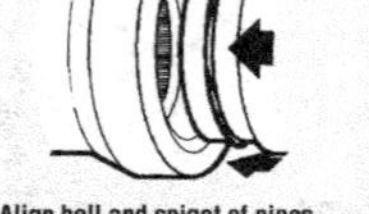

Doing	Carefully clean all dirt and foreign substances from the joining surfaces of the bell or groove end of pipe.	Lubricate bell jointing surface liberally. Use a brush, cloth, sponge or gloves to cover entire surface. Only approved lubricant should be used.	Carefully clean spigot or tongue end of pipe, including the gasket recess.	Lubricate the spigot or tongue end of the pipe, including the gasket recess.	Lubricate the gasket thoroughly (unless it is self lubricating) before it is placed on the spigot or tongue.	Fit the gasket carefully. Equalize the rubber gasket stretch by running a smooth, round object, inserted between gasket and spigot, around the entire circumference several times.	Align bell and spigot of pipes to be joined. Before homing the joint, check that the gasket is in contact with the entry taper around the entire circumference. Make sure pipe is aligned.
Prevents	Improper prepared bell jointing surface may prevent homing of the pipe.	A bell not lubricated or improperly lubricated may cause gasket to roll and possibly damage the bell.	Improperly prepared spigot and gasket recess may prevent gasket from sealing properly.	Gasket may twist out of recess if lubricant in recess is lacking or insufficient.	Excessive force will be required to push the pipe to the home position if gasket is not well lubricated.	Unequal stretch could cause bunching of gasket and may cause leaks in the joint or crack the bell.	Improper alignment can dislodge gasket causing leaks or possibly break the bell.

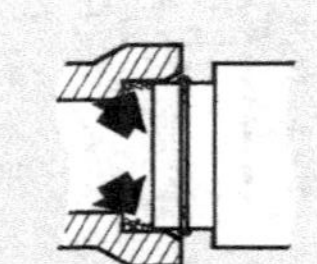
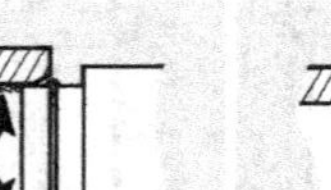

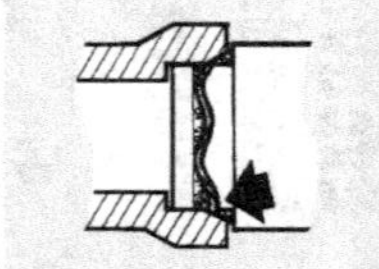

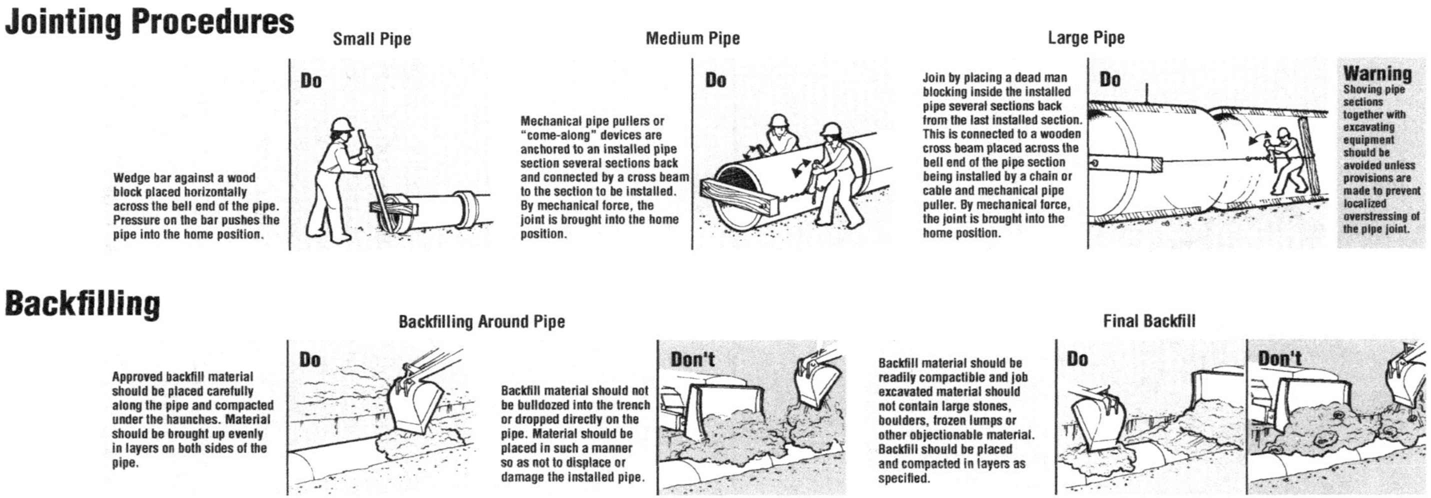

Jointing Procedures

Small Pipe

Do

Wedge bar against a wood block placed horizontally across the bell end of the pipe. Pressure on the bar pushes the pipe into the home position.

Medium Pipe

Do

Mechanical pipe pullers or "come-along" devices are anchored to an installed pipe section several sections back and connected by a cross beam to the section to be installed. By mechanical force, the joint is brought into the home position.

Large Pipe

Do

Join by placing a dead man blocking inside the installed pipe several sections back from the last installed section. This is connected to a wooden cross beam placed across the bell end of the pipe section being installed by a chain or cable and mechanical pipe puller. By mechanical force, the joint is brought into the home position.

Warning
Shoving pipe sections together with excavating equipment should be avoided unless provisions are made to prevent localized overstressing of the pipe joint.

Backfilling

Backfilling Around Pipe

Do

Approved backfill material should be placed carefully along the pipe and compacted under the haunches. Material should be brought up evenly in layers on both sides of the pipe.

Don't

Backfill material should not be bulldozed into the trench or dropped directly on the pipe. Material should be placed in such a manner so as not to displace or damage the installed pipe.

Final Backfill

Do

Don't

Backfill material should be readily compactible and job excavated material should not contain large stones, boulders, frozen lumps or other objectionable material. Backfill should be placed and compacted in layers as specified.

Figure 6.13 *(Continued)* Concrete pipe installation procedures. *American Concrete Pipe Association.*

Figure 6.14 An elastomeric seal through a concrete containment. *Artwork provided by psi–Pipeline Seal & Insulator, Inc. Link-Seal® Modular seals is a registered trademark of psi.*

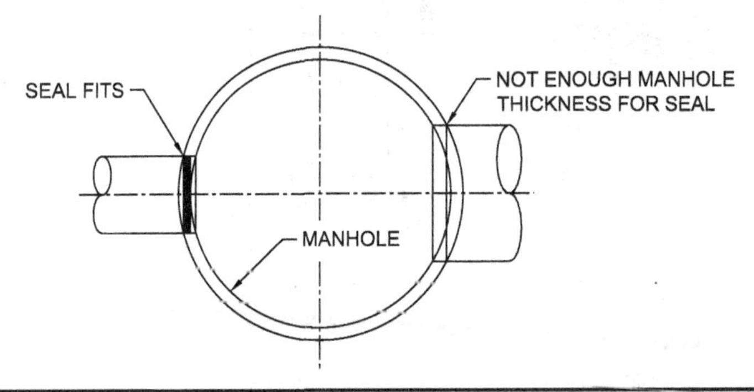

Figure 6.15 Pipe penetrations in a manhole.

to asbestos and evident health effects. Many substitutes exist that do not fall under suspicion. And the inhalation exposure may still be present if fibers are present in the water.

Other Composites

Composites are materials that are made of two or more dissimilar materials that are combined in a macroscopic matrix to form another material. The composite is designed to have better properties than any of the individual materials used to make it, and these usually center around resistance to chemicals or a high strength-to-weight ratio.

The individual materials may exist in more-or-less discrete layers in the composite, as in PEX-Aluminum-PEX tubing, or they may be more uniform, as in reinforced concrete pipe.

There are many different composites available for piping. We have already discussed some of these, but a few others are of particular interest.

Centrifugally Cast Glass-Fiber Reinforced, Polymer Mortar Pipe

This material is built up into layers, and offers a high-strength material that is resistant to corrosion and possesses a smooth interior surface for low fluid friction. See Figure 6.16.

These pipes are available in pressure and non-pressure classifications, with the wall thicknesses built to match the requirements. Sizes are available from 18 to 110 in

FIGURE 6.16 CCFRPM sample. Note the layering, with a smooth resin on the inside, a heavily reinforced chopped glass and mortar layer, a core of polymer mortar, followed by another heavily reinforced layer of chopped glass and resin, and finally, an exterior layer of sand and resin.

diameter, and pressures up to 250 psi. Connections are made with filament wound sleeves that contain an internal elastomeric gasket, or with bell-and-spigot joints. Low-profile bell-and-spigot joints are available for slip-lining applications, and flush bell-and-spigot joints are available for non-pressure jacking applications.

Lined Piping Systems

Steel or stainless steel piping systems are available with materials like fluoropolymers bonded to the interior. Other systems use thermoplastic-lined FRP piping to combine the chemical resistance and lightweight strength. These systems are often used in Pulp & Paper for fume collection, especially where chlorine is handled. Such systems can be rated for 150 psi (10 bar) at 300°F (149°C).

Elastomers

Material selection is not always about piping. Elastomers are used for gaskets, seals, and coatings, and there are other materials that are applied to valves to resist wear. A brief description of some of the common materials not described above follows.

Polyvinylidene Fluoride (PVDF)

PVDF is also known by the trade names Kynar®, Hylar®, and Sygef®. It has a low melting point, but possesses high strength, and exhibits high resistance to solvents, acids, and bases.

Polytetrafluoroethylene (PTFE)

PTFE is also known by the trade name Teflon®. This is a chemically inert and non-toxic compound, but chemical decomposition occurs at temperatures of 660°F (350°C). Because of its extremely low coefficient of friction, and because it creeps at low temperatures, it is often used in ball valve seals. PTFE in fact has the lowest coefficient of friction of any known solid[10]. PTFE is suitable for applications up to approximately 500°F (260°C).

Nitrile Rubber, or Buna-N

This is a synthetic rubber that is resistant to aliphatic hydrocarbons like methane and acetylene. It is not resistant to ozone or aromatic hydrocarbons. It is the most popular material for O-rings.

[10] PTFE is the only known substance to which a gecko cannot stick.

Ethylene Propylene Diene Monomer (EPDM) Rubber

EPDM is compatible with fireproof hydraulic fluids, ketones, bases, and water. It is incompatible with oils, gasoline, and concentrated acids. It is often used in hoses, tubing, and seals.

Polychloroprene

Polychloroprene is better known by the trade name Neoprene®. It is a synthetic rubber that exhibits resistance to many chemicals, making it a good choice for many gasket and hose applications.

Fluoropolymer

Viton® is the popular trade name of this synthetic rubber which smells like cinnamon. It has excellent heat resistance up to 400°F (200°C), and resists hydrocarbons. It is often used for seals and gaskets.

Polyetheretherketone (PEEK) or Polyketones

PEEK is a thermoplastic that has a very high stiffness (522,000 psi, or 3.6 GPa) relative to other thermoplastics, and a very high tensile strength (13 ksi, or 90 Mpa). It resists alkalis, aromatic hydrocarbons, alcohols, and oils, but is not recommended for most acids.

Insulating Materials

Fiberglass

Preformed fiberglass pipe insulation is available in sizes up to 34 in. It is also available with a factory-installed jacket bonded to aluminum foil, and reinforced with a fiberglass scrim. The jacket may be kraft paper or polyethylene, and serves as a vapor retarder and a barrier to physical damage of the fiberglass. The pieces are cut with a knife and taped together.

This is a good choice for dry locations, but if the insulation ever becomes wet it must be replaced.

Fiberglass is appropriate for pipe temperatures from -50°F to 250°F (-46°C to 121°C) and has a thermal conductivity of approximately 0.30 BTU-in/hr-ft^2-°R (0.04 W/m K).

Calcium Silicate

Calcium silicate is a lightweight, porous, chalky insulator that is cut with a hand saw and wired to the pipe. It is available in pre-formed shapes to fit pipes and a variety of fittings, and is then covered with a metal jacket, usually aluminum.

Calcium silicate is used to insulate hot pipes up to 1200°F (649°C). It has a thermal conductivity of approximately 0.50 BTU-in/hr-ft^2-°R (0.072 W/m K) at 400°F (204°C), and a density of 14.5 lb/ft^3 (232 kg/m^3).

Where gaps exist at fittings (and especially when flat boards are fitted around vessels) a lightweight insulating cement is applied.

Calcium silicate must be protected from moisture, since once wet its insulating properties are greatly diminished.

Cellular Glass

Cellular glass is a closed-cell material that is used in applications from -450°F to 900°F (-268°C to 482°C). Being a closed-cell material, it is an excellent choice for locations subject to moisture. It is impermeable to liquids and it does not burn. It may also be used for underground applications, and is available in a wide range of preformed shapes.

Cellular glass has a thermal conductivity of 0.29 BTU-in/hr-ft^2-°R (0.039 W/m K) at 75°F (24°C) and a density of 7.5 lb/ft^3 (120 kg/m^3).

Foam Synthetic Rubber

Preformed foam synthetic rubber insulation is available in sizes up to 6 in NPS. It is used primarily for plumbing and hydronics applications, and has a recommended temperature range of -297°F to 220°F (-183°C to 105°C). It is often used to prevent condensation on cold service piping.

It is slipped over or slit and wrapped around the pipe. Butt joints and seams are sealed with a special adhesive. Foam synthetic rubber is designed for above ground installations only. If exposed to weather, a protective finish must be applied to protect it from ultraviolet degradation.

Foam synthetic rubber has a thermal conductivity of 0.27 BTU-in/hr-ft^2-°R (0.039 W/m K) at 75°F (24°C) and a density of 3 to 6 lb/ft^3 (48 to 96 kg/m^3).

Polyisocyanurate

This is a cellular polymer insulation for use at temperatures from -297°F to 300°F (-183°C to 149°C). It may be used for general industrial piping, chilled water, or tank and vessel insulation. Polyisocyanurate will degrade when exposed to the ultraviolet spectrum, so a covering is required for locations exposed to sunlight. Polyisocyanurate is combustible, but certain manufacturers are able to attain a Flame Spread Index of 25 or less, and a Smoke Developed Index of 50 or less.

Polyisocyanurate has a thermal conductivity of 0.19 BTU-in/hr-ft^2-°R (0.027W/m K) at 75°F (24°C) and a density of 2.05 lb/ft^3 (32.8 kg/m^3).

Mineral Wool

Mineral wool insulation is suitable for temperature ranges up to 1200°F (1177°C). It is made from basalt rock and steel slag with an organic binder. The binder requires that a heat-up schedule be maintained for service temperatures above 450°F (232°C). During this heat-up phase, some of the resin will start a controlled decomposition so adequate ventilation is required to vent vapors.

Mineral wool is water repellant, yet vapor-permeable, and it may be used outdoors. A metal jacket is recommended to protect the insulation from physical damage as well as weatherproofing. The insulation is cut with a knife and wired to the pipe. Nesting schedules are available to provide increased insulation thicknesses.

While there is no known carcinogenic risk of inhaling the fibers, OSHA has established exposure limits for inhalation, and respiratory protection is required during installation.

Mineral wool has a thermal conductivity of 0.25 BTU-in/hr-ft^2-°R (0.035 W/m K) at 100°F (24°C) and a density of 4.4 lb/ft^3 (70 kg/m^3).

Extruded Polystyrene

Extruded polystyrene is a rigid thermoplastic foam insulator that is used for piping in the range of -297°F to 165°F (-183°C to 74°C). It is most often used for cold lines to prevent heat gain and surface condensation or sweating. It will degrade when exposed to sunlight if not covered.

At least one manufacturer includes a fire retardant to inhibit accidental ignition, but the Smoke Developed Index may be as high as 165.

Extruded polystyrene has a thermal conductivity of 0.259 BTU-in/hr-ft^2-°R (0.037W/m K) at 75°F (24°C) and a density of 1.6 lb/ft^3 (26 kg/m^3).

CHAPTER 7

Fittings

Piping systems require fittings in order to change direction and connect to all of the equipment and devices required to make them function. These fittings are manufactured in standard dimensions referred to as the fitting "take-out." See Table 7.1. Virtually any piping component that attaches permanently to a pipe can be considered a fitting.

Fittings must obviously be manufactured for every type of pipe and every type of connection, i.e., threaded, flanged, or welded. Socket welded and threaded fittings are available in sizes up to 4 in diameter, but are not commonly used in sizes above 2 in diameter.[1]

Fittings may be classified according to their material of construction or according to their function. The various standard specifications for fittings are generally classified according to MOC. In this chapter however, we will examine the fittings by function and purpose. The variety of fittings is staggering since there are so many combinations of materials and functions. A trip through the plumbing section of a home improvement center will impress upon the reader the number of choices that exist in connecting two pieces of pipe together. And that is just residential plumbing.

Common fittings include:

- Flanges
- Elbows (90 and 45°) and reducing elbows
- Tees, reducing tees, and cleanouts
- Unions
- Laterals
- Reducers (concentric, eccentric)
- Caps
- Plugs
- Nipples, couplings, and half-couplings.

Less common fittings are:

- Swages
- Bull plugs
- Crosses
- Wyes

[1] This coincides with the typical sizes regarded as field-routed piping, so "field-routed" generally implies threaded or socket welded pipe.

NPS	OD	90° Ell Long R A	90° Ell Short R A	180° Bends Long R K	180° Bends Short R K	45° Ell B	Tees C	Caps E	Crosses C	Stub Ends Lap OD G	Stub Ends ANSI L	Stub Ends MSS L
1/2	0.840	1 1/2		1 7/8		5/8	1	1		1 3/8	3	2
3/4	1.050	1 1/8		1 11/16		7/16	1 1/8	1		1 11/16	3	2
1	1.315	1 1/2	1	2 3/16	1 5/8	7/8	1 1/2	1 1/2		2	4	2
1 1/4	1.660	1 7/8	1 1/4	2 3/4	2 1/16	1	1 7/8	1 1/2	1 7/8	2 1/2	4	2
1 1/2	1.900	2 1/4	1 1/2	3 1/4	2 7/16	1 1/8	2 1/4	1 1/2	2 1/4	2 7/8	4	2
2	2.375	3	2	4 3/16	3 3/16	1 3/8	2 1/2	*1 1/2*	2 1/2	3 5/8	6	2 1/2
2 1/2	2.875	3 3/4	2 1/2	5 3/16	3 15/16	1 3/4	3	*1 1/2*	3	4 1/8	6	2 1/2
3	3.500	4 1/2	3	6 1/4	4 3/4	2	3 3/8	*2*	3 3/8	5	6	2 1/2
3 1/2	4.000	5 1/4	3 1/2	7 1/4	5 1/2	2 1/4	3 3/4	*2 1/2*	3 3/4	5 1/2	6	3
4	4.500	6	4	8 1/4	6 1/4	2 1/2	4 1/8	*2 1/2*	4 1/8	6 3/16	6	3
5	5.563	7 1/2	5	10 5/16	7 3/4	3 1/8	1 7/8	*3*	4 7/8	7 5/16	8	3
6	6.625	9	6	12 5/16	9 5/16	3 3/4	5 5/8	*3 1/2*	5 5/8	8 1/2	8	3 1/2
8	8.625	12	8	16 5/16	12 5/16	5	7	*4*	7	10 5/8	8	4
10	10.750	15	10	20 3/8	15 3/8	6 1/4	8 1/2	*5*	8 1/2	12 3/4	10	5
12	12.750	18	12	24 3/8	18 3/8	7 1/2	10	*6*	10	15	10	6
14	14.000	21	14	28	21	8 3/4	11	*6 1/2*	11	16 1/4	12	6
16	16.000	24	16	32	24	10	12	*7*	12	18 1/2	12	6
18	18.000	27	18	36	27	11 1/4	13 1/2	*8*	13 1/2	21	12	6
20	20.000	30	20	40	30	12 1/2	15	*9*	15	23	12	6
22	22.000	33	22	44		13 1/2	16 1/2	10	16 1/2			
24	24.000	36	24	48	36	15	17	10 1/2	17	27 1/4	12	6
26	26.000	39	26	52		16	19 1/2	10 1/2				
30	30.000	45	30	60	45	18 1/2	22	10 1/2				
34	34.000	51	34			21	25	10 1/2				
36	36.000	54	36		54	22 1/4	26 1/2	10 1/2				
42	42.000	63	42			26		12				

TABLE 7.1 Fitting Take-outs. Take-outs for Caps in italics apply only to Std and XS.

Applicable Specifications

ANSI	B16.1	Cast Iron Pipe Flanges and Flanged Fittings, Class 25, 125, 250, and 800
ANSI	B16.3	Malleable Iron Threaded Fittings
ANSI	B16.4	Cast Iron Threaded Fittings
ANSI	B16.5	Pipe Flanges and Flanged Fittings: NPS 1/2 through 24
ANSI	B16.9	Factory-Made Wrought Steel Buttwelding Fittings
ANSI	B16.11	Forged Steel Fittings, Socket-Welding and Threaded
ANSI	B16.12	Cast Iron Threaded Drainage Fittings
ANSI	B16.14	Ferrous Pipe Plugs, Bushings and Locknuts with Pipe Threads
ANSI	B16.15	Cast Bronze Threaded Fittings
ANSI	B16.18	Cast Copper Alloy Solder Pressure Fittings
ANSI	B16.20	Metallic Gaskets for Pipe Flanges – Ring-Joint, Spiral-Wound, and Jacketed
ANSI	B16.21	Nonmetallic Flat Gaskets for Pipe Flanges
ANSI	B16.22	Wrought Copper and Copper Alloy Solder Joint Pressure Fittings
ANSI	B16.23	Cast Copper Alloy Solder Joint Drain Fittings (DWV)
ANSI	B16.24	Cast Copper Alloy Pipe Flanges and Flanged Fittings
ANSI	B16.25	Buttwelding Ends
ANSI	B16.26	Cast Copper Alloy Fittings for Flared Copper Tubes
ANSI	B16.28	Wrought Steel Buttwelding Short Radius Elbows and Returns
ANSI	B16.36	Orifice Flanges
ANSI	B16.39	Malleable Iron Threaded Pipe Unions
ANSI	B16.42	Ductile Iron Pipe Flanges and Flanged Fittings, Classes 150 and 300
ANSI	B16.45	Cast Iron Fittings for Sovent® Drainage Systems
ANSI	B16.47	Large Diameter Steel Flanges: NPS 26 through NPS 60 (Welding Neck and Blind Flanges only)
ANSI	B16.48	Steel Line Blanks
ANSI	B16.49	Factory-Made Wrought Steel Buttwelding Induction Bends for Transportation and Distribution Systems
ANSI	F704	Standard Practice for Selecting Bolting Lengths for Piping System Flanged Joints
ANSI	F2015	Standard Specification for Lap Joint Flange Pipe End Applications
API	6A	Specification for Wellhead and Christmas Tree Equipment
API	17D	Specification for Subsea Wellhead and Christmas Tree Equipment
MSS	SP-6	Standard Finishes for Contact Faces of Pipe Flanges and Connecting-End Flanges of Valves and Fittings
MSS	SP-9	Spot Facing for Bronze, Iron and Steel Flanges

MSS	SP-44	Steel Pipeline Flanges
MSS	SP-60	Connecting Flange Joint Between Tapping Sleeves and Tapping Valves
MSS	SP-65	High Pressure Chemical Industry Flanges and Threaded Stubs for Use with Lens Gaskets
MSS	SP-95	Swaged Nipples and Bull Plugs
MSS	SP-97	Integrally Reinforced Forged Branch Outlet Fittings - Socket Welding, Threaded and Buttwelding Ends
SAE	J518	Hydraulic Flanged Tube, Pipe, and Hose Connections, Four-Bolt Split Flange Type

Flanges

There are several styles of flanges in use. For hydraulic service, SAE J518 4-bolt flanges are common. However, for industrial, commercial, and institutional applications the most commonly used flanges conform to the requirements of ANSI B16.5. These flanges are available in a variety of styles and pressure classes. The dimensions of these flanges are tabulated in Tables 7.2 through 7.5.

Flanges are not recommended for use in underground service where the flange will be in direct contact with soil, since this opens the possibility for corrosion of bolted connections. Where it is necessary to use a flange in underground service, as at a valve, it is preferred to have the flanged connection exposed in a valve pit for maintenance access.

Flange Ratings

The allowable pressures for a given class are material and temperature dependent. Thus, for a 150 lb carbon steel flange, the allowable pressure is 275 psi at 100°F, 150 psi at 500°F, and 40 psi at 1000°F. The only temperature at which the flange is rated for 150 psi is 500°F. Heavier flanges are rated at their nominal pressure rating only at 850°F.

Pressure classes of cast iron flanges are established in ANSI B16.1; pressure classes for other flanges are established by ANSI B16.5. The pressure classes of flanges are commonly referred to in terms of "pounds" rather than "pounds per square inch." But because the allowable pressures are the same as the nominal pressure classes only at the elevated temperatures noted above, it is best to think of the pressure classes as nominal values, and not as "pressures." The pressures and temperatures do not correspond to saturated steam properties, as is sometimes thought.

Although ANSI now refers to the pressure ratings as "classes," the use of the term "pound" remains prevalent in the industry. In describing the pressure class of flanges, the terms "pound" and "class" may be considered interchangeable.

Further, it is important to note that the pressures listed in the ANSI B16.1 and B16.5 tables are for non-shock pressures. Tables 7.6 and 7.7 show the non-shock pressure and temperature ratings for cast iron and carbon steel flanges respectively.

Note that ANSI B16.1 permits cast iron flanges to be used in gaseous and steam service. This is a carry-over from the days when cast iron piping was prevalent in infrastructure and heating. Cast iron would not be the choice for steam or gaseous service today. Always use carbon or stainless steel for gaseous or steam services.

Materials for flanges are grouped into 24 categories in an effort to combine materials that are likely to be joined together. However, if two dissimilar groupings are mated, the lower rating of the two must be used.

Nominal Pipe Size	Flange OD	Dia of Bolt Circle	Dia of Bolts	No. of Bolts	Length of Studs 1/16 in Raised Face	Bolt Length	Length Thru Hub			Thickness
							Weld Neck[1]	SO Threaded SW[1]	Lap Joint	Blind[1]
in	in	in	in		in	in	in	in	in	in
1/2	3 1/2	2 3/8	1/2	4	2 1/4	1 3/4	1 7/8	5/8	5/8	7/16
3/4	3 7/8	2 3/4	1/2	4	2 1/4	2	2 1/16	5/8	5/8	1/2
1	4 1/4	3 1/8	1/2	4	2 1/2	2	2 3/16	11/16	11/16	9/16
1 1/4	4 5/8	3 1/2	1/2	4	2 1/2	2 1/4	2 1/4	13/16	13/16	5/8
1 1/2	5	3 7/8	1/2	4	2 3/4	2 1/4	2 7/16	7/8	7/8	11/16
2	6	4 3/4	5/8	4	3	2 3/4	2 1/2	1	1	3/4
2 1/2	7	5 1/2	5/8	4	3 1/4	3	2 3/4	1 1/8	1 1/8	7/8
3	7 1/2	6	5/8	4	3 1/2	3	2 3/4	1 3/16	1 3/16	15/16
3 1/2	8 1/2	7	5/8	8	3 1/2	3	2 13/16	1 1/4	1 1/4	15/16
4	9	7 1/2	5/8	8	3 1/2	3	3	1 5/16	1 5/16	15/16
5	10	8 1/2	3/4	8	3 3/4	3 1/4	3 1/2	1 7/16	1 7/16	15/16
6	11	9 1/2	3/4	8	3 3/4	3 1/4	3 1/2	1 9/16	1 9/16	1
8	13 1/2	11 3/4	3/4	8	4	3 1/2	4	1 3/4	1 3/4	1 1/8
10	16	14 1/4	7/8	12	4 1/2	3 3/4	4	1 15/16	1 15/16	1 3/16
12	19	17	7/8	12	4 1/2	4	4 1/2	2 3/16	2 3/16	1 1/4
14	21	18 3/4	1	12	5	4 1/4	5	2 1/4	3 1/8	1 3/8
16	23 1/2	21 1/4	1	16	5 1/4	4 1/2	5	2 1/2	3 7/16	1 7/16
18	15	22 3/4	1 1/8	16	5 3/4	4 3/4	5 1/2	2 11/16	3 13/16	1 9/16
20	17 1/2	25	1 1/8	20	6	5 1/4	5 11/16	2 7/8		1 11/16
22	29 1/2	27 1/4	1 1/4	20	6 1/2	5 1/2	5 7/8	3 1/8		1 13/16
24	32	29 1/2	1 1/4	20	6 3/4	5 3/4	6	3 1/4		1 7/8
26	34 1/4	31 3/4	1 1/4	24	7	6	5	3 3/8		2
30	38 3/4	36	1 1/4	28	7 1/4	6 1/4	5 1/8	3 1/2		2 1/8
34	43 3/4	40 1/2	1 1/2	32	8	7	5 5/16	3 11/16		2 5/16
36	46	42 3/4	1 1/2	32	8 1/4	7	5 3/8	3 3/4		2 3/8
42	53	49 1/2	1 1/2	36	8 3/4	7 1/2	5 5/8	4		2 5/8

[1] The 1/16" raised face is included in the "Length thru Hub" dimension of the Weld Neck (WN), Slip On (SO), Threaded, and Socket Weld (SW) flange, and also in the "Thickness" dimension of the Blind flange.

Table 7.2 ANSI B16.5 Flange dimensions for 150 lb flanges. Bolting arrangement for 125 lb cast iron (ANSI B 16.1) flanges are the same as for 150 lb ANSI B16.5 steel flanges.

							Length Thru Hub			Thickness
Nominal Pipe Size	Flange OD	Diam of Bolt Circle	Diam of Bolts	No. of Bolts	Length of Studs 1/16" Raised Face	Bolt Length	Weld Neck[1]	SO Threaded SW[1]	Lap Joint	Blind[1]
in	in	in	in		in	in				
1/2	3 3/4	2 5/8	1/2	4	2 1/2	2	2 1/16	7/8	7/8	9/16
3/4	4 5/8	3 1/4	5/8	4	2 3/4	2 1/2	2 1/4	1	1	5/8
1	4 7/8	3 1/2	5/8	4	3	25 1/2	2 7/16	1 1/16	1 1/16	11/16
1 1/4	5 1/4	3 7/8	5/8	4	3	2 3/4	2 9/16	1 1/16	1 1/16	3/4
1 1/2	6 1/8	4 1/2	3/4	4	3 1/2	3	2 11/16	1 3/16	1 3/16	13/16
2	6 1/2	5	5/8	8	3 1/4	3	2 3/4	1 5/16	1 5/16	7/8
2 1/2	7 1/2	5 7/8	3/4	8	3 3/4	3 1/4	3	1 1/2	1 1/2	1
3	8 1/4	6 5/8	3/4	8	4	3 1/2	3 1/8	1 11/16	1 11/16	1 1/8
3 1/2	9	7 1/4	3/4	8	4 1/4	3 3/4	3 3/16	1 3/4	1 3/4	1 3/16
4	10	7 7/8	3/4	8	4 1/4	3 3/4	3 3/8	1 7/8	1 7/8	1 1/4
5	11	9 1/4	3/4	8	4 1/2	4	3 7/8	2	2	1 3/8
6	12 1/2	10 5/8	3/4	12	4 3/4	4 1/4	3 7/8	2 1/16	2 1/16	1 7/16
8	15	13	7/8	12	5 1/4	4 3/4	4 3/8	2 7/16	2 7/16	1 5/8
10	17 1/2	15 1/4	1	16	6	5 1/4	4 5/8	2 5/8	3 3/4	1 7/8
12	20 1/2	17 3/4	1 1/8	16	6 1/2	5 3/4	5 1/8	2 7/8	4	2
14	23	20 1/4	1 1/8	20	6 3/4	6	5 5/8	3	4 3/8	2 1/8
16	25 1/2	22 1/2	1 1/4	20	7 1/4	6 1/2	5 3/4	3 1/4	4 3/4	2 1/4
18	28	24 3/4	1 1/4	24	7 1/2	6 3/4	6 1/4	3 1/2	5 1/8	2 3/8
20	30 1/2	27	1 1/4	24	8	7	6 3/8	3 3/4	5 1/2	2 1/2
22	33	29 1/4	1 1/2	24	8 3/4	7 1/2	6 1/2	4	6	2 5/8
24	36	32	1 1/2	24	9	7 3/4	6 5/8	4 3/16		2 3/4
26	38 1/4	34 1/2	1 5/8	28	10	8 3/4	7 1/4	7 1/4		3 1/8
30	43	39 1/4	1 3/4	28	11 1/4	10	8 1/4	8 1/4		3 5/8
34	47 1/2	43 1/2	1 7/8	28	12 1/4	10 3/4	9 1/8	9 1/8		4
36	50	46	2	32	12 3/4	11 1/4	9 1/2	9 1/2		4 1/8
42	57	52 3/4	2	36	13 3/4	13 1/2	10 7/8	10 7/8		4 5/8

[1] The 1/16" raised face is included in the "Length thru Hub" dimension of the Weld Neck (WN), Slip On (SO), Threaded, and Socket Weld (SW) flange, and also in the "Thickness" dimension of the Blind flange.

TABLE 7.3 ANSI B16.5 Flange dimensions for 300 lb flanges. Bolting arrangement for 250 lb cast iron (ANSI B 16.1) flanges are the same as for 300 lb ANSI B16.5 steel flanges.

Nominal Pipe Size	Flange OD	Diam of Bolt Circle	Diam of Bolts	No. of Bolts	Length of Studs 1/4" Raised Face	Length Thru Hub			Thickness
						Weld Neck[1]	SO Threaded SW[1]	Lap Joint	Blind[1]
in	in	in	in		in	in	in	in	in
1/2	3 3/4	2 5/8	1/2	4	3	Use 600 lb			
3/4	4 5/8	3 1/4	5/8	4	3 1/4				
1	4 7/8	3 1/2	5/8	4	3 1/2				
1 1/4	5 1/4	3 7/8	5/8	4	3 3/4				
1 1/2	6 1/8	4 1/2	3/4	4	4				
2	6 1/2	5	5/8	8	4				
2 1/2	7 1/2	5 7/8	3/4	8	4 1/2				
3	8 1/4	6 5/8	3/4	8	4 3/4				
3 1/2	9	7 1/4	7/8	8	5 1/4				
4	10	7 7/8	7/8	8	5 1/4	3 1/2	2	2	1 3/8
5	11	9 1/4	7/8	8	6 1/2	4	2 1/8	2 1/8	1 1/2
6	12 1/2	10 5/8	7/8	12	5 3/4	4 1/16	2 1/4	2 1/4	1 5/8
8	15	13	1	12	6 1/2	4 5/8	2 11/16	2 11/16	1 7/8
10	17 1/2	15 1/4	1 1/8	16	7 1/4	4 7/8	2 7/8	4	2 1/8
12	20 1/2	17 3/4	1 1/4	16	7 3/4	5 3/8	3 1/8	4 1/4	2 1/4
14	23	20 1/4	1 1/4	20	8	5 7/8	3 5/16	4 5/8	2 3/8
16	25 1/2	22 1/2	1 3/8	20	8 1/2	6	3 11/16	5	2 1/2
18	28	24 3/4	1 3/8	24	8 3/4	6 1/2	3 7/8	5 3/8	2 5/8
20	30 1/2	27	1 1/2	24	9 1/2	6 5/8	4	5 3/4	2 3/4
22	33	29 1/4	1 5/8	24	10	6 3/4	4 1/4		2 7/8
24	36	32	1 3/4	24	10 1/2	6 7/8	4 1/2	6 1/4	3
26	38 1/4	34 1/2	1 3/4	28	11 1/2	7 5/8	7 5/8		3 1/2
30	43	39 1/4	2	28	13	8 5/8	8 5/8		4
34	47 1/2	43 1/2	2	28	13 3/4	9 1/2	9 1/2		4 3/8
36	50	46	2	32	14	9 7/8	9 7/8		4 1/2
42	57	52 3/4	2 1/2	36	16 1/4	11 3/8	11 3/8		5 1/8

[1] The 1/16" raised face is included in the "Length thru Hub" dimension of the Weld Neck (WN), Slip On (SO), Threaded, and Socket Weld (SW) flange, and also in the "Thickness" dimension of the Blind flange.

Table 7.4 ANSI B16.5 Flange dimensions for 400 lb flanges.

Nominal Pipe Size	Flange OD	Diam of Bolt Circle	Diam of Bolts	No. of Bolts	Length of Studs 1/4" Raised Face	Length Thru Hub			Thickness
						Weld Neck[1]	SO Threaded SW[1]	Lap Joint[1]	Blind[1]
in	in	in	in		in	in	in	in	in
1/2	3 3/4	2 5/8	1/2	4	3	2 1/16	7/8	7/8	9/16
3/4	4 5/8	3 1/4	5/8	4	3 1/4	2 1/4	1	1	5/8
1	4 7/8	3 1/2	5/8	4	3 1/2	2 7/16	1 1/16	1 1/6	11/16
1 1/4	5 1/4	3 7/8	5/8	4	3 3/4	2 5/8	1 1/8	1 1/8	13/16
1 1/2	6 1/8	4 1/2	3/4	4	4	2 3/4	1 1/4	1 1/4	7/8
2	6 1/2	6	5/8	8	4	2 7/8	1 7/16	1 7/16	1
2 1/2	7 1/2	5 7/8	3/4	8	4 1/2	3 1/8	1 5/8	1 5/8	1 1/8
3	8 1/4	6 5/8	3/4	8	4 3/4	3 1/4	1 13/16	1 13/16	1 1/4
3 1/2	9	7 1/4	7/8	8	5 1/4	3 3/8	1 15/16	1 15/16	1 3/8
4	10 3/4	8 1/2	7/8	8	5 1/2	4	2 1/8	2 1/8	1 1/2
5	13	10 1/2	1	8	6 1/4	4 1/2	2 3/8	2 3/8	1 3/4
6	14	11 1/2	1	12	6 1/2	4 5/8	2 5/8	2 5/8	1 7/8
8	16 1/2	13 3/4	1 1/8	12	7 1/2	5 1/4	3	3	2 3/16
10	20	17	1 1/4	16	8 1/4	6	3 3/8	4 3/8	2 1/2
12	22	19 1/4	1 1/4	20	8 1/2	6 1/8	3 5/8	4 5/8	2 5/8
14	23 3/4	20 3/4	1 3/8	20	9	6 1/2	3 11/16	5	2 3/4
16	27	23 3/4	1 1/2	20	9 3/4	7	4 3/16	5 1/2	3
18	29 1/4	25 3/4	1 5/8	20	10 1/2	7 1/4	4 5/8	6	3 1/4
20	32	28 1/2	1 5/8	24	11 1/4	7 1/2	5	6 1/2	3 1/2
22	34 1/4	30 5/8	1 3/4	24	12	7 3/4	5 1/4		3 3/4
24	37	33	1 7/8	24	12 3/4	8	5 1/2	7 1/4	4
26	40	36	1 7/8	28	13 1/4	8 3/4	8 3/4		4 1/4
30	44 1/2	40 1/4	2	28	14	9 3/4	9 3/4		4 1/2
34	49	44 1/2	2 1/4	28	15	10 5/8	10 5/8		4 3/4
36	51 3/4	47	2 1/2	28	15 3/4	11 1/8	11 1/8		4 7/8
42	58 3/4	53 3/4	2 3/4	28	17 1/2	12 3/4			5 1/2

[1] The 1/16"raised face is included in the"Length thru Hub"dimension of the Weld Neck (WN), Slip On (SO), Threaded, and Socket Weld (SW) flange, and also in the"Thickness"dimension of the Blind flange.

Table 7.5 ANSI B16.5 Flange dimensions for 600 lb flanges.

Pressure Class	25		125				250				800
Material	ASTM A 126		ASTM A 126				ASTM A 126				ASTM A 126
Class	A		A	B			A	B			B
NPS (in)	4-36	42-96	1-12	1-12	14-24	30-48	1-12	1-12	14-24	30-48	2-12
Service Temp (°F)	Maximum Non-Shock Pressure (psig)										
-20 to 150	45	25	175	200	150	150	400	500	300	300	800
200	40	25	165	190	135	115	370	460	280	250	
225	35	25	155	180	130	100	355	440	270	225	
250	30	25	150	175	125	85	340	415	260	200	
275	25	25	145	170	120	65	325	395	250	175	
300			140	165	110	50	310	375	240	150	
325			130	155	105		295	355	230	125	
375				145			265	315	210		
425				130				270			
450				125				250			

Limitations:

Class 25: Maximum pressure shall be limited to 25 psig when Class 25 cast iron flanges and flanged fittings are used for gaseous service. Tabulated pressure-temperature ratings above 25 psig for Class 25 cast iron flanges and flanged fittings are applicable for non-shock hydraulic service only.

Class 250: When used for liquid service the tabulated pressure-temperature ratings in sizes 14 in and larger are applicable to Class 250 flanges only and not to Class 250 fittings.

Class 800: The tabulated rating is not a steam rating and applies to non-shock hydraulic pressure only.

Table 7.6 ANSI B16.1 Cast Iron Flange Temperature and Pressure Ratings.

Pressure Class	150	300	400	600	900	1500	2500
PN Number	20	50	68	100	150	250	420
Service Temperature (°F)	Maximum Non-Shock Pressure (psig)						
100	285	740	990	1480	2200	3705	6170
200	260	675	900	1350	2025	3375	5625
300	230	655	875	1315	1970	3280	5470
400	200	635	845	1270	1900	3170	5280
500	170	600	800	1200	1795	2995	4990
600	140	550	730	1095	1640	2735	4560
650	125	535	715	1075	1610	2685	4475
700	110	535	710	1065	1600	2665	4440
750	95	505	670	1010	1510	2520	4200
800	NOT RECOMMENDED ABOVE 800						

Table 7.7 ANSI B16.5 Temperature and Pressure Ratings for ASTM A105 Carbon Steel Flange.

Most of what the average engineer deals with is at moderate temperatures and pressures, say up to 300°F and 200 psig. That is, it falls into the realm of Class 150, ASTM A105 flanges.

But sometimes we are faced with more demanding services, whether they are related to elevated temperatures, pressures, or corrosive fluids that require different metallurgy. In those cases, the reader is advised to obtain access to ANSI B16.5 (or B16.47) to determine the pressure rating at the particular temperature for that service. Do not rely on tables gleaned from generic sources, since these tend to generalize the term "carbon steel." Carbon steel covers a lot of territory, and the difference in pressure rating for Class 300 flanges at 100°F can swing from 620 to 750 psig. Some tables list pressures for carbon steel flanges up to 1000°F, but ANSI B16.5 suggests that carbon steels not be used above 800°F. At those elevated temperatures, higher grades of stainless or the Group 3 nickel-molybdenum alloys should be used.

Pressure classes of cast iron flanges are established by ANSI B16.1. These are rated at Classes 25,125, 250, and 800. The Class 125 flanges are always flat faced, and can be mated to 150 lb steel flanges. When this is done, there are two choices to prevent the cast iron flange from cracking due to the stress of bolting to a raised face flange:

1. Use low-strength bolting (less than 30 ksi minimum yield strength).
2. Machine off the raised face of the steel mating flange, and use a full face gasket as shown in Figure 7.1 with intermediate or high-strength bolts.

The same holds true for mating Class 250 cast iron flanges with 300 lb steel raised face flanges. Class 250 CI flanges have a raised face, and according to ANSI B16.5, both mating flange faces should be machined flat. In practice, sometimes you will see Class 250 CI raised face flanges installed against Class 300 steel flanges with ring gaskets as shown in Figure 7.2 and no machining. In these cases, the low-strength bolting should be used.

Flange Facings

The interface between a pair of flanges is certainly the most critical aspect of the flange, since this is what seals the fluid inside the system. Many configurations are available, but the most common for industrial services utilizes a gasket that compresses between the flange facing surfaces.

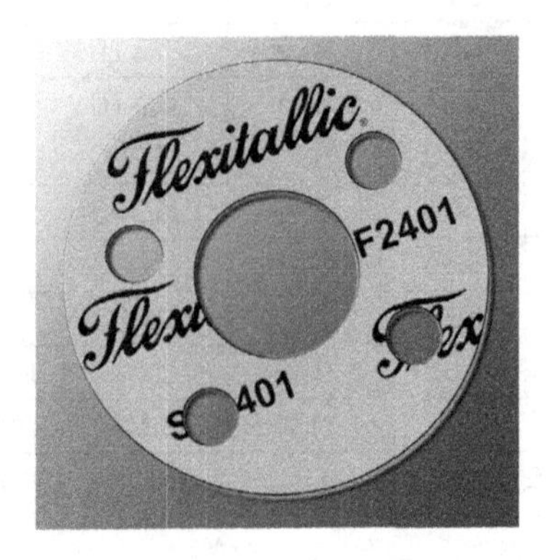

FIGURE 7.1 A full face gasket. *Flexitallic.*

FIGURE 7.2 A spiral-wound ring gasket. *Flexitallic.*

Engineers often collect the detritus of projects, and years ago my boss reached under his review table and produced a compressed fiber gasket that showed almost no gasket compression on one side. It seems a flange was leaking in a pulp mill, and the fitters were continually trying to seal the leak. In spite of the advice my boss gave them, they proceeded to remove the flanges and mill the raised face off the CS flanges, thinking that the increased surface of the sealing area would be the solution.

Obviously, the smaller sealing surface of a raised face flange provides a much higher compressive pressure when the bolts are torqued down on the flange. It is this compressive pressure that seals the fluid inside the flange. The solution to the mill's problem was to remove the flanges and align them better. It is imperative that flange faces are installed parallel to one another.

Flange faces may be of the following types:

- Plain straight face
- Plain face corrugated or scored
- Male-and-female
- Tongue-and-groove
- Raised face
- Ring Type Joint

Plain Face, Straight, Corrugated, and Scored

These flanges are machined with the sealing surfaces lying in the same plane. They may be smooth or have concentric grooves cut into the face to help grip the gasket to keep it from blowing out.

Plain face flanges may use either ring or full face gaskets.

Male-and-Female and Tongue-and-Groove

Male and female flange faces consist of a recess in one flange and a protruding portion in the other which engages the recess and compresses a gasket. This style has the advantage of capturing the gasket and restraining it from blowing out the sides. It has the disadvantage of requiring the joint to be sprung in order to remove a valve or fitting.

Similarly, the tongue-and-groove faces mate with a male and female connection. The difference between the tongue-and-groove and the male-and-female facings is the

FIGURE 7.3 A ring-type joint gasket. *Flexitallic.*

tongue-and-groove are narrow rings whose ID is larger than the pipe bore. The ID of the male portion of a male-and-female facing is the same ID as the pipe bore.

These styles are most often used in hydraulic applications.

Raised Face

The raised face flange has a ring on the face that compresses the gasket. An advantage to this style is that the lines need not be sprung to remove a gasket, valve, or fitting.

Ring Type Joint (RTJ)

This style is used in the oilfield with metallic ring-type gaskets as shown in Figure 7.3.

Types of Flanges

Welding Neck (WN) Flange

The welding neck flange (or more commonly "weld neck" flange) is illustrated in Figure 7.4. Weld neck flanges are attached to the adjoining pipe with a circumferential butt weld. Because the ID of heavy schedule pipe may be smaller than the bore of a weld neck flange, it is important to ensure that the bore of the flange matches the ID of the adjoining pipe. If a shoulder is present inside the flange due to the bore being smaller than the pipe, turbulence could result from high velocities. If the velocities were very high, erosion could result, but this does not seem to be a common problem.

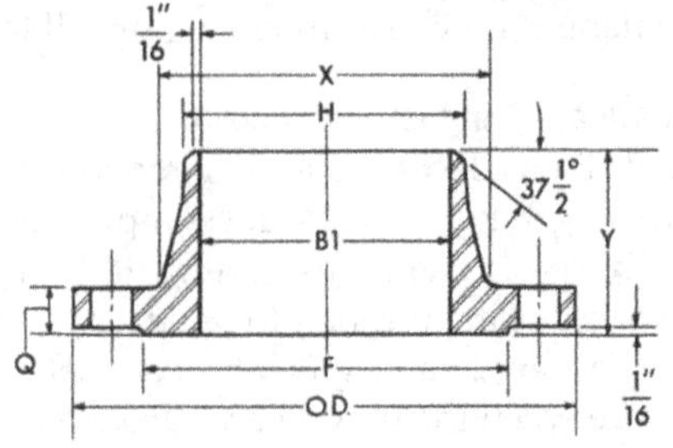

FIGURE 7.4 A weld neck flange, 150 lb pressure class.

The matching of the flange bore appears to fall under the category of "good practice," so most piping specifications require the flange to be bored to match the pipe ID.

Weld neck flanges have a tapered hub, which adds significantly to their strength and rigidity, as well as to their weight. Weld neck flanges are suitable for severe services where temperature extremes or high pressures may be encountered.

Slip-On (SO) Flanges

The slip-on flange is shown in Figure 7.5. This flange is essentially a ring that is placed over the pipe end, with the flange face extending from the end of the pipe by enough distance to apply a weld bead on the inside diameter. The ODs are also welded on the back side of the flange.

Slip-on flanges have a lower material cost than weld neck flanges, and are more easily aligned. For this reason they find favor among some engineers and contractors, but they are not as strong as weld neck flanges. Due to their lower strength, they are only available in sizes up to 2½ in in 1500 lb. They are not available in 2500 lb. Most specifications limit the use of slip-on flanges to 300 lb.

Slip-on flanges may also be used as lap joint flanges if Type B or Type C stub ends are used. Sometimes the designer will encounter a configuration in which the slip-on flange is to be welded directly to an elbow. This presents a problem in that unless the hole through the flange is relieved (ground) on the inside radius of the elbow bend, the flange will not weld squarely to the axis of the ell. The flange may be ground and fitted, but this is a situation that is best avoided.

Socket Weld (SW) Flanges

Shown in Figure 7.6, socket weld flanges contain a shoulder on the inside of the flange that acts as a guide to set the depth at which the pipe is welded to the flange. They are fabricated by inserting the pipe end into the flange until it bottoms out against the shoulder, and then retracting the pipe 1/8 in before welding it in place. This practice was originally employed to reduce cracking due to thermal stresses in stainless steel superheaters, but over the years has become standard practice for the installation of all socket weld flanges.

Sometimes, an internal weld bead is applied to seal the annulus between the pipe OD and the socket ID. This also imparts additional strength to the flange assembly, but this practice is not in common use for general industrial applications. One can appreciate the value of the internal weld for pharmaceutical, food and beverage, and some chemical applications, although it should be noted that the clearance available for performing such a weld on small diameter pipes may limit the ability to achieve it.

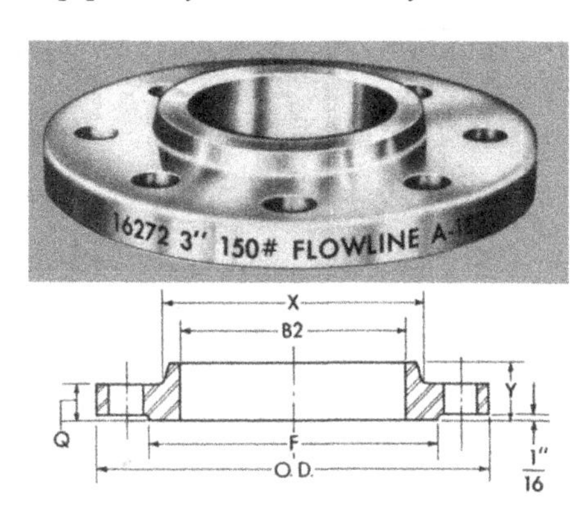

FIGURE 7.5 A slip-on flange, 150 lb pressure class.

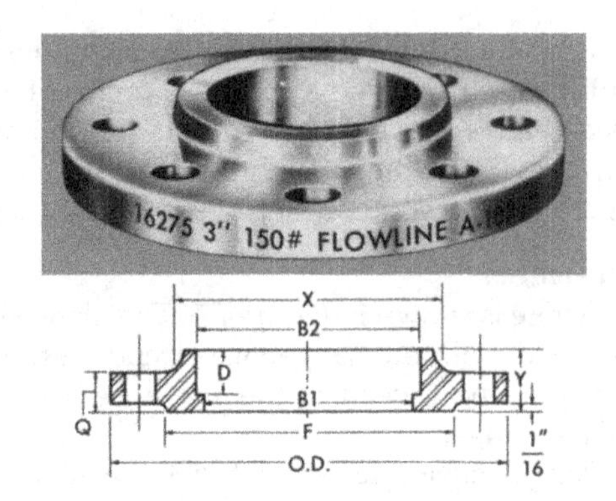

FIGURE 7.6 A socket-weld flange, 150 lb pressure class.

The number of bolts used to join any standard flange pair is always divisible by four. It is important to ensure that when piping is fabbed in a shop that it fit in the field. A common designation in pipe spool drawings is "2HU," which means to the fabricator that there should be "two holes up" at the vertical quarter point of a flange. Another way of saying this on piping drawings is to note that "bolt holes straddle centerline." Because the number of holes is divisible by four, if any of the bolt holes of a flange straddle the centerline, then they all do. If your spool was simply a straight length of pipe with a flange on each end, you would still want the flanges to have the same orientation with respect to each other in order to ensure that there was consistency among the spools. This will assure fit-up in the field.

Lap Joint (LJ) or Van Stone Flanges

Lap joint flanges use a stub end that is welded to the pipe. A ring flange fits loosely around the stub end, permitting easy flange alignment and joint disassembly. This obviates the need to provide careful alignment of the bolt holes. A lap joint flange is shown in Figure 7.7. "Lap joint" and "Van Stone" are interchangeable terms.

The stub ends are available in three styles (Types A, B, and C) and two lengths (Short and Long). See Figure 7.8.

The short length stub ends are manufactured in accordance with MSS SP-43 and are available in all three types. Long length stub ends are manufactured in accordance with ANSI B16.9 and are only available for Types A and B.

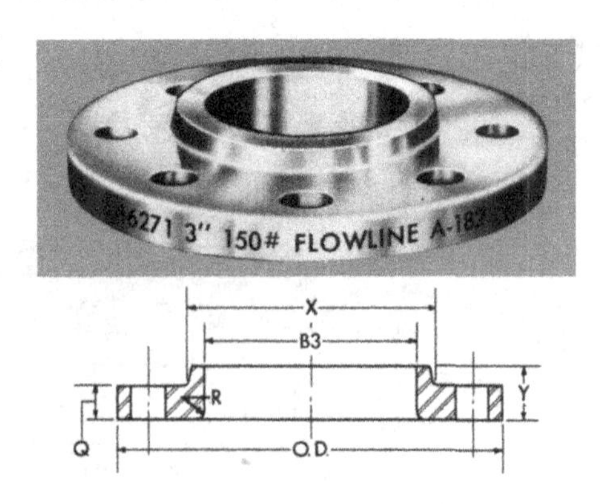

FIGURE 7.7 A lap-joint flange, 150 lb pressure class.

FIGURE 7.8 Stub ends for lap joint flanges.

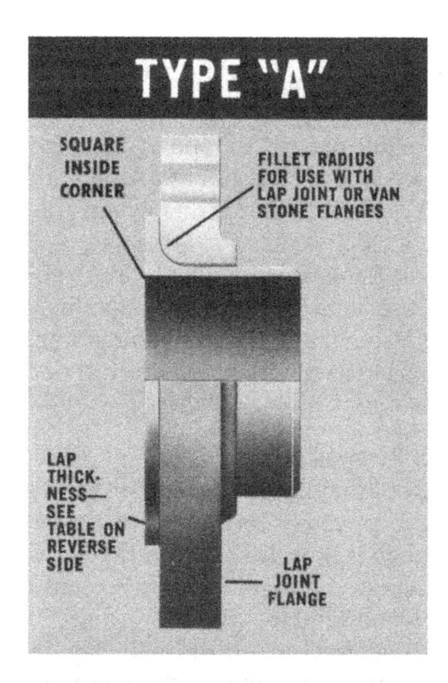

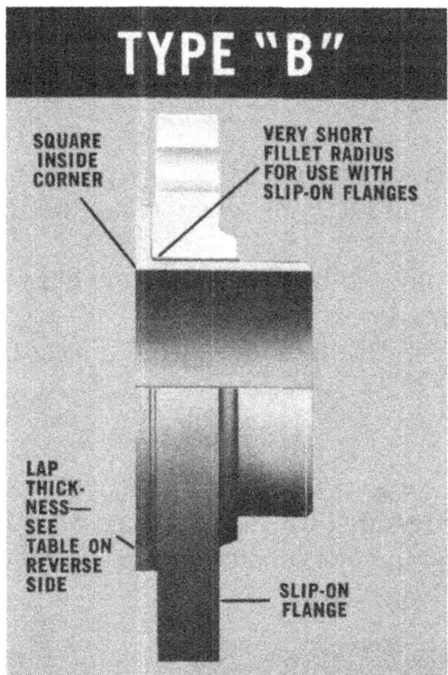

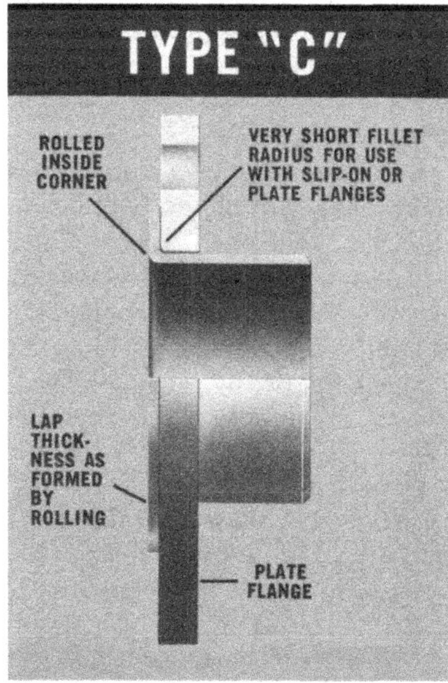

Type A stub ends have a generous fillet radius on the OD, and must be used with lap joint flanges. Slip-on or plate flanges cannot be used with Type A stub ends because the fillet radius of the stub end will prevent the slip-on or plate from laying flat against the lip of the stub end. The Short length is commercially available in Schedule 40 in sizes up to 24 inch. The Long length is commercially available in Schedule 80 up to 24 in, Schedule 160 up to 12 in, and XXS up to 8 in.

Type B stub ends have a shorter fillet radius than Type A stub ends on the OD, and are used with slip-on flanges. Type B stub ends are commercially available only up to Schedule 40 in sizes up to 24 in.

Type C stub ends have the same fillet radius on the OD as Type B stub ends, and are used with slip-on or plate flanges. Plate flanges do not offer much strength, and so Type C stub ends are relegated to low pressure applications, and are commercially available only in Schedules 5 and 10. Type C stub ends also have a rounded inside edge, whereas the other two stub end types have squared inside edges.

The lip of the stub end forms what is essentially a raised face that seals against the gasket.

Lap joint flanges are available in up to 2500 lb pressure class. In addition to the ability to ease alignment and disassembly, another advantage of these flanges is the low cost of using them in corrosion resistant pipe construction. Low-cost carbon steel flanges may be used with stainless steel or other high alloy hubs to reduce the cost of the flange assembly.

Threaded or Screwed Flanges

These flanges are suitable only for low-pressure systems in which there are no thermal cycles that could cause the threads to loosen. See Figure 7.9.

The flange is supplied with a tapered internal thread that the pipe screws into. Threaded flanges find applications in areas in which hot work (welding, burning, and grinding) is undesirable.

As with other threaded fittings, they are sometimes backwelded (seal welded) to prevent leaking. This is obviously only to be used where disassembly is not required. Furthermore, no credit is taken for increased strength whenever a threaded fitting is backwelded.

Blind Flanges

Blind flanges are used whenever a line must be capped off at a flange. It is good practice to install blind flanges at the ends of headers or at locations where future tie-ins are anticipated.

Blind flanges are also used extensively for manways, in which case a davit is recommended for ease in handling the unbolted manway cover.

A blind flange is illustrated in Figure 7.10. Blind flanges for sizes 26 through 48 in are specified by ANSI B16.47.

Orifice Flanges

A special type of flange is the orifice flange, which always occurs as a matched pair of flanges. These flanges are used to measure the pressure drop across a fixed orifice. The orifice is specially sized for the expected flow and fluid parameters and is located at the center of a plate which is inserted between the flange faces.

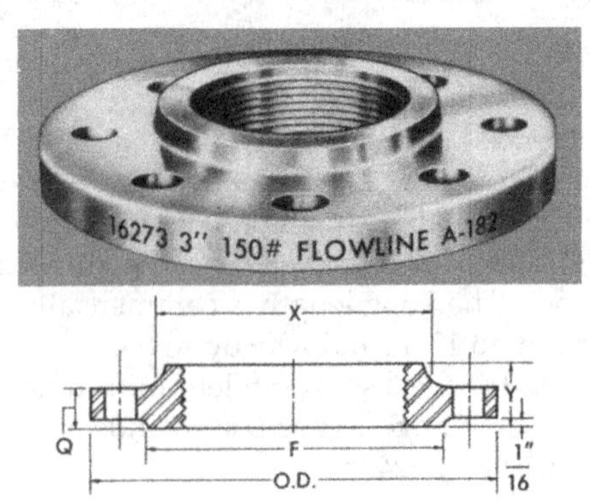

Figure 7.9 A threaded flange, 150 lb pressure class.

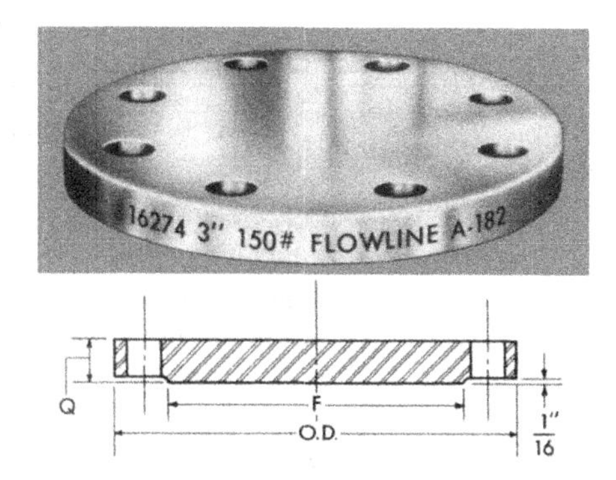

Figure 7.10 A blind flange, 150 lb pressure class.

These are never less than 300 lb flanges, with the increase in flange thickness provided to accommodate a radial hole that runs from the flange OD to the flange bore. The inside edge of the hole must be square to the pipe bore and free of burrs. The outer edge of the hole is threaded to permit a tubing bushing to be attached. A manometer or pressure differential device can then measure the difference in pressure across the orifice and the flow can then be calculated.

Jack screws are also provided to assist in spreading the flanges so that the orifice plate may be more easily removed.

This is a very simple and inexpensive way to measure fluid flow, but certain geometries must be observed in order to attain accurate measurements. The locations of the orifice flange must always remain flooded in liquid service (a vertical drop of the fluid through the orifice flange would be incorrect) and typically the orifice plate manufacturer requires straight pipe runs of at least 5 diameters upstream and 10 diameters downstream in order to minimize turbulent flow and achieve reproducible measurements.

API Flanges

API flanges are used primarily in the oil fields where pressures can be quite high. The best strategy in dealing with these flanges is to avoid altogether any effort to mix and match them with ANSI flanges.

> A change order was processed during the installation of some basket strainers on a cooling water system. The contractor had pre-fabbed the connecting spools with ANSI flanges, but when the strainers arrived from the vendor they were supplied with API flanges. None of the bolt holes lined up. This predated the discontinuation of API 605 (Large Diameter Carbon Steel Flanges).
>
> It turned out that a single mention of API appeared in the reference specs for the basket strainer specification. The Owner tried to no avail to recover the cost of refitting the API companion flanges on the adjoining pipe. Even though the vendor's approval drawing submittal did not make clear that they planned on using the API flanges, the Owner had to admit that the spec allowed API flanges. This was a case of the spec writer getting carried away and trying to cover every contingency, which is not always good practice.

See Table 7.8 for a summary of API flanges.

API Spec	Type	Gasketing	Pressures	Face
605	Discontinued			
6A	6B Weld Neck or Segmented	R (Oval or Octagonal)	Up to 10,000 psi	May or may not have raised face
		RX (Oval or Octagonal) Interchangeable with R gaskets	Up to 15,000 psi	
	6BX Weld Neck or Segmented	BX (Octagonal)	Up to 20,000 psi	Always have raised face
17D	SS Weld Neck or Blind Flange	BX	Up to 10,000 psi	May or may not have raised face
	SV Swivel (similar to Lap Joint)	BX	Up to 10,000 psi	May or may not have raised face

Table 7.8 API Flanges.

Studding Outlets

These are essentially metal blocks contoured to fit onto tank and vessel shells, heads and bottoms. The contoured side is welded to the vessel and the other end is provided with a flange facing and holes tapped for the flange bolts.

SAE Flanges

SAE 4-bolt flanges are often used on high pressure services like hydraulics or high pressure water systems. SAE Code 61 is for pressure ratings between 500 and 5000 psi in sizes from ½ to 5 in diameter. SAE Code 62 are rated for 6000 psi for all sizes (½ to 2 in).

Both styles of flanges are available in "captive" or "split" styles. The captive flange fits over the hydraulic tube and the split flange is in two parts that clamp the tube on either side. It is important to note that Code 61 and Code 62 are dimensionally similar (see Table 7.9), and so great care must be exercised to prevent using a flange rated for a lower pressure on a high-pressure application. Adding further to the confusion is the fact that at least one manufacturer of hydraulic equipment has developed its own standard for four bolt flanges that is also dimensionally similar to Code 61 and Code 62.

Dielectric Connections

Stray electrical currents can accelerate corrosion, so it is desirable to electrically isolate underground piping using dielectric connections at the flanges. See Figure 7.11. A dielectric sleeve covers the bolts, and dielectric washers electrically insulate the nuts from the flange surfaces. The gasket separates the flange faces eliminating metal-to-metal contact between the flanges. Thus isolated, the underground portion of the pipe is usually coated and wrapped to further shield it from contact with earth. The next step in protection is to provide a sacrificial anode that is bonded to the pipe electrically. The anode corrodes instead of the pipe. This is called "cathodic protection." Zinc is commonly used as the sacrificial anode, with the pipe acting as the cathode.

Gaskets

Full face gaskets (Figure 7.1) extend to the outer diameter of the flange hub. In addition to the hole for the pipe bore, the bolt holes are also cut into the gasket. These are used

Nominal Pipe Size	SAE CODE 61						SAE CODE 62					
	Flange Length	Flange Width	Flange Thickness	C-to-C Long Dim	C-to-C Short Dim	Pressure Rating	Flange Length	Flange Width	Flange Thickness	C-to-C Long Dim	C-to-C Short Dim	Pressure Rating
in	in	in	in	in	in	psi	in	in	in	in	in	psi
1/2	2.12	1.188	0.265	1.500	0.688	5000	2.20	1.25	0.305	1.594	0.718	6000
3/4	2.56	1.500	0.265	1.875	0.876	5000	2.79	1.625	0.345	2	0.938	6000
1	2.75	1.750	0.315	2.062	1.030	5000	3.18	1.875	0.375	2.25	1.094	6000
1 1/4	3.12	2.000	0.315	2.312	1.188	4000	3.74	2.125	0.405	2.625	1.25	6000
1 1/2	3.69	2.375	0.315	2.750	1.406	3000	4.45	2.5	0.495	3.125	1.438	6000
2	4.00	2.812	0.375	3.062	1.688	3000	5.23	3.125	0.495	3.812	1.75	6000
2 1/2	4.50	4.280	0.375	3.500	2.000	2500						
3	5.31	4.000	0.375	4.188	2.438	2000						
3 1/2	5.98	4.500	0.442	4.750	2.750	500						
4	6.37	5.000	0.442	5.125	3.062	500						
5	7.24	6.000	0.442	6.000	3.624	500						

TABLE 7.9 SAE four bolt flanges. Note how similar the dimensions are between the two classes for a single nominal pipe size.

primarily for cast iron flanges so that there is no moment exerted around the gasket (as would happen with a ring gasket) that could snap the brittle flange.

Ring gaskets (Figure 7.2) are used with raised face flanges, and are designed so that the ID of the gasket matches the bore of the pipe and the OD of the gasket fits into the circle described by the inner edges of the bolts. This locates the gasket radially in the flange interface and also permits easier replacement of the gasket.

Another type of gasket is the "ring-type" (Figure 7.3), not to be confused with the ring gasket above. Ring-type gaskets are made for ring-type joints, most commonly used in oil fields with API 6A flanges. These gaskets have oval or octagonal cross sectional profiles.

Gaskets must be selected based on compatibility with the fluid, operating temperature and pressure, and performance (compressibility) of the material. Generally, the higher the force that may be applied to the gasket, the longer the gasket will last in service.

Note that asbestos is a common gasket material because it performs well at high temperatures and is compressible. The use of asbestos is to be avoided, since it is a carcinogen. Removal of old asbestos gaskets requires asbestos containment procedures.

FIGURE 7.11 A dielectric connection on a flange pair.

Gasket Material	Application	Maximum Temperature (°F)	(C)	Maximum Pressure (psi)	(bar)
Synthetic rubbers	Water, Air	250	121	60	4
Vegetable fiber	Oil	250	121	160	11
Synthetic rubbers with inserted cloth	Water, Air	250	121	500	34
Solid Teflon	Chemicals	500	260	300	20
Compressed Asbestos	Most applications	750	399	333	23
Carbon Steel	High pressure fluids	750	399	2 133	145
Stainless Steel	High pressure or corrosive fluids	1200	649	2 500	171
Spiral wound SS/ Teflon	Chemicals	500	260	500	34
Spiral wound CS/ Asbestos	Most applications	750	399	333	23
Spiral wound SS/ Asbestos	Corrosive	1200	649	208	14
Spiral wound SS/ Ceramic	Hot Gases	1900	1038	132	9

Table 7.10 Gasket Materials.

Sixty countries, including Australia and those countries in the European Union, have banned the use of asbestos in whole or in part.

Table 7.10 lists a variety of gasket materials and their suitable applications.

Bolting

Flanges may be joined with either hex head bolts or studs nutted on each end. The length of the bolt or stud should be limited so that only two threads protrude beyond the nut. This prevents exposed threads from corroding which permits easier removal of the nut for future disassembly.

A common chore for the pipefitters in one steel mill was to "rehearse" the bolts of flanges that were to be opened during an upcoming outage. The lines were in operation, so this involved removing the bolts one at a time, either with wrenches or, in case they were highly corroded, by burning them off with an oxy-fuel torch. New bolts or studs would then be installed with a liberal coating of anti-seize applied to the threads.

Available downtime is always limited during an outage. Rehearsing the bolts minimizes the time required to break open a flange, but the fluid service and pressure must be innocuous enough that slight leakage could be tolerated.

Cast iron flange bolting is given by ANSI B16.1. Steel flange bolting is listed in ANSI B16.5.

In general, the flange faces must be brought together in a parallel fashion. This is especially critical for corrugated metal gaskets to prevent the corrugations from being deformed unevenly as the bolts are tightened. The procedure for bolting up flanges is generally:

1. Visually inspect the flanges to ensure that the faces are flat, with no burrs or nicks.
2. Check the threads of the fasteners by ensuring that the nuts can be turned by hand onto the bolt or stud. Replace defective fasteners.
3. Lubricate bolt or stud threads, as well as the contact surfaces of the washers and nuts. Do not apply lubricant to flange or gasket faces, and never use any liquid or metallic based lubricants on the gaskets, as these could deteriorate the gasket material. The lubricant must be approved by the gasket manufacturer and should be spread evenly and thinly. Hardened washers should be used under nuts.
4. Center the gasket on the flange. For a raised face flange, use several of the bolts to help align the gasket. The gasket OD will closely match the ID of the circle formed by the bolts.
5. Draw the flanges together evenly using a star bolting pattern. The bolts should be torqued to 30 percent of the final torque value. The idea is to get the flanges to compress the gasket as evenly as possible. Torque wrenches should be used for this and the wrenches need to be calibrated periodically.
6. Following the same star pattern, the bolts should be torqued to 60 percent of their final torque value.
7. Using the same star pattern, torque the bolts to the final torque value.
8. Complete one more pass at torquing the bolts using an adjacent bolt-to-bolt sequence.
9. Re-torque the bolts 12 to 24 hours after initial installation when possible to minimize any creep in the joint.

Bolt Torques

It is the internal stress of the bolt that keeps the gasket in compression and provides a leak-tight joint. Most gasket manufacturers recommend a stress of 45,000 psi (310 N/mm^2). Flange bolt torques are given in Table 7.11.

Other Fittings

Elbows

Elbows ("ells") are used more than any other fitting. They are used to maneuver around obstacles and they introduce flexibility into piping systems.

Just like flanges, elbows and other steel fittings can be obtained with ends suitable for butt-welding, socket-welding, or threading, and cast and ductile iron elbows are available with flanged ends.

Butt-Weld Elbows

Forged steel buttwelding elbows are manufactured in accordance with ANSI B16.9. Long radius elbows are preferred to short radius elbows due to their lower pressure drops, which when referring to fittings are called "minor losses."

Nominal Bolt Dia	Number of Threads per Inch	Thread Root Dia	Thread Root Area	STRESS					
				30,000 psi		45,000 psi		60,000 psi	
				Torque	Clamping Force	Torque	Clamping Force	Torque	Clamping Force
(in)		(in)	(sq. in.)	(ft lb)	(lb/bolt)	(ft lb)	(lb/bolt)	(ft lb)	(lb/bolt)
1/4	20	0.185	0.027	4	810	6	1,215	8	1,620
5/16	18	0.24	0.045	8	1,350	12	2,025	16	2,700
3/8	16	0.294	0.068	12	2,040	18	3,060	24	4,080
7/16	14	0.345	0.093	20	2,790	30	4,180	40	5,580
1/2	13	0.4	0.126	30	3,780	45	5,670	60	7,560
9/16	12	0.454	0.162	45	4,860	68	7,290	90	9,720
5/8	11	0.507	0.202	60	6,060	90	9,090	120	12,120
3/4	10	0.62	0.302	100	9,060	150	12,590	200	18,120
7/8	9	0.731	0.419	160	12,570	240	18,855	320	25,140
1	8	0.838	0.551	245	16,530	368	24,795	490	33,060
1 1/8	8	0.963	0.728	355	21,840	533	32,760	710	43,680
1 1/4	8	1.088	0.929	500	27,870	750	41,805	1,000	55,740
1 3/8	8	1.213	1.155	680	34,650	1,020	51,975	1,360	69,300
1 1/2	8	1.338	1.405	800	42,150	1,200	63,225	1,600	84,300
1 5/8	8	1.463	1.68	1,100	50,400	1,650	75,600	2,200	100,800
1 3/4	8	1.588	1.98	1,500	59,400	2,250	89,100	3,000	118,800
1 7/8	8	1.713	2.304	2,000	69,120	3,000	103,680	4,000	138,240
2	8	1.838	2.652	2,200	79,560	3,300	119,340	4,400	159,120
2 1/4	8	2.088	3.423	3,180	102,690	4,770	154,035	6,360	205,380
2 1/2	8	2.338	4.292	4,400	128,760	6,600	193,140	8,800	257,520
2 3/4	8	2.58	5.259	5,920	157,770	8,880	236,655	11,840	315,540
3	8	2.838	6.324	7,720	189,720	11,580	284,580	15,440	379,440

TABLE 7.11 Bolting torques for alloy steel bolts. *Garlock Sealing Technologies.*

The radius of the elbow is described in terms of the nominal pipe diameter. See Figure 7.12. This is the "take-out" of the fitting. Short radius elbows have a takeout dimension equal to 1D where D is the nominal diameter of the pipe (the pipe size). The single exception to this rule is that 42 in SR ells have a takeout of 48 in.

Short radius (SR) elbows are generally only used where space constraints prevent the use of long radius elbows. They are only available in 90° bend configurations, but could be cut on an angle for use as a 45° or other angled SR elbow.

FIGURE 7.12 Long and short radius elbows. "D" equals the nominal diameter.

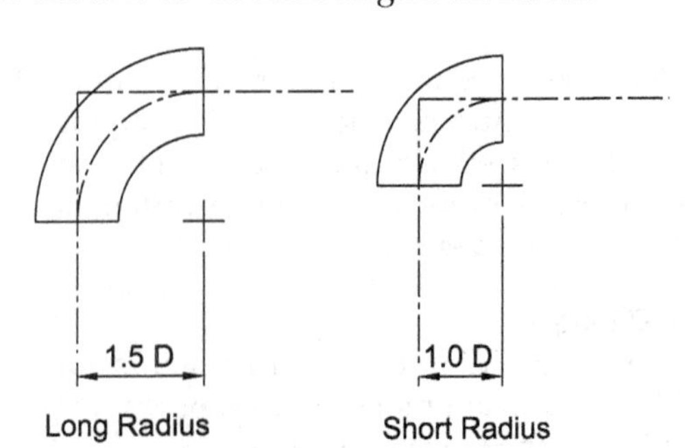

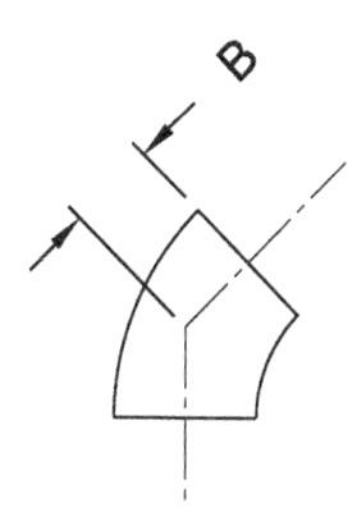

Figure 7.13
Forty-five degree elbow. "B" is the take-out dimension.

Long radius (LR) elbows have a takeout equal to 1½ D. Thus, the takeout of a 2 in LR ell is 3 in; the takeout of a 36 in ell is 54 in. Long radius ells are available in 90° or 45° configurations. Sometimes one may encounter data for r/D ratios of other than 1.0 or 1.5. Tables for minor losses sometimes indicate that ratios of two or three are available, but these are not common in general industrial piping.

For pneumatic conveying applications, 5-D bends are often used to allow the material to flow smoothly without an abrupt change in direction. In plants that convey abrasive materials pneumatically, the outside curves of the bends are sometimes reinforced with a structural channel. The legs of the channel are welded to the outside curvature of the pipe, and the space between the channel and the pipe is grouted. This reinforcement reduces vibration and noise transmission.

Forty-five degree elbows (Figure 7.13) are commonly used. These have the same radius of curvature as LR elbows. The take-out dimension for 45° ells between 4 and 20 in inclusive is equal to 0.625 times the nominal pipe size.

Rolling Offsets

A common configuration of 45s and other ells is the "rolling offset." Consider the box shown in Figure 7.14 in which an elbow pair is rolled in three dimensions to produce an exit pipe that remains parallel to the inlet, but is offset in two directions, the Y and Z directions. This forms a right triangle inside the box whose hypotenuse L is the centerline of the skewed pipe. The length of the hypotenuse is given by

$$L = \sqrt{X^2 + Y^2 + Z^2} \qquad \text{Equation 7.1}$$

If one needs to find the location of the intersection of the skewed pipe as it exits the box, then

$$X = (\tan\theta)(Y^2 + Z^2)^{1/2} \qquad \text{Equation 7.2}$$

Where θ is the angle of the elbow.

Socket-Weld Elbows

Forged steel socket-welded elbows (and other socket-welded fittings) are covered under ANSI B16.11. Socket welded fittings are available in pressure ratings of 3000, 6000, and 9000 psi. Their geometry is somewhat compact and the right angle formed by the fitting is more abrupt than with a short radius elbow.

Tees

Tees are used to form branch connections. If the size of the branch is the same size or one or two sizes smaller than the run pipe, then a tee or reducing tee is usually used, since the cutting of a fishmouth[2] can weaken the pipe and requires the application of the

[2] The welding of a smaller branch onto a larger run is also called a "nozzle weld."

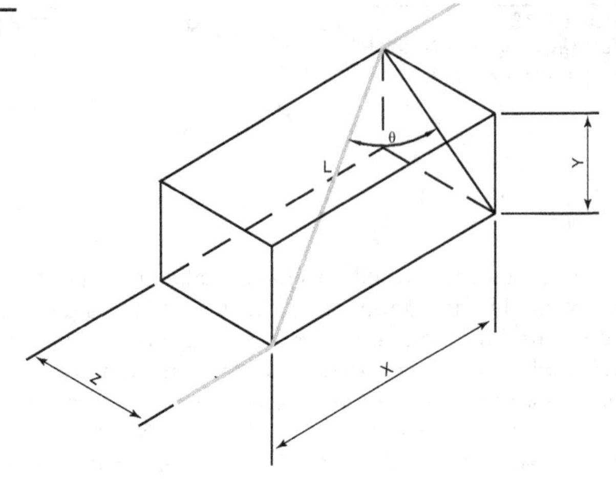

FIGURE 7.14 Rolling offset.

appropriate branch connection code calculations discussed in Chapter 4. Installation of a butt welding tee manufactured in accordance with ANSI B16.9 ensures that the joint will comply with the rated pressure of the adjoining pipe, without the need for further calculation. See Figure 7.15.

Tees are often used for cleanout access or for the installation of components like thermowells. The tee is installed where an elbow would otherwise be used and the thermowell is inserted into the run of the pipe at the end of the tee.

Cleanouts

While tees and laterals are often used for cleanouts in industrial settings, the term "cleanout" refers to fittings used to snake sanitary sewer lines.

Laterals

Not to be confused with wyes, laterals are special fittings that are essentially tees with the branch connection formed at 45°. See Figure 7.16.

Threaded Fittings

Forged steel threaded fittings are standardized in ANSI B16.11, which specifies pressure ratings, dimensions, tolerances, markings, and material requirements. They are available in pressure ratings of 2000, 3000, and 6000 psi.

FIGURE 7.15 A butt-weld tee fitting.

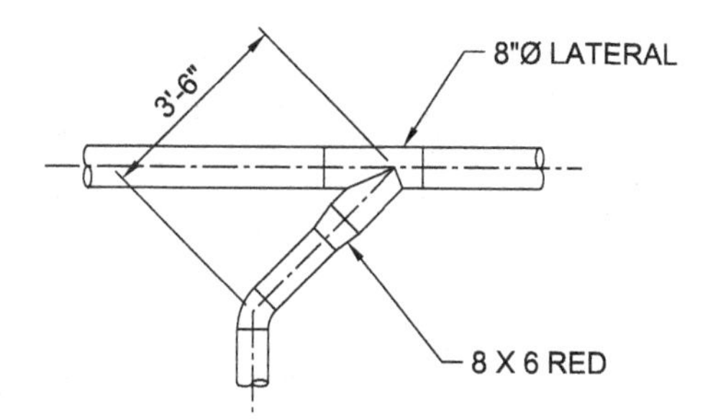

Figure 7.16 Lateral.

Malleable iron fittings are standardized in ANSI B16.3, which specifies pressure ratings, dimensions, tolerances, markings, material requirements, threading, and coatings. They are available in pressure classes 150 and 300. See Table 7.12.

Cast iron threaded fittings are available in pressure classes 125 and 250 and are covered under ANSI B16.4. See Table 7.13.

Threaded connections are commonly available in sizes up to 2 in. Although ANSI B1.20.1 specifies threads up to 24 in, sizes above 2 in are rarely threaded due to the difficulty in assembly and the chance of leakage.

Reducers

Reducers are used to change diameters between adjoining pipe sections. Whether or not the flow of fluid is from the large diameter to the small diameter or vice versa, the correct term is reducer, not "enlarger" or "increaser." Reducers are always designated such that the larger diameter precedes the smaller diameter. Reducers are available in two configurations: concentric and eccentric. Eccentric reducers are handy for establishing a constant BOP elevation on headers that must reduce. They are also used extensively at pump suctions to prevent gas bubbles from being sucked into the pump intake. In this case the eccentric reducer is installed with the flat on top of the run of pipe. Whenever

		Maximum Working Pressure (psig)			
		Class			
		150	300		
Temperature		All Sizes	1/4–1	1 1/4–2	2 1/2–3
deg F	deg C	in	in	in	in
–20 to 150	–29 to 66	175	2000	1500	1000
200	93	265	1785	1350	910
250	121	185	1575	1200	825
300	149	150	1360	1050	735
350	177	—	1150	900	650
400	204	—	935	750	560
450	232	—	725	600	475
500	260	—	510	450	385
550	288	—	300	300	300

Table 7.12 Malleable iron fitting pressure ratings per ANSI B16.3.

Maximum Working Pressure (psig)			
Temperature		Pressure Class	
deg F	deg C	125	250
–20 up to 150	–29 to 66	175	400
200	93	165	370
250	121	150	340
300	149	140	310
350	177	125	280
400	204	—	250

Table 7.13 Cast iron fitting pressure ratings per ANSI B16.4.

eccentric reducers are used, the designer must always specify where the flat side should be. "FOT" means "flat on top"; "FOB" means "flat on bottom."

Reducers may be fabricated by hand from a straight piece of pipe by cutting wedges out of one end of the pipe segment. This practice is uncommon, although it is sometimes performed as part of an apprenticeship test or in extreme cases when a commercial fitting is not readily available. Concentric and eccentric reducers are available in a wide range of sizes and schedules, as indicated in Table 7.14. Manufacture of butt-welding reducers is governed by ANSI B16.9.

Threaded reducers have a female thread on each end. These are available in forged steel and malleable iron. They are distinguished from bushings in that bushings have one male and one female thread.

Bushings

As noted above, bushings are a type of reducing fitting with both male and female ends. See Figure 7.17. They may be threaded or plain for use as a socket weld fitting. Further, the socket weld fitting is available with a female thread for transitioning between a socket weld connection and a threaded connection.

Bushings are available in cast iron, malleable iron, cast steel, brass, copper, and forged carbon and stainless steel.

The International Fuel Gas Code does not permit the use of cast iron bushings for flammable gas piping. Like most instances of the codes, no explanation is offered. Presumably, there is a suspicion or past evidence of these bushings being over-tightened and cracking. Because the bushing fits inside a larger diameter pipe, it would be difficult to see a crack. This code specifically prohibits cast iron bushings from being used, but

Figure 7.17 SS threaded reducing bushing.

Nominal Pipe Size			Length	Avg Dia	Average Wall Thickness			
					Std	XS	Sch 160	XXS
in			in	in	in	in	in	in
0.75	x	0.375	1.5	0.8625	0.1020	0.1400	0.1095	0.1540
0.75	x	0.5	1.5	0.9450	0.1110	0.1505	0.2035	0.3010
1	x	0.375	2	0.9950	0.1120	0.1525	0.1250	0.1790
1	x	0.5	2	1.0775	0.1210	0.1630	0.2190	0.3260
1	x	0.75	2	1.1825	0.1230	0.1665	0.2345	0.3330
1.25	x	0.5	2	1.2500	0.1245	0.1690	0.2190	0.3380
1.25	x	0.75	2	1.3550	0.1265	0.1725	0.2345	0.3450
1.25	x	1	2	1.4875	0.1365	0.1850	0.2500	0.3700
1.5	x	0.5	2.5	1.3700	0.1270	0.1735	0.2345	0.3470
1.5	x	0.75	2.5	1.4750	0.1290	0.1770	0.2500	0.3540
1.5	x	1	2.5	1.6075	0.1390	0.1895	0.2655	0.3790
1.5	x	1.25	2.5	1.7800	0.1425	0.1955	0.2655	0.3910
2	x	0.75	3	1.7125	0.1335	0.1860	0.2815	0.3720
2	x	1	3	1.8450	0.1435	0.1985	0.2970	0.3970
2	x	1.25	3	2.0175	0.1470	0.2045	0.2970	0.4090
2	x	1.5	3	2.1375	0.1495	0.2090	0.3125	0.4180
2.5	x	1	3.5	2.0950	0.1680	0.2275	0.3125	0.4550
2.5	x	1.25	3.5	2.2675	0.1715	0.2335	0.3125	0.4670
2.5	x	1.5	3.5	2.3875	0.1740	0.2380	0.3280	0.4760
2.5	x	2	3.5	2.6250	0.1785	0.2470	0.3595	0.4940
3	x	1.25	3.5	2.5800	0.1780	0.2455	0.3440	0.4910
3	x	1.5	3.5	2.7000	0.1805	0.2500	0.3595	0.5000
3	x	2	3.5	2.9375	0.1850	0.2590	0.3910	0.5180
3	x	2.5	3.5	3.1875	0.2095	0.2880	0.4065	0.5760
3.5	x	1.25	4	2.8300	0.1830	0.2545	0.1250	0.1910
3.5	x	1.5	4	2.9500	0.1855	0.2590	0.1405	0.2000
3.5	x	2	4	3.1875	0.1900	0.2680	0.1720	0.2180
3.5	x	2.5	4	3.4375	0.2145	0.2970	0.1875	0.2760
3.5	x	3	4	3.7500	0.2210	0.3090	0.2190	0.3000
4	x	1.5	4	3.2000	0.1910	0.2685	0.4060	0.5370
4	x	2	4	3.4375	0.1955	0.2775	0.4375	0.5550
4	x	2.5	4	3.6875	0.2200	0.3065	0.4530	0.6130
4	x	3	4	4.0000	0.2265	0.3185	0.4845	0.6370

Table 7.14 Commercially-available reducer sizes and take-outs. The average wall thicknesses and diameters are provided to aid in modeling reducers in some stress analysis software packages.

(continued on next page)

Nominal Pipe Size			Length	Avg Dia	Average Wall Thickness			
					Std	XS	Sch 160	XXS
in			in	in	in	in	in	in
4	x	3.5	4	4.2500	0.2315	0.3275	0.2655	0.3370
5	x	2	5	3.9690	0.2060	0.2965	0.4845	0.5930
5	x	2.5	5	4.2190	0.2305	0.3255	0.5000	0.6510
5	x	3	5	4.5315	0.2370	0.3375	0.5315	0.6750
5	x	3.5	5	4.7815	0.2420	0.3465	0.3125	0.3750
5	x	4	5	5.0315	0.2475	0.3560	0.5780	0.7120
6	x	2.5	5.5	4.7500	0.2415	0.3540	0.5470	0.7080
6	x	3	5.5	5.0625	0.2480	0.3660	0.5785	0.7320
6	x	3.5	5.5	5.3125	0.2530	0.3750	0.3595	0.4320
6	x	4	5.5	5.5625	0.2585	0.3845	0.6250	0.7690
6	x	5	5.5	6.0940	0.2690	0.4035	0.6720	0.8070
8	x	3.5	6	6.3125	0.2740	0.4090	0.4530	0.4375
8	x	4	6	6.5625	0.2795	0.4185	0.7185	0.7745
8	x	5	6	7.0940	0.2900	0.4375	0.7655	0.8125
8	x	6	6	7.6250	0.3010	0.4660	0.8125	0.8695
10	x	4	7	7.6250	0.3010	0.4185	0.8280	0.8370
10	x	5	7	8.1565	0.3115	0.4375	0.8750	0.8750
10	x	6	7	8.6875	0.3225	0.4660	0.9220	0.9320
10	x	8	7	9.6875	0.3435	0.5000	1.0155	0.9375
12	x	5	8	9.1565	0.3165	0.4375	0.9685	0.8750
12	x	6	8	9.6875	0.3275	0.4660	1.0155	0.9320
12	x	8	8	10.6875	0.3485	0.5000	1.1090	0.9375
12	x	10	8	11.7500	0.3700	0.5000	1.2185	1.0000
14	x	6	13	10.3125	0.3275	0.4660	1.0625	0.4320
14	x	8	13	11.3125	0.3485	0.5000	1.1560	0.4375
14	x	10	13	12.3750	0.3700	0.5000	1.2655	0.5000
14	x	12	13	13.3750	0.3750	0.5000	1.3590	0.5000
16	x	8	14	12.3125	0.3485	0.5000	1.2500	0.4375
16	x	10	14	13.3750	0.3700	0.5000	1.3595	0.5000
16	x	12	14	14.3750	0.3750	0.5000	1.4530	0.5000
16	x	14	14	15.0000	0.3750	0.5000	1.5000	—
18	x	10	15	14.3750	0.3700	0.5000	1.4530	0.5000
18	x	12	15	15.3750	0.3750	0.5000	1.5465	0.5000

Table 7.14 *(continued).*

Nominal Pipe Size			Length	Avg Dia	Average Wall Thickness			
					Std	XS	Sch 160	XXS
in			in	in	in	in	in	in
18	x	14	15	16.0000	0.3750	0.5000	1.5935	—
18	x	16	15	17.0000	0.3750	0.5000	1.6875	—
20	x	12	20	16.3750	0.3750	0.5000	1.6405	0.5000
20	x	14	20	17.0000	0.3750	0.5000	1.6875	—
20	x	16	20	18.0000	0.3750	0.5000	1.7815	—
20	x	18	20	19.0000	0.3750	0.5000	1.8750	—
22	x	14	20	18.0000	0.3750	0.5000	1.7655	—
22	x	16	20	19.0000	0.3750	0.5000	1.8595	—
22	x	18	20	20.0000	0.3750	0.5000	1.9530	—
22	x	20	20	21.0000	0.3750	0.5000	2.0470	—
24	x	14	20	19.0000	0.3750	0.5000	1.8750	—
24	x	16	20	20.0000	0.3750	0.5000	1.9690	—
24	x	18	20	21.0000	0.3750	0.5000	2.0625	—
24	x	20	20	22.0000	0.3750	0.5000	2.1565	—
26	x	18	24	22.0000	0.3750	0.5000	0.8905	—
26	x	20	24	23.0000	0.3750	0.5000	0.9845	—
26	x	22	24	24.0000	0.3750	0.5000	1.0625	—
26	x	24	24	25.0000	0.3750	0.5000	1.1720	—
28	x	18	24	23.0000	0.3750	0.5000	0.8905	—
28	x	20	24	24.0000	0.3750	0.5000	0.9845	—
28	x	24	24	26.0000	0.3750	0.5000	1.1720	—
28	x	26	24	27.0000	0.3750	0.5000	—	—
30	x	20	24	25.0000	0.3750	0.5000	0.9845	—
30	x	24	24	27.0000	0.3750	0.5000	1.1720	—
30	x	26	24	28.0000	0.3750	0.5000	—	—
30	x	28	24	29.0000	0.3750	0.5000	—	—
32	x	24	24	28.0000	0.3750	0.5000	1.1720	—
32	x	26	24	29.0000	0.3750	0.5000	—	—
32	x	28	24	30.0000	0.3750	0.5000	—	—
32	x	30	24	31.0000	0.3750	0.5000	—	—
34	x	24	24	29.0000	0.3750	0.5000	1.1720	—
34	x	26	24	30.0000	0.3750	0.5000	—	—
34	x	30	24	32.0000	0.3750	0.5000	—	—

Table 7.14 *(continued).* *(continued on next page)*

Nominal Pipe Size			Length	Avg Dia	Average Wall Thickness			
					Std	XS	Sch 160	XXS
in			in	in	in	in	in	in
34	x	32	24	33.0000	0.3750	0.5000	—	—
36	x	24	24	30.0000	0.3750	0.5000	1.1720	—
36	x	26	24	31.0000	0.3750	0.5000	—	—
36	x	30	24	33.0000	0.3750	0.5000	—	—
36	x	32	24	34.0000	0.3750	0.5000	—	—
36	x	34	24	35.0000	0.3750	0.5000	—	—
42	x	24	24	33.0000	0.3750	0.5000	1.1720	—
42	x	26	24	34.0000	0.3750	0.5000	—	—
42	x	30	24	36.0000	0.3750	0.5000	—	—
42	x	32	24	37.0000	0.3750	0.5000	—	—
42	x	34	24	38.0000	0.3750	0.5000	—	—
42	x	36	24	39.0000	0.3750	0.5000	—	—

Table 7.14 *(continued).*

one could argue that malleable iron or cast steel bushings, being less brittle than cast iron, would be acceptable[3].

Nipples

Short segments of threaded pipe are called nipples. They are called out by their diameter and their overall length. Sometimes the make-up (assembled) length required is very short. In such cases the threads from both ends meet in the middle and the nipple is referred to as "close-by-close." A close-by-close nipple is shown in Figure 7.18.

Unions

If you were to construct a closed loop of threaded pipe, as the final joint was tightened, the other end would be loosened. Unions are used to make such connections. They are also used to enable a pipe to be separated for removal of in-line components.

Unions form a metal-to-metal seal. See Figure 7.19. These surfaces must never be coated with any sealing compound or tape. The seal derives from the precision of the mating profiles and the smooth surface finish.

The end connections may be socket welded or threaded.

Crosses

Similar to tees, crosses are available to form branch connections on both sides of the run. See Figure 7.20. They are available with butt weld, socket weld, and threaded end connections.

[3] Unfortunately, the bins in hardware or "big box" stores are not always particularly well labeled, and they certainly do not go to the trouble of specifying that the fitting is ANSI B16.4 Cast Iron vs. ANSI B16.3 Malleable Iron. More likely, the bins at the hardware store will be labeled "black iron," and then it's anyone's guess as to what the material is. Imprecise terms such as "black iron" are to be avoided in engineering and construction. You might as well say it's "metal."

So the safer route is to use a reducer for fuel gas reducing fittings since the difference between MI, CI, and cast steel is not easily discerned.

FIGURE 7.18 SS close-by-close nipple.

FIGURE 7.19 A disassembled threaded union showing ground seat and metal-to-metal seal.

FIGURE 7.20 A threaded cross.

In practice, crosses are not often used. Pressure loss data are rare, but one could expect the flow of fluids to be preferentially straight through the cross.

> Fabricated crosses should also be avoided if possible. Even if not size-on-size, the piping codes do not address stress intensification factors for crosses. The SIFs must be resolved through finite element analysis.

U-Bends or 180° Returns

The ANSI B16.9 U-bend has the same take-out dimension as two long radius 90° elbows. ANSI B16.28 U-bends are available in extra-long radius (corresponding to two 90° elbows with r/D = 2) and short radius (corresponding to two 90° elbows with r/D = 1).

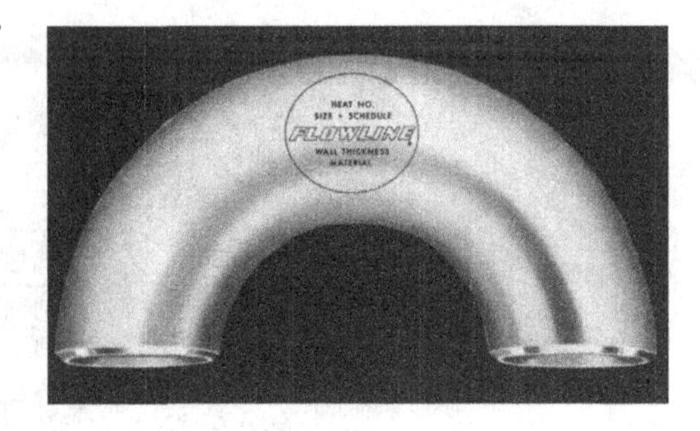

FIGURE 7.21 A butt-weld U-bend.

U-bends are most often used in heat exchange applications, where a grid of piping is required. See Figure 7.21.

Caps and Plugs

Caps and plugs are used to terminate and seal the end of a run of pipe. Caps are available as butt weld or threaded connections and plugs are available as threaded connections.

Bull Plugs and Swaged Nipples

These are manufactured in accordance with MSS-SP-95. Swaged nipples are male end reducing fittings and are available in sizes NPS 1/4 through NPS 12 with both concentric and eccentric patterns. Bull plugs are hollow or solid male closures available in sizes NPS 1/8 through NPS 12. These fittings are available with ends that are threaded, beveled, square cut, grooved, or any combination of these.

The applications of these fittings are identical to other plugs and reducers, but bull plugs and swages are used primarily in oil country piping, and do not find use in general industry.

Couplings and Half-Couplings

Couplings are used to join two male pipe ends together. Half-couplings are used to provide a female threaded branch connection by welding the unthreaded end to the adjoining pipe. This may be done in-line (coaxially) with a socket weld half-coupling, or it may be done by fillet welding the half-coupling to the OD of the header pipe. For critical applications, a reinforced branch connection fitting is preferred.

Couplings may be threaded, socket welded, or for larger bore piping, of the mechanical joint variety.

Integrally Reinforced Forged Branch Outlet Fittings

MSS SP-97 standardizes integrally reinforced forged branch connections for socket welding, threaded, and butt-welding ends. The term "OLET®" applied generically to this type of fitting is incorrect, since this suffix is a trademark of Bonney Forge. Several varieties of these fittings are shown in Figures 7.22 through 7.25.

Wyes

Wyes are more common in material handling or sewage applications than in general industrial piping. For diverting fluid flow, it is more common to use a tee or a lateral. Wyes are distinguished from tees and laterals in that the branches diverge from the axis of the main run an equal angular amount on each side of the main axis. Like tees and laterals, all three pipe sections are always coplanar.

Figure 7.22 A Weldolet® branch connection fitting.[1]

Figure 7.23 A Thredolet® branch connection fitting for threading.[1]

Figure 7.24 A Sockolet® branch connection fitting for attaching a socket weld joint.[1]

Figure 7.25 A Latrolet® branch connection fitting for turning a straight run into a lateral.[1] The attached fitting may be butt-welded, socket welded, or threaded.

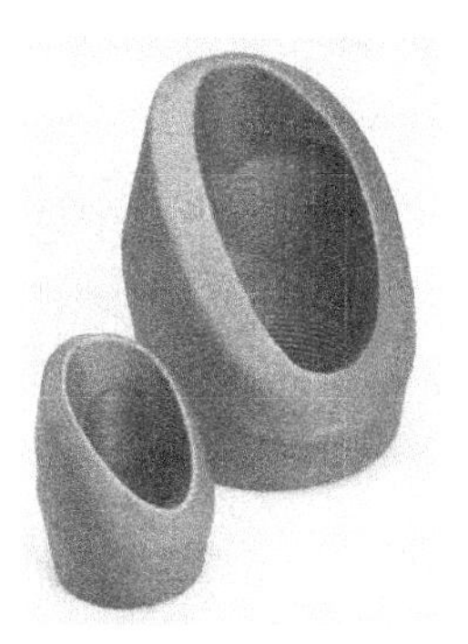

[1] OLET® is a well-known and registered trademark of Bonney Forge Corporation and its affiliated companies. Bonney Forge Corporation and its affiliated companies are the owners of all rights, title, and interest in and to the OLET® products and fittings, and representations and depictions thereof in any and all media are copyrights of Bonney Forge Corporation and its affiliated companies. OLET® and the representations and depictions specifically appearing in this text authored by Brian Silowash are authorized by Bonney Forge Corporation and its affiliated companies.

Material	Fitting Connection	Nominal Pressure Ratings	Specification
Carbon and Stainless Steel	Butt Weld	Consult applicable ASME Pipe Code	ANSI B16.9
	Socket Weld	3000, 6000, 9000 psi	ANSI B16.11
	Threaded	2000, 3000, 6000 psi	ANSI B16.11
	Flange	Consult ANSI B16.5 or B16.47	ANSI B16.5 or B16.47
Cast Iron	Threaded	125, 250	ANSI B16.4
	Flanged	125, 250	ANSI B16.4
Cast Bronze	Threaded	250 Steam or 400 WOG	ANSI B16.15
Malleable Iron	Threaded	150, 300	ANSI B16.3

TABLE 7.15 Typical fitting pressure ratings. Note that the ratings are temperature-dependent, as indicated in the respective specifications.

Ratings of Fittings

Note that the ratings of fittings are temperature and pressure dependent. While the pressure ratings are classed by nominal pressures, the reader should refer to the specifications that govern the manufacture of the fitting for specific pressure ratings at the design temperature.

Some common fitting ratings are summarized in Table 7.15.

CHAPTER 8

Valves and Appurtenances

As part of a flow control strategy, I once designed a water system with a pressure bleed valve to dump excess water when the process did not require it. A bypass valve was to open and dump water into a reservoir in order to prevent the pump from deadheading when a discharge valve closed.

Due to the particular organizational structure of the company, the valve sizing and selection took place in the Process Control Department, since this was a "control valve."

Prior to my departure to another assignment, the controls designer and I were alerted by the valve vendor to the possibility of cavitation. His recommendation was to provide a means for aspirating air into the valve intake to aid in relieving the vapor pressure bubbles that might form under the cavitation conditions. The recommendation was to install valved branch openings at carefully selected points upstream of the control valve. If necessary, these valves would be manually opened, permitting air to aspirate into the valve inlet.

We had tried this once on a smaller scale for a minor cavitation issue on another project and found this approach to be unreliable. I had advised the project team about my skepticism prior to departing for the next project.

Upon system start-up the valve did in fact cavitate. It was a large butterfly valve and the enormous vibration due to the cavitation broke the pipe and flooded the basement of the facility.

There are several lessons to be learned from this:

1. Cavitation is a dangerous condition that must be avoided.
2. Lack of continuity of personnel within a project may result in poor follow-through.
3. Dumping large quantities of water to prevent deadheading may work under the correct circumstances, but more elegant and energy-efficient choices (such as taking advantage of Variable Frequency Drives) may present a better technical solution, even if first-cost is higher.

The variety of valves available to employ on a project is staggering, and usually the choice of which to use is based on the technical application, the ease of use by operators and maintenance personnel, and the cost.

In application, valves are used for two purposes:

1. On/off service, in which the valve is intended to be full open to permit flow, or full-closed to stop flow.
2. Throttling service, in which the valve is required to modulate the flow through it in order to satisfy a process condition such as regulating flow rate or pressure.

Not all valves can be used in throttling service, but many valves suitable for throttling are also used for on/off service.

Valve Trim

Trim refers to certain internal parts of a valve, including:

- Seat rings
- Disk or facing
- Stem
- Stem guide sleeves

The selection of suitable trim materials is always an important consideration for valve manufacturers. The challenge is to find wear-resistant materials that resist corrosion and galling[1], are tough, and possess coefficients of expansion that approximate that of the valve body material.

One common trim material is known by the trade name Stellite®. This is a hardfacing alloy of cobalt, chromium, and sometimes tungsten. The sealing surfaces for metal-to-metal seals must be machined very smoothly, so hard, wear-resistant trim is quite important.

Gate Valves

Gate valves (Figure 8.1) utilize a flat insert that travels perpendicular to the flow stream. The diameter through the valve is essentially the same diameter as the pipe, and since the insert or gate lies entirely outside the flow stream when the valve is full open, gate valves have very low-pressure drops.

Gate valves are used for making hot taps, for general on/off service, and are useful for handling slurries. The gate is out of the slurry flow path when open, and slices through the slurry to close. They are also applied in viscous liquid service. They are not a good choice for applications that require cleanliness or sanitary conditions.

Gate valves are available in two basic designs:

- Double Disc Type gate valves have parallel seats against which a wedge is driven when the gate is closed. This achieves a tight seal between the disc and seat. Variations on this design include valves that depend on fluid pressure to force the disc against the seat.

[1] Galling is the seizing of metal surfaces as they slide against each other.

Figure 8.1 Outside Stem & Yoke flanged gate valve with rising stem and epoxy coated cast iron body. *Photo provided by Watts Regulator Company.*

- Wedge Type gate valves have a disc that is in the shape of a wedge that seats against two inclined seats. Solid wedges are most often used and excel in high flow or turbulent applications like steam service, since the solid design minimizes vibration and chatter. Split wedge discs are more flexible and may be used where pipeline strains may distort the valve seats. But pipeline stresses are difficult to predict without a stress analysis, and the operating stresses may be increased by improper installation.

There are two types of gate valve stems; Rising Stem and Non-Rising Stem. Rising Stem valves are preferred because the position of the gate is readily discernible by observing the position of the valve stem. If the stem is up, the gate is open. If the stem is down, then the valve is closed. Figure 8.2 shows an open gate valve in industrial water service. Figure 8.3 shows a closed rising stem gate valve.

Non-Rising Stems are generally only used where a space conflict may arise when the stem is extended. Where Non-Rising Stem valves are used, it may be advisable to post a sign to alert plant personnel to not rely only on looking at the stem to determine the valve position. The Rising Stem design keeps the stem threads out of contact with the fluid. With a Non-Rising Stem design, the threads remain internal to the valve

Figure 8.2 A rising stem gate valve in the open position. Note the full face gasket on the near flange pair.

Figure 8.3 A 1 in gate valve used as a steam-out connection.

body and may corrode, erode, or accumulate deposits on the stem threads, rendering operation difficult.

Gate valves are equipped with an outside stem and yoke, giving rise to the term "OS&Y Gate Valve." At the top of the yoke, a stationary yoke nut advances the lead screw when the handwheel is turned. This raises or lowers the gate.[2]

Naturally there must be a means to prevent the fluid inside the valve from leaking around the stem as it passes through the valve body. This is accomplished in gate valves through the use of packing that is compacted around the stem with the use of a packing gland. As the packing wears and the valve leaks around the stem, the gland can be tightened progressively to increase the pressure on the packing, deforming it against the sealing surfaces.

Gate valves are only used in on/off applications and should never be applied in situations where the flow of fluid is to be modulated. The reason for this is that a gate valve that is partially open behaves more or less as though it is fully open. There is no control range to speak of. Another reason is that the gate and seats can erode if exposed for extended periods to the flow of fluid, as it is when in the partially open position. Once erosion has occurred a positive seal can never be achieved.

Gate valves may be used for liquids or gases.

Globe Valves

Globe valves (Figure 8.4) have a rounded body through which the fluid traverses a circuitous path. This results in more pressure drop through a fully open globe valve than a similarly sized gate or ball valve.

Globe valves may be used for on/off or throttling applications of liquids or gases. These valves are constructed with a disk that translates against a seat. The flow of fluid is usually directed up through the seat and around the disk. For this reason, most large globe valves have a direction arrow embossed on the body casting, as shown in Figure 8.5. Sometimes, however, an engineer will desire to install the valve backward. If the screw that attaches the disk to the stem were to fail, they reason that in some circumstances it would be desirable for the disk to fall against the seat, thus stopping the flow. In this case, the valve would fail closed. This configuration also uses system

[2] "Righty-tighty/lefty-loosey" applies, but handwheels are often cast with "Open" arrows pointing counterclockwise so that we don't get confused, especially when they are mounted in positions other than upright.

FIGURE 8.4 A threaded bronze globe valve with non-rising stem and threaded bonnet. *Photo provided by Watts Regulator Company.*

FIGURE 8.5 A globe valve with a union bonnet. Note the direction arrow embossed on the body.

pressure to help seat the valve, but exposes the stem threads to the fluid in the closed position. The circumstances under which such a scheme is desirable are probably infrequent, but the opportunity exists to apply globe valves in this fashion, assuming of course that the disk seats properly when it fails.

Y-Pattern globe valves are most often used as shutoff valves on boilers.

Globe valves are available in sizes up to 12 in.

Check Valves

Excessive use of check valves should be avoided where possible. Inexperienced engineers and designers feel that check valves offer some means of foolproofing a piping system, ensuring that the flow is checked if an unbalanced pressure condition occurs.

When troubleshooting a piping system, too many check valves can introduce too many variables. The internal checking devices often fail open or closed, stuck with the accumulated goo that the system is trying to move. In fact, in many installations the internals of the check valve have been removed. A good starting point in such a troubleshooting effort is to ask the maintenance personnel whether the check valve still has its internals.

Check valves are used to prevent backflow of a fluid, as on the discharge of a pump. Backflowing fluid may cause a column of liquid to drain after energy has been expended to lift it to a higher elevation. It may also cause damage to a pump by turning the impeller in the wrong direction.

All check valves are supplied with a direction arrow on the valve body to indicate the direction of flow.

Swing Check Valves

The most common type of check valve operates by gravity. See Figure 8.6. The velocity pressure of the fluid exerts a force over the area of the disk and overcomes the force of gravity, unseating the disk. It is imperative that these valves be oriented in a position in which the disk seats itself when no fluid flows. Figure 8.7 shows a swing check valve mounted in the vertical position.

As might be expected, there is a minimum flow required to unseat the disk. Further, and less obvious, there is a minimum velocity that must be maintained in order to prevent the valve from chattering, or opening and closing rapidly. This is an unstable

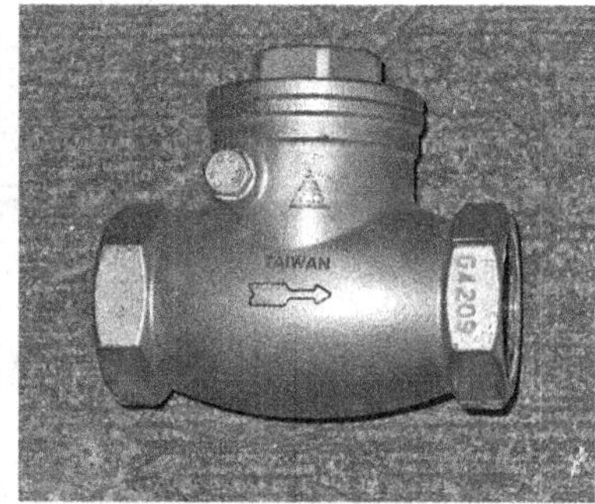

FIGURE 8.6 Swing check valves.

FIGURE 8.7 A cast iron swing check valve, 125 lb, with full-face gasket on upstream end, and lug-type butterfly valve downstream.

Figure 8.8 A lift check valve.

Figure 8.9 A cast iron silent check valve used to prevent water hammer due to pump shut down. *Photo provided by Watts regulator Company.*

condition that at best will result in noise, and may lead to valve component failure. Therefore, the valve must sometimes be downsized in order to keep the velocity high enough to keep the disk lifted. For steam, the recommended velocity is 14,000 to 15,000 FPM (71 to 76 m/sec).

The conventional swing check valve pivots on a pin attached to the top of the disk, rotating the disk up and toward the edge of the flow stream when pressure is applied in the intended flow direction by the velocity of the fluid. Should the flow direction happen to be reversed due to a pressure imbalance, the disk swings back down against the valve seat, sealing the valve closed.

Some manufacturers offer a disk whose pin is located closer to the center of the disk. This balances the disk and is alleged to provide better flow characteristics and non-slam closing.

Lift Check Valves

Lift check valves are available in straight-through, tee, or wye patterns. This type of check valve may be installed horizontally or vertically, but those lift checks relying solely on gravity must be installed so that the disc travel is vertical. Other valves contain a spring which must be forced open under normal flow conditions. See Figure 8.8.

Ball Check Valves

Another type of check valve operates using gravity as the checking means. A lightweight polymer ball is captured within the valve body, and at rest lays against a spherical profile seat. These valves must be oriented vertically. They are often used in air-operated diaphragm pumps to prevent the fluid from backflowing into the pump.

Silent Check Valves

The silent check valve (Figure 8.9) is a spring-loaded straight-through design. These may be constructed with a solid disk that lifts off the seat against inlet pressure or with a twin disc design that is hinged in the center of the disc.

These quick acting valves are intended for liquid service. They are promoted as reducing water hammer when pumps shut down and flow is reversed in the liquid column.

These are available in sizes up to 24 in.

Foot Valves

This is a special type of check valve that is designed for use at the submerged bottom of a pump suction. The purpose of the foot valve is to retain a column of water inside the vertical portion of the pump suction, thus retaining the prime on the pump. If the pump is not self-priming, then a fluid inlet tap on the suction line in combination with the foot valve enables the pump to be primed.

Foot valves are usually equipped with suction strainers to prevent large, low-density particles from being drawn into the suction line and damaging the pump. If a wrench were to fall into a pump wet well, it would not cause as many problems as something like a plastic bag, which could get trapped on the suction strainer and starve the pump.

Ball Valves

Ball valves (Figure 8.10) may be used for liquid or gas service, and may also be used for on/off or throttling applications. For critical throttling applications however, the profile of the opening in the valve is designed for controllability over its range of positions. The orifice cut into the ball is machined in the shape of a triangle, leading to the name "V-port ball valve."

Standard ball valves are quarter-turn valves. That is, the stem is rotated through 90 degrees in going from full open to full closed positions. The ball is metal, with a hole drilled through it through which the fluid flows when the hole is oriented along the axis of the valve. The position of the ball is easily discernible to the operator, since the lever which fastens to the stem is parallel to the hole in the ball. Thus, when the lever is perpendicular to the pipe, the valve is closed. An open valve is indicated when the handle is parallel to the valve, as in Figure 8.11.

Ball valves are easily turned with levers in sizes up to about 3 in. Beyond that, manually operated valves are usually equipped with geared handwheels which indicate the position of the valve with an arrow located atop the operator. See Figure 8.12.

Care must be exercised in the specification of ball valves to ensure that the bore size is appropriate for the application. The standard bore is actually one size smaller than the nominal ball valve size. That is, a 6 in standard bore ball valve has a 4 in bore and is therefore smaller than the inside diameter of the adjoining pipe. If the valve must accept the full line size, as in certain cleaning operations like rodding out or pigging, then the valve must be specified as "full-port."

FIGURE 8.10 A two-piece threaded carbon steel ball valve with lockable handle. *Photo provided by Watts Regulator Company.*

Figure 8.11 A ball valve used to isolate a pressure gauge. Note the lack of a pressure bleed valve on the pressure gauge, as well as the inconvenient handle position of the ball valve.

Figure 8.12 A bank of 8 in SS ball valves with manual right-angle gear operators with handwheels inside a chemical plant.

"Double-reduced" bores are also available for sizes 2 in and under. These have a bore that is two pipe sizes smaller than the nominal size.

The sealing surface between the ball and the seat is critical to the successful application of the ball valve. This sealing surface relies on both the sealing method and the seat material.

Manufacturers employ the following four methods for achieving a good seal between the ball and seat:

1. Mechanical compression of the ball and seat.
2. A floating ball which is pressed into the seat when high pressure is applied to the opposite side.
3. A ball held in place with a trunnion, with the seats compressed against the ball via line pressure or springs.
4. A floating ball with pressurized seats.

The seats are often made with a low-friction material like TFE, but these have an upper temperature limit of 350°F (177°C) at 100 psi. At lower pressures 400°F (204°C) can be achieved, but higher temperatures are only possible through the use of other materials employing glass-fiber matrices, polyimides, or PEEK. Above 650°F the only suitable sealing surfaces are metal-to-metal or metal-to-ceramics. Care must be exercised in selecting metal materials to prevent galling. Dissimilar metals must be used and the differential hardnesses generally must exceed 10 Rockwell C. Temperatures up to 1500°F (816°C) can be achieved.

A variety of other configurations of ball valves are available other than flowing straight through the valve. Where a flow needs to be diverted to multiple locations, 3-way or 4-way valves may be used.

Ball valves may be used in flow control applications. In these cases, the port of the ball is often shaped like a "V" rather than a circle in order to achieve better variability of flow across the lower range of valve positions.

Note that when closed, a small amount of fluid is trapped inside the ball. In special instances, manufacturers can drill a hole into the side of the valve body to permit venting this fluid to atmosphere.

The term "car-seal open/closed" refers to a metal tab that is crimped around a wire to indicate that a valve is to be operated only with the proper authorization. This is a system that is often used in the process industries to warn operators against inadvertently placing a valve (usually a quarter-turn valve) in the wrong position.

Many quarter-turn valves are manufactured with the ability to physically lock them open or closed. Placing a padlock through the handle is a more positive physical barrier than a car-seal. But the car-seal remains an effective administrative control. The padlock serves as a more rigid barrier to changing the position of a valve however.

Car-seals originated in railcar shipping. In order to prove that the contents have not been tampered with, a wire is passed through the handle (of a door or a valve) and a lead ball is crimped around the wire with the seal of the shipping entity applied to the lead seal. These are often seen on utility meters to provide indication to the meter reader of tampering.

Butterfly Valves

Butterfly valves consist of a disk which rotates inside the body, forming a seal against elastomeric or metal seats. These valves are very versatile, being used for manual or process control, on/off or throttling, and liquids or gases. They are often useful due to their short take-outs, and comparatively lower costs.

Butterfly valves are quarter-turn valves. Note that lever operators are often equipped with a squeeze handle that disengages a pin from a round rack of teeth, as shown in Figure 8.15. The pin is used to set the position of the disk inside the valve for throttling applications. The valve position may also be locked into place with a thumbscrew.

Wafer-Type

Wafer-type ball valves fit between two flanges, with studs passing from one flange through the other. The valve is held in place and sealed with gaskets by the tension of the studs. These are the lowest cost butterfly valves. See Figure 8.13.

Lug-Type

Lug-type valves (Figure 8.14) must be used where the butterfly valve is at the end of a pipeline since there would be no second flange to secure studs. Instead, lugs are cast on the valve body with tapped holes that match the bolt hole pattern for the size and pressure class of the flange. Bolts are then passed through the flange holes and are threaded into the tapped holes in the lug, as shown in Figure 8.15.

Figure 8.13 A wafer-type butterfly valve with cast iron body, stainless steel shaft, 10-position lever, and elastomeric seats. The valve is sandwiched between two flanges. Note that the elastomeric seat serves as a flange gasket. No additional gasket is required. *Photo provided by Watts Regulator Company.*

Figure 8.14 A lug-type butterfly valve with cast iron body, stainless steel shaft, 10-position lever, and elastomeric seats. The seat acts as a flange gasket. *Photo provided by Watts Regulator Company.*

Figure 8.15 A lug-type butterfly valve downstream of a detonation arrestor. Note that the position of the handle indicates that this valve is open.

In many cases, the pressure rating of a butterfly valve at end-of-line service must be derated by a factor of two since there is no companion flange to hold the assembly together. Other manufacturers offer full-pressure ratings, so it is best to verify the rating with the valve manufacturer.

High-Performance Butterfly Valves (HPBV)

Unless otherwise specified, butterfly valves will be supplied with elastomeric seals. These valves are suitable only for infrequent positioning, such as on/off service, or those instances when the valve position is set and left alone for long periods.

In modulation service, where the valve position must be adjusted often to satisfy a process condition, the elastomeric seals are unsuitable, since they would wear quickly and render the valve inoperable. In those cases, a precision metal seal is required. These valves are called "High-Performance Butterfly Valves" (HPBV). Some are designed to reduce wear and friction even further through the use of an offset that drives the disk into the seat with a translational motion in addition to rotational motion.

Fluid Velocities through Control Valves

While the only practical limits for velocity of gases in pipelines depend on noise generation and choked flow, liquids at high velocities can cavitate, leading to noise, vibration, and damage to the valve. Cavitation may occur if the pressure drop through the valve exceeds 15 to 20 psi (1 to 1.4 bar).

Table 8.1 gives maximum water velocities through control valves. The manufacturer of the control valve should be consulted if the velocities approach these maximum values.

The valve should be located lower than the discharge if acting as a dump valve in order to reduce the possibility of cavitation. This provides some backpressure on the valve.

Severe duty control valves are available to meet high-pressure drop/high flow demands. These break the pressure drop down across multiple stages of specially designed trim inside the valve. Such valves are able to safely drop the pressure from pressures in the range of 5500 psi (379 bar) to atmospheric pressure.

Valve Type	Service	Maximum Velocity	
		ft/sec	m/sec
Butterfly	Continuous	23	7.0
	Infrequent	27	8.2
Ball	Continuous	32	9.8
	Infrequent	39	11.9

TABLE 8.1 Maximum liquid velocities through control valves.

Needle Valves

Needle valves are used wherever a small amount of fluid must be metered from a system. The operation of such a valve requires many turns of the handle to open the valve all the way. This implies a high degree of adjustability. The prudent engineer will choose such valves carefully, since not all are of high quality.

I was on top of a calender during engineering checkouts ("commissioning") of a system that heated and compressed polypropylene fibers, effectively welding the fibers together at discrete spots so that when it exited the calender, it formed a fabric with considerable strength. The calender consisted of two very heavy rolls that were heated to a temperature that would melt the polypropylene, and squeezed the fabric between the nip of the rolls as it passed through. The subsystem I was working on was the hydraulics system that did the squeezing.

My English boss was observing me from the floor below as I climbed 10 ft to the top of the structure to bleed air out of the system. Another engineer had designed the hydraulics system with needle valves so that the air could be bled off. Hydraulics systems do not operate properly if air is present in the circuit, and the place to locate air vents is at the very top of the circuit.

The system was still under pressure as I turned the handle on the needle valve. The handle consisted of a bent piece of thin-gage metal with two holes on the sides through which a thin steel pin was crimped. In other words, the handle kept falling off the valve as I worked to open the valve.

Suddenly, a pencil-thick stream of hydraulic fluid shot out the top of the valve, struck the concrete ceiling 8 ft above me, and rained down on my hard hat, my shoulders, and all of the adjoining equipment. My unflappable English boss stood on the floor with crossed arms, shaking his head as I desperately tried to close the valve with the broken handle.

Pressure Regulating Valves

Figure 8.16 shows a pressure regulating valve. Figure 8.17 shows the same valve in a cooling water line. These are self-regulating valves that are used to reduce the line pressure without the need for external feedback control. On a system that requires fine pressure control, a pressure transmitter would be installed downstream of a control valve and would relay an electrical signal to a programmable logic controller that would operate a pressure control valve. Self-regulating pressure reducing valves cannot perform this function, but instead are useful for reducing the pressure downstream of the valve. This is desirable to reduce water hammer, wear on orifices, or simply to achieve a lower pressure for components not rated for the higher pressure. In the latter case, a safety valve would also have to be added downstream of the pressure-regulating valve to protect the lower-rated components in case the regulating valve failed.

These valves are also available for steam or gas service. Figure 8.18 shows a valve listed for steam service. Note that each of these valves is equipped with an adjusting

Figure 8.16 A pressure regulating valve. Note the adjusting screw at the top of the valve. *Photo provided by Watts Regulator Company.*

Figure 8.17 A pressure regulating valve on a cooling water line.

Figure 8.18 A pressure regulating valve for steam service. *Photo provided by Watts Regulator Company.*

screw that is used to place more or less tension on an internal spring that fits against a diaphragm. The diaphragm raises or lowers a valve stem, which increases or decreases the orifice size, changing the pressure setting. A pressure gauge located downstream of the valve is required in order to check the set pressure. Note also that the direction in which the screw is turned is counter-intuitive: to increase the pressure, the screw must be turned clockwise; to decrease the pressure, the screw is turned counterclockwise.

Some pressure regulating valves are available with a compensator adjustment control that is used to eliminate hum or chatter within the valve. This may be an important feature for commercial, institutional, or residential applications.

Pressure Relief Valves (PRVs)

Pressure vessels are required by the ASME Boiler and Pressure Vessel Code to be protected from overpressurization by relief devices. An overpressure situation may arise from a pump operating out of its expected range of operation, from a fire that raises the temperature and hence the pressure of a closed system, or from a chemical reaction that creates high pressures due to vapor generation. Figure 8.19 shows some typical pressure relief valves.

The purpose of these valves is to relieve pressure in systems in order to prevent damage from overpressure situations. The Boiler and Pressure Vessel Code[3] distinguishes among several types of relief valves:

- "Pressure Relief Valve" is defined as a pressure relief device that is designed to reclose and prevent the further flow of fluid after normal conditions have been restored. This is the generic term that encompasses all four of the following terms.

[3] See ASME Section VIII – Division I, UG-125 and UG-126.

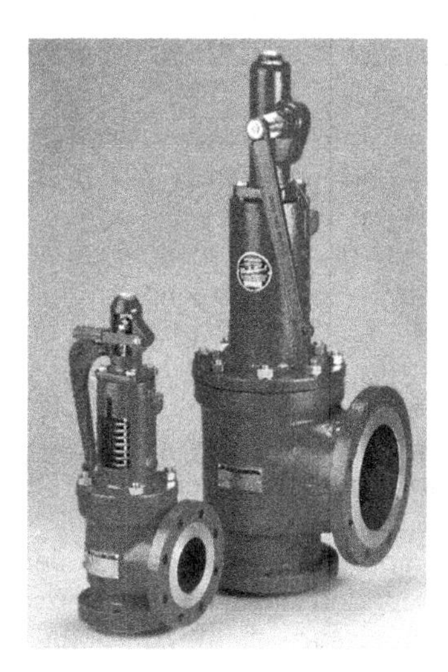

Figure 8.19 Pressure relief valves. *Supplied courtesy of Farris Engineering, division of Curtiss-Wright Flow Control Corporation.*

- "Safety Valve" is defined as a pressure relief valve that is actuated by inlet static pressure and is characterized by rapid opening or "pop" action. These are also referred to as "pop-off" valves.
- "Relief Valve" is defined as a pressure relief valve that is actuated by inlet static pressure and which opens in proportion to the increase in pressure over the opening pressure.
- "Safety Relief Valve" is defined as a pressure relief valve that may be either of the pop-off or proportional type.
- "Pilot Operated Pressure Relief Valve" is defined as a pressure relief valve in which the major relieving device (the disk lifting from the seat) is combined with and is controlled by a self-actuated auxiliary pressure relief valve.

In any case, the relief of pressure above the MAWP in a rated vessel must be provided by purely mechanical means. While electrical devices may be used to monitor and control pressure, the protection against overpressurization must be provided by mechanical devices. When installed on a boiler or pressure vessel these pressure relief valves must be identified with the appropriate ASME stamp:

- V – Power Boiler Safety Valve
- HV – Heating Boiler Safety Valve
- UV – Pressure Vessel Safety Valve
- UV3 – High-Pressure Vessel Safety Valve
- TV – Transport Tank Safety Valve

Manufacturers of pressure relief valves also provide non-coded valves for relief applications for other than boiler and pressure vessels.

No valves may be placed between a boiler and the code-required relief device. No valves may be placed between a pressure vessel and the code-required relief device, although there are some exceptions described in ASME Section VIII, UG-135 (d) (1) and (2).

ASME Boiler and Pressure Vessel Code - Section I Requirements

The number of PRVs on a boiler is set by ASME BPV Section I. Each boiler, regardless of size, shall have at least one safety valve or safety relief valve. If the boiler has over 500 SF (46 m^2) of bare tube water heating surface, or if it is an electric boiler with a power input greater than 1100 kW, then it must have two or more safety valves or safety relief valves.

At least one of the safety valves shall be set at or below the MAWP. If additional valves are used, the highest maximum set pressure shall not exceed the MAWP by more than 3 percent. The total range of set pressures of saturated steam on a boiler is not permitted to exceed 10 percent of the highest pressure to which any valve is set. However, the pressure setting of safety relief valves on high-temperature (superheated) water boilers may exceed this 10 percent range.

There are other rules regarding boiler PRVs, and the reader is referred to ASME BPV Code – Section I, PG-67 for details.

In practice, one or more relief valves on a boiler are usually set to a pressure lower than the rest, since this provides early warning (and an additional measure of safety) during a potential overpressure event. Although no isolation valves may be placed upstream of a PRV used to prevent overpressure, a gate valve is sometimes used upstream of an extra PRV known as a "power-control valve." The valve is set to blow before any of the others, and because high-pressure steam throttling through a valve orifice can sometimes cut the valve seat, this valve is the "sacrificial" valve and can be isolated and removed for repairs while the system is still in service. The capacity of a power-control valve cannot be used in the relief calculations however, since it may be valved off or out of service.

Section I requires lifting levers on all relief valves except those used in organic fluid vaporizer service. See Figure 8.20 for an example of a relief valve equipped with a test lever.

ASME Boiler and Pressure Vessel Code - Section VIII Requirements

Section VIII of the ASME Code requires pressure relief valves on pressure vessels to have lifting levers for air, steam, and hot water (over 140°F) service, except when all of the following conditions are met:

1. The user has a documented procedure and an associated implementation program for the periodic removal of the relief valve for inspection, testing, and repair.
2. The user specifies that no test lever be supplied.
3. The user obtains permission to omit the lifting lever from the authority having jurisdiction over the installation of a pressure vessel.

Operation

Relief valves have a spring that presses the disc against the seat, as shown in the cutaway section of Figure 8.21. When the pressure of the fluid under the seat lifts the disc, the spring force increases in proportion to the amount of lift. Because of this, PRVs are permitted an overpressure allowance of 10 percent of the set pressure to achieve full lift.

Most PRVs are equipped with a secondary chamber, called a "huddling chamber" which permits the pressurized fluid to act over the total surface area of the disc, rather than the portion exposed by the valve orifice. This permits maximum lift and higher

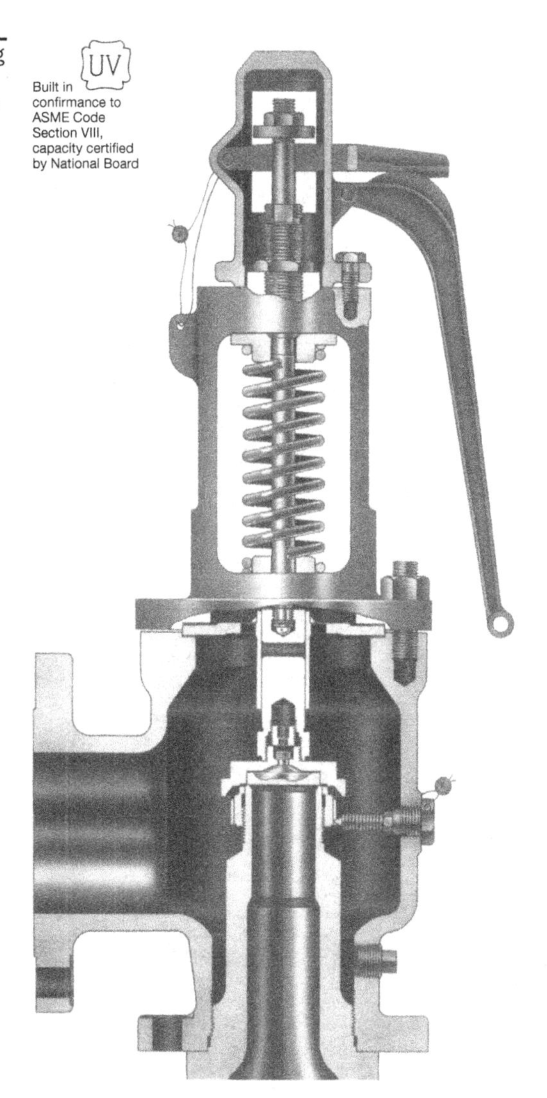

Figure 8.20 PRV with lifting lever for Section I service. Note the car-seal devices to indicate that the valve settings have not been tampered with. *Supplied courtesy of Farris Engineering, division of Curtiss-Wright Flow Control Corporation.*

flow within the allowable overpressure limits. Once the disc has lifted from the seat, the valve will not close until the system pressure has fallen below the set pressure, since the force acting on the spring is equal to the pressure times the larger area exposed to the system pressure. The difference between the set pressure and the closing re-seating pressure or closing point pressure is called the blowdown.

Most manufacturers recommend that system operating pressures are not more than 90 percent of the set pressure in order to maintain proper seating of the disc.

Pilot-Operated Valves

Pilot-operated valves are used where closer tolerances between the set pressure and the operating system pressure must be achieved. See Figure 8.22. They are recommended only for relatively clean services however, since the internal clearances may be less than those inside a conventional PRV.

Built in conformance to ASME Code Section VIII, capacity certified by National Board (air, gas, steam[6])

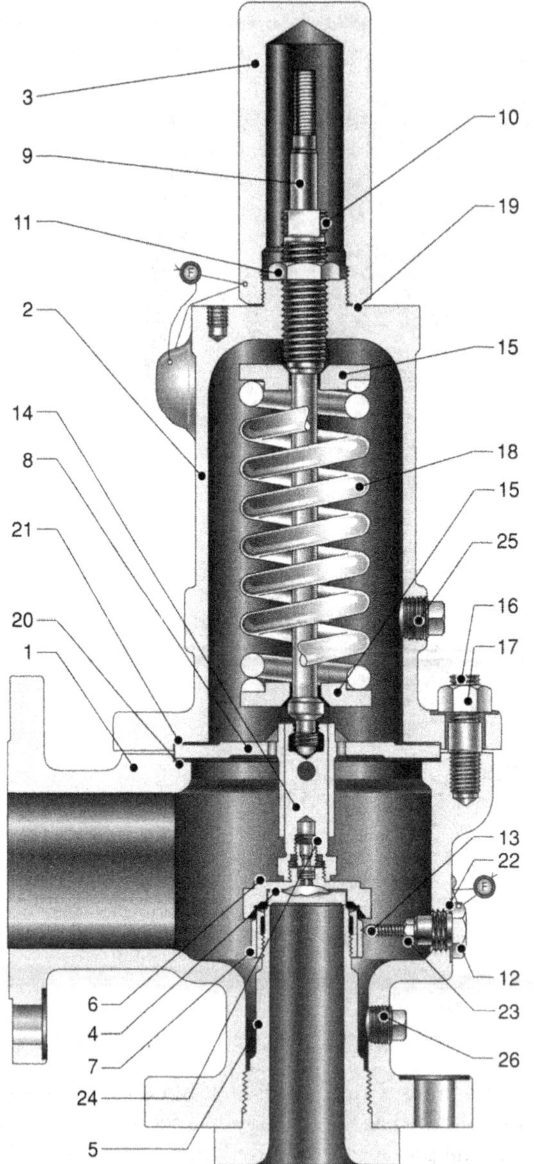

Item	Part Name		Material
1	Body	26( )A10 thru 26( )A16	SA-216 GR. WCB Carbon Steel
		26( )A32 thru 26( )A36	SA-217 GR. WC6, Alloy St. (1¼ CR-½ Moly)
2	Bonnet	26( )A10 thru 26( )A16	SA-216 GR. WCB Carbon Steel
		26( )A32 thru 26( )A36	SA-217 GR. WC6, Alloy St. (1¼ CR-½ Moly)
3	Cap. Plain Screwed		Carbon Steel
4	Disc		316 St. St.
5	Nozzle		316 St. St.
6	Disc Holder		316 St. St.
7	Blow Down Ring		316 St. St.
8	Sleeve Guide		316 St. St.
9	Stem		316 St. St.
10	Spring Adjusting Screw		Stainless Steel
11	Jam Nut (Spr. Adj. Screw)		316 St. St.
12	Lock Screw (B.D.R.)		316 St. St.
13	Lock Screw Stud		316 St. St.
14	Stem Retainer*		17-4-PH St. St.
15	Spring Button		Carbon St., Rust proofed or 316 St. St
16	Body Stud		ASME SA-193 GR. B7 Alloy St.
17	Hex Nut (Body)		ASME SA-194 GR. 2H Alloy St.
18	Spring	26( )A10 thru 26( )A16	Chrome Alloy Rust Proofed
		26( )A32 thru 26( )A36	High Temperature Alloy Rust Proofed
19	Cap Gasket		316 St. St.
20	Body Gasket		316 St. St.
21	Bonnet Gasket		316 St. St.
22	Lock Screw Gasket		316 St. St.
23	Hex Nut (B.D.R.L.S.)		Stainless Steel
24	Lock Screw (D.H.)		Stainless Steel
25	Pipe Plug (Bonnet)		Steel
26	Pipe Plug (Body)		Steel

General Notes:

1. Parentheses in type number indicate orifice designation, as in 26FA10.
2. For corrosive and low temperature materials, see pages 17 through 21.
3. For open and packed lever materials and test gags, see accessories on pages 68 & 69.
4. For capacities, see pages 39-42 U.S. Units, 57-60 Metric Units.
5. For dimensions and weights, see pages 72-75.
6. Also suitable for liquid service where ASME Code Certification is not required. For ASME Code Certified liquid service, use the 2600L Series as illustrated on page 14.

*For 316 Stem Retainer add S1 suffix to Type #.

FIGURE 8.21 A pressure relief valve for air, gas, or steam service. *Supplied courtesy of Farris Engineering, division of Curtiss-Wright Flow Control Corporation.*

The pilot-operated valve has a main valve with a piston or diaphragm operated disc and a pilot line, which allows system pressure into the piston chamber. Because the piston area is larger than the disc seat area, the disc is held closed. When the set pressure is reached, the pilot shuts off the system fluid to the piston chamber and vents the piston chamber. The system pressure can then force the disc to open.

Pilot-operated valves permit the operating pressure to be within 5 percent of the set pressure without increased seat leakage.

Figure 8.22 Pilot-operated relief valves. *Supplied courtesy of Farris Engineering, division of Curtiss-Wright Flow Control Corporation.*

Design Considerations

Because the relieved fluid exits such valves at high velocity and high mass flow rates, it is imperative that:

1. The exit point be directed in a safe direction to avoid damage to equipment and personnel in the event that the valve opens.
2. The valve be properly anchored, since the combination of velocity and mass flow will generate a reaction thrust on the valve. The longer the exhaust piping is away from the valve or anchor point, the more moment will be placed on that anchor point. Therefore great care must be exercised in the design of the relief exhaust. Some valves are equipped with dual exhaust ports located 180° from each other. But unless the exhaust piping lies entirely in the vertical plane formed by these two ports, and are symmetrical, thrust loads will resolve back to the vessel.
3. The exhaust pipe must be designed to eliminate the entry of precipitation. Imagine a relief exhaust pipe that has filled with rain or snow, and now imagine a relief scenario. The whole idea is to relieve the system pressure as quickly and safely as possible. Any foreign objects inside the relief exhaust will impede the flow and will be ejected at a high rate of speed, creating a dangerous condition.

One method frequently employed to reduce the moment on a relief anchor is to provide an "umbrella fitting." This is a sleeve into which the exit pipe from the relief valve is inserted. There is no hard pipe connection, and the exhaust pipe may be anchored without placing any other reaction forces onto the relief valve.

The umbrella fitting must still drain precipitation, condensate, and be properly supported (keeping in mind also that it may see large thermal displacements depending on the fluid relieved). See Figure 8.23.

Some pressure relief valves are equipped with a threaded tap at the top of the cap. A "test gag" (Figure 8.24) is threaded finger tight into this opening in order to prevent the seat from lifting during hydrotesting of the system. The hydrotest is at 1.5 times the design pressure (an alternate pneumatic test may be made at 1.1 times the design pressure for power boilers[4] or 1.25 times the design pressure for pressure vessels[5] if "appropriate safety measures are taken").

[4] See ASME Boiler and Pressure Vessel Code, Section I (Power Boilers), PG-73.4.1(b).

[5] See ASME Boiler and Pressure Vessel Code, Section VIII (Pressure Vessels), UG-136(d)(2)(b).

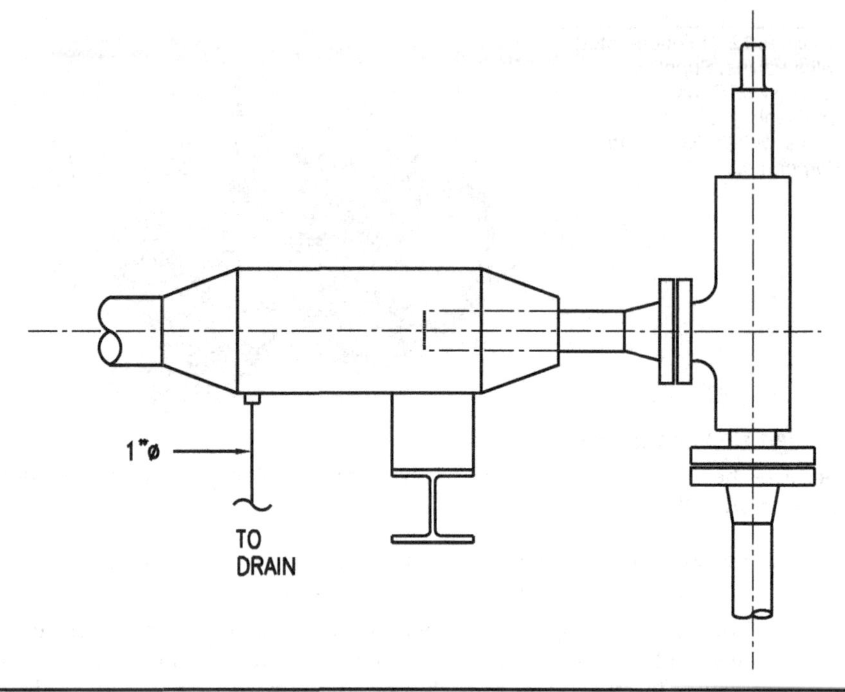

Figure 8.23 An umbrella fitting for relieving pipe support thrust loads at a PRV.

Because the test gag defeats the relief valve, it is imperative that these gags are removed immediately after the hydrotest.

Sizing of relief valves must take the fluid characteristics into account. Engineers must note that two-phase flow problems are especially critical since gases expand at reduced pressures downstream of the relief device. Relief device sizing is a critical aspect of pressure relief problems, and the reader is referred to the Design Institute for Emergency Relief Systems (DIERS) as well as the ASME code.

Temperature and Pressure (T&P) Valves

Temperature and Pressure valves such as the one shown in Figure 8.25 are required on all storage water heaters that operate above atmospheric pressure. The temperature setting must not exceed 210°F (99°C) and the pressure setting must not exceed the tank's rated working pressure or 150 psi, whichever is less.

These devices should be tested periodically to ensure their safe operation. Care must be taken in areas that have hard or mineral-laden water, since the minerals can deposit on the valve seat in a short time and cause nuisance leaks that require valve replacement. The discharge from these valves must be hard-piped to a safe location that will avoid damage or injury in the event of a discharge.

Rupture Disks

Rupture disks (Figure 8.26) are non-reclosing devices which serve the same purpose as safety valves. They are designed to relieve pressure rapidly. Rupture disks may be used as stand-alone relief devices or placed upstream of pressure relief valves to prevent the fluid from contacting the internals of the relief valve. This may be advantageous if the

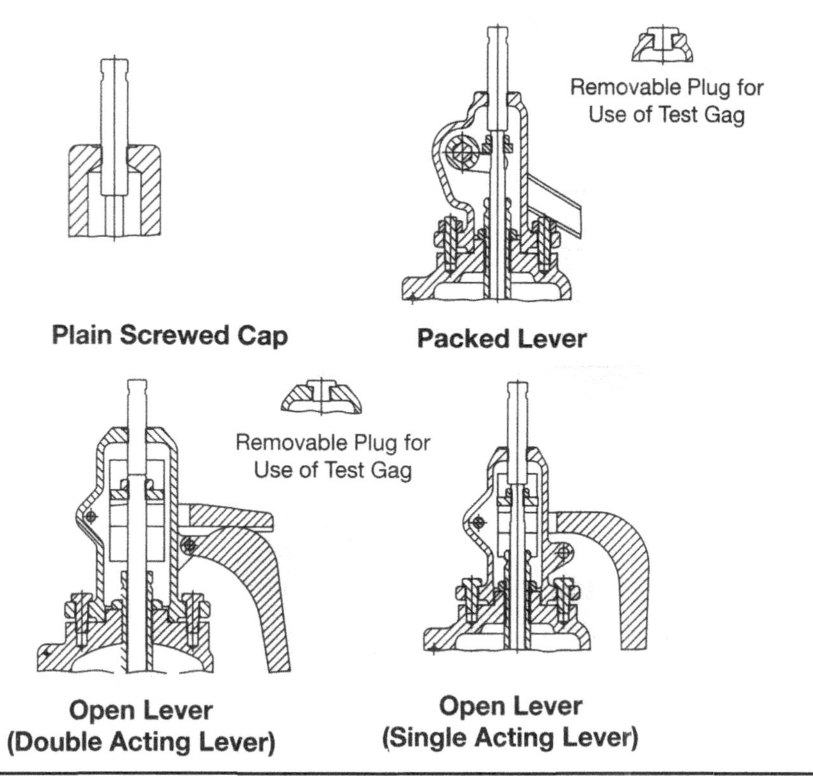

FIGURE 8.24 Test gags. *Supplied courtesy of Farris Engineering, division of Curtiss-Wright Flow Control Corporation.*

FIGURE 8.25 A temperature and pressure relief valve removed from a hot water heater tank. Accumulations of corrosion such as shown here may prevent proper operation of the valve.

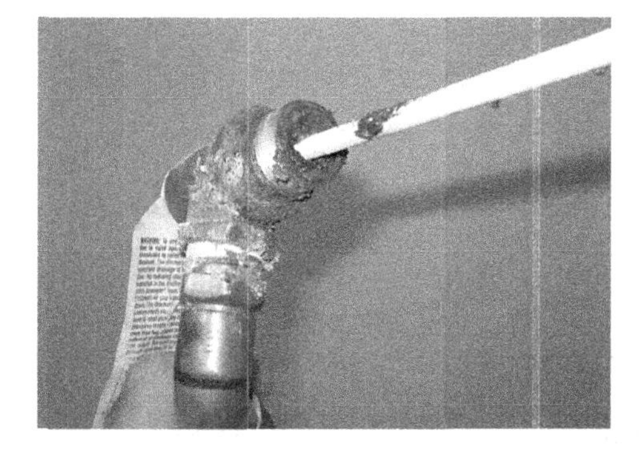

fluid would be corrosive to the relief valve internals or if the fluid might be expected to collect on the relief valve seat and interfere with its operation should the valve be required to relieve. Where rupture disks are located upstream of a relief valve, they must be of the non-fragmenting type so that bits of the rupture disk do not jam inside the relief valve and interfere with its proper operation.

Refineries place a rupture disk upstream of a relief valve to control fugitive emissions from leaking relief valves. This approach may be part of a Leak Detection and Repair Program that is required by the EPA under the US Clean Air Act[6].

[6] 40 CFR Part 60, New Source Performance Standards (NSPS), and 40 CFR Parts 61 and 63, National Emission Standards for Hazardous Air Pollutants (NESHAP).

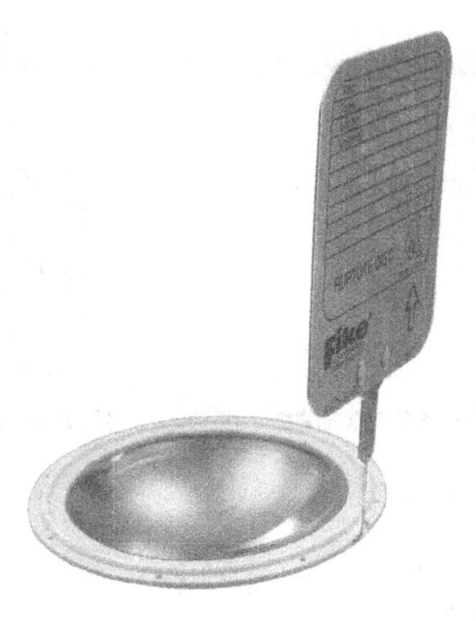

Figure 8.26 A reverse-acting rupture disk has a convex bulge facing into the fluid to be relieved. *Fike Corporation.*

Because rupture disks are engineered to relieve at specific pressures, they must be installed correctly. Flow arrows are provided to indicate the proper orientation. This is important since there are forward-acting and reverse-acting designs, with the bulge either facing into or away from the fluid to be relieved. Special holders for the disk are also required to ensure that the disks are properly centered, sealed, and may be properly torqued. The disk manufacturers have specific torquing requirements, and some disks may be ordered as a pre-torqued unit that eliminates some of the variability in installation. The holders may be designed to be inserted between flanges, as in Figure 8.27, or may be installed in threaded pipe, as in Figure 8.28.

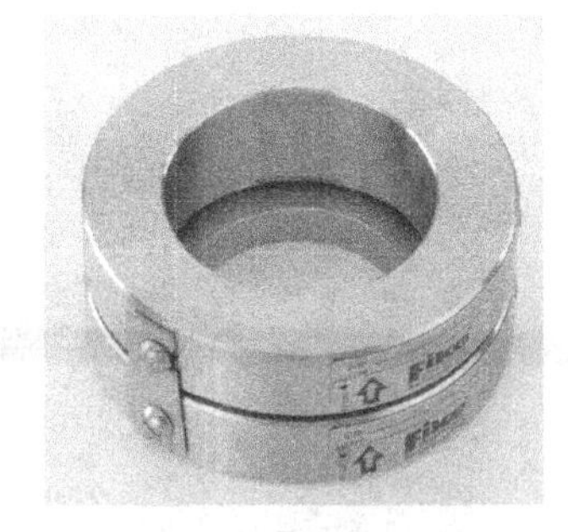

Figure 8.27 A rupture disk holder designed to be installed between two flanges. *Fike Corporation.*

Figure 8.28 A union-type rupture disk holder for threaded pipe. *Fike Corporation.*

ASME Requirements for Rupture Disks

The tolerance for burst pressure must not exceed ± 2 psi for marked pressures up to and including 40 psi, and ± 5 percent for marked pressures above 40 psi.

There are three methods permitted by ASME for acceptance testing of rupture disks. The most common method requires that at least two disks from each manufacturing lot be burst tested at the specified disk temperature. The disks must burst within the rupture tolerance.

Rupture disks are marked with the ASME UD symbol.

Design Considerations

The following parameters must be considered when specifying a rupture disk:

- Normal operating pressure
- Maximum Allowable Working Pressure of the vessel to be protected
- Cyclic pressure duty
- Exposure to vacuum
- Chemical compatibility between the disk and the fluid
- Normal operating temperature
- Maximum operating temperature
- Fluid characteristics such as polymerization, liquid, gas, or two-phase flow

Just like other safety relief devices, the exhaust from a rupture disk must be vented to a safe location.

Valve Leakage

No valve is completely leak-proof, and leakage is a relative term. The industry has adopted standards to quantify the amount of leakage through control valves. These standards are set forth in ANSI/FCI 70-2.

The extent to which a particular valve leaks depends on several variables:

- The number of seats. A butterfly valve has two seats and this reduces its ability to seal completely around the axis of rotation of the disk.
- Greater actuator thrust (or torque) will seal more tightly.
- Higher pressure drop across the valve sealing surface will promote leakage.
- Low surface tension liquids will leak more easily, as will gases.

ANSI/FCI 70-2 defines six classifications of valve seat leakage, but for most applications only two are commonly used: Class IV is a metal-to-metal seal and Class VI is a soft-seat seal (in which either the plug[7] or seat or both are made of a resilient material).

The complete classification system is:

- Class I is identical to Classes II, III, and IV in construction and design, but no shop tests are performed if the user and the supplier agree. There is no maximum leakage allowable set for this classification of valves.

[7] In this context, "plug" may refer to a disk, a ball, or a plug (as in a plug valve).

- Class II is for metal-to-metal seats. Air or water may be the test fluid, at 45 to 60 psig (3 to 4 bar) and 50 to 125°F (10 to 52°C). The maximum allowable leakage is 0.5 percent of the full-open capacity of the valve. To test such a valve, pressure is applied to the inlet, with the valve outlet open to atmosphere or connected to a low pressure measuring device. Full normal closing thrust is provided by an actuator.
- Class III is identical to Class II (metal-to-metal seats), except the maximum allowable leakage is limited to 0.1 percent of the full-open rated capacity.
- Class IV is identical to Class II (metal-to-metal seats), except the maximum allowable leakage is limited to 0.01 percent of the full-open rated capacity.
- Class V is also for metal-to-metal seats, but the test fluid is water at the maximum service pressure drop across the valve plug and at 50 to 125°F (10 to 52°C). The test pressure may not exceed the ANSI valve body rating. The maximum allowable leakage is 0.0005 ml per minute of water per inch of orifice diameter per psi differential. The closing thrust is set by the net specified thrust of the actuator.
- Class VI is for resilient-seating valves. The test fluid is air or nitrogen at 50 to 125°F (10 to 52°C) at a test pressure of 50 psig (3.5 bar), or the maximum rated pressure differential across the valve plug, whichever is lower. The closing thrust is set by the full normal closing thrust of the actuator. The maximum allowable leakage is measured in "bubbles per minute," or ml per minute, and must not exceed the values given in Table 8.2. These values are standardized in ANSI/FCI 70-2.

The term "bubble-tight" derives from the testing of Class VI valves, but as can be seen from Table 8.2, even at small diameters some amount of leakage, however small, is permitted.

Nominal Port Diameter	Allowable Leakage	
(Inches)	(ml Per Minute)	(Bubbles Per Minute*)
1	0.15	1
1.5	0.30	2
2	0.45	3
2.5	0.60	4
3	0.90	6
4	1.70	11
6	4.00	27
8	6.75	45
10	9.00	63
12	11.5	81

* Bubbles per minute as tabulated are a suggested alternative based on a suitable calibrated measuring device, in this case a 0.25-inch O.D. X 0.032-inch wall tube submerged in water to a depth of from 1/8 to 1/4 inch. The tube end shall be cut square and smooth with no chamfers or burrs. The tube axis shall be perpendicular to the surface of the water. Other measuring devices may be constructed and the number of bubbles per minute may differ from those shown as long as they correctly indicate the flow in milliliters per minute.

Table 8.2 Valve leakage rates for Class VI valves, from ANSI/FCI 70-2.

Plug Valves

Plug valves are quarter-turn valves similar to ball valves in operation and function, but instead of a ball, plug valves utilize a frustum or a cylinder to control the flow through the valve.

The plug rests in a cavity inside the valve body that is ground to the same profile as the plug. These valves are often supplied with lubrication ports through which grease may be injected to lift the plug from the seat, and to apply a film of lubrication to aid in turning the plug within the valve.

Plug valves are used in both liquid and gas service, but are best used for on/off applications rather than throttling, since only crude control can be obtained in partially-opened positions.

Diaphragm Valves

Not as common as other valves in general industry, the diaphragm valve offers tight closure in gas and liquid service. It may be especially useful in liquids containing grit or suspended particles, or those that may be corrosive or form scale. It finds wide use in the chemical process industry, electronic component manufacturing, pharmaceutical manufacturing, and in ion-exchange water demineralizers.

Such valves are constructed with a resilient diaphragm which is compressed by the translating stem against a weir. Rubber based diaphragms are suitable for temperatures up to 220°F (104°C), but PTFE diaphragms permit operating temperatures up to 300°F (149°C). Diaphragm Valves are available in sizes from ½ to 16 in, with lower pressure ratings as the size increases. See Table 8.3.

Note that unlike other valves, closing a diaphragm valve significantly decreases the liquid volume inside the valve body. This means that if the pipeline is blocked on both sides of the valve, closing it could increase the pressure enough that damage may occur (rupturing the diaphragm for instance).

Since an elastomeric diaphragm yields as it contacts the seat, closing a diaphragm valve has a different feel to the operator than closing a gate valve. Overtightening the valve may be avoided by providing diaphragm valves with an indicator to let the operator know that the valve is adequately closed.

Triple-Duty Valves

The triple-duty valve (Figure 8.29) is a combination valve that functions as a shut-off valve, a check valve, and a flow control valve for balancing flows. They are often employed in hydronic applications on the discharge of a pump. These valves save cost and space, but are rarely used in industrial applications.

Size		Pressure Rating	
NPS (in)	DN (mm)	psi	bar
1/2 to 4	15 to 100	150	10.3
5 to 6	125 to 150	125	8.6
8	200	100	6.9
10 to 12	150 to 300	65	4.5
14 to 16	350 to 400	50	3.4

TABLE 8.3 Sizes and pressure ratings of diaphragm valves.

Fig 8.29 A triple-duty valve. This valve functions as a shutoff, check, and balance valve. *Photo provided by Watts Regulator Company.*

Backflow Preventers

Cross connections exist whenever a potable water system may be contaminated by chemicals or microbes. Air gaps are one very reliable method of ensuring that the contaminant cannot reach the potable water system. Consider a fire protection system utilizing a tank and fire pump. The system feeds a carbon steel network of sprinklers in a plant. City water is introduced into the tank as shown in Figure 8.30. The overflow to drain tank is the flood level rim, and the distance between the discharge of the potable water line and the flood level rim is the air gap distance. One can see that there is no way for the process fluid to enter the potable water system. Because the fire protection system is constructed of carbon steel, with lots of stagnant dead legs, it must not be connected to the potable water system since it could contaminate the city water supply with dissolved iron, and possibly harmful bacteria.

Another way to design such a system might be to use the city water to directly feed the fire protection system as shown in Figure 8.31. This would have the advantage of eliminating the tank and pump (assuming that the city water pressure is sufficient). But if there were ever a drop in pressure in the city water line, as might occur if a line ruptures, the water in the fire protection system could get sucked back into the potable water main. In Figure 8.32 the potable water line should be protected with a backflow preventer.

Backflow preventers are devices that protect against backflow by checking the flow in the reverse direction. There are four types:

1. Atmospheric-Type Vacuum Breakers are the least expensive type of backflow preventer. They are available in sizes up to 3 inches, and must be installed vertically. They are suitable only for preventing backsiphonage.

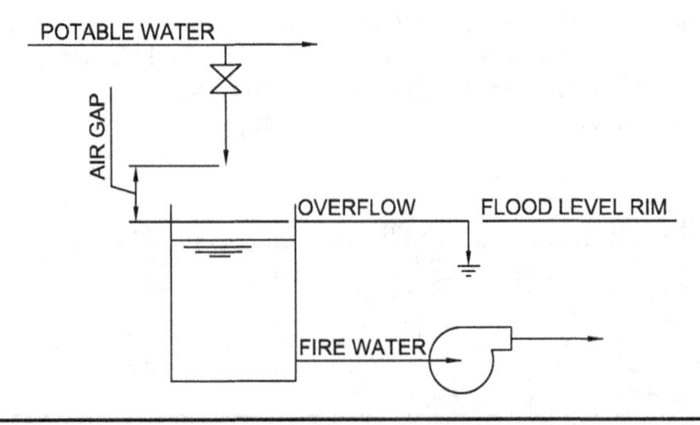

Figure 8.30 An air gap to prevent contamination of the potable water source.

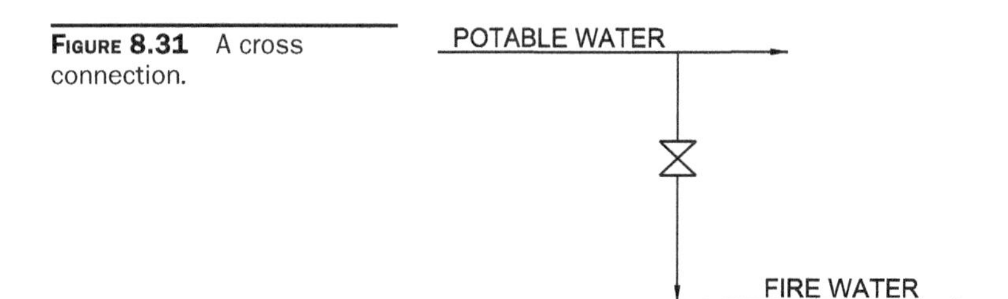

Figure 8.31 A cross connection.

2. Pressure-Type Vacuum Breakers are available in sizes from ½ in to 10 in. Pressure Vacuum Breakers are intended to prevent backsiphonage. They are therefore not appropriate for conditions that might induce a backpressure.
3. Double Check Backflow Prevention Assemblies are available in sizes up to 12 in. They consist of two check valves located between two full-port shutoff valves. There are also four test cocks on an Double Check BFP assembly. They are intended only for low hazard applications. Double Check Detector Assemblies are equipped with meters to assist in detecting leaks or unauthorized taps. See Figure 8.33.
4. Reduced Pressure Principle Backflow Prevention Assemblies are available in sizes up to 10 in. They consist of two check valves with a pressure differential relief valve between them, and the check valves are located between two full-port shutoff valves. There are also four test cocks on an RP assembly. The relief valve ensures that the pressure downstream of the first check valve is always less than the inlet pressure to the assembly. In normal operation, the relief is closed and no water is discharged. This assembly is considered to offer the highest degree of backflow prevention. See Figure 8.34. When equipped with a meter to detect leaks or unauthorized taps, these are referred to as Reduced Pressure Detector Assemblies. See Figure 8.35.

Note that a cross connection exists even if there is a *potential* for contamination. Figure 8.36 shows a potential cross connection in which a hose bibb has been attached to the test cock of a Reduced Pressure BFP.

Figure 8.32 A cross connection between a potable water line and a fire protection system. The second riser from the left is potable water from the city supply; the third riser supplies the fire protection system. If the potable water pressure drops, nothing prevents the stagnant and iron-laden fire protection water from entering the potable supply.

Figure 8.33 A double-check detector assembly. *Photo provided by Watts Regulator Company.*

Figure 8.34 A reduced pressure BFP in a utility service entrance room.

Figure 8.35 A reduced pressure backflow prevention detector assembly. *Photo provided by Watts Regulator Company.*

Figure 8.36 A reduced pressure backflow preventer with a hose bibb attached. The hose bibb constitutes a cross connection since the potential exists to attach a hose upstream of the backflow preventer.

ASSE Standard	1016	1017	1062	1066	1070
Title	Performance Requirements for Automatic Compensating Valves for Individual Showers and Tub/Shower Combinations	Performance Requirements for Temperature Actuated Mixing Valves for Hot Water Distribution Systems	Performance Requirements for Temperature Actuated, Flow Reduction (TAFR) Valves for Individual Fixture Fittings	Performance Requirements for Individual Pressure Balancing In-Line Valves for Individual Fixture Fittings	Performance Requirements for Water Temperature Limiting Devices
Location	Point-of-Use	Hot Water Source	Point-of-Use	Point-of-Use	Point-of-Use
Application	Individual Tubs and Tub/ Shower Combinations	Control of in-line domestic hot water temperature	Bath, Shower, Sink	Bath, Shower, Sink	Bath, Sink
Protects Against	Scalding and Thermal Shock	Scalding	Scalding	Thermal Shock	Scald
Regulation	±3.6°F	±3°F to ±7°F, depending on valve capacity	Reduce flow to 0.25 GPM within 5 seconds after outlet temperature reaches 120°F	±3°F with 50% change in water supply pressure and reduces flow to 0.5 GPM when cold water supply fails	±7°F

Table 8.4 ASSE Standards for scalding and thermal shock.

ASSE Valves

Water for bathing that is too hot can cause scalding. Water that is too cold can cause thermal shock, which in a shower could cause injury from slipping or falling.

Studies performed by the Harvard Medical School show that the time to produce injury to tissue decreases rapidly with only slight increases in temperature. Such increases in temperature could occur when cold-water pressure drops suddenly due to opening a valve on a washing machine, toilet, or other fixture. It only takes three seconds to cause a first-degree burn on adult skin tissue at 140°F (60°C) water temperature, and five seconds to cause a second degree burn at the same temperature[8].

One way to avoid scalding is to set the hot water temperature to a lower value, say 120°F (49°C). While this avoids scalding (it takes eight minutes at this temperature to produce a first degree burn in adults), it is also a temperature that supports the growth of Legionella bacteria. Because one is still faced with the prospect of thermal shock due to sudden cold water temperatures, a better solution is to install devices that are designed to control the water temperature.

A comparison of different ASSE devices appears in Table 8.4.

[8] Viola, David W. 2002, "Water Temperature Control and Limitation," Plumbing Manufacturers Institute.

ASSE 1016 Control Valves

These devices are used to provide scald and thermal shock protection at showers and bathtubs. In order to be listed as an ASSE 1016 valve, the device must be controllable by the bather and have an adjustable maximum temperature limit that is set by the installer (and may be later adjusted by the owner). The flow rate must also reduce to 0.5 GPM (1.9 l/min) if the cold water stops flowing to the valve.

The valve must maintain the desired outlet temperature within a range of 3.6°F (±2.0°C).

ASSE 1017 Control Valves

Valves manufactured to this standard are used to reduce temperature throughout the hot water distribution system. These are mixing valves located at the water-heating source rather than at the end use fixture. In the absence of other preventive measures, one should keep in mind that the Legionella bacteria do not die rapidly until temperatures reach 131°F (55°C).

ASSE 1016 valves should be used in conjunction with ASSE 1017 valves to prevent scalding.

ASSE 1062 Temperature Actuated Flow Reduction (TAFR) Valves

TAFR valves reduce the risk of scalding by restricting flow when the temperature exceeds a preset limit. The maximum limit is 120°F (49°C). These are point-of-use devices either integral to the fixture or available as an add-on after installation.

ASSE 1066 Pressure Balancing In-Line Valves

These valves are for preventing thermal shock on individual fixtures. The discharge temperature is permitted to vary 3 degrees from the set temperature when incoming water pressure changes 50 percent. While not designed specifically to prevent scalding, the flow through these devices is to be reduced to 0.5 GPM or less whenever the cold water supply fails.

ASSE 1070 Water Temperature Limiting Devices

These valves are for individual or multiple fixtures and protect against scalding. ASSE 1070 valves do not apply to showers.

Steam Traps

Steam traps remove condensate and noncondensibles (air) from steam lines. If sufficient pressure is available in the live steam line, the condensate can be pushed back to the boiler room without the need for condensate pumps. If the condensate is sufficiently clean, it is preferred to return it to the boiler for heat reclaim and re-use as boiler feedwater. Boiler feedwater is costly to create due to both the heat-up of the water and the treatments and chemicals required to maintain a clean steam system. Demineralization and/or anti-scaling treatments are required since minerals inside a steam system are essentially thermal insulators. Thus, an excellent place to save money in a plant is in the return of clean condensate back to the boiler.

Float Traps

A chamber contains a float that is on a linkage. As the chamber fills with condensate, the level rises until the float trips open a valve that expels the condensate. These are equipped with a thermostatic air vent to vent noncondensibles that become trapped within the chamber. Noncondensibles would displace steam inside the trap and prevent the formation of condensate.

Advantages

- Able to handle a wide range of condensate loads across a wide range of pressures and flows
- Large load capacity for its size
- Discharges condensate at essentially the same temperature as the steam
- Resistant to water hammer

Disadvantages

- Susceptible to freezing

Inverted Bucket Traps

The condensate flow into an inverted bucket trap is from the bottom. Inside the trap chamber an upside-down bucket hangs from a lever that is attached to a valve internal to the trap. When steam enters the trap, the bucket becomes buoyant and floats up to close the valve until the steam condenses.

A small hole at the top of the bucket permits air to be vented through the trap also.

Advantages

- Can withstand high pressure
- Resistant to water hammer
- Fails open, so it is safer for certain applications like turbine drains
- May be used on superheated steam by adding a check valve on the inlet

Disadvantages

- Bleeds-off noncondensibles very slowly
- Must use an inlet check valve if steam pressure fluctuates
- If trap loses water seal around the rim of the inverted bucket, steam is wasted. This can happen if there is a sudden loss of steam pressure that flashes the condensate to steam

Liquid Expansion Trap

This type of trap contains an oil-filled element that expands when steam enters the chamber. This seats a valve until the steam condenses, and then the oil-filled element contracts, pulling the disk away from the seat to allow the condensate to be expelled.

Advantages

- Can be manufactured to discharge at lower temperatures
- Excellent at discharging air and condensate during warm-up
- Withstands vibration and water hammer

Disadvantages

- Flexible element is susceptible to corrosion
- Not quick-acting

- May freeze
- Must usually be applied in parallel with another type of trap

Balanced Pressure Trap

Similar to the liquid expansion trap, the balanced pressure trap contains a sealed compartment that is filled with a solution with a boiling point lower than that of water. During start-up the valve is wide open so that air is easily vented. When condensate surrounds the sealed compartment the liquid contained inside vaporizes and expands the compartment, which seats an internal valve. Once the condensate cools, the vapor in the sealed compartment condenses and opens the valve to expel the condensate. This in turn allows steam to enter the trap and the liquid inside the sealed compartment again vaporizes.

Advantages

- Full open on start-up permits venting of noncondensibles and large condensate loads
- Not susceptible to freezing
- Suitable for varying steam pressure
- Can withstand superheat
- Easily maintained.

Disadvantages

- Bellows style traps are prone to failure from corrosion and water hammer. (Sealed compartments of welded stainless steel are better under these conditions).
- Does not evacuate condensate until condensate temperature drops below the steam temperature. This can lead to waterlogging.

Bimetallic Traps

Bimetallic discs inside these traps deflect due to temperature changes. This movement lifts a valve from the seat causing the condensate to be expelled. Because the forces exerted by bimetallic strips are small, a large mass must be used to exert enough force to lift the valve from the seat. The increased mass makes the bimetallic strip slow to react.

Some bimetallic traps are manufactured to produce a quick blast discharge, which helps to clear dirt from the internals.

Advantages

- Large condensate capacity
- Full open on start-up permits venting of noncondensibles and large condensate loads
- Not susceptible to freezing
- Withstands water hammer, corrosive condensate, and high steam pressures
- Easy to maintain without removing trap from the line.

Disadvantages

- A separate check valve may be required downstream of the trap if reverse flow is possible
- Waterlogging of the steam space will occur unless the steam trap is installed at the end of a 3 to 10 ft (1 to 3 m) long uninsulated cooling leg
- May be blocked by dirt due to low flows through the trap, although good practice indicates that traps be installed with strainers upstream
- Bimetallic steam traps do not respond quickly to changes in load or pressure due to the slow reaction time of the thermal mass of the bimetallic element.

Thermodynamic Traps

The thermodynamic trap has a small condensing chamber above a single disc that is permitted to rise and fall above another chamber that admits condensate. The hot condensate flashes to steam as it leaves the trap and some of this steam remains above the disc, seating it against the condensate chamber below. When this steam condenses, the condensate pressure below forces the disc open, and the cycle repeats.

Advantages

- Large condensate capacity
- Not susceptible to water hammer or vibration
- Not susceptible to corrosion or superheat
- Not damaged by freezing, and can even be freeze-proofed if installed with the disc in the vertical plane
- Easily maintained with trap in-line
- Clicking noise of disc opening and closing indicates proper operation.

Disadvantages

- Pressure drop across the trap must be sufficient for condensate to flash to steam.
- High-velocity air on start-up will cause trap to close and air-bind. A separate thermostatic air vent must be piped in parallel with the trap to alleviate this condition or the trap can be ordered with an integral anti-air-binding disc.
- Noise caused by clicking of the trap may prevent its use in some environments, although diffusers that reduce the noise can be fitted to the trap.
- Oversizing the trap will increase cycle time and subsequent wear.

Prior to start-up, steam lines must be blown to atmosphere, bypassing the steam traps. It is imperative that the oils, lacquers, mill scale, etc. that are present in an installed steam line are eliminated by bringing the lines to temperature and blowing them down with steam. If this is not accomplished, then the traps will clog and be rendered useless. This could lead to other problems such as water hammer.

Figure 8.37 A basket strainer with 125 lb flanged cast iron. *Photo provided by Watts Regulator Company*

Figure 8.38 A Y-type strainer. The flange at the bottom is removed to remove the strainer screen. *Photo provided by Watts Regulator Company.*

Strainers

Strainers may be cleaned manually or by automatically flushing. They are available as basket strainers in which the strainer screen is shaped like a bucket (Figure 8.37), and is removed from the top of the strainer through a flanged connection. Y-type strainers (Figure 8.38) have a screen that is open on both ends. The flow of fluid is through the center of the screen and radially through it. In order for this screen to be effective, both ends must be sealed. This strainer has the advantage of being able to be blown down (cleaned) in operation through the addition of a valve at the end of the strainer body, where the solids would collect. Most of these screens however are very flimsy, and care must be taken to avoid even slight amounts of backflow through the strainer that could cause the screen to collapse.

On larger strainers the screen is supported by a perforated metal liner that imparts more strength. This may not prevent the screens from collapsing on backflow however. Some larger strainers are shown in Figure 8.39.

Some self-cleaning strainers require a motor to rotate the screen, which is scraped. The solids are then automatically blown down. Still other strainers operate using the pressure differential across the screen to rotate internal components and flush out trapped solids. These are self-cleaning and require no external power.

The installation of strainers on water intakes for plants is imperative in areas affected by zebra mussels. These mussels are extremely prolific, and can completely clog pipelines and heat exchangers. While chemical treatments can be effective, their larvae can also be filtered from the water using strainers. These are tenacious creatures, being able to live out of water for up to 14 days.[9]

[9] Ironically, their filter-feeding has led to improved water quality in many bodies of water. Were it not for their tendency to attach to virtually any substrate, efforts to control them might not be so vigorous.

FIGURE 8.39 A Y-type strainer protecting the inlet of a centrifugal pump. The check valve at the pump discharge should protect backflow through this strainer. Note that the strainers are rotated horizontally due to a lack of elevation which would prevent the strainer screen from being removed. The blowdown lines are still located at the lowest point, although we recognize that this is not an optimal arrangement for blowing down particulates trapped by the strainer screen.

Instrumentation

Temperature Elements and Indicators

Temperature elements are often protected inside a pipe with a metal sheath that is inserted into the pipe. This sheath, called a thermowell, also permits the temperature element to be removed while the pipe is under pressure. The additional mass of the thermowell and the gap between the thermowell and the element tend to reduce the reaction time of the element to changes in fluid temperature.

Temperature elements are often inserted into a tee where the piping changes direction (where ordinarily an elbow would be used). This allows longer elements to be inserted into the fluid stream.

Pressure Transmitters and Indicators

Pressure transmitters and indicators are usually threaded into a coupling. It is good practice to install these devices so that they can be valved out-of-service for removal while the pipe is under pressure. It is even better practice to provide a bleed valve to safely relieve the pressure in the nipple connected to the PT or PI after the isolation valve is closed.

Flow Meters

The measurement of flow is very important to anyone who attempts to perform a heat balance on a system. Unfortunately, flow meters are often a rare commodity in a piping system, even though there are quite economical methods of capturing this information.

Orifice Plates

Probably the easiest way to measure flow is by installing an orifice plate. By measuring the pressure drop across a known geometry, the flow of the fluid can be estimated. These require full pipe flow, as well as a certain amount of stable flow, usually created by allowing five diameters of straight pipe downstream of the orifice plate, and a number of diameters of straight pipe upstream of the orifice. The number of diameters required upstream of the plate depends on:

- The ratio of the orifice diameter to the pipe diameter (called the "beta factor" of the orifice plate)
- Whether or not any bends exist in the vicinity of the orifice plate, and whether they are in-plane or out-of-plane
- Whether another flow disturbing component like a valve exists upstream of the orifice plate

The number of upstream diameters may range from 10 to 55 or more, depending upon the application.

Orifice plates are quite accurate but the turndown in flow rates may be only 4:1 or 5:1. They do not perform well at flows lower than 20 percent of the rated capacity.

Tubing is routed from the taps of an orifice flange to a differential pressure cell which converts the differential pressure into an electronic signal for use in monitoring or control. The pressure taps may also be located in the piping rather than in an orifice flange, but great care must be taken to ensure that the holes are square to the pipe axis, on center, and that burrs do not exist. All of these conditions could influence the accuracy of the resulting differential pressure.

Orifice plates can be used in liquids or gases, but dirty fluids containing entrained particles will erode the orifice over time. Like the taps, the orifices must be precisely machined, since any erosion or corrosion will affect the accuracy.

Annubar®

Annubar® is a registered trademark of Emerson Process Management/Rosemount. These devices function like a series of pitot tubes to measure the differential pressure in a fluid stream. They have multiple orifices across the profile of the pipe cross-section so that an average is obtained[10].

Turbine Flow Meters

A propeller or turbine can be mounted inside the flow stream. The velocity of the fluid imparts a rotation to the propeller and the rotational speed of the propeller can be related to the fluid velocity, and hence to the volumetric flow.

Turbine flow meters are accurate and do not require the long straight runs of pipe upstream and downstream of the device that some other instruments require.

Magnetic Flow Meter

Magnetic flow meters have no moving parts, so they are often used for measuring the flow of a dirty fluid stream. Mag flow meters take advantage of Faraday's Law, which states that the induced voltage across a conductor as it moves at right angles through a magnetic field is proportional to the velocity of the conductor. In this case, the conductor is the fluid itself, so fluids with poor conductance like hydrocarbons and distilled water are not good candidates for measuring with a mag flow meter.

Because mag flow meters are free of internal obstructions, they are excellent where low pressure drop and low maintenance are desired.

[10] Recall that the fluid stream exhibits a variable velocity profile across the pipe cross section with zero flow occurring at the surface of the pipe due to the no-slip condition at the fluid/pipe interface.

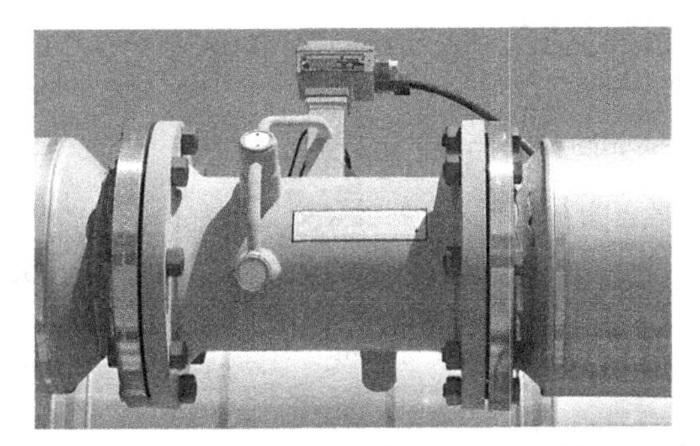

FIGURE 8.40 An ultrasonic flowmeter installed in a straight run of pipe. Note the sensors which pick up the Doppler shift of the discontinuities in the fluid.

Magmeters are accurate to within 1 percent, but they are sensitive to air bubbles. If air is entrained in the fluid, the flow will read higher than it really is.

Ultrasonic Flow Meters

Ultrasonic flow meters, such as shown in Figure 8.40, use a Doppler shift of reflected sound waves to measure the flow of a fluid. The fluid must contain discontinuities such as particulates of air bubbles to reflect the sound waves, and the liquid must be conductive.

Ultrasonic flow meters also are useful when low-pressure drop and low maintenance are desired. Clamp-on ultrasonic flow meters are available to measure flow inside a pipe that was installed without flow measurement devices. These have limitations however, and their accuracy is generally much worse than an in-line device and may be within only ±20 percent, depending on the pipe materials.

Hoses and Expansion Joints

Hoses

A wide variety of hoses are available, made of corrugated metal or various reinforced rubber or plastic materials. These may be fitted with any combination of end connections. Unless the hose is to be supplied with plain ends for fitting the end connection in the field, hoses are always special ordered.

The available pressure ratings can be quite high even for bores up to 2 in. Metal corrugated hoses are available with braiding on the exterior (similar to "Chinese handcuffs"). This braiding is flexible and helps to protect the hose, as well as assist in limiting the axial displacement due to internal pressure. See Figure 8.41. In the corrugations there exists two different areas over which internal pressure acts, as shown in Figure 8.42. The internal pressure acts on the additional area created by the annulus of the corrugations. This exerts a force that acts to stretch out the corrugated hose. The value of that thrust may be approximated by:

$$F = P\frac{\pi}{4} D_m^2 \qquad \text{(Equation 8.1)}$$

where

F = Thrust [lb]

P = Gauge pressure [lb/in^2]

D_m = Mean diameter = $\dfrac{OD + ID}{2}$

The thrust generated is independent of the number of corrugations.

Smooth bore hoses do not have this problem, but these are only available in reinforced rubber or plastic materials, and so are unable to resist higher pressures.

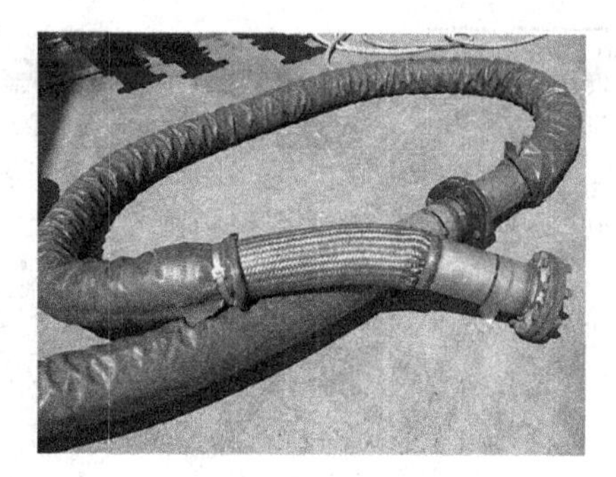

FIGURE 8.41 A corrugated stainless steel hose. Note the exterior braiding which acts as an axial restraint.

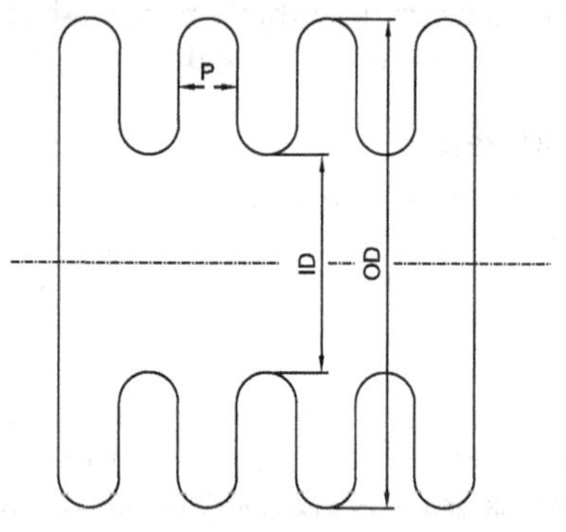

FIGURE 8.42 The pressure inside expansion joint bellows acts to stretch the bellows.

The value of hoses lies in their ability to account for movement, and to limit mechanical vibrations. But there are limits on how much the hose can move, and these limits must be accounted for in the design of the piping configuration. The minimum bend radius must be observed to prevent kinking, and twisting about the axis of the hose is to be avoided.

Expansion Joints

Expansion joints are used to take up displacements caused by thermal expansion of the adjoining pipe. Expansion joints should probably be avoided if possible, but sometimes the displacements and subsequent forces are too great to be absorbed by either the inherent flexibility of the piping or the ability of the supporting structure to absorb the forces and moments imposed by the thermal growth.

Expansion joints for pressure piping are always fabricated of corrugated metal, and the additional forces imposed by the internal pressure acting on the annulus of the corrugations are absorbed by tie rods that bind the flanges together across the expansion joint. Cloth expansion joints are available for high temperature ductwork.

The amount of movement absorbed by expansion joints is actually not very much. And the movements are cumulative, so that if an expansion joint is rated to absorb ½ in of axial movement, that amount of axial displacement will be de-rated if it also has to account for any parallel displacement between the adjoining pipes.

Due to the potentially large pressure thrusts produced by the corrugations of expansion joints, most applications require that the joints be restrained with tie rods. This means that the best location for an expansion joint to accommodate thermal growth is probably in a line that is at right angles to the line that experiences the growth. In that way the expansion joint can deflect laterally rather than axially.

CHAPTER 9

Pipe Supports

My first exposure to stress analysis came when I received a call from a contractor who was installing a 3 in continuous blowdown line on a boiler at a paper mill. "These hangers ain't loading up," he complained. My company had hired a contract engineer to perform a stress analysis for this system. He had performed the analysis and moved on to his next job, leaving me with a computer analysis with which I was unfamiliar and a contractor who couldn't get the spring hangers to budge under the weight of the 3 in pipe. When I visited the site, I was impressed with the size of the hangers purchased for the installation. Sifting through the analysis input, I learned that the contractor had overridden the default insulation densities, electing to enter them manually. Which would have been okay if he hadn't moved the decimal point to the right by three extra places. The program sized hangers for insulation that would be 1000 times as dense as the calcium silicate that was installed.

There may be no other area within piping design that causes as much difficulty as proper pipe support. The designer must be cognizant of multiple issues:

- Do any lines require stress analysis?
- How far apart should the supports be spaced?
- What type of supports are appropriate for the service?
- Is there enough space to install the hangers, cans, supports, etc.?
- Will the lines be restrained adequately in the horizontal plane?

Reference Standards

MSS	SP-58	Pipe Hangers and Supports – Materials, Design, and Manufacture
MSS	SP-69	Pipe Hangers and Supports – Selection and Application
MSS	SP-89	Pipe Hangers and Supports – Fabrication and Installation Practices
MSS	SP-90	Guidelines on Terminology for Pipe Hangers and Supports
ASCE	7	Minimum Design Loads for Buildings and Other Structures
ASME	B31.1	Power Piping
ASME	B31.3	Process Piping
IBC		International Building Code (from International Code Council)

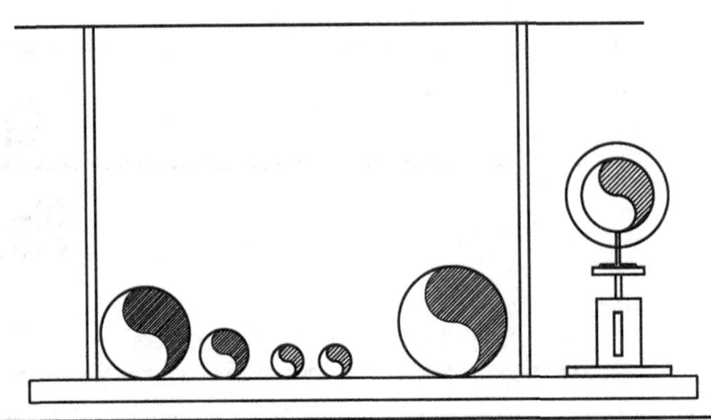

Figure 9.1 A trapeze support. Note smaller lines in center, heavier lines near vertical supports.

Pipe Routings

A "rookie" mistake is to lay out a pipe route without regard for how it is to be supported. Piping is rarely self-supporting. Provisions must almost always be made for supporting piping in order to keep loads off of equipment flanges or to prevent excessive sagging in the pipe that at worst interferes with drainage and at best looks bad.

If pipe racks are not available, the designer must look for overhead steel[1] or walls with which to support the lines. Of course looking for steel is not always enough. The steel must be adequate to support the static and dynamic loads that may be imposed by the piping.

On a new project this steel may be provided in the form of pipe racks. These racks are specially designed to support the piping and the expected loads need to be conveyed to the structural engineer early in the project. On an existing installation, where a new line is to be run, the steel must often be checked to ensure that it is able to support the additional loads. In a world filled with "pre-engineered buildings," the ability of the structure to handle additional piping loads is not a foregone conclusion. Pre-engineered buildings with tapered columns, purlins, and bar joists are designed to be economical by minimizing the amount of steel required for the anticipated design conditions. Once the clever structural engineer accounts for wind loads, snow loads, dead and live loads, there is often very little left for piping. The piping designer must then often post up from the floor to prevent overstressing the structure.

Common sense dictates that larger or heavier pipes be located closest to the vertical supports holding them. Lighter loads can be placed toward the center of a horizontal support. See Figure 9.1.

Support Considerations

Lines supported by rods from above will be more flexible than lines supported by shoes from below. Rods can pivot about their attachment point, and so they are not the best choice for lines that are not otherwise restrained horizontally. Struts or snubbers may be required to prevent the pipes from moving excessively in the horizontal plane. Pipe

[1] "Steel" in this sense refers to structural steel. Any other structure besides structural steel may be used if it is adequate for the loads imposed by the piping, but structural steel is the most common in industrial settings.

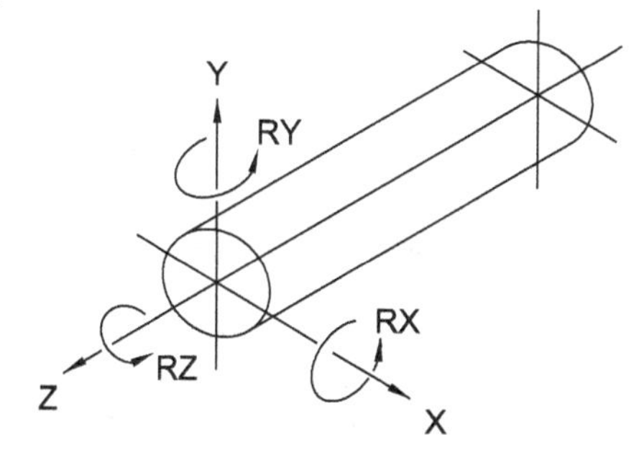

Figure 9.2 Pipe support orientation.

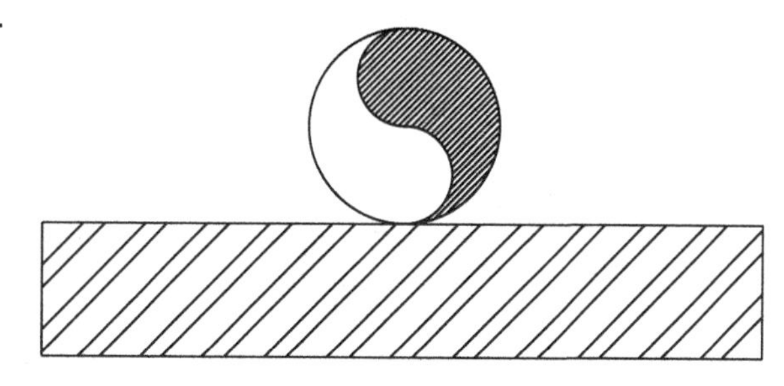

Figure 9.3 Single restraint, 5 DOF.

supported from below can take advantage of the friction between the pipe and the support (if it is rigid enough) to limit horizontal movement.

Degrees-of-Freedom

Consider the pipe segment shown in Figure 9.2. The axis of the centerline lies along the Z-axis, the vertical direction lies along the Y-axis, and the direction transverse to the pipe centerline and in the horizontal plane is the X-axis. Each of these axes can also be used to describe rotational parameters by applying the right-hand rule[2] around the axis.

This convention can be used to describe displacements, forces, and moments. Because we have three linear axes and three rotational axes, we have six degrees-of-freedom (DOF). An unrestrained point on a pipe floating in space would have six DOF. We selectively remove some of these DOFs when we apply restraints or supports to the pipe.

A simple pipe support is shown in Figure 9.3. We lay the pipe across a piece of supporting steel and rely on gravity to hold it in place. The steel provides a vertical restraint, and since the pipe is free to move in the positive vertical direction (up), if a force so acted upon it, we say that this support is a +Y restraint.

Perhaps we don't want the pipe to ever lift off of the supporting steel. We can install a U-bolt to secure it as shown in Figure 9.4. This is a Y-restraint, but on closer inspection, we can argue that since the U-bolt traps the pipe in the transverse direction, there is also an X restraint. Further, if the nuts are very tight, there may also be an axial restraint acting along the Z-axis. This Z-axis restraint is provided only at two points—the 12 and 6 o'clock positions of the pipe. This is not much to hold the pipe in the axial direction, and it relies entirely on the friction force produced by the tightening of the nuts. Still, if the supporting steel is stiff in the axial direction and if the nuts are tight, we can say that the pipe is restrained axially.

[2] Place your right thumb in the direction of the axis, and make a fist. The rotational parameter is oriented in the same direction as your fingers.

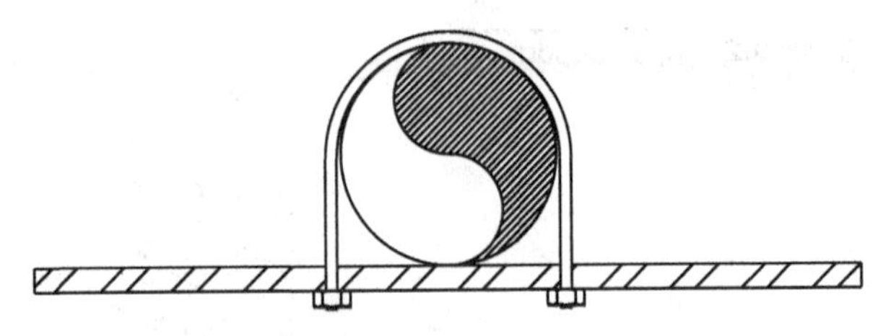

FIGURE 9.4 A U-bolt with single nuts at bottom.

We also note that with this U-bolt support, the pipe is free to rotate in the RY and RX directions. This support acts with two DOF if the nuts are tight. If the U-bolt is double nutted as in Figure 9.6 and some play is permitted between the U-bolt and the pipe, then it becomes a four DOF support, with only the vertical and transverse directions restrained (after some small permitted translation in those directions).

Types of Supports

The support specified for a given location depends on the desired amount of movement and the space available for the support. Many supports and support components are available as manufactured items, but in practice the simpler components are usually fabricated in the field, especially on industrial projects. The reason for this is that much of the design for industrial projects is custom, due to the location or load requirements of the particular support. Unless the project requires dozens of "medium steel brackets," they are more likely to be field fabbed than purchased.

Vendor catalogs and MSS Standard Practices do a good job of showing various components in some of their appropriate configurations, but careful examination of the notes accompanying such information indicates that the responsibility for the correct selection of components for a given application belongs to the design engineer. You will find that many supports will require custom designs rather than standard details.

Rack Piping

Consider a long mill building of several hundred feet, with structural framing along the walls to support piping and cable trays. Of course we will need to account for thermal expansion of hot lines, but what kind of care should we exercise to ensure that headers that are operating at more-or-less ambient do not impose unreasonable forces on the building structure? After all, even the building has expansion joints designed into it to account for seasonal thermal growth.

As a general rule, we are not concerned with thermal expansion, regardless of length, until the temperature *difference* between the installed condition and the operating condition approaches 150°F (approximately 80°C) delta T. In the case of long utility headers, because they are fabricated of materials very similar to those used in the buildings, we expect them to grow at the same rate as the building (at ambient temperatures). The building has expansion joints for the skin and even for major horizontal members such as crane girders. The columns are definitely designed to remain vertical.

As stipulated above, we recognize that there may be lines that operate at process temperatures different from ambient. The most obvious of these is steam and condensate, and expansion of these will have to be engineered into the system. Warm services such as Cooling Water Return, may have operating temperatures in the 135 to 140°F (57 to 60°C) range. The ambient design temperature in a cool climate may be 40°F (4°C), yielding a delta T of 95°F. Not really a large temperature difference, but will it move relative to the building steel? Yes. Will it matter? That will depend on how the headers are restrained.

Utility rack piping is often restrained with U-bolts. As happens so often in engineering, we can take a seemingly simple tool like a U-bolt and turn it into a complicated analysis. How should the nuts be applied to the U-bolt?

- A single nut on the bottom of each leg as shown in Figure 9.4 will snug the U-bolt onto the pipe, and grab the pipe at two points. Without some other means to hold the nut in place, it is easy to envision vibrations and slight thermal growth working the nut loose. The nuts could be tack-welded, peened, pinned, wired, or installed with lock washers to prevent them from working loose. Or, the nuts could be double-nutted as shown in Figure 9.5.
- Applying the nuts to both the top and bottom of the steel as shown in Figure 9.6 also solves the problem of the nuts working loose. It also offers an opportunity to use the U-bolt to guide the pipe rather than hold it fast.

If the pipe is actually clamped by the U-bolt as in Figures 9.4 and 9.5, we expect it to move through the U-bolts. The U-bolt support grabs the pipe at only two points, top and bottom, and the friction usually isn't sufficient to stop the pipe from moving. In an extreme case, some of the U-bolts may start to lean. When the pipe cools, they should return to their original position.

In practice, most rack piping will not be a straight shot through the length of the plant. There are obstacles to be avoided such as equipment, overhead doors, etc. And it may be a good idea to include a few bends in such piping anyway, since the effect of the pipe moving would be ameliorated by bends designed into the headers, which add flexibility.

It is sometimes suggested that it would be easy to install rack piping through the end of the building, by leaving the siding off, and then just sliding the lengths into place and performing the field welds outside the building. This may be possible if:

1. There are limited branch connections.
2. There are no in-line valves.
3. There are no bends.
4. The locations of the branch connections is not critical.

FIGURE 9.5 A double-nutted U-bolt. This configuration permits movement in all directions and rotations, but limits displacement in the vertical and transverse directions.

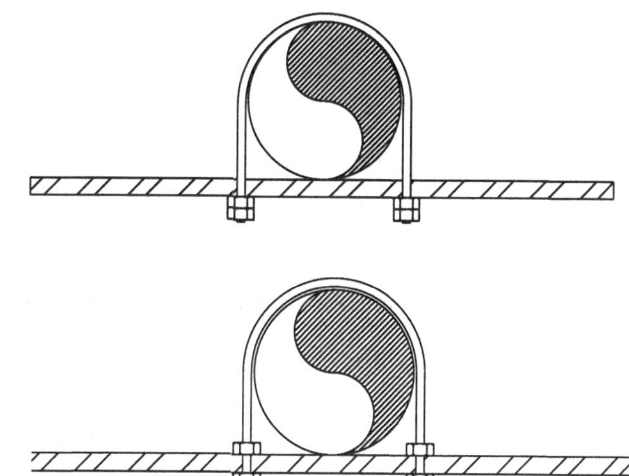

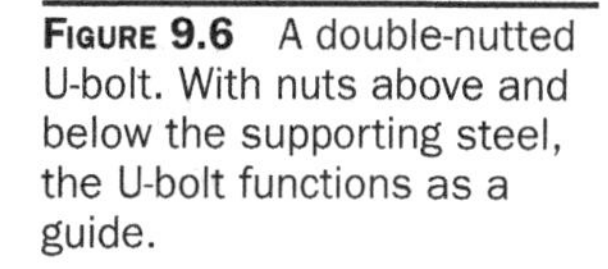

FIGURE 9.6 A double-nutted U-bolt. With nuts above and below the supporting steel, the U-bolt functions as a guide.

In other words, this is not a feasible construction method. A better method if rigging space permits is to prefab the piping and structural frame in sections. The piping can be mounted in the rack frames in the shop, and painted or insulated at ground level. At the construction site the entire frame can be lifted into place and the piping connections completed between sections. This can be done by welding, but flanging is easier. Some consideration must be made for proper field alignment.

Sloping of headers is often advantageous to facilitate drainage, but on long rack piping, this may prove more trouble than it is worth. The height at the end of a 500 ft run would have to be 62.5 in to accommodate even a modest slope of 1/8 in/ft. And the trouble in sloping such a line has to be weighed against the effectiveness of such sloping. No line is going to drain completely anyway, and if the process demands such levels of cleanliness or recovery, then you probably need to consider a pigging system. Intermediate drains from slopes every 50 or 100 ft may be possible, but again, the cost and constructability will probably rule these options out as well.

Finally, rack piping should be designed using Flat-On-Bottom (FOB) reducers as the header size decreases so that the piping does not require shimming. This permits a constant top of steel elevation along the rack.

Structural Supports

Throughout the analysis of pipe supports there must be considerable coordination between the piping engineer and the structural engineer. It is advisable to solicit input from both the structural and the electrical engineers at the beginning of the project, since cable trays and conduits will likely be supported from the same steel as the piping. The territory should be carved out at the beginning to avoid later interferences, and to arrive at an intelligent design that meets the requirements of all of the disciplines.

There is naturally some overlap between the responsibilities of the piping and structural engineer with regard to supports. It is a good idea to arrive at an understanding of when help will be needed with structural supports. Trivial supports certainly fall under the purvey of the piping engineer, but the cut-off of where help should be solicited really depends on the confidence of the piping engineer in distributing high loads to the structure. One rule-of-thumb is to seek assistance for loads greater than 10,000 lb (or perhaps 4500 kg) or moments greater than 10,000 ft-lb (approximately 1400 kgf-m).

While it is not possible to know what loading conditions may be in the future, there are several considerations that should be recognized:

- A corrosion allowance of 1/16 to 3/16 in is often taken on the steel members for harsh environments. Many plants have deferred maintenance, and unless you specify hot dipped galvanized steel or a superior coating system, a liberal corrosion allowance is appropriate.
- Foundations are much less expensive when initially installed and should be designed conservatively because modifications to foundations are usually very expensive. Foundation design is best left to a qualified structural engineer.
- The maximum tension or compressive stresses in the steel should be in the range of 12 to 16 ksi for A36, or 16 to 22 ksi for A992 Grade 50, as long as these values do not exceed the allowable stresses calculated based on the support geometry.

For stand-alone posts that are subjected to eccentric loads, consider using a closed section, like a pipe or structural steel tube (Hollow Structural Sections, or "HSS"). Not only do these sections offer superior resistance to axial torque, they are easier to paint than wide flange beams. A slender wide flange used as a column has very little resistance to torque, and can be deflected significantly under light loads.

Support Spacing

Years ago at a consulting firm we decided to save money by using PVC instead of stainless steel for a deionized water line. Deionized water is not corrosive, but it is something of a solvent in that it is "hungry for ions." Carbon steel would be a poor choice for DI water since the iron would dissolve.

Unfortunately, the support spacing in the rack piping was too long for the PVC, and when the client complained that the piping sagged, a retrofit with steel channels was made to help support the PVC.

There are many tables available to the engineer with recommendations on the spacing of supports. Some of these are more useful than others. The International Mechanical Code provides a table that merely provides a maximum horizontal spacing without regard for diameter. Clearly such a table is geared toward hydronics and plumbing rather than industrial applications.

ASME B31.1 provides guidelines in Table 121.5 for maximum spacing based on certain typical criteria such as:

- Horizontal pipe runs
- Maximum operating temperature of 750°F (400°C)
- No concentrated loads such as valves, flanges, specialties (or a vertical run)
- Insulated carbon steel pipe with a bending stress not greater than 2300 psi
- Standard weight or heavier
- Permissible sag is 0.1 in (2.5 mm)

The ASME B31.1 recommended support spans are provided in Table 9.1, along with additional sizes not included the ASME table. In addition to the conditions above, note the following:

1. No consideration is made for seismic conditions. Seismically sensitive areas should always have a stress analysis performed for critical lines.
2. Even though the table groups steam together with air and gas, the prudent designer will consider steam lines to be water-filled. This will certainly be the case for the hydrotest, and substantial water may exist in steam lines during other periods such as start-up.
3. The 0.1 in sag is a very common criterion for support spacings. This is considered to allow adequate drainage, but one must recognize that there will always be some pocketing of liquids in any pipeline installed with normal construction tolerances. Think of the 0.1 in sag as a design guideline, rather than an acceptable limit for drainage.[3]

[3] If wind loads are a serious consideration, a separate analysis may be undertaken to ensure that the natural frequency of the pipe segments is greater than 4 Hz. One source states that if the sag exceeds 5/8 in, the natural frequency may exceed 4 Hz. But normal design practice of 0.1 in sag would eliminate this altogether.

Nominal Pipe Size	Suggested Maximum Span				Minimum Threaded Rod Dia	Safe Rod Load
	Water Service		Steam, Gas, or Air Service			
	ft	m	ft	m	in	lb
1	7	2.1	9	2.7	0.375	610
2	10	3	13	4	0.375	610
3	12	3.7	15	4.6	0.500	1130
4	14	4.3	17	5.2	0.635	1810
6	17	5.2	21	6.4	0.750	2710
8	19	5.8	24	7.3	0.875	3770
10	21	6.4	27	8.2	0.875	3770
12	23	7	30	9.1	0.875	3770
14	25	7.6	32	9.8	1.000	4960
16	27	8.2	35	10.7	1.000	4960
18	28	8.5	37	11.3	1.125	6230
20	30	9.1	39	11.9	1.250	8000
24	32	9.8	42	12.8	1.500	11630
30	33	10	44	13.4	1.750	15700
36	35	10.7	48	14.6	2.000	20700

TABLE 9.1 Recommended pipe support spacing for horizontal carbon steel pipe with no concentrated loads, and temperatures below 750°F (400°C).

4. A conservative rule-of-thumb up to and including 20 in diameter pipe is that the support span in feet should not exceed 10 ft plus 12 times the nominal pipe size. For example, the span for 4 in pipe should not exceed 10 ft plus 4 ft equals 14 ft.

Suggested pipe spacings for various plastic piping materials are shown in Table 9.2.

Shoes

Shoes are among the most common restraints, and are used extensively for insulated pipe as they provide clearance for the insulation between the pipe and the support steel. They are most often fabricated from WT4s for insulation up to 3 1/2 in and WT6s for insulation thicknesses between 4 and 5 1/2 in. Manufactured shoes are also available with slots cut to accommodate the banding of insulation and jackets.

Shoes may be guided by welding strips parallel to the sole plate. This constitutes a transverse restraint. Clips (angles) over the sole plate can further be used to limit the amount of movement up as shown in Figure 9.7.

Shoes should be designed so that they extend over both sides of the bearing surface of the supporting steel. This is to prevent dirt or water from accumulating on the bearing surface. The length of the shoe must also be sufficient to prevent the shoe from falling off the support steel. If this happens the pipe will be prevented from returning to its

NPS	DN	Suggested Maximum Span					
		Fiberglass		PVC Schedule 80		CPVC Schedule 80	
		ft	m	ft	m	ft	m
0.5	15	—	—	—	—	3	0.9
0.75	20	—	—	—	—	3	0.9
1	25	12	3.7	6	1.8	3	0.9
1.5	40	14	4.3	6.5	2.0	4	1.2
2	50	15	4.6	7	2.1	4	1.2
3	80	17	5.2	8	2.4	4	1.2
4	100	19	5.8	9	2.7	4	1.2
6	150	22	6.7	10	3.0	4	1.2
8	200	24	7.3	11	3.4	4	1.2
10	250	27	8.2	12	3.7	4	1.2

TABLE 9.2 Recommended pipe support spacing for plastics for water service.

neutral position upon cool-down. See Figure 9.8. This will place an inordinate amount of stress on the piping (and the support steel) upon cooling.

Example 9.1 Force produced by axial strain of piping
Given: A 6 in Sch 40 carbon steel pipe, 10 ft long, anchored at both ends
Find: The amount of force generated by the expansion of the pipe from 70°F to 250°F

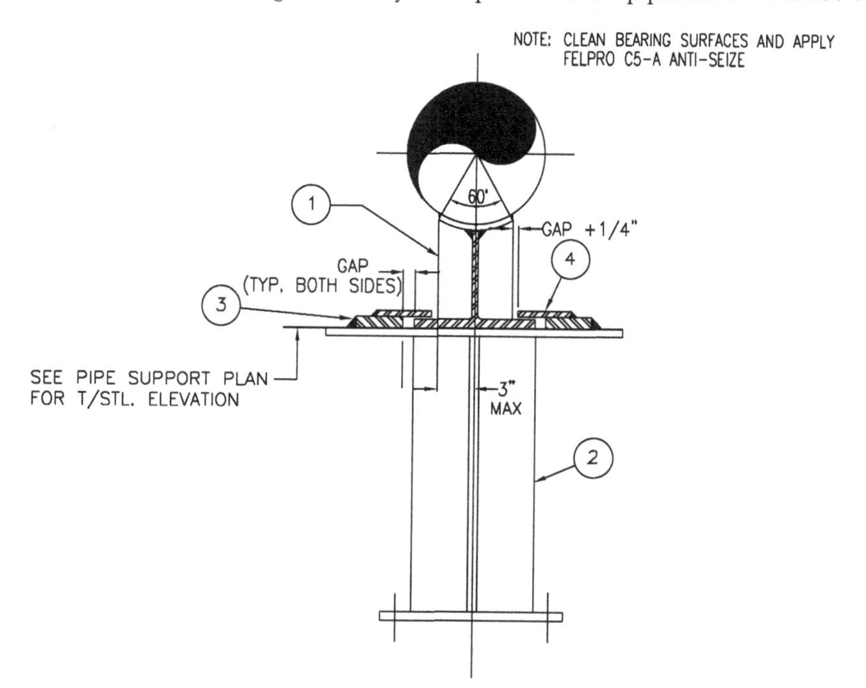

FIGURE 9.7 A guided shoe.

FIGURE 9.8 A shoe that has slipped off of support steel will place high stress on pipe and support steel upon cooling.

Solution: Let

F = Force [lb]
A = Metal cross-sectional area [in²] = 5.581 in² for 6 in Sch 40 pipe
E = Young's Modulus = 30×10^6 lb/in²
ϵ = Strain [in/in]
σ = Stress [lb/in²]
δ = Displacement due to thermal growth [in]
L = Length of pipe [in]
A = Coefficient of Linear Expansion = 0.000 0065 in/in/°F for carbon steel

The thermal growth is given by

$$\delta = \alpha L(\Delta T) \quad \text{Equation 9.1}$$
$$= (0.0000065 \text{ in/in/°F})(120 \text{ in})(250°\text{F} - 70°\text{F}) = 0.14 \text{ in}$$

Strain is defined as

$$\epsilon = \delta/L \quad \text{Equation 9.2}$$
$$= 0.14 \text{ in}/120 \text{ in} = 0.0012 \text{ in/in}$$

We recall that Young's modulus is equal to the slope of the stress-strain curve in the elastic region, so

$$E = \sigma/\epsilon$$

Rearranging,

$$\sigma = E\epsilon \quad \text{Equation 9.3}$$
$$= (30 \times 10^6 \text{ lb/in}^2)(0.0012 \text{ in/in}) = 36{,}000 \text{ lb/in}^2$$
$$\sigma = F/A \quad \text{Equation 9.4}$$

so

$$F = \sigma A = (36{,}000 \text{ lb/in}^2)(5.581 \text{ in}^2) = 200{,}000 \text{ lb}$$

Obviously, something is going to move. If one of the "anchors" is really not an anchor in the theoretical sense, but deflects 0.02 in (about half of a millimeter), then the strain reduces to

$$\epsilon = (0.14\ \text{in} - 0.02\ \text{in})/120\ \text{in} = 0.001\ \text{in/in and}$$

$$\sigma = E\epsilon = (30 \times 10^6\ \text{lb/in}^2)\ (0.001\ \text{in/in}) = 30{,}000\ \text{lb/in}^2,\ \text{yielding}$$

$$F = \sigma A = (30{,}000\ \text{lb/in}^2)\ (5.581\ \text{in}^2) = 167{,}000\ \text{lb, a difference of 33,000 lb due to the movement of one anchor by half of a millimeter.}$$

The forces generated by even moderately-sized pipes at modest temperature changes are tremendous, but even slight deflection of anchors alleviates the strain and subsequent forces.

The engineer must keep in mind that these forces will be transmitted to the supporting steel, which very often is the building steel (which may be supporting heavy equipment, cranes, etc.). If the displacements are large enough, hot piping is certainly capable of stressing steel into the plastic range, and thermal forces must be accounted for in the design.

Anchors

Anchors restrict displacements in all three axes, and rotation in all three axes. The simplest anchor is a double U-bolt as shown in Figure 9.9. For small loads, this may be adequate, but again, one must recognize that a U-bolt will grab the pipe only at the top and bottom. This cannot be considered a rigid connection for anything other than light loads. Other, more rigid anchors will be of welded construction, as illustrated in Figure 9.10. This anchor utilizes a C4x7.25 with the legs welded to a pipe between NPS 4 and 8 in. This provides a very good bearing surface for the pipe, permits an easy weld, and grabs the pipe securely. The C4 can be welded to any convenient (and rigid) structure.

Other more complex anchors are limited only by the imagination of the engineer, or the economics of the project.

Trapezes

Trapezes are erected below overhead steel to provide horizontal members on which to support pipes or electrical utilities. The principals of design are similar to rack piping, namely, to locate the larger or heavier lines near the vertical members.

Light duty services may take advantage of a variety of manufactured clamps similar to Figures 9.11 and 9.12 to attach to overhead steel. Medium duty services may use welding lugs similar to Figure 9.13 to attach the vertical members to the overhead steel. Neither of these trapezes will resist horizontal forces however.

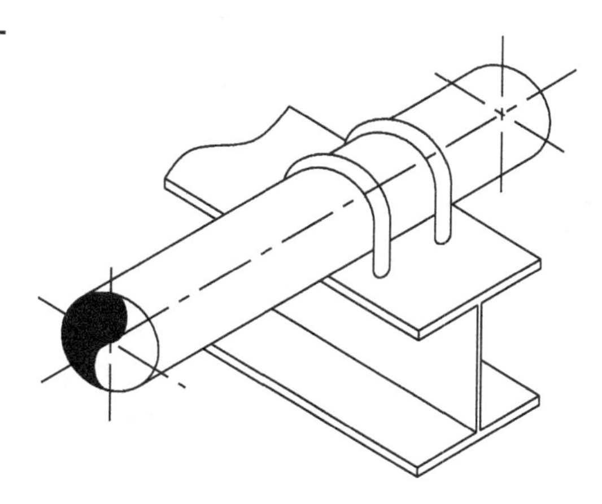

FIGURE 9.9 Double U-bolts acting as an anchor.

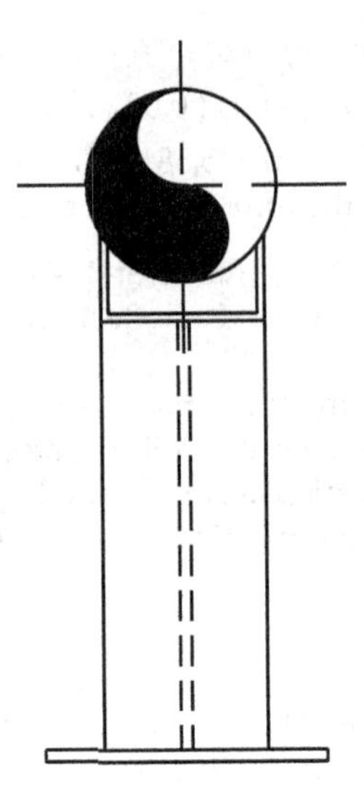

Figure 9.10 A small anchor for NPS 4 to 8 in, using a C4.

Figure 9.11 Anvil Figure 133 standard duty beam clamp. *Courtesy of Anvil International.*

Figure 9.12 Anvil Figure 86 C clamp. *Courtesy of Anvil International.*

Figure 9.13 Anvil Figure 55 welding lug. *Courtesy of Anvil International.*

Heavy duty trapezes that must resist horizontal forces are welded to the overhead structural steel so that moments are transferred into the joint between the overhead steel and the vertical members of the trapeze.

Rods

Single rods may be dropped from overhead steel to support a single pipe as shown in Figure 9.14. Note that rods will swing if subjected to horizontal forces. If the horizontal displacement is large, considerable force can be generated in the horizontal direction. Figure 9.15 shows a rod hanger that has been subjected to horizontal thermal growth. In practice, rods are limited to applications in which the expected angle from vertical is less than 4° (sin 4° = 0.07).

Note that erection is made easier by using a turnbuckle to connect rods with left- and right-hand threads. This simplifies the adjustment of elevation. Simply using a connecting coupling on right-hand threaded rods implies that the rods must be cut to the exact length.

Rollers

Where significant axial growth will occur, rollers are commonly used. This is more typical of cross-country transmission lines (especially where they cross bridges) than it is of industrial lines. The rollers prevent the expansion loads from being transferred into the bridge structure. For industrial lines shoes are more typical.

Since the axial growth is caused by thermal expansion, and since hot lines require insulation, it is necessary to install a crush-proof saddle over the roller to protect the insulation. Some manufacturers provide insulated saddles, and others are plain, with the insulation abutting the saddle.

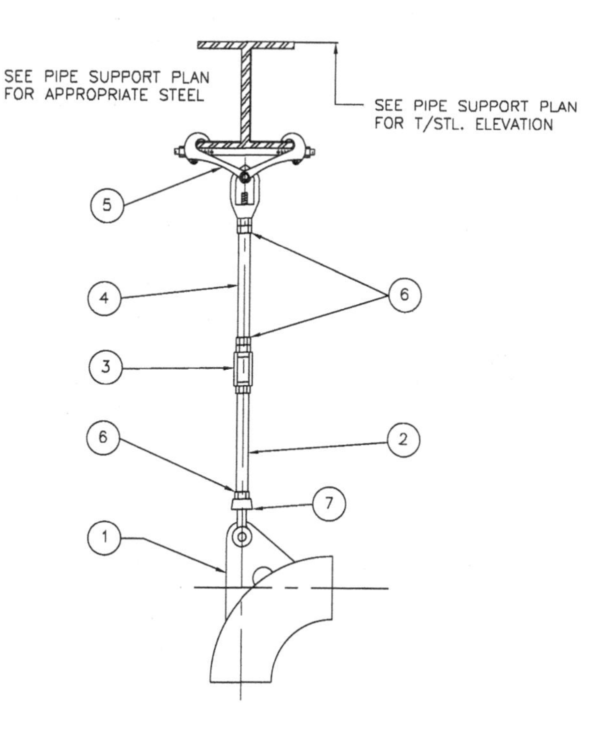

FIGURE 9.14 A rod hanger.

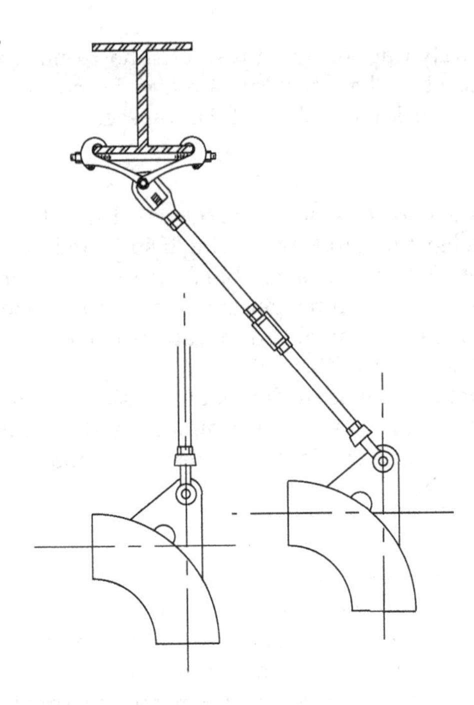

Figure 9.15 A displaced rod hanger (greatly exaggerated).

Spring Hangers

Spring hangers are available as variable spring hangers or constant support hangers. The variable spring hanger compresses a spring inside a can, and depending on the piping configuration, the reaction of the spring is either greater or less in the cold position than it is in the hot position. In most cases, the movement, and hence the spring compression, is slight, and the difference in the force applied by the spring is tolerable. Hooke's Law states that the force exerted by a spring is proportional to the deflection of the spring.

In other cases, and especially where the calculated vertical growth is large, a constant support hanger is used. These also use springs, but through the clever design of levers, the reaction on the pipe remains a constant force.

Great care must be taken in selecting spring hangers to ensure that sufficient vertical space exists either above or below the piping to fit not only the spring hanger, but the rods, shoes, or other components required to tie the pipe and spring hanger to the structure.

See Figure 9.16 for an example of a spring hanger installation.

Spring hangers are shipped with travel stops that must be removed prior to start-up.

Stress Analysis

Piping that requires a stress analysis generally meets one of the following conditions:

1. Lines 3 in (DN 80) or larger, whose operating temperature is 150°F (80°C) higher than the installed temperature[4]

[4] Note again that the temperatures cited are delta T.

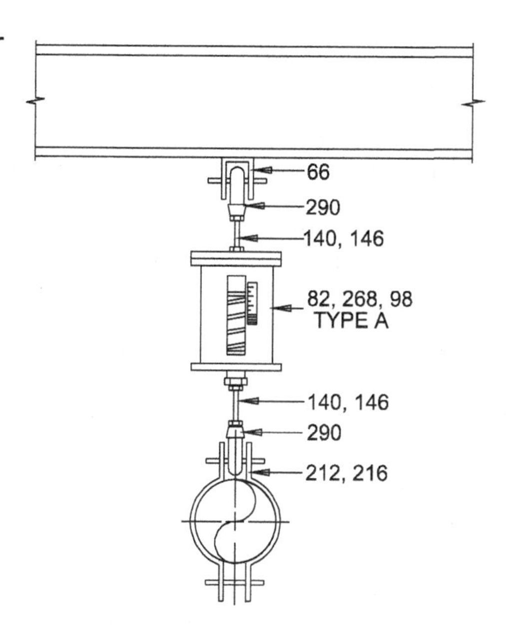

FIGURE 9.16 Sample spring hanger detail with call-outs for manufacturer's part numbers. *Courtesy of Anvil International.*

2. Suction and discharge lines from reciprocating or rotating equipment
3. Relief valve stacks with an inlet pressure greater than 150 psig (10 bar)

It is up to the Mechanical Engineer to determine when it is appropriate to perform an analysis. Usually, this will be stated in a Design Basis, but often it is overlooked.

One method of determining if an analysis is required is to apply the flexibility criteria given in ASME B31.1, 119.7.1, and ASME B31.3, 319.4.1 and discussed in Chapter 4.

$$Dy/(L - U)^2 \leq 30\, S_A/E_C \qquad \text{Equation 9.5}$$

Stress analysis is sometimes referred to as "flexibility analysis." The two terms are synonymous, since if a line is sufficiently flexible it will not be overstressed. While stress analyses had been performed by hand for many years, the availability of software to perform the complex calculations has rendered hand calculations obsolete. Further, the analysis of complex configurations is simplified considerably by use of the stress analysis software.

Stress Analysis Software

A variety of software packages are commercially available to aid in the analysis of piping stresses. Most of these programs convert the analog concept of a drawing into discrete digital data. This becomes an exercise in entering the data in a manner that accurately represents the piping system. One popular program is CAESAR II®, a registered trademark of COADE, Inc. The discussion that follows is based on CAESAR II®, but much of it will also be applicable to other pipe stress analysis packages.

Understanding Stress Analysis Software

What information does one get from a stress analysis?

- Confirmation that the internal stresses are less than the code allowable stresses
- Displacements at each of the node points

- Reactions at pipe supports
- Specification of spring hangers, if they were modeled in the system

It is important to note that the software does not locate or suggest locations of pipe supports. Nor does it suggest when to use spring hangers. These decisions are left up to the analyst.

The analyst builds a computer model of the pipe system using a set of rules that the software can understand. The software can be very picky about understanding these rules, and this often leads to a lot of time spent debugging the model. The results of the analysis are only as good as the accuracy of the model. The analyst must always exercise good engineering judgment in developing the model, and in interpreting the results.

The software works by breaking the model down into a series of segments. Then it analyzes each segment, considering thermal expansion and beam equations. Any loads are applied as concentrated or uniform loads. These loads include:

- The weight of the pipe itself
- The weight of fluid inside the pipe
- The weight of any insulation
- The weight of flanges or valves
- Any uniform applied loads such as snow loads or wind loads
- Any concentrated loads
- Any dynamic loads such as earthquake loads, relief valve reactions, water hammer.

Types of Analysis

Two types of analysis are available through CAESAR II®: static and dynamic. Dynamic analysis includes loads imposed by transient occurrences, such as earthquakes, water hammer, and relief valves firing. Some of these loads, like earthquake loads, can be approximated using static analysis. Others must use dynamic analysis.

A static analysis considers only loads that are applied steadily, such as loads due to thermal expansion or the dead weight of the pipe system.

Load Cases

There are several load cases that must be understood in order to perform a stress analysis. These load cases may appear in static or dynamic analyses.

- The Sustained Case includes dead weight of pipe, fluid, insulation, and pressure of the fluid in the pipe.
- The Occasional Case includes snow, wind, relief valve discharge, and earthquake.
- The Operating Case includes all of the sustained load cases, plus the thermal loads due to pipe expansion.
- The Expansion Case is the algebraic difference between the Operating Case and the Sustained Case

There is no code stress check for the Operating Case for analyses performed under B31.1. Similarly, B31.3 does not process a code stress check for the Operating case under normal circumstances[5]. If a stress analysis report is prepared for the Operating Case, it

[5] ASME B31.3 Appendix P offers alternative rules for flexibility analysis, and a code stress for the Operating Case is possible under those rules.

will yield zeroes for the code stresses or give the user an error message. This is often a source of great confusion for those who read the stress analysis reports. Code stress checks are performed only for the Sustained, Occasional, and Expansion cases. The user should consult the specific pipe code for the analysis to determine which load cases are appropriate for stress checks.

Data Entry

Before the analysis can begin, the following conditions must be met:

- The design must be frozen (no further changes)
- The analyst must have all of the design drawings
- The analyst must decide on a coordinate system
- The analyst must identify the nodes

The coordinate system is a right-hand coordinate system, with +Y always in the "up" direction. +X usually points south, with +Z pointing west. This is a common configuration, but others may be used if they are more convenient. The restraints and the displacements also follow the right-hand rule.

Nodes are what the software uses to identify points in the model. They are always along the centerline of the pipe.

It is up to the analyst to determine where the nodes go and what to call them. Some software has a feature that automatically numbers the nodes, with a user-defined increment. This feature may be used or may be overwritten. The feature does not locate the nodes. If you wanted to model a piece of straight piping that was 20,000 ft long, you could number the first node "10" and the end node "20," and your job would be done. But you would soon realize that you need to locate all of the supports along the 20,000 ft length.

Nodes should be located wherever something "interesting" is happening to the pipe. "Interesting" means any place where you need to describe a change in the pipe, like a new fitting, or where you want the output to give you information about that point of the pipe. So nodes will be required:

- At every support
- At every change in direction
- Wherever there is a spec change (material, wall thickness)
- Wherever there is a change in the stiffness or loading of the pipe (as at a valve)
- Wherever there is a change in the properties of whatever is inside the pipe (temperature, pressure, density)

It is sometimes convenient to name the nodes according to the pipe size. For instance, the first node of a 6 in diameter line might be 6000, the second node 6010. It is always a good idea to leave 5 or 10 numbers between nodes in case you have to break the segment and add something in between.

Nodes are labeled on the piping drawings. For rigorously documented projects, stress isometrics are prepared. While this produces a very nice document package, it requires the additional step of drawing the iso. It is easier to simply label the nodes on the piping drawings, if that level of documentation can be tolerated.

The software builds an isometric of your piping system using vectors. A graphical representation of the system is available for viewing during the data entry.

One must be very careful to not duplicate any nodes unintentionally. There will be times when you need to make reference to a node that has previously been located. Tees, for example, require that the node at the intersection be specified in three places. But if you unintentionally duplicate a node, expect to spend some time debugging.

Because the data entry process is tedious and requires much precision, you should save your data frequently.

The following items are typical of the various fields that must be entered into the software to fully describe the piping system:

- From and To describe the beginning and ending nodes of the element you are about to describe.
- DX, DY, DZ are direction vectors. These are used to describe the geometry of the element. If the element begins at node 10 and ends at node 20, you have to tell the program how the pipe gets from node 10 to node 20. Note that such entries do not describe thermal growth. They are only lengths in the cold position.
- Diameter is the OD of the pipe.
- Weight or Schedule is the schedule of the pipe.
- Mill Tolerance is the standard mill tolerance on wall thickness. It is always taken as 12.5 percent less than the wall thickness unless the tolerance is specified as something else during the purchase of the piping. CAESAR II® uses the mill tolerance to perform a minimum thickness calculation based on the pressure and the piping code, so only the negative value of the mill tolerance is useful to us. (The + mill tolerance is usually disabled). If the pipe wall is not thick enough, CAESAR II® will generate a warning.
- The corrosion allowance is usually specified as 1/16 in, but may be as high as 0.1 in. The CA is generally only active when used in conjunction with ASME B31.3. The B31.3 code requires that CA be accounted for in the Sustained and Occasional load stress calculations only. But the Expansion case may also be enabled to include CA, and this is thought to be more conservative.

  Corrosion allowance may also be activated for ASME B31.1 jobs, even though the default is to not consider corrosion allowance in B31.1 stress calculations.
- Insulation thickness
- Temperature and pressure can be stated for various operating cases. The software usually selects an ambient installed temperature of 70°F (21°C) to assess the amount of thermal growth that will occur. The installed temperature must be known, but really only the delta T is crucial to the analysis. The pressure should be entered, but is usually not an important factor unless expansion joints are modeled, or unless the system is a plastic piping system (e.g., FRP). In the case of plastic piping, Bourdon Pressure Effects come into play. This is the phenomenon of internal pressure acting to straighten out a piping system that contains bends. When pipes are bent, the cross section is ovalized. It is thought that the internal pressure tends to distend such a cross section from an oval into a circle. In doing so, the bend tends to "rotate" or open up. Forged fittings or molded FRP elbows have a circular cross section and should not be subject to Bourdon Pressure Effects. Normally, the Bourdon Pressure Effects must be turned OFF, which is the default.
- Bend is used to designate elbows and bends. Certain programs will automatically assign intermediate nodes along the elbow, at the upstream weld line and at the midpoint. The important thing to note is that the TO node ends up at the

downstream weld line of the elbow, although it is always modeled as being at the intersection of the centerlines. There must always be another element following an elbow in order to designate what the orientation of the ell is. Elbow definition is often a primary cause of fatal errors during the run.

- Valves or flanges are known by the generic term "rigids." Selectable fields permit the user to enter the weight of the rigid element. Valve and flange databases are also available. These invoke a selection of valves and flanges, with various end connections and classes. The length and weight are then automatically entered into the appropriate fields. The user should check the valve values against catalog data to ensure that the database values are accurate.
- Expansion joints may be modeled though their use in steam and condensate lines is generally to be avoided. Some projects, especially those involving gases may require expansion joints that must be modeled.
- Concentric and eccentric reducers are selectable elements. The length must be entered (and an axial offset if the reducer is eccentric), as well as the diameters and wall thicknesses. If your software does not include a database for reducers, Table 7.14 contains average diameters and wall thicknesses that may be used to approximate them.

  Additionally, if you are using ASME B31.1, the program will apply a Stress Intensification Factor (SIF) at each end of the reducer. ASME B31.3 is silent regarding SIFs at reducers.
- SIFs for configurations other than elbows and bends are also selectable. Common branch connections may be specified and are calculated internally by the software. Others, including reducers under B31.1 or some flanges under B31.3, must be entered manually. The analyst should refer to Tables D-1 (B31.1) or D300 (B31.3).

Available selections for specifying SIFs include:

1. "Reinforced fabricated tees" are tees such as may be field-fabricated. In this case, a "fish mouth" is cut on the end of the branch pipe and the opening is traced onto the header and cut out. The branch is welded to the header, and a curved reinforcing pad is placed around the branch, and welded to the header. Reinforcing pads are always welded to the header; not to the branch.
2. "Unreinforced fabricated tees" are identical to the reinforced fabricated tee, except that the repad is omitted.
3. "Welding tees" are ASME B16.9 fittings.
4. "Sweepolet®" is a Bonney Forge trademark. These are referred to as contoured, integrally reinforced, butt-weld branch connections by the piping codes, with low stress intensification factors for low stresses and long fatigue life. The user should note that there may be a different SIF if using a manufacturer other than Bonney Forge.
5. "Weldolet®" is also a Bonney Forge trademark. These area butt-weld branch connections, designed to minimize stress concentrations and provide integral reinforcement.
6. "Extruded welding tees" are special fittings made by pulling a die through the wall of a pipe. These tees match the requirements set forth in the piping codes (see ASME B31.1, 104.3.1(G), or ASME B31.3, 304.3.4(a)). These tees are prevalent in the power industry.

7. "Butt-weld" is not a tee at all, but rather gives the analyst an opportunity to determine the SIF at a girth butt weld, which normally has an SIF of 1.0, which means we don't worry about it. However, if there is a dramatic change in the wall thickness at a point, then you could specify it here.
8. "Socket weld" is a socket weld fitting. Because socket welds are almost exclusively used on small bore lines, and because small bore lines do not usually require analysis, the SIF is rarely required to be specified at socket welds.
9. "Tapered transition" is not a tee either, but is another way to examine SIFs at adjoining pipes with different wall thicknesses.
10. "Threaded" is not a tee either. According to ASME B31.1 and B31.3, the SIF at threaded fittings is always 2.3. This is rarely encountered in normal industrial stress analyses.
11. "Double welded" pertains to double-welded slip on flanges. ASME B31.3 assigns an SIF of 1.2.
12. "Lap joint flanges" with ASME B16.9 LJ stubs are assigned an SIF of 1.6 by ASME B31.3.
13. "Latrolet®" is another Bonney Forge fitting used for 45° lateral connections. It is available as a butt-weld, 3000# or 6000# classes for socket weld, and threaded applications.
14. "Bonney Forge Insert Weldolet®" is another contoured butt-weld branch connection used in less critical applications. Like the Sweepolet, the attachment welds are easily examined by radiography, ultrasound, and other standard non-destructive techniques. The user should note that there may be a different SIF if using a manufacturer other than Bonney Forge.
15. "Full encirclement tees" are special tees applicable only for the IGE/TD-12 Code.

It is important to note that some configurations are not addressed specifically in the piping codes. One such arrangement is a fabricated cross. Because the codes are silent regarding fabricated crosses, the stress analysis software does not address these either. An attempt may be made to model the cross as two tees, with the intersections sharing a common node. But the result will likely be a warning that indicates that the user has attempted to frame four pipes into a common node. The SIFs will be ignored, yielding a lower stress than required. In cases like this, a Finite Element Analysis (FEA) should be run to determine the actual SIF. This value may then be plugged into the stress analysis software to determine the true stresses. Figure 9.17 illustrates the FEA wireframe model for a cross.

- Restraints are for anchors, supports, guides, and so on. This is probably the most critical area of the data entry because the user must not only determine where the restraints will go (which nodes will be restrained), but also how the restraint will act.

1. "Anchors" are fixed restraints that permit no movement at all. This means no translational movement and no rotational movement.
2. "X, Y, Z" are translational restraints acting along the axis named. If there is no sign preceding the letter, the restraint acts in both directions. If there is a "+" or

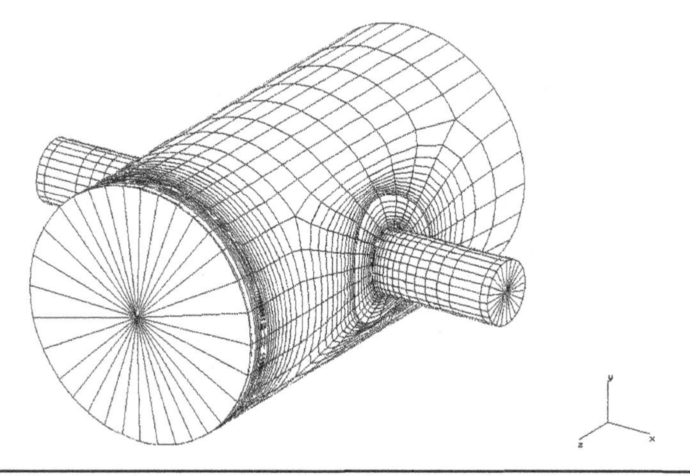

Figure 9.17 FEA wireframe model of a fabricated cross.

"–" sign, the sign indicates the direction of FREE movement. The user can also indicate whether there is a gap. The gap is how far the pipe may move until it encounters the restraint. Gapping the restraint produces a "nonlinear restraint."

Nonlinear restraints are restraints in which the force applied to the restraint does not vary in a linear relationship with the displacement of the pipe. Examples are when gaps close, or when pipes lift off their supports.

3. "RX, RY, RZ" are rotational restraints that function similarly to the translational restraints.
4. "Guides" are translational double-acting restraints that act perpendicular to the pipe and in the horizontal plane. If the axis of the pipe is vertical, then two perpendicular horizontal restraints are generated. Guides are really no different than translational restraints, but are useful for defining such translational restraints on a pipe that is skewed from the orthogonal coordinates
5. "Limits" are translational restraints that act parallel to the axis of the pipe. Again, this type of restraint can be described easily by using the translational restraints.
6. "XROD, YROD, ZROD" are large-rotation rod or hanger restraints. The analysis time increases dramatically when rod restraints are defined, because they introduce many more variables into the equations and the solutions take a long time to converge. It is best to avoid these. Even if you have a rod hanger, you can probably define it as a +Y restraint, since if the rotation is small (and it should be) the pipe will be free to move in the X and Z directions, as well as the RX, RY, and RZ directions.
7. "X2, Y2, Z2" designate translational bi-linear restraints used to model soil support (buried pipe). These can also be used for elastic/plastic action of the piping system
8. "RX2, RY2, RZ2" designate rotational bi-linear restraints, similar to X2, Y2, and Z2.

9. "XSPR, YSPR, ZSPR" designate "bottom out" springs, and occur infrequently in analysis.
10. "XSNB, YSNB, ZSNB" designate translational snubbers, and again are not often encountered in general industry. They are used where displacements due to occasional loads must be controlled.

- "Hangers" specifically means spring hangers.
- If the element you are defining begins or ends at a vessel or tank, you can introduce nozzle flexibility into the model. Nozzles are often modeled as anchors. Allowing for nozzle flexibility often helps in reducing the calculated loads on nozzles. This is important if you are faced with an allowable nozzle load on a tank or vessel. Modeling the nozzle as an anchor is more conservative, but not necessarily more accurate.
- Displacements define any initial displacements, such as may occur when defining a tie-point to an already hot system.
- Forces and moments define any initial forces or moments. These values must usually be calculated. Forces and moments may also be used to model the weight of a valve operator, or occasional maintenance loads.
- Uniform loads may be used to define a load beyond the weight of the pipe, fluid, and insulation. This may be a dust load inside a large-diameter duct, a snow load, or perhaps a walkway mounted off of a large diameter duct. The uniform loads are described by vectors that act along the centerline of the pipe.

  Uniform loads may also be used to calculate seismic reactions in a static analysis. You could perform a dynamic analysis to determine seismic reactions, but this is a complex analysis. Adequate results can often be achieved by simulating the seismic loads imposed on a piping system in the static analysis. In this case, a factor is applied to the weight of the system, and this weight is then applied as a uniform load to the system in one of the horizontal directions. The factors can be derived from procedures provided in the ASCE 7 "Minimum Design Loads for Buildings and Other Structures."

  The interested reader is referred to ASME B31E "Standard for the Seismic Design and Retrofit of Above-Ground Piping Systems" and The American Lifelines Alliance "Seismic Design and Retrofit of Piping Systems"[6]. The American Lifelines document offers an excellent treatment of seismic analysis.
- Wind loads on piping exposed to wind may be modeled by including a Wind Shape Factor from ASCE 7. Another way of handling the wind load is to calculate it by hand and apply it to the piping as a horizontal uniform load.
- Material properties are used to establish the allowable stress for the piping material. Cold and hot allowable stresses are either entered from the code appendices or selected from a database of common materials. The hot allowable stress entered must correspond to the operating temperatures entered earlier. Selecting the material also automatically fills in the values for Elastic Modulus, Poisson's Ratio, and pipe density.

[6] http://www.americanlifelinesalliance.org/pdf/Seismic_Design_and_Retrofit_of_Piping_Systems.pdf

- Longitudinal weld joint efficiency is usually 1.0 but may be as low as 0.60 for furnace butt-welded pipe.
- Fluid density is usually entered in units of lb/in^3 (kg/m^3), or in terms of the Specific Gravity of the fluid.
- Insulation densities are available as default values (usually taken to be calcium silicate) through a material database, or may be entered manually. If you manually enter a value, be certain that the value is correct. This is one place where missing a decimal point can have a DRAMATIC impact on the results.

An unfortunate tendency of many who use computer programs is to place too much faith in the results, simply because the output is a neatly organized report, with results out to 4 decimal places. Users of such programs must constantly be aware that the results are only as good as the input. In the case of CAESAR II®, the input may be checked in a variety of ways. A graphic isometric may be manipulated to provide views from virtually any angle. Input data may be reviewed in a spreadsheet format to check for consistency among the various elements and properties. And while errors will prevent a job from being run, the warnings and notes are also important and will be reviewed by the prudent analyst.

These tasks should be ongoing through the data entry process. The model must be checked for accuracy before any results are examined.

Interpreting the Results

The first thing an analyst does after checking the accuracy of the model and completing a stress run is to check if the Sustained, Occasional, and Expansion load cases passed the code stress checks. Even if they have passed, that is not the end of the story. The displacements and the loads at the restraints must also be checked to make sure that the results are reasonable. Large displacements may indicate that some data entry is incorrect, or that the pipe is insufficiently restrained. High loads or moments on the restraints may indicate that the pipe is not sufficiently flexible or is overly restrained.

Very often the stress checks fail in some area. In those cases, the analyst must dig deeper. A good place to start is with the piping geometry to make sure that the input is correct. Graphic representations or animations are useful for spot checking this. Checking the restraint reports is also very useful. If a restraint is showing zero load in the operating case near an overstressed node, this indicates that the pipe is rising off the support. This may be a good candidate for a spring hanger. Remember that the software does not locate supports or recommend hangers; that is left to the analyst.

Also look for restraints with no or very low loads. These can often be eliminated, resulting in a cost savings.

CHAPTER 10

Drafting Practice

Engineering curricula are not usually loaded with practical information. If a current engineering program even offers engineering graphics, it is often biased toward learning CAD rather than in the clear presentation of information on a sheet of bond.

I once had a junior engineer working for me who was billed as good at producing CAD drawings, which made me feel inadequate because at the time I had not yet learned any of the CAD programs.

Not long afterward, I was reviewing a piping drawing by the engineer. He had located a pump baseplate spanning between two columns of a pipe bridge. Worse yet, he called out the baseplate as checker plate, and none of the valving was accessible from platforms.

The Purpose of Piping Drawings

It might be advisable to remind ourselves what the purpose of a piping drawing is. Like other engineering drawings, piping drawings have multiple purposes, depending on who is using them. If a contractor could be relied upon to build a system or building to satisfy the customer's requirements and to comply with the applicable codes, one could argue that drawings are not necessary at all. Except for "field-routed" piping, the Owner could simply turn the contractor loose and get the finished product. Obviously, for anything but the simplest project, planning is required to ensure that all of the systems operate as they should, and everything fits together with no interferences.

The Contractor

The contractor relies on the piping drawings to explain how the piping and equipment must be installed. This will ensure that everything "fits" (no interferences), and that there are adequate maintenance and operational clearances, which may not otherwise be obvious to an installer.

The contractor must also determine what material is required to be furnished so that he can order the appropriate quantities and types of materials for the project. The "shopping list" that he prepares is called a Material Take-Off (MTO).

Further, the prospective contractor will review the drawings during the bid phase, so that he can submit a competitive bid for the project. In that case, he will need to understand the scope of work so that he can estimate his material costs (developed from the MTO) as well as the manpower required to erect the system in the schedule set forth by the Owner.

The Owner

The owner may use piping plans and sections to review operational requirements and clearances. Sophisticated owners may review P&IDs to determine if the control schemes are satisfactory. Some owners have simple requirements, relying more on manual controls. More complicated systems require more complicated controls, with Distributed Controls Systems and Programmable Logic Controllers, so that the systems are automated, with minimal operator intervention.

P&IDs are usually posted in control rooms so that operators can see schematically how the process is intended to behave without having to trace down every foot of piping. These become part of the record documents for the project, and a savvy owner will ensure that these records are maintained to reflect any revisions.

The piping plans and sections should also be kept up-to-date to reflect the latest revisions. This will minimize the cost of further revisions by reflecting accurate conditions in the field.

Drawing Sizes

Drawings appear in many nonstandard sizes, especially during the concept phase of planning a project. Some of these are very long and are limited in size only by the width of the roll of paper and the area of the lay-down table. However, there are standard sizes for both U.S. and metric drawings. See Table 10.1.

Note that even though there is a standard (ANSI Y14.1) for U.S. drawing sizes, there are two common sizes that are not represented in the ANSI standard. These are 24 in × 36 in and 30 in × 42 in. Another size that is sometimes used is 30 in × 40 in although this is not common. All of the other sizes are in multiples of 8 1/2 in × 11 in that can be folded down to this size. This is convenient, as it permits easy transport and storage of folded drawings.

Note that the metric sizes all have the same aspect ratio (or very close) that approximates the square root of 2. Again, successive folds reduce the final size down to 210 mm × 297 mm. The largest standard metric size A0 has an area very close to 1 m^2.

Sizes		Area	Aspect Ratio
Metric	**mm × mm**	**mm^2**	
A4	210 × 297	62370	1.4143
A3	297 × 420	124740	1.4141
A2	420 × 594	249480	1.4143
A1	594 × 841	499554	1.4158
A0	841 × 1189	999949	1.4138
US	**in × in**	**in^2**	
A	8.5″ × 11″	93.5	1.2941
B	11″ × 17″	187	1.5455
C	17″ × 22″	374	1.2941
D	22″ × 34″	748	1.5455
E	34″ × 44″	1496	1.2941
D1	24″ × 36″	864	1.5000
E1	30″ × 42″	1260	1.4000

Table 10.1 Standard drawing sizes.

Drawing Scales

Scales commonly used in engineering drawings are listed in Table 10.2. In the U.S., there are two drawing scale systems: Architectural and Engineering. In industrial piping projects the architectural scales are more commonly used. Piping drawings usually look best at 1/4″ = 1′-0″ or 3/8″ = 1′-0″, but that does not obviate the possibility of other scales. Long pipe routes may be drawn at any convenient scale since they merely map the route. The smaller 1/4 in or 3/8 in scales are used for piping details and interference checking.

It is a good idea for the scales used between disciplines to remain consistent. This permits interference checking to be more easily conducted from paper drawings. But if the entire project is performed with a CAD package in which the disciplines are referenced to each other, this requirement may be eliminated if there is a reason to prepare the drawings at a different scale. The development of 3-D drawing packages, as well as smart interference checking, may also eliminate the need for drawings to remain at the same scale across disciplines. Given a choice though, the structural, electrical, and piping plans should be in the same scale.

Note that structural plans frequently employ stick figures to represent the structure. This is a common source of piping/structural interferences since there is no scaled width of the structural member. See Figure 10.1. The detail should get better on structural elevations and details, depending on the conventions of the structural department.

Architectural		**Scale**	**3/32″=1′**	**1/8″=1′**	**3/16″=1′**	**1/4″=1′**	**3/8″=1′**	**1/2″=1′**	**3/4″=1′**	**1″=1′**	**1 1/2″=1′**	**3″=1′**
Actual scale			0.00781	0.01042	0.01563	0.02083	0.03125	0.04167	0.06250	0.08333	0.12500	0.25000
Border insertion			128	96	64	48	32	24	16	12	8	4
Feature sizes in inches												
	1/2″	0.5	64.00	48.00	32.00	24.00	16.00	12.00	8.00	6.000	4.000	2.000
Heading	**3/8″**	0.375	48.00	36.00	24.00	18.00	12.00	9.00	6.00	4.500	3.000	1.500
	1/4″	0.25	32.00	24.00	16.00	12.00	8.00	6.00	4.00	3.000	2.000	1.000
Text	**1/8″**	0.125	16.00	12.00	8.00	6.00	4.00	3.00	2.00	1.500	1.000	0.500
Standard text and DIM arrows	**3/32″**	0.09375	12.00	9.00	6.00	4.50	3.00	2.25	1.50	1.125	0.750	0.375
Ext. line offsets	**1/16″**	0.0625	8.00	6.00	4.00	3.00	2.00	1.50	1.00	0.750	0.500	0.250

Engineering		**Scale**	**1″=100′**	**1″=60′**	**1″=50′**	**1″=40′**	**1″=30′**	**1″=20′**	**1″=10′**	**1″=1′**
Actual scale			0.00083	0.00139	0.00167	0.00208	0.00278	0.00417	0.00833	0.08333
Border insertion			1200	720	600	480	360	240	120	12
Feature sizes in inches										
	1/2″	0.5	600.00	360.00	300.00	240.00	180.00	120.00	60.00	6.000
Heading	**3/8″**	0.375	450.00	270.00	225.00	180.00	135.00	90.00	45.00	4.500
	1/4″	0.25	300.00	180.00	150.00	120.00	90.00	60.00	30.00	3.000
Standard	**1/8″**	0.125	150.00	90.00	75.00	60.00	45.00	30.00	15.00	1.500
Standard text and DIM arrows	**3/32″**	0.09375	112.50	67.50	56.25	45.00	33.75	22.50	11.25	1.125
Ext. line offsets	**1/16″**	0.0625	75.00	45.00	37.50	30.00	22.50	15.00	7.50	0.750

TABLE 10.2 CAD scale factors. CAD packages that use "paper space" obviate the need for scaling text.

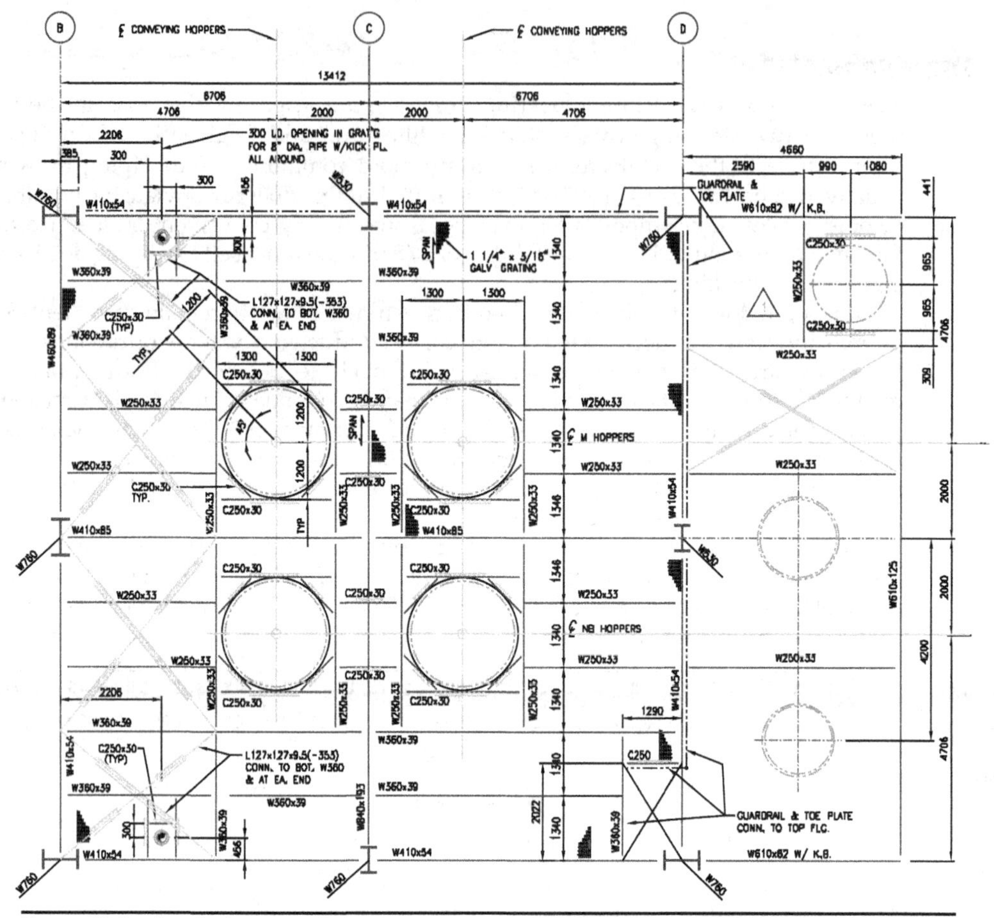

FIGURE 10.1 A typical structural plan.

See Figure 10.2 for an elevation of a structure that displays the members to scale. If someone has to route piping through this, or support piping from it, they at least stand a chance of not encountering an interference.

Another good idea is to incorporate a graphic scale on the drawings. This is especially useful when working with drawings that are reproduced at a size smaller than the original scale. See Figure 10.3. It is good practice to indicate on the General Notes that drawings are not to be scaled. The resolution is simply not adequate for an accurate location to be determined at the common scales (1/4″ = 1′-0″ is 48 to 1). Still it is useful for planning purposes to include graphic scales, just so an idea of the true scale can be obtained.

Consistent presentation of information on drawings demands that a consistent approach is maintained throughout a drawing set. CAD greatly assists in this effort, but as with most every other aspect of engineering, there are many details to pay attention to. CAD standards, like the size and style of fonts, the dimension styles, and layer (or level) conventions need to be set up and followed, but in the world of consulting, individual clients may have their own standards.

Table 10.2 provides a convenient reference for inserting borders and scaling text and dimensions. Text and dimensioning at 3/32″ is generally adequate for readability without taking up too much space on the drawing.

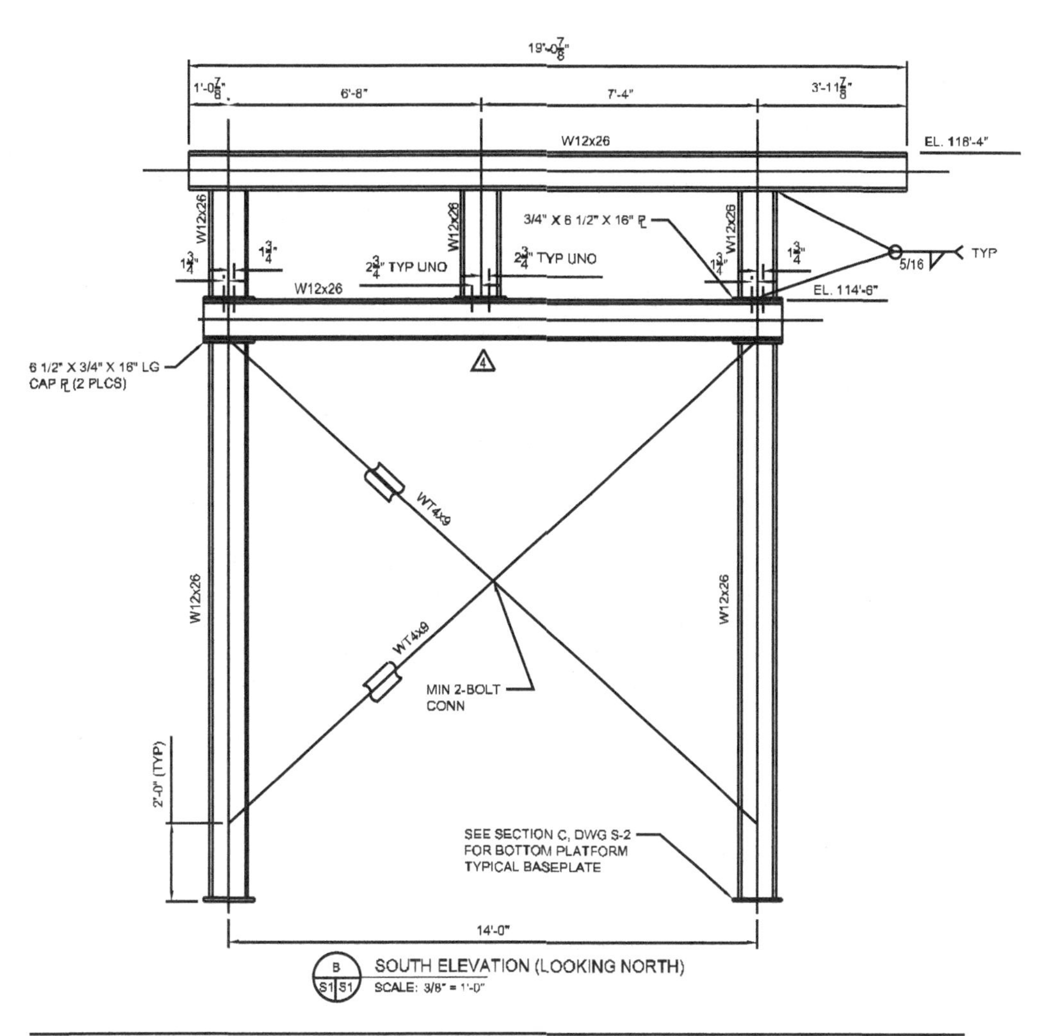

FIGURE 10.2 Structural elevation with to-scale structural members.

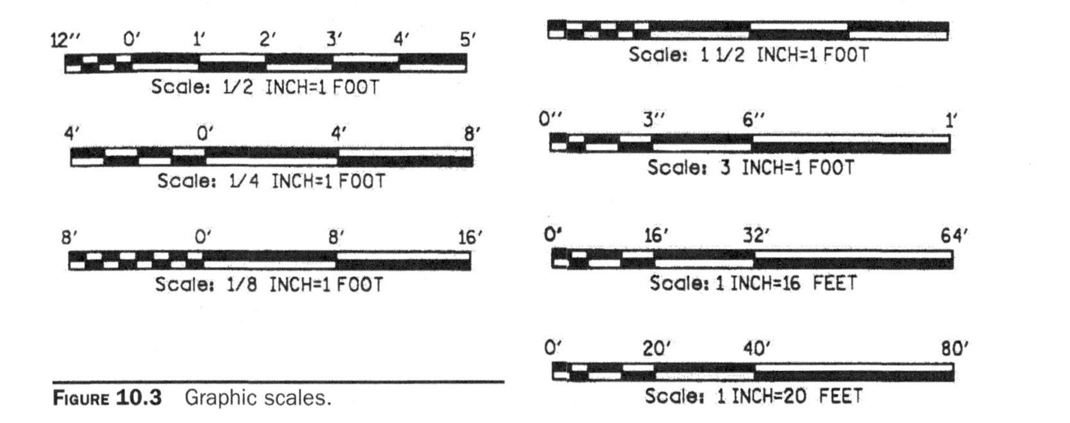

FIGURE 10.3 Graphic scales.

Symbology

Valves and Piping

Especially in flow sheets and P&IDs, piping components are represented by symbols rather than by representations of their physical appearance. P&IDs should be accompanied by a legend to describe the symbols used to create the P&ID.

Figure 10.4 shows valve and piping symbols. Many of these will also be used in the piping plans and sections. On plans, sections, elevations and details, many of these symbols are used as well, especially if the object in question is a valve. It is sometimes important to show the orientation of a valve handwheel on a gate valve, and in those cases, the yoke and handwheel will be shown to scale on the drawing. See Figure 10.5.

Some valves are normally closed during operation. These are identified on P&IDs as shaded, as shown in Figure 10.6. Although relief valves are normally closed, they are never shaded, since it is understood that they are normally closed.

Figure 10.7 shows some common control valve operators. The devices that actually control operators such as these, and provide remote feedback to control rooms are the instruments shown in Figure 10.8.

More-or-less standard abbreviations are used throughout industry to describe the instruments and controls. The abbreviations link together with up to four letters describing the function of the device. The system is depicted in Table 10.3.

Suppose there is a tank that is filled by a process, and pumped to another process. If the pump were to fail, then the fluid would continue to rise inside the tank. A high level switch, or LSH (Level Switch High) would be designed to announce a Level Alarm High (LAH) at the control room. If the alarm were to go unnoticed, then additional action would be required. Another higher level switch (a Level Switch High High, or LSHH) would be used to close the fill valve, preventing an overflow.

Process Symbols

Engineers use the process symbology shown in Figure 10.9 to develop the P&IDs. Symbology developed by the International Society of Automation (ISA) is normally used to portray the instrumentation and logic on the P&ID's. Different linework is used to distinguish process lines from control lines. This enables the control loops to be graphically depicted. See Figure 10.10.

Process-intensive industries like chemical, petrochemical, pharmaceutical, and pulp & paper use line numbers to identify the individual pipelines. A convention for calling these out is illustrated in Figure 10.11. The fluid service designations must be standard throughout the project and are also included in the P&ID legend as shown in Table 10.4.

Finally, some general symbols are shown in Figure 10.12. Of particular value is the phantom line that indicates the limits of scope of supply.

> The notion of defining the scope of supply is immensely important to establish contractual obligations. Sometimes this requires additional effort for the engineer, as when two or more different piping contractors may be needed for a very large project. In that case, care must be taken to clearly identify the scope of each contractor. The sequence of construction also must be taken into account and that may require the ability to independently pressure test two segments of the same line. Flanges or valves may be required at the scope interface; not for any process reason, but rather to facilitate the construction or the contract definition. Anchors may also be required at the interface if the particular line is hot and requires a stress analysis to be run by two different consultants.

(text continues on p. 255)

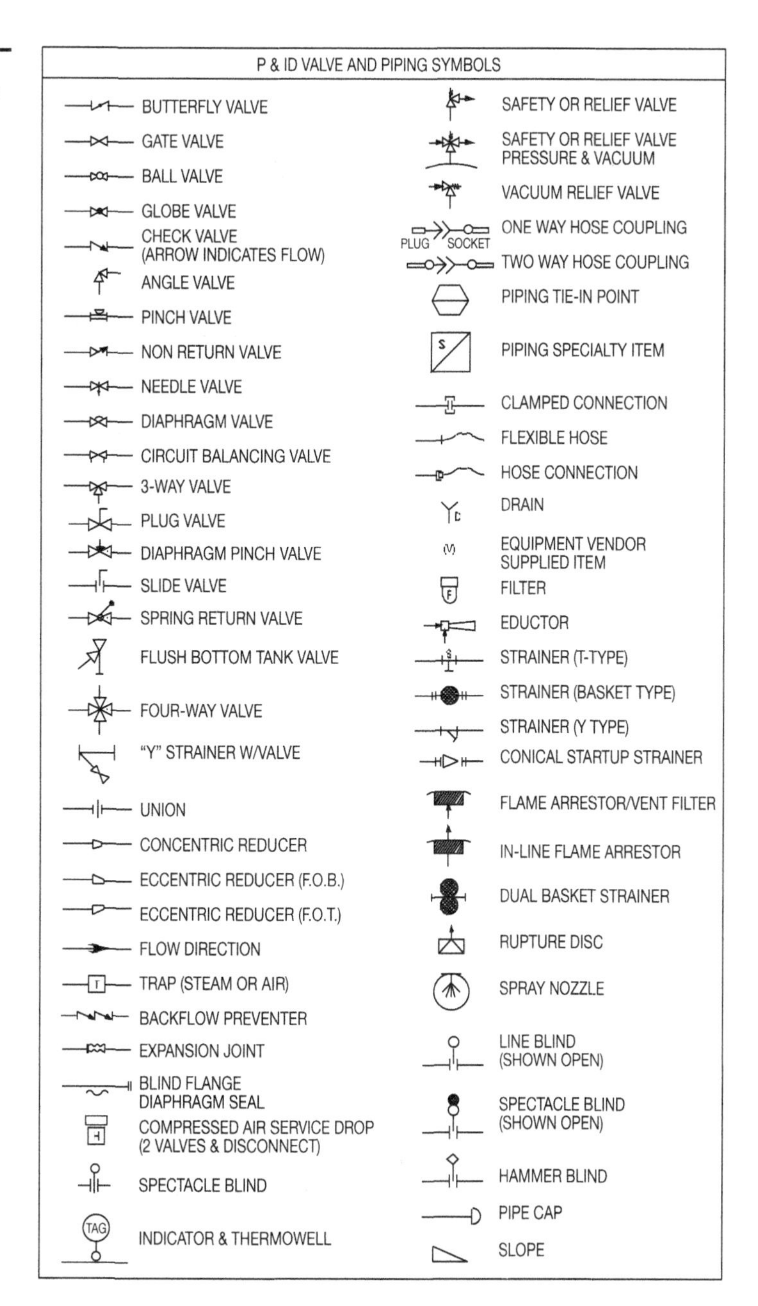

Figure 10.4 Valves and instruments for P&IDs. This will form part of the P&ID legend.

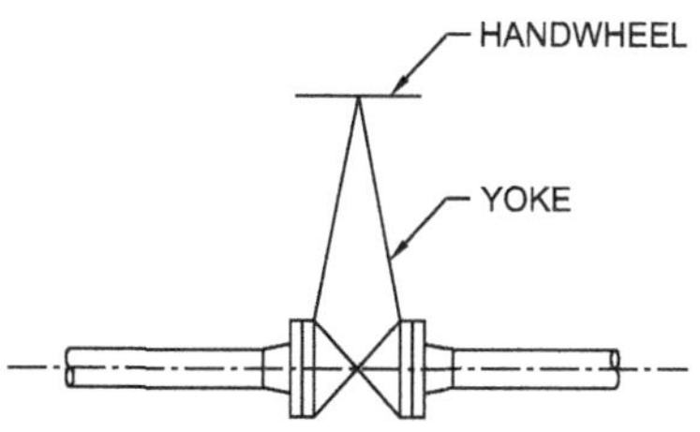

Figure 10.5 A gate valve showing the handwheel orientation.

Figure 10.6 Normally closed valves are represented with the valve outline shaded.

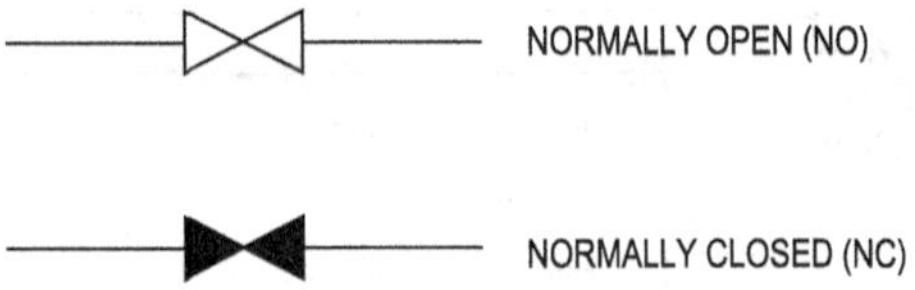

NORMALLY OPEN (NO)

NORMALLY CLOSED (NC)

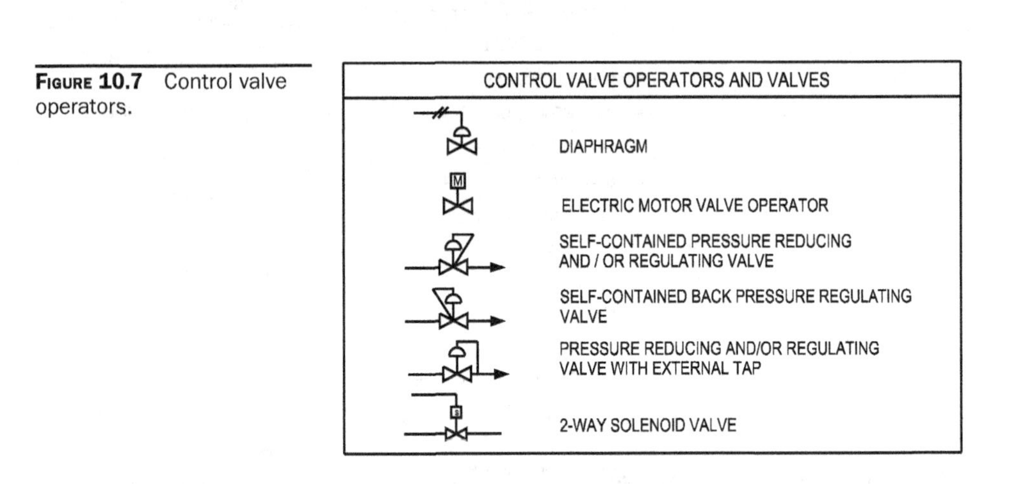

Figure 10.7 Control valve operators.

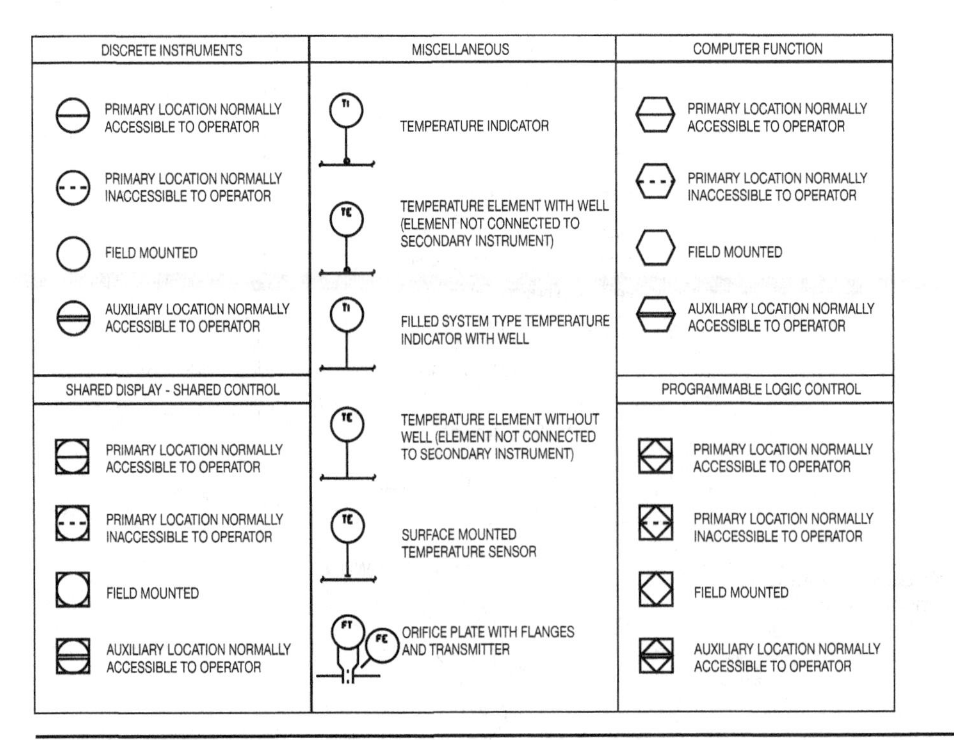

Figure 10.8 Instrumentation symbols.

LETTERS OF INSTRUMENT IDENTIFICATION			
LETTER	FIRST LETTER	SECOND LETTER	THIRD LETTER
A	ANALYTICAL	ALARM	ALARM, POSITION 'A'
B	BURNER	BATCH BIAS	BOX, POSITION "B"
C	CONDUCTIVITY	CONTROL; CONTROLLER CONVERTER	CONTROLLER, CLOSED CONVERTER, POSITION 'C'
D	DENSITY	DIFFERENTIAL	DRIVE DISC
E	VOLTAGE	ELEMENT	ELEMENT
F	FLOW	RATIO(FRACTION)	
G	THICKNESS (GAGING)	GAGE	
H	HAND	HAND	HIGH
I	CURRENT	INDICATOR INDICATING	INDICATOR
J	POWER	SCAN	
K	TIME	CONTROL STATION	CONTROL STATION
L	LEVEL	LOGGER, LIGHT	LOW
M	MOISTURE OR HUMIDITY	MOTOR	METER, MIDDLE OR INTERMEDIATE
O	OPTICAL	ORIFICE	OPEN
P	PRESSURE		PROGRAMMER
Q	EVENT NUMBER	INTEGRATOR; TOTALIZER	INTEGRATOR; TOTALIZER
R	RESTRICTION	RECORDER	RECORDER
S	SPEED OR FREQUENCY	SWITCH; SAFETY	SWITCH
T	TEMPERATURE	TRANSMITTER	TRANSMITTER
U	MULTIVARIABLE	MULTIFUNCTION	MULTIFUNCTION
V	VISCOSITY; VACUUM VIBRATION; VIDEO	VALVE	VALVE
W	WEIGHT OR FORCE	WELL	WELL
X	MISCELLANEOUS	MISCELLANEOUS	
Y	RELAY OR COMPUTE	RELAY OR COMPUTE	RELAY OR COMPUTE
Z		UNCLASSIFIED FINAL	UNCLASSIFIED FINAL

Table 10.3 Instrument identifications.

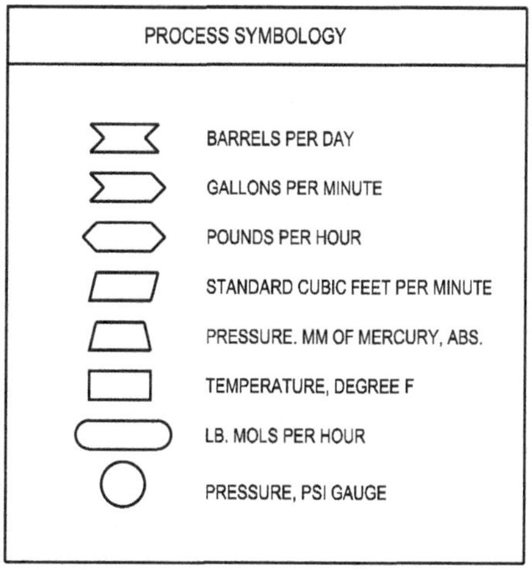

Figure 10.9 Process symbology. The symbols are often omitted on less complex projects.

PROCESS SYMBOLOGY

- BARRELS PER DAY
- GALLONS PER MINUTE
- POUNDS PER HOUR
- STANDARD CUBIC FEET PER MINUTE
- PRESSURE. MM OF MERCURY, ABS.
- TEMPERATURE, DEGREE F
- LB. MOLS PER HOUR
- PRESSURE, PSI GAUGE

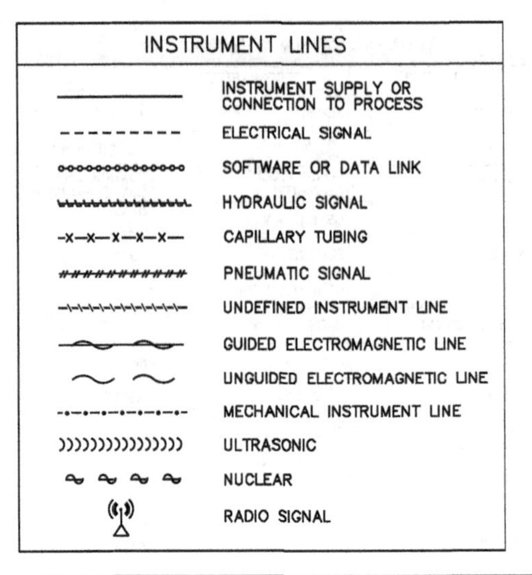

FIGURE 10.10 Linework used to depict process and control schemes.

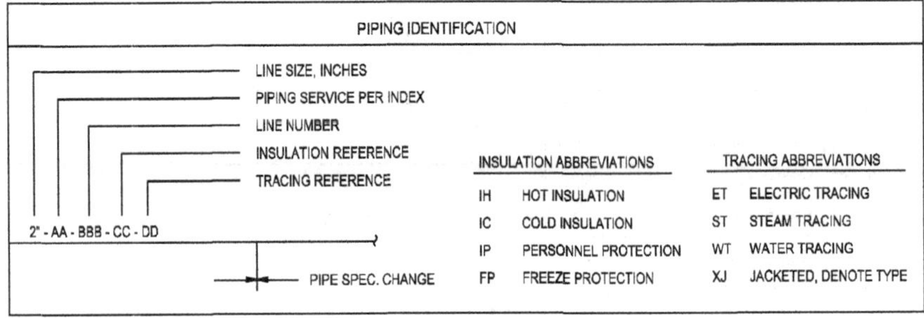

FIGURE 10.11 Typical pipeline identification callout.

PIPING SERVICE DESIGNATION

BFW	BOILER FEED WATER	MC	MEDIUM PRESSURE STEAM CONDENSATE
CHWR	CHILLED WATER RETURN	MS	MEDIUM PRESSURE STEAM
CHWS	CHILLED WATER SUPPLY	N	NITROGEN
CWR	COOLING WATER RETURN	NG	NATURAL GAS
CWS	COOLING WATER SUPPLY	PA	PLANT AIR
DG	DRAINS-NON CORROSIVE	PR	PRODUCT-SANITARY PIPING
DIW	DEIONIZED WATER	PS	PROCESS-STAINLESS
DW	POTABLE WATER	PW	PROCESS WATER
FO	FUEL OIL	SL	SLUDGE
FW	FIRE WATER	SO	SEAL OIL
HBS	HIGH BOILING SOLVENT	SW	SOFTENED WATER
HC	HIGH PRESSURE STEAM CONDENSATE	TW	TREATED WATER
HO	HYDRAULIC OIL	V	PROCESS VENT
HPCS	HIGH PRESSURE CLEANING SOLVENT	VA	PROCESS VAPOR
HS	HIGH PRESSURE STEAM	VG	VENTS-NON CORROSIVE
HW	HOT WATER	VS	VAPOR SAMPLE
IA	INSTRUMENT AIR	WAW	WASH WATER
IG	INERT GAS	WEW	WELL WATER
LBS	LOW BOILING SOLVENT	WW	WASTE WATER
LC	LOW PRESSURE STEAM CONDENSATE		
LO	LUBE OIL		
LS	LOW PRESSURE STEAM		

TABLE 10.4 Fluid service designations.

FIGURE 10.12 General symbols.

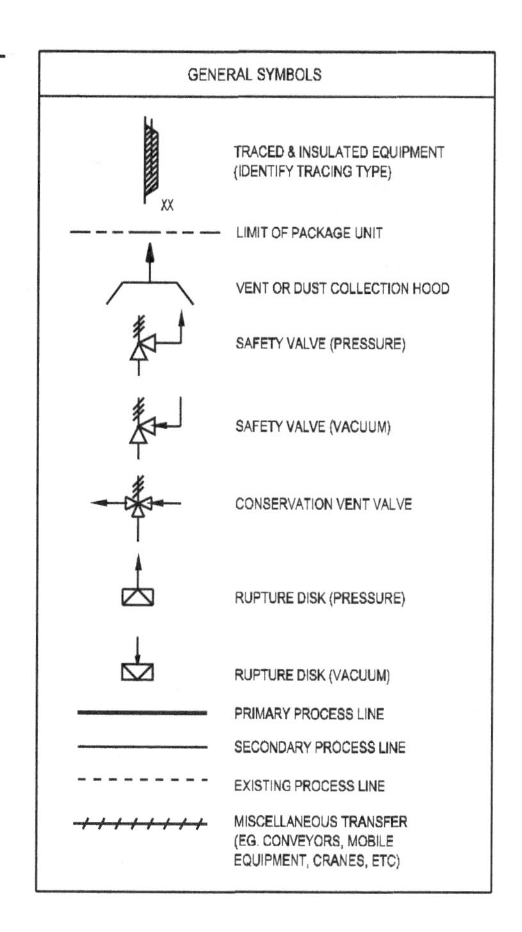

Welding Symbols

Welding has evolved into a highly technical science, and information is conveyed to the welder via the standard symbols shown in Figures 10.13 and 10.14. Note that the reference line and the arrow are the only required elements of the symbol. The arrow may be placed on the left or right hand side of the reference line, but the other elements, when used, must be placed in the locations specified.

The welding symbols are intended as a convenient "short-hand" method of conveying a lot of information in a little space. Welding information can also be indicated via notes, details, specifications, and so on.

Drafting Practices for Piping

Piping Plans and Elevations

The most common piping drawing is the plan view, which shows an overhead layout of the piping and its relationship to the surrounding structure and equipment. A good scale for piping plans is 1/4" = 1'-0", but other scales are useful. For projects other than cross-country piping, scales smaller than 1/8" = 1'-0" are to be avoided. Too much detail is lost at these small scales and elevations and details must be relied on to fill in the missing information.

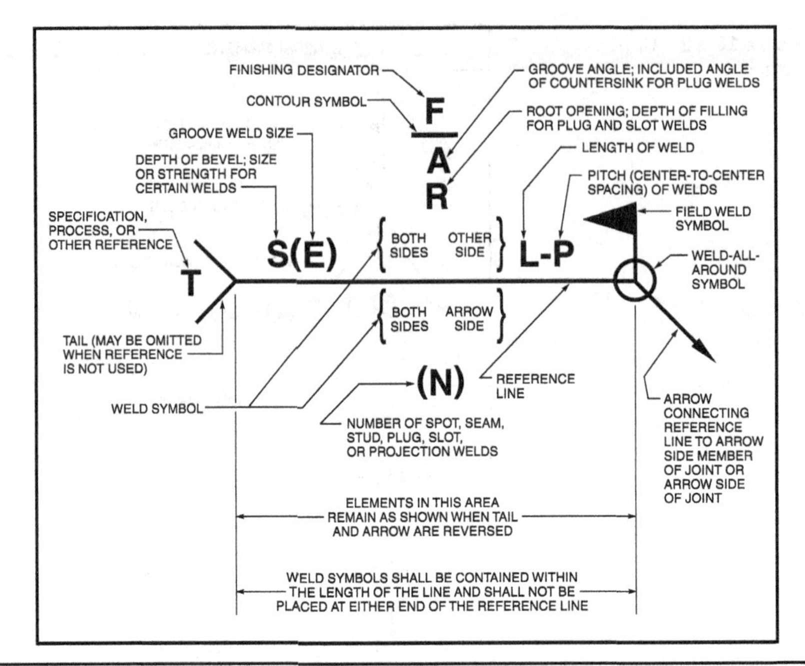

FIGURE 10.13 From AWS A2.4:2007, Standard Symbols for Welding, Brazing, and Nondestructive Examination, Figure 3 – Standard Location of the Elements of a Welding Symbol. *Reproduced with permission from the American Welding Society (AWS), Miami, FL USA.*

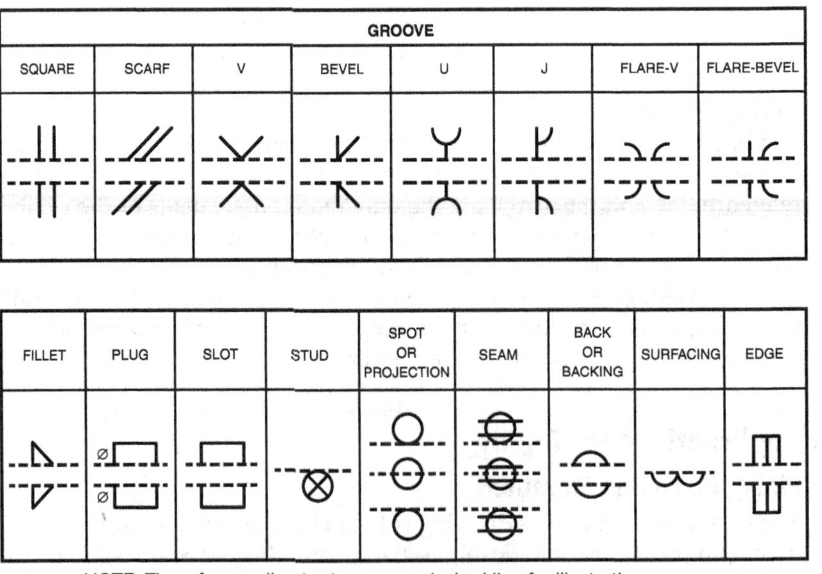

FIGURE 10.14 From AWS A2.4:2007, Standard Symbols for Welding, Brazing, and Nondestructive Examination, Figure 1 – Weld Symbols. *Reproduced with permission from the American Welding Society (AWS), Miami, FL USA.*

Every plan must have a North arrow. The preferred orientation of the drawing is with the North arrow pointing to the top or left of the sheet. Most sites have a convenient North reference that is orthogonal to the plant layout. This is known as "Plant North", and may deviate from "true" North by up to 45°.

Column bubbles should ideally be located along the top and right-hand side of the drawing. Some company standards call for the column bubbles to be placed on the left-hand side. Anyone who has had to search for a plant area in a stick file knows that the search is much easier if the column bubbles are on the open side of the stick file.

For large plans in which a grid of drawings is required to cover the extent of the design, a key plan is an extremely useful guide to orient the reader to the drawing area.

When a pipe crosses (at an angle) over another in plan, the general rule is to show the top pipe as continuous, with the pipe below it broken. See Figure 10.15.

When pipes run parallel, the lines are broken above in enough areas to show the location of the pipes at the lower elevations. See Figure 10.16.

The elevation of piping run in pipe racks is generally denoted as Bottom of Pipe (BOP). This will be the same elevation as the Top of Steel (TOS) unless a pipe support or guide is required, as is often the case with steam and condensate piping. Rack piping that changes size usually does so with an eccentric reducer, oriented with the flat portion on the bottom (FOB) to maintain a constant BOP elevation.

The elevation of underground piping is denoted as BOP also. Unless the piping is a gravity drain line, reducers are usually concentric. Invert elevations are only called out to designate the elevation of the inside of the pipe used in a gravity drain. Inverts are often specified for trench drains, but are rarely used for pipe. The BOP designation is more common and convenient. Figure 10.17 shows the difference between "invert" and "BOP" elevations.

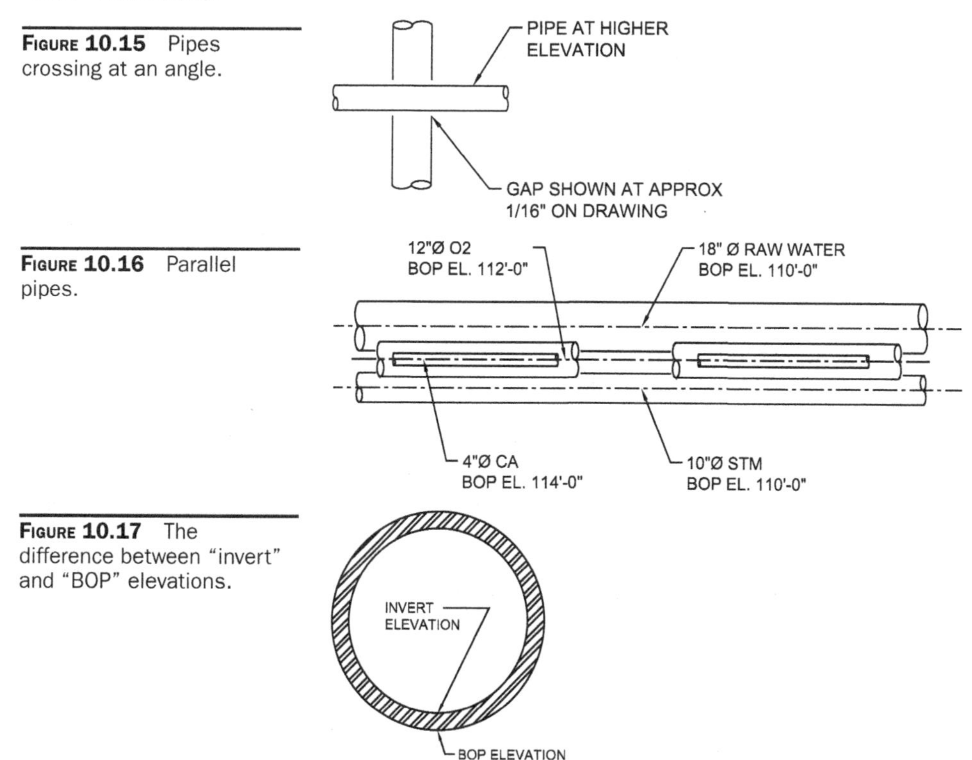

Figure 10.15 Pipes crossing at an angle.

Figure 10.16 Parallel pipes.

Figure 10.17 The difference between "invert" and "BOP" elevations.

Many projects require that the trenches for underground piping be "laid back" at an angle less than the angle of repose of the soil, rather than using a trench box to protect the workers. This is especially true if:

- The soil is sandy (low angle of repose)
- There is plenty of room, as on a "Greenfield" project
- The burial depth is not great

Obviously, the cover must exceed the frost line for a given area in order to prevent both freezing and heaving of the pipe. But because excavation is a significant cost, the designer must be aware of locating the lines for maximum economy.

The angle of repose is the angle between the horizontal and the side of a conical pile of material (in this case, soil).

If space does not permit the trench to be laid back, then trench boxes or shoring must be used for trench depths of 5 ft or more[1].

Piping is normally dimensioned to the centerline of the pipe. Recommended spacing between centerlines of pipes of various dimensions for both insulated and uninsulated lines is shown in Table 10.5. This takes into account flange diameters, but special care must be exercised where large diameter branch lines must come off of headers in dense rack piping.

Clearances

Minimum clearances for overhead piping (including insulation) should be as shown in Table 10.6. It is advisable to contact the operating rail service to verify clearances from railroads.

Suitable maintenance clearance around pumps and equipment is also necessary. A good rule-of-thumb is 3 ft (1 m) of clearance. The National Electric Code specifies clearances required around electrical equipment. These clearances depend on the voltage of the equipment, and are referred to as "working space." See NEC Articles 110.26 and 110.32.

Careful planning by the Piping Department is sometimes for naught. In order to protect the intent of the design, it may be necessary to review the requirements with the Electrical Department.

In one instance, a careful layout of cooling tower coldwell piping was rendered moot when a wall of conduit 3 ft high blocked access to the pumps and valves. This "hurdle" had to be cleared literally during commissioning. In another occurrence, a potable water pump room had floor access blocked when conduit was clamped to Unistrut® which was bolted to the floor. This prevented any access to the pumps with rolling equipment.

Just because you've carefully designed the layout, it doesn't mean that your efforts will bear fruit. Take the extra step and coordinate access with the other disciplines.

[1] See 29 CFR 1926.652(a): Protection of employees in excavations.

Uninsulated

NPS	1	1.5	2	3	4	6	8	10	12	14	16	18	20	24	30
30	19	19	20	22	23	25	25	29	30	30	31	32	33	35	38
24	18	18	18	21	21	22	23	25	26	27	28	29	30	32	
20	15	15	17	17	18	20	21	22	23	24	25	26	27		
18	14	14	16	16	17	18	20	21	22	23	24	25			
16	13	14	15	15	16	17	19	20	21	22	23				
14	13	13	14	14	15	16	18	19	19	20					
12	12	12	13	14	14	15	16	18	19						
10	11	11	12	12	13	14	15	16							
8	10	10	11	11	12	13	14								
6	9	9	9	10	10	11									
4	8	8	8	9	9										
3	7	7	7	8											
2	6	6	7												
1.5	6	6													
1	6														

Insulated

NPS	1	1.5	2	3	4	6	8	10	12	14	16	18	20	24	30
30	20	20	21	23	25	26	28	30	31	32	33	34	35	37	40
24	19	19	19	22	23	24	26	27	28	29	30	31	32	34	
20	16	16	18	19	20	21	22	24	25	26	27	28	29		
18	15	15	17	18	19	20	22	23	24	25	26	27			
16	14	15	16	17	18	19	21	22	23	24	25				
14	14	14	15	17	18	19	20	21	22	23					
12	13	13	15	16	17	17	19	20	21						
10	12	12	14	15	16	17	18	19							
8	11	11	13	13	14	15	16								
6	10	10	10	11	12	13									
4	9	9	9	10	11										
3	8	8	9	10											
2	7	7	8												
1.5	7	7													
1	7														

TABLE 10.5 Recommended spacing between pipe centerlines.

Description	Feet-Inches	mm
Over Roadways	17′-0″	5200
Over Walkways	6′-8″	2035
In Confined Spaces	6′-8″	2035
Yard Piping	17′-0″	5200
Over Railroads (Top of Rail)	23′-0″	7010
Adjacent to Railroads (From Centerline)	12′-0″	3660

TABLE 10.6 Minimum pipeline clearances.

Field Routed Piping and Single-Line versus Double-Line Piping

Usually 2 in diameter and smaller piping is represented on drawings as a single line and 2 1/2 in and larger is shown as double-line. That is, the OD of the pipe is represented at full size.

The exact split between single-line and double-line representations may vary somewhat, depending on the scale of the project or the standards prepared by the Chief Engineer.

The split at 2 in is a good rule-of-thumb however, and it also designates a good split between field-routed piping and dimensioned piping. This is a convenient coincidence, because it quickly shows what may be field-routed.

Field-routed piping is a concept that saves design time. The idea is that small-bore piping is easy enough to route from Point A to Point B (the "Tie Points") around the obstacles in between that an inordinate amount of time would be spent in designing and dimensioning the routing. Piping of such small sizes is very flexible and so pipe stresses would rarely need to be considered. It is more economical to allow the fitters to establish the route. Sometimes a list is simply provided with the tie points and the entire route is left to the contractor. Sometimes, however, it may be advantageous to show the anticipated routing, and then this is represented with the single line piping, which may also be required to show valve, equipment, or instrumentation locations.

Several years ago, a consulting firm developed a concept for quickly designing steel mills that relied on the contractors to route the piping and electrical lines. They marketed it as a cost savings for the Owner, as well as a means to compress the schedule.

While this approach may have some merit on the surface, most engineers would object to the shift of responsibility to the contractor. The contractor would have to have somebody perform material take-offs and plan the routing anyway. Many companies would prefer to have that work performed by qualified personnel, whether they work for the engineer or the contractor.

From a technical, operational, and maintenance standpoint, it is better to have the layout performed by engineers and designers. The "minimal design" practice has not been popular in the U.S. recently, and has given way to the more traditional and proven design practices.

Piping Details

Detail drawings are often prepared at 3/8″ = 1′-0″ since this scale offers room to provide lots of information on the drawing sheet. Engineering companies take advantage of re-usable standard details for common equipment connections, steam trap installations, pipe supports, and the like. Many of these are available as pre-drawn CAD details through third party suppliers, and these save considerable time and effort.

CHAPTER 11

Pressure Drop Calculations

The term "fast track" as it is applied to engineering projects is defined as a project in which design is taking place before all of the information is available. In fact, portions of the project may be under construction while other portions are still in design. One of the first efforts in a piping project is to calculate pressure drops. This exercise is often required even before adequate information is available. Welcome to fast track design.

The reason that the pressure drop calculations are so important at this stage is because long-lead items such as pumps, motor starters, and switchgear must be specified, bid, and purchased. The fact that this must all occur while the vendor is fine-tuning energy balances (and hence flow rates) only adds to the stress under which the engineer must operate. Fortunately, there are some good guesses that we can make to assure that the final result satisfies the requirements of the equipment, the plant, the project schedule, and the budget.

These guesses may relate to anticipated life of the equipment (as in the degree to which internal surfaces may become rougher as they corrode), acceptable loss coefficients through valves or fittings (since there is some variation in the published data), and of course, the intelligent selection of a contingency factor, or "factor of safety," to account for variations in the unknowns and to provide a margin of error.

A careful accounting of the lengths and fittings is achievable by applying a methodical approach to the problem. In spite of all our care, a reasonable factor of safety of 15 percent is often applied to the end result since the last thing anyone on a piping project wants is to start a pump and have it "churn" the fluid without delivering the proper capacity at the proper pressure. For those who are worried about the inherent waste of energy that may result in a slightly oversized pump, there are always variable speed drive technologies that may be applied to the pump in order to take advantage of the affinity laws. A small reduction in pump speed can lead to significant energy savings.

Certainly a huge part of engineering is the economical application of resources to a problem, and so it is unacceptable to grossly oversize a piece of equipment due to the large capital cost as well as the domino effect of installation costs, taxes, and the cost of the necessarily larger electrical gear that would be required to keep a larger-than-necessary pump operating. Hence it is necessary to determine as closely as possible the requirements of the pump. But in no case may the pump or driver be undersized.

There are software programs available that model the piping system and flows and produce the required pump performance characteristics. These are certainly of value, but especially for simple single-line flow, hand calculations are easy and may be performed quickly without the expense of software or the all-too-frequent foibles with computers[1].

[1] Even though they have become an inextricable part of the engineer's life.

Concepts Involved in Pressure Drop

Fluids flow through pipes due to a difference in pressures within the piping system. The pressure forces the fluid from high-pressure regions into low pressure regions.

Most of our efforts as piping engineers revolve around moving water through pipes. But if we have to move a gypsum slurry that is 15 percent denser than water, or ethanol which is 21 percent less dense, then we will need more or less pressure to accomplish the same task than if we were using water with a specific gravity of 1.0.

The calculations for pressure drops in fluid dynamics problems derive from the First Law of Thermodynamics. The resulting equation is known as Bernoulli's Equation[2]. While we will not undertake the derivation of the equation here, it is important to note the following:

- The various terms of the equation are referred to as "heads," and are distinguished from pressures in that the head of a fluid at a given point is equal to the pressure divided by the density of the fluid.
- "Head" has units of distance, either in feet or meters.
- Care must be taken to apply the gravitational acceleration in order to get the units to balance. This may be taken to be $g_c = 32.2\ ft/sec^2$ ($9.81\ m/sec^2$) for all locations, since as we will see, any local variation in gravity will be subsumed in the safety factors and rounding that will be applied in the calculations.

Bernoulli's Equation

Arguably the most important equation to the piping engineer is Bernoulli's Equation. This relationship is valid only for incompressible fluids such as liquids, and for gases at low Mach numbers ($M < 0.3$). Fortunately, most industrial piping for compressible gases is designed for velocities less than 0.3M.

The Mach number is defined as the ratio of the actual speed of a gas to the speed of sound in that gas, or

$$M = V/c \quad \text{(Equation 11.1)}$$

where

V = Speed of the gas

c = Speed of sound in the gas

$$c = \sqrt{kRT} \quad \text{(Equation 11.2)}$$

where

k = Ratio of Specific Heats (See Table 11.1)

R = Gas Constant (See Table 11.1)

T = Absolute Temperature [°R or K][3]

R may be determined for any gas using the relationship

$$R = R_u/M_m \quad \text{(Equation 11.3)}$$

[2] Named after Swiss mathematician Daniel Bernoulli, 1700–1782.

[3] Degrees Kelvin do not show the degree symbol superscript, thus the freezing point of water is correctly written "273.15 K."

Gas	Chemical Symbol	Molecular Mass, M_m	R			c_p		c_v		k
			J/kg K	ft lbf/ lbm °R	ft^2/sec^2 °R	J/kg K	BTU/lbm °R	J/kg K	BTU/lbm °R	
Air		28.98	286.9	53.33	1717	1.004	0.2399	717.4	0.1713	1.40
Carbon Dioxide	CO_2	44.01	188.9	35.11	1130	840.4	0.2007	651.4	0.1556	1.29
Carbon Monoxide	CO	28.01	296.8	55.17	1776	1039	0.2481	742.1	0.1772	1.40
Helium	He	4.003	2077	386.1	12,432	5225	1.248	3147	0.7517	1.66
Hydrogen	H_2	2.016	4124	766.5	24,681	14,180	3.388	10,060	2.402	1.41
Methane	CH_4	16.04	518.3	96.32	3102	2190	0.5231	1672	0.3993	1.31
Nitrogen	N_2	28.01	296.8	55.16	1776	1039	0.2481	742.0	0.1772	1.40
Oxygen	O_2	32.00	259.8	48.29	1555	909.4	0.2172	649.6	0.1551	1.40
Steam	H_2O	18.02	461.4	85.78	2762	2000	0.478	1540	0.368	1.30

Table 11.1 Properties of some common gases.

where

R_u = Universal Gas Constant = 1544 ft lbf / lbmole °R (8314 N m / kgmole K)

M_m = Molecular Mass of the gas

We note that the speed of sound is a function of temperature only for an ideal gas.

Example 11.1

Given: Air at 100 psig flowing through a 4 in Sch 40 steel pipe at 100 ft/sec

Find: Is Bernoulli's Equation valid for this flow condition?

First we find the velocity of sound in the air. We will assume that the air flowing through the pipe is 85°F. Compressed air is generally cooled after compression to drop out excessive moisture.

Applying Equation 11.2 yields

$$\begin{aligned} c &= \sqrt{kRT} \\ &= \sqrt{(1.40)\,(1717\ \text{ft}^2 / \text{sec}^2\ {}^\circ\text{R})\,(85^\circ\text{F} + 460^\circ\text{R})} \\ &= 1145\ \text{ft/sec} \end{aligned}$$

$$M = \frac{100\dfrac{\text{ft}}{\text{sec}}}{1145\dfrac{\text{ft}}{\text{sec}}} = 0.09$$

Therefore, Bernoulli's Equation is applicable to this flow condition.

To further our discussion of Bernoulli's Equation we define the two points in a piping system in which we are interested. We identify the first point with the logical subscript 1 and the second point with the subscript 2. We further stipulate that we will apply the equation in English units.

Bernoulli's equation is given by

$$\frac{p_1}{\rho_1} + \frac{v_1^2}{2g_c} + z_1 + h_A = \frac{p_2}{\rho_2} + \frac{v_2^2}{2g_c} + z_2 + h_E + h_f \qquad \text{(Equation 11.4)}$$

where

p = pressure [lb/ft^2]

ρ = density [lb/ft^3]

v = velocity [ft/sec]

g_c = the gravitational constant = 32.2 ft/sec^2 (9.81 m/sec^2)

z = elevation [ft]

h_A = Head added to the process, as by a pump [ft]

h_E = Head extracted from the process, as from a turbine [ft]

h_f = Head loss due to friction [ft]

In trying to properly size a pump, the bulk of the effort lies in the last term, determining the head loss due to friction.

The equation is usually rewritten so that similar terms can be subtracted from one another.

$$h_A = \left(\frac{p_2}{\rho_2} - \frac{p_1}{\rho_1}\right) + \left(\frac{v_2^2}{2g_c} - \frac{v_1^2}{2g_c}\right) + (z_2 - z_1) + h_E + h_f \quad \text{(Equation 11.5)}$$

Because the density is constant for an incompressible fluid, this may be written as

$$h_A = \left(\frac{p_2 - p_1}{\rho}\right) + \left(\frac{v_2^2 - v_1^2}{2g_c}\right) + (z_2 - z_1) + h_E + h_f \quad \text{(Equation 11.6)}$$

Pressure Head

The term p/ρ is referred to as the "pressure head". This has the dimensions of length and is calculated in units of feet or meters. Note that a conversion factor must be applied if the pressure is given in psi and the density is in lb/ft^3. For example, water at 65 psi and a density of 62.4 lb/ft^3 has a pressure head of

$$p/\rho = (65 \text{ lb/in}^2)\,(12 \text{ in/ft})^2 \,/\, (62.4 \text{ lb/ft}^3) = 150 \text{ ft}$$

In reality, the pressures are subtracted prior to performing the conversion from psi into PSF.

Velocity Head

The term $v/2g_c$ is referred to as the "velocity head." This is the head due to kinetic energy. This term is usually very small, and while it would be technically incorrect to disregard it, for practical piping problems its contribution is often negligible. Because the first point of the system often refers to a lake, tank, or similar large container, the v_1 term can be taken to be zero in such cases, since the velocity of the surface of a lake or tank does not appreciably change due to pumping out of it. The same may be true of the discharge point 2.

The velocity head plays a role in calculating friction head losses through fittings.

Elevation Head

The z term is the "elevation head." This is the head due to potential energy, and accounts for that portion of the head required to move the fluid from one elevation to another.

Friction Losses

While there are several software products available to aid in computing friction losses, the ability to perform hand calculations should not be underestimated. To begin with, most of the effort centers around obtaining a fair count of the lengths of pipe and the number of fittings. Once that is accomplished, the rest is accounting, and a manual calculation system is usually no more difficult than modeling the pipe system in a format that can be understood by the piping software.

Friction losses depend on several variables:

- Flow rate
- Diameter
- Type of pipe (Roughness)
- Length of pipe (Major Losses)
- Number and sizes of fittings and valves (Minor Losses)
- Entrance and Exit Losses (Minor Losses)

The diameter of the pipe is determined based on velocity. The velocity is usually based on recommended rules-of-thumb that have been used with success over the years. Some of these recommended velocities appear in Table 11.2. These recommended velocities represent a compromise between what is thought to be a reasonable energy cost in overcoming the friction balanced against the first cost of the installed pipe. When sizing pipe it is useful to consider that the incremental cost of increasing the pipe by one size is usually minimal, and the increase in cost may often be justified by the reduced energy costs in pumping the fluid, or the ability to increase capacity at some future time without having to add another separate pipe. If, however, the application involves solids in suspension, then all bets are off: minimum velocities will be required in order to keep the solids from dropping out of the flow stream.

Flow rates in GPM for common pipe sizes and practical velocities are given for various pipe schedules in Tables 11.3 through 11.7. These tables can be used to quickly find the diameter of pipe if the desired flow in GPM is known.

Water

Depending on the particular application, water velocities should be between 3 and 10 fps (1 to 3 m/sec). The danger of creating water hammer increases above 10 fps.

Pump suctions operate in a region of low pressure, and high velocities can cause a reduction in the vapor pressure, creating local vaporization of the water. This leads to small vapor bubbles which then collapse inside the pump, and the result is an erosive condition that can lead to early pump failure. This phenomenon may occur with any liquid and is known as "cavitation." Pump suction lines should always be sized for low velocities between 2 and 5 fps (0.6 to 1.5 m/sec) to reduce the chance of creating cavitation in the pump.

Hydronic piping should be sized for a maximum pressure loss of 4 ft of head per 100 ft of pipe. This is a requirement of ASHRAE 90.1 Energy Standard for Buildings Except Low-Rise Residential Buildings. The intent is to limit the friction loss, and hence, the energy consumption required to pump water through a building. Because hydronic piping is usually closed-loop, most of the energy consumption derives from friction losses as opposed to moving the fluid to a higher elevation.

(text continues on page 270)

Fluid	Application	Range			
		Low	High	Low	High
		ft/sec	ft/sec	m/sec	m/sec
Acetylene		67		20.4	
Air	Air or Flue Gas Ducting	10	35	3.0	10.7
Air	Centrifugal Compressor - All Piping	50	100	15.2	30.5
Air	Piston Compressor Discharge	70	100	21.3	30.5
Air	Piston Compressor Suction	50	70	15.2	21.3
Ammonia	Gaseous	33	100	10.1	30.5
Ammonia	Liquid	2	6	0.6	1.8
Benzene		6		1.8	
Bromine	Gaseous	33		10.1	
Bromine	Liquid	4		1.2	
Calcium Chloride		4		1.2	
Carbon Tetrachloride		6		1.8	
Chlorine	Dry Gas	33	83	10.1	25.3
Chlorine	Dry Liquid	5		1.5	
Chloroform	Gaseous	33		10.1	
Chloroform	Liquid	6		1.8	
Ethylene	Gaseous	100		30.5	
Ethylene Dibromide		4		1.2	
Ethylene Dichloride		6		1.8	
Ethylene Glycol		6		1.8	
Hydrochloric Acid	Gaseous	67		20.4	
Hydrochloric Acid	Liquid	5		1.5	
Hydrogen		67		20.4	
Methyl Chloride	Gaseous	67		20.4	
Methyl Chloride	Liquid	6		1.8	
Natural Gas	75 psig and Below Main Lines	35	115	10.7	35.1
Natural Gas	Cross-Country	80	250	24.4	76.2
Natural Gas	Low Pressure Main Lines	3	6	0.9	1.8
Oil	Gravity Flow	2	3	0.6	0.9
Oil	Heavy Viscosity	2	3	0.6	0.9
Oil	Light Viscosity	3	6	0.9	1.8
Oil	Suction Lines	3	4	0.9	1.2
Oxygen	Up to 200 psig	30	100	9.1	30.5
Paper Stock	2% to 2.5% A.D Consistency	3	10	0.9	3.0
Paper Stock	3% to 6% A.D. Consistency	1	8	0.3	2.4
Perchlorethylene		6		1.8	
Propylene Glycol		5		1.5	
Sand	5 to 25% by Volume	12		3.7	

TABLE 11.2 Recommended velocities of fluids in pipelines.

Fluid	Application	Range			
		Low	High	Low	High
		ft/sec	ft/sec	m/sec	m/sec
Sand	Coarse & Granulated Slag	12	13	3.7	4.0
Sand	Fine Graded	8	10	2.4	3.0
Sand	Gravel up to 1/2"	14		4.3	
Sand	Ordinary	11		3.4	
Sewage	Slurry	2.5	3	0.8	0.9
Sodium Chloride	No Solids	5	8	1.5	2.4
Sodium Chloride	With Solids	6	15	1.8	4.6
Sodium Hydroxide	0% - 30%	6		1.8	
Sodium Hydroxide	30% - 50%	5		1.5	
Sodium Hydroxide	50% - 73%	4		1.2	
Steam	Boiler to Turbine Cold Reheat	100	135	30.5	41.1
Steam	Boiler to Turbine Hot Reheat	135	170	41.1	51.8
Steam	HP Bypass of Turbine	200	270	61.0	82.3
Steam	HP District Heating	833	1250	253.9	381.0
Steam	Long Run	135	200	41.1	61.0
Steam	LP Bypass of Turbine	270	335	82.3	102.1
Steam	Saturated up to 15 psig for Heating	17	70	5.2	21.3
Steam	Saturated, 50 psig and Higher	100	167	30.5	50.9
Steam	Superheated 200 psig and Higher	167	300	50.9	91.4
Steam	Superheated Main	100	200	30.5	61.0
Styrene		6		1.8	
Sulfur Dioxide		67		20.4	
Sulfuric Acid		4		1.2	
Tar	Discharge Lines	2	2.5	0.6	0.8
Tar	Gravity Flow	1	1.5	0.3	0.5
Tar	Suction Lines	1	2	0.3	0.6
Trichlorethylene		6		1.8	
Vinyl Chloride		6		1.8	
Vinylidene Chloride		6		1.8	
Water	Boiler Feedwater Discharge	10	17	3.0	5.2
Water	Centrifugal Pump Discharge	5	12	1.5	3.7
Water	Centrifugal Pump Suction	2	5	0.6	1.5
Water	City Water/Service Mains	2	5	0.6	1.5
Water	Fire Hose		10		3.0
Water	General Service	4	10	1.2	3.0
Water	Gravity Flow	2	3	0.6	0.9
Water	Hot Water Recirc		3		0.9
Water	Reciprocating Pump Discharge	5	10	1.5	3.0
Water	Reciprocating Pump Suction	2	5	0.6	1.5
Water	Sea Water	5	12	1.5	3.7

Nominal Pipe Size	Inside Area (Sq ft)	3 fps	4 fps	5 fps	6 fps	7 fps	8 fps	9 fps	10 fps
1/2"	0.0025	3.3	4.4	5.6	6.7	7.8	8.9	10.0	11.1
3/4"	0.0043	5.7	7.7	9.6	11.5	13.4	15.3	17.2	19.1
1"	0.0066	8.8	11.8	14.7	17.7	20.6	23.6	26.5	29.5
1 1/4"	0.0113	15.3	20.4	25.4	30.5	35.6	40.7	45.8	50.9
1 1/2"	0.0154	20.8	27.7	34.6	41.5	48.5	55.4	62.3	69.2
2"	0.0254	34.2	45.6	56.9	68.3	79.7	91.1	102.5	113.9
2 1/2"	0.0379	51.0	68.0	85.0	102.0	119.0	136.0	153.0	169.9
3"	0.0580	78.0	104.0	130.1	156.1	182.1	208.1	234.1	260.1
4"	0.0990	133.3	177.7	222.1	266.5	310.9	355.3	399.8	444.2
6"	0.2204	296.7	395.7	494.6	593.5	692.4	791.3	890.2	989.1
8"	0.3784	509.4	679.2	849.0	1018.8	1188.6	1358.4	1528.2	1698.0
10"	0.5922	797.3	1063.0	1328.8	1594.5	1860.3	2126.0	2391.8	2657.6
12"	0.8373	1127.2	1503.0	1878.7	2254.5	2630.2	3005.9	3381.7	3757.4
14"	1.0124	1362.9	1817.3	2271.6	2725.9	3180.2	3634.5	4088.8	4543.1
16"	1.3314	1792.5	2390.0	2987.5	3584.9	4182.4	4779.9	5377.4	5974.9
18"	1.6941	2280.7	3041.0	3801.2	4561.5	5321.7	6082.0	6842.2	7602.5
20"	2.0876	2810.5	3747.3	4684.2	5621.0	6557.8	7494.7	8431.5	9368.3
24"	3.0121	4055.1	5406.8	6758.5	8110.2	9461.9	10813.7	12165.4	13517.1

Table 11.3 Flow rates in GPM through Schedule 10 pipe.

Nominal Pipe Size	Inside Area (Sq ft)	3 fps	4 fps	5 fps	6 fps	7 fps	8 fps	9 fps	10 fps
1/2"	0.0021	2.8	3.8	4.7	5.7	6.6	7.5	8.5	9.4
3/4"	0.0037	5.0	6.6	8.3	10.0	11.6	13.3	14.9	16.6
1"	0.0060	8.1	10.8	13.5	16.2	18.8	21.5	24.2	26.9
1 1/2"	0.0141	19.0	25.3	31.6	38.0	44.3	50.6	56.9	63.3
2"	0.0233	31.4	41.8	52.3	62.7	73.2	83.6	94.1	104.6
2 1/2"	0.0333	44.8	59.8	74.7	89.7	104.6	119.6	134.5	149.4
3"	0.0513	69.1	92.1	115.1	138.1	161.2	184.2	207.2	230.2
4"	0.0884	119.0	158.7	198.4	238.0	277.7	317.4	357.0	396.7
6"	0.2006	270.1	360.1	450.1	540.1	630.2	720.2	810.2	900.2
8"	0.3474	467.7	623.6	779.5	935.4	1091.3	1247.2	1403.1	1559.0
10"	0.5476	737.2	983.0	1228.7	1474.5	1720.2	1966.0	2211.7	2457.4
12"	0.7854	1057.4	1409.8	1762.3	2114.8	2467.2	2819.7	3172.1	3524.6
14"	0.9575	1289.1	1718.8	2148.5	2578.2	3007.9	3437.5	3867.2	4296.9
16"	1.2684	1707.6	2276.9	2846.1	3415.3	3984.5	4553.7	5122.9	5692.1
18"	1.6230	2185.0	2913.4	3641.7	4370.1	5098.4	5826.8	6555.1	7283.5
20"	2.0211	2721.0	3628.0	4535.0	5442.0	6349.0	7256.0	8163.0	9070.0
24"	2.9483	3969.3	5292.4	6615.5	7938.6	9261.7	10584.8	11907.9	13231.0
30"	4.6640	6279.1	8372.2	10465.2	12558.3	14651.3	16744.4	18837.4	20930.4
36"	6.7770	9123.9	12165.1	15206.4	18247.7	21289.0	24330.3	27371.6	30412.9
40"	8.4020	11311.6	15082.1	18852.7	22623.2	26393.7	30164.2	33934.8	37705.3
42"	9.2810	12495.0	16660.0	20825.0	24990.0	29155.0	33320.0	37485.0	41650.0
48"	12.1770	16393.9	21858.5	27323.1	32787.7	38252.4	43717.0	49181.6	54646.2
60"	19.1470	25777.6	34370.1	42962.6	51555.1	60147.6	68740.2	77332.7	85925.2
96"	49.4830	66618.8	88825.1	111031.4	133237.7	155444.0	177650.3	199856.5	222062.8

Table 11.4 Flow rates in GPM through Schedule 40 pipe.

Nominal	Inside Area	3	4	5	6	7	8	9	10
Pipe Size	(Sq ft)	fps	fps	fps	fps	fps	fps	fps	fps
1/2"	0.0016	2.2	2.9	3.6	4.3	5.0	5.7	6.5	7.2
3/4"	0.0030	4.0	5.4	6.7	8.1	9.4	10.8	12.1	13.5
1"	0.0050	6.7	9.0	11.2	13.5	15.7	18.0	20.2	22.4
1 1/2"	0.0123	16.6	22.1	27.6	33.1	38.6	44.2	49.7	55.2
2"	0.0205	27.6	36.8	46.0	55.2	64.4	73.6	82.8	92.0
2 1/2"	0.0294	39.6	52.8	66.0	79.2	92.4	105.5	118.7	131.9
3"	0.0459	61.8	82.4	103.0	123.6	144.2	164.8	185.4	206.0
4"	0.0798	107.4	143.2	179.1	214.9	250.7	286.5	322.3	358.1
6"	0.1810	243.7	324.9	406.1	487.4	568.6	649.8	731.0	812.3
8"	0.3171	426.9	569.2	711.5	853.8	996.1	1138.4	1280.7	1423.0
10"	0.5185	698.1	930.7	1163.4	1396.1	1628.8	1861.5	2094.2	2326.9
12"	0.7530	1013.8	1351.7	1689.6	2027.5	2365.4	2703.4	3041.3	3379.2
14"	0.9213	1240.3	1653.8	2067.2	2480.7	2894.1	3307.6	3721.0	4134.5
16"	1.2272	1652.2	2202.9	2753.6	3304.4	3855.1	4405.8	4956.5	5507.3
18"	1.5762	2122.0	2829.4	3536.7	4244.1	4951.4	5658.8	6366.1	7073.4
20"	1.9689	2650.7	3534.3	4417.9	5301.5	6185.0	7068.6	7952.2	8835.8
24"	2.8852	3884.3	5179.1	6473.9	7768.7	9063.5	10358.2	11653.0	12947.8
30"	4.5869	6175.3	8233.8	10292.2	12350.7	14409.1	16467.6	18526.0	20584.4
36"	6.6810	8994.6	11992.8	14991.0	17989.2	20987.4	23985.6	26983.8	29982.0
40"	8.2960	11168.9	14891.8	18614.8	22337.8	26060.7	29783.7	33506.7	37229.6
42"	9.1680	12342.9	16457.1	20571.4	24685.7	28800.0	32914.3	37028.6	41142.9
48"	12.0480	16220.2	21626.9	27033.7	32440.4	37847.1	43253.9	48660.6	54067.3
60"	18.9860	25560.8	34081.1	42601.3	51121.6	59641.9	68162.2	76682.4	85202.7
96"	49.2240	66270.2	88360.2	110450.3	132540.3	154630.4	176720.4	198810.5	220900.5

Table 11.5 Flow rates in GPM through XS pipe.

Nominal	Inside Area	3	4	5	6	7	8	9	10
Pipe Size	(Sq ft)	fps	fps	fps	fps	fps	fps	fps	fps
1/2"	0.0004	0.5	0.7	0.9	1.1	1.3	1.4	1.6	1.8
3/4"	0.0010	1.3	1.8	2.2	2.7	3.1	3.6	4.0	4.5
1"	0.0020	2.7	3.6	4.5	5.4	6.3	7.2	8.1	9.0
1 1/2"	0.0066	8.9	11.8	14.8	17.8	20.7	23.7	26.7	29.6
2"	0.0123	16.6	22.1	27.6	33.1	38.6	44.2	49.7	55.2
2 1/2"	0.0171	23.0	30.7	38.4	46.0	53.7	61.4	69.1	76.7
3"	0.0289	38.9	51.9	64.8	77.8	90.8	103.8	116.7	129.7
4"	0.0542	73.0	97.3	121.6	145.9	170.3	194.6	218.9	243.2
6"	0.1308	176.1	234.8	293.5	352.2	410.9	469.6	528.3	587.0
8"	0.2578	347.1	462.8	578.5	694.2	809.8	925.5	1041.2	1156.9
10"	0.4176	562.2	749.6	937.0	1124.4	1311.8	1499.2	1686.6	1874.0
12"	0.6303	848.6	1131.4	1414.3	1697.1	1980.0	2262.9	2545.7	2828.6

Table 11.6 Flow rates in GPM through XXS pipe.

Nominal Pipe Size	Inside Area (Sq ft)	3 fps	4 fps	5 fps	6 fps	7 fps	8 fps	9 fps	10 fps
1/2"	0.0000	0.0	0.0	0.0	0.0	0.0	0.0	0.0	0.0
3/4"	0.0020	2.7	3.6	4.5	5.4	6.3	7.2	8.1	9.0
1"	0.0036	4.8	6.5	8.1	9.7	11.3	12.9	14.5	16.2
1 1/2"	0.0098	13.2	17.6	22.0	26.4	30.8	35.2	39.6	44.0
2"	0.0155	20.9	27.8	34.8	41.7	48.7	55.6	62.6	69.6
2 1/2"	0.0246	33.1	44.2	55.2	66.2	77.3	88.3	99.4	110.4
3"	0.0376	50.6	67.5	84.4	101.2	118.1	135.0	151.9	168.7
4"	0.0645	86.8	115.8	144.7	173.7	202.6	231.6	260.5	289.5
6"	0.1467	197.5	263.3	329.2	395.0	460.8	526.7	592.5	658.3
8"	0.2532	340.9	454.5	568.1	681.8	795.4	909.0	1022.6	1136.3
10"	0.3941	530.6	707.4	884.3	1061.2	1238.0	1414.9	1591.7	1768.6
12"	0.5592	752.8	1003.8	1254.7	1505.7	1756.6	2007.6	2258.5	2509.5
14"	0.6827	919.1	1225.5	1531.9	1838.2	2144.6	2451.0	2757.4	3063.7
16"	0.8953	1205.3	1607.1	2008.9	2410.7	2812.5	3214.2	3616.0	4017.8
18"	1.1370	1530.7	2041.0	2551.2	3061.5	3571.7	4082.0	4592.2	5102.5
20"	1.4071	1894.4	2525.8	3157.3	3788.8	4420.2	5051.7	5683.1	6314.6
24"	2.0342	2738.6	3651.5	4564.4	5477.3	6390.2	7303.0	8215.9	9128.8

TABLE 11.7 Flow rates in GPM through Schedule 160 pipe.

Steam

Steam flows are often stated in terms of pounds per hour. In order to determine the velocity of steam in a pipe, it is necessary to know its temperature and pressure. From there, one can look up the specific volume to calculate the diameter based on the desired velocity.

Example 11.2
Given: Saturated steam at 100 psig, at a flow rate of 8000 lb/hr
Find: A suitable carbon steel pipe diameter

We need to first determine the volume occupied by saturated steam at 100 psig. We do this by interpolating the specific volume from the data shown in Table A.8 in the Appendix.

$$v_g = 3.889 \text{ ft}^3/\text{lb}$$

$$\text{Volumetric Flow Rate} = \dot{V} = (3.889 \text{ ft}^3/\text{lb})\ (8000 \text{ lb/hr})\ (\text{hr}/3600 \text{ sec}) = 8.64 \text{ ft}^3/\text{sec}$$

From Table 11.1 we note that the recommended range of velocities for saturated steam above 50 psig is 100 to 167 fps. We choose 150 fps.

$$\text{ID} = (8.64 \text{ ft}^3/\text{sec})\ /\ (150 \text{ ft/sec}) = 0.058 \text{ ft}^2$$

From Appendix 1 we see that 3 in Schedule 40 carbon steel has a flow area of 0.051 ft^2; 4 in Schedule 40 has a flow area of 0.088 ft^2. Let's examine the actual velocities of the commercially available pipe:

$$\text{Velocity of 3 in Sch 40} = (8.64 \text{ ft}^3/\text{sec})\ /\ 0.051 \text{ ft}^2 = 169 \text{ fps}$$

$$\text{Velocity of 4 in Sch 40} = (8.64 \text{ ft}^3/\text{sec})\ /\ 0.088 \text{ ft}^2 = 98 \text{ fps}$$

The 3 in Sch 40 pipe exceeds the maximum recommended velocity, and the 4 in produces a velocity less than the low end of the recommended range. We could use 3 1/2 in diameter pipe, but that is an unusual size, so we select the 4 in diameter because it will

- Provide for possible future capacity
- Decrease noise generation
- Provide less friction loss

Roughness of the pipe is a function of the pipe material as well as its age. Ferrous pipes that have been in service many years may be corroded, scaled, or tubercular, and the relative roughness will have increased due to these surface irregularities. However, for low flow rates it turns out that the roughness does not matter at all. This occurs only for laminar flow, which is a special case since most industrial flow problems lie within the turbulent regime.

The number of fittings and other factors such as entrance and exit losses that contribute to the minor losses may be converted into an "Equivalent Length," which is then added to the length of straight pipe for a total length. This exercise must be performed for each segment that has a different diameter in the system.

Much attention has been given to pressure drop calculations throughout the years, and the result is that there are several good methods to compute pressure drops. Some are certainly more convenient than others, and engineers know a good shortcut when they see one. Once again, because the majority of piping systems pertain to water, much effort has been devoted to the calculation of pressure drops through common water pipe materials.

Major Losses

There are four methods used to calculate the head loss due to friction:

1. Darcy Weisbach Equation
2. Hazen Williams Formula
3. Fanning Friction Factor
4. Tabular Methods

Darcy Weisbach Equation

The head loss due to friction can be found by the Darcy Weisbach Equation:

$$h_f = f\frac{L_e}{D}\frac{V^2}{2g_c} \quad \text{(Equation 11.7)}$$

where

f = The Darcy or Moody Friction Factor [dimensionless]
L_e = Equivalent Length of the pipe [ft or m]
D = ID of the pipe, or alternately, the Hydraulic Diameter [ft or m]
V = Velocity of the fluid [ft/sec or m/sec]
g_c = The gravitational constant = 32.2 ft/sec^2 (9.81 m/sec^2)

The velocity of the fluid is understood to be the average velocity, and is best described as the volumetric flow rate divided by the cross-sectional wetted area of the pipe. This distinction is made due to the varying velocity profile that results from the non-slip condition of the fluid adjacent to the inside wall of the pipe.

The Equivalent Length term is used so that the friction losses of both the straight pipe and the fittings may be determined simultaneously.

The Hydraulic Diameter is a concept that is used for less-than-full pipe flow, non-circular ducts, or open channel flow. It is defined as

$$D_h = 4A/U \qquad \text{(Equation 11.8)}$$

where

A = The cross-sectional area

U = The wetted perimeter of the cross section

Thus, for a full pipe,

$$D_h = 4\,\pi\,r^2 \,/\, 2r\,\pi = 2\,r$$

as we would expect.

The f term, the Darcy Friction Factor, may be determined by applying one of the following methods. The most common method is to enter the Moody Diagram. While the other methods are valid, their use is included here for the sake of completeness. Most engineers will find their application to be too complex in practice.

- Moody Diagram
- Colebrook Equation
- Haaland Equation
- Swamee-Jain Equation
- Serghide's Solution.

The latter three methods provide approximations of the Colebrook equation.

Moody Diagram

The Moody Diagram (Figure 11.1) is an extremely useful tool that is used to determine the friction factor based on pipe roughness and flow conditions. The flow conditions are described by the Reynolds Number, which is the ratio of inertial forces to viscous forces. The Reynolds Number is defined as

$$Re = \rho v D_h/\mu \qquad \text{(Equation 11.9)}$$

where

ρ = Density [lbm/ft³ or kg/m³]

v = Velocity [ft/sec or m/sec]

D_h = Hydraulic Diameter as described above [ft or m]

μ = Dynamic Viscosity [lbm/sec ft or N sec/m²]

There are several "types" of viscosities. The "Absolute Viscosity" is the same as the Dynamic Viscosity. Dynamic Viscosity data are often given in "centipoises" or "micropoises," where 1 poise is 0.1 N sec/m². Dynamic viscosities of water are tabulated in Table 11.8.

Another viscosity term that is often encountered is the Kinematic Viscosity, which is equal to the Dynamic Viscosity divided by the density. Water densities are given in Table 11.9.

Kinematic Viscosity is represented by the Greek letter nu (ν), and has the units [ft²/sec or m²/sec]. In the SI system, 1 m²/sec is defined as a "Stoke," but because this represents a large quantity, the more practical unit "centistoke" [cSt] is used.

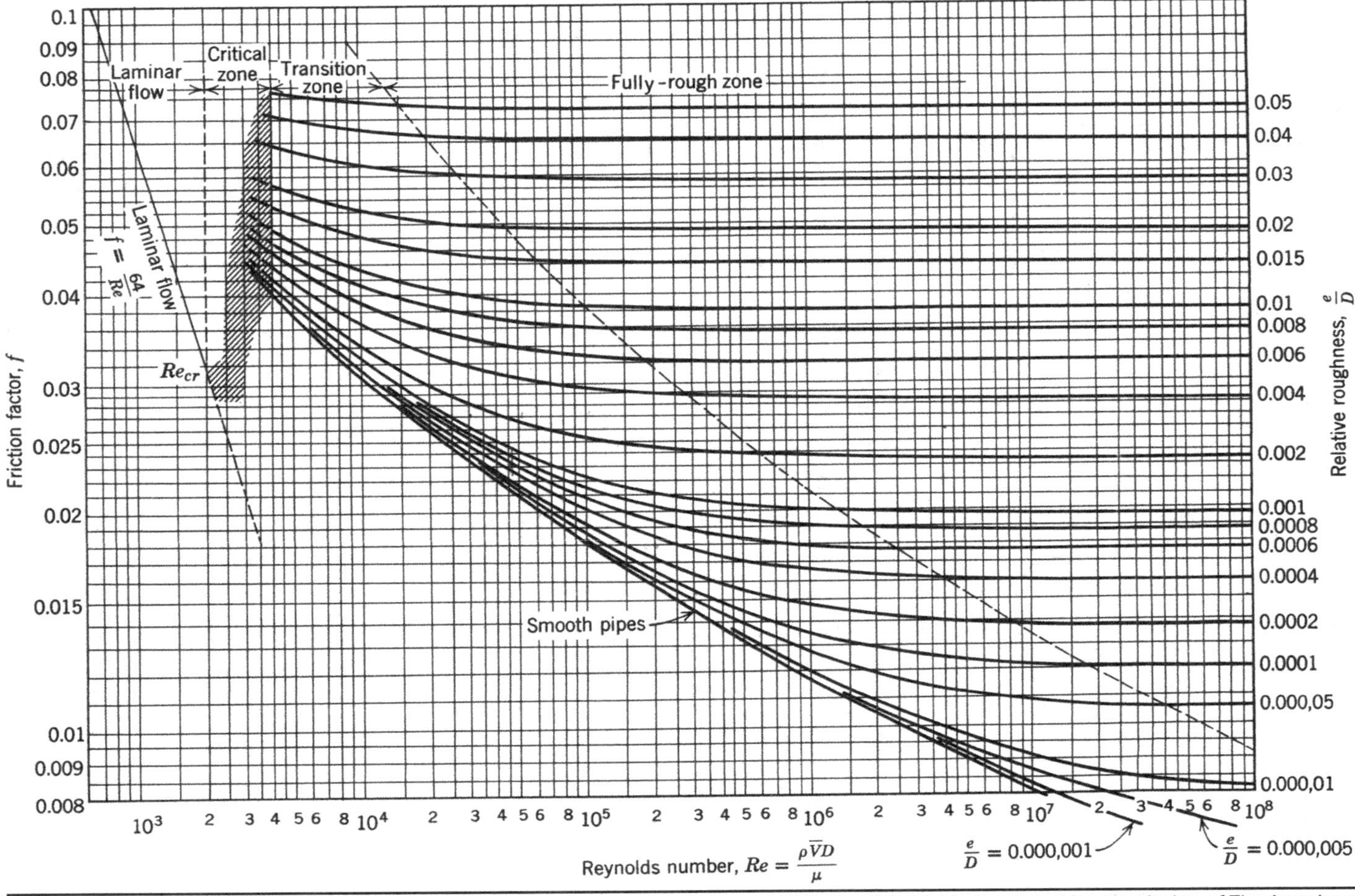

Figure 11.1 Moody Diagram, L.F. Moody, "Friction factors for pipe flow," Trans. ASME Vol. 66. 1944. *Reprinted by permission of The American Society of Mechanical Engineers. All rights reserved.*

Temperature		Dynamic Viscosity μ		Temperature		Dynamic Viscosity μ		Temperature		Dynamic Viscosity μ		Temperature		Dynamic Viscosity μ	
°F	°C	lbm/ft sec	kg/m sec	°F	°C	lbm/ft sec	kg/m sec	°F	°C	lbm/ft sec	kg/m sec	°F	°C	lbm/ft sec	kg/m sec
32.0	0	0.001204	0.001792	78.8	26	0.000585	0.000871	123.8	51	0.000362	0.000538	168.8	76	0.000251	0.000373
33.8	1	0.001163	0.001731	80.6	27	0.000573	0.000852	125.6	52	0.000355	0.000529	170.6	77	0.000248	0.000369
35.6	2	0.001125	0.001674	82.4	28	0.000560	0.000833	127.4	53	0.000350	0.000521	172.4	78	0.000245	0.000364
37.4	3	0.001089	0.001620	84.2	29	0.000548	0.000815	129.2	54	0.000344	0.000512	174.2	79	0.000241	0.000359
39.2	4	0.001054	0.001569	86.0	30	0.000536	0.000798	131.0	55	0.000339	0.000504	176.0	80	0.000239	0.000355
41.0	5	0.001021	0.001520	87.8	31	0.000525	0.000781	132.8	56	0.000333	0.000496	177.8	81	0.000236	0.000351
42.8	6	0.000990	0.001473	89.6	32	0.000514	0.000765	134.6	57	0.000329	0.000489	179.6	82	0.000233	0.000346
44.6	7	0.000960	0.001429	91.4	33	0.000503	0.000749	136.4	58	0.000323	0.000481	181.4	83	0.000230	0.000342
46.4	8	0.000931	0.001386	93.2	34	0.000493	0.000734	138.2	59	0.000319	0.000474	183.2	84	0.000227	0.000338
48.2	9	0.000904	0.001346	95.0	35	0.000484	0.000720	140.0	60	0.000314	0.000467	185.0	85	0.000224	0.000334
50.0	10	0.000879	0.001308	96.8	36	0.000474	0.000705	141.8	61	0.000309	0.000460	186.8	86	0.000222	0.000330
51.8	11	0.000854	0.001271	98.6	37	0.000465	0.000692	143.6	62	0.000304	0.000453	188.6	87	0.000219	0.000326
53.6	12	0.000831	0.001236	100.4	38	0.000456	0.000678	145.4	63	0.000300	0.000447	190.4	88	0.000216	0.000322
55.4	13	0.000808	0.001202	102.2	39	0.000448	0.000666	147.2	64	0.000296	0.000440	192.2	89	0.000214	0.000319
57.2	14	0.000786	0.001170	104.0	40	0.000439	0.000653	149.0	65	0.000292	0.000434	194.0	90	0.000212	0.000315
59.0	15	0.000765	0.001139	105.8	41	0.000431	0.000641	150.8	66	0.000288	0.000428	195.8	91	0.000209	0.000311
60.8	16	0.000745	0.001109	107.6	42	0.000423	0.000629	152.6	67	0.000284	0.000422	197.6	92	0.000207	0.000308
62.6	17	0.000726	0.001081	109.4	43	0.000415	0.000618	154.4	68	0.000280	0.000416	199.4	93	0.000204	0.000304
64.4	18	0.000708	0.001054	111.2	44	0.000408	0.000607	156.2	69	0.000276	0.000410	201.2	94	0.000202	0.000301
66.2	19	0.000691	0.001028	113.0	45	0.000400	0.000596	158.0	70	0.000271	0.000404	203.0	95	0.000200	0.000298
68.0	20	0.000674	0.001003	114.8	46	0.000394	0.000586	159.8	71	0.000268	0.000399	204.8	96	0.000198	0.000295
69.8	21	0.000658	0.000979	116.6	47	0.000387	0.000576	161.6	72	0.000265	0.000394	206.6	97	0.000196	0.000291
71.6	22	0.000642	0.000955	118.4	48	0.000380	0.000566	163.4	73	0.000261	0.000388	208.4	98	0.000194	0.000288
73.4	23	0.000627	0.000933	120.2	49	0.000374	0.000556	165.2	74	0.000257	0.000383	210.2	99	0.000192	0.000285
75.2	24	0.000612	0.000911	122.0	50	0.000368	0.000547	167.0	75	0.000254	0.000378	212.0	100	0.000189	0.000282
77.0	25	0.000599	0.000891												

TABLE 11.8 Dynamic Viscosity of Water (1 kg/m sec = 1000 cp). *Jean Yves Messe*

Temperature		Density		Vapor Pressure		
°F	°C	lb/ft³	kg/m³	Pa	m of water	ft of water
32.0	0.0	62.42	999.8	611.0	0.0623	0.204
41.0	5.0	62.43	1000.0	872.0	0.0889	0.292
50.0	10.0	62.41	999.8	1227.0	0.1251	0.411
59.0	15.0	62.38	999.2	1704.0	0.1738	0.570
68.0	20.0	62.32	998.3	2337.0	0.2383	0.782
77.0	25.0	62.25	997.1	3166.0	0.3228	1.059
86.0	30.0	62.16	995.7	4242.0	0.4326	1.419
95.0	35.0	62.06	994.1	5622.0	0.5733	1.881
104.0	40.0	61.94	992.3	7375.0	0.7521	2.467
113.0	45.0	61.82	990.2	9582.0	0.9771	3.206
122.0	50.0	61.68	988.0	12335.0	1.2578	4.127
131.0	55.0	61.53	985.7	15740.0	1.6051	5.266
140.0	60.0	61.37	983.1	19919.0	2.0312	6.664
149.0	65.0	61.21	980.5	25008.0	2.5502	8.367
158.0	70.0	61.03	977.6	31161.0	3.1776	10.425
167.0	75.0	60.85	974.7	38548.0	3.9309	12.897
176.0	80.0	60.66	971.6	47359.0	4.8294	15.845
185.0	85.0	60.45	968.4	57803.0	5.8944	19.339
194.0	90.0	60.25	965.1	70108.0	7.1492	23.455
203.0	95.0	60.03	961.6	84525.0	8.6194	28.279
212.0	100.0	59.81	958.1	101325.0	10.3325	33.899

TABLE 11.9 Properties of Water. *Jean Yves Messe*

Therefore $\nu = \mu/\rho$ and so the Reynolds Number may also be written as

$$Re = vD_h/\nu \qquad \text{(Equation 11.10)}$$

The pipe roughness used in the Moody Diagram is called the "Relative Roughness" and it is the ratio of the pipe roughness ϵ divided by the pipe diameter D.

Pipe Roughness There is wide variation in values of pipe roughness as published in various sources. For instance, some sources indicate the roughness of copper and brass to be 0.000003 ft, while others give it as high as 0.003 ft. This will obviously affect the value of the friction factor, and it is assumed that the higher value is for a very rough casting. Table 11.10 provides a consensus sampling of published roughness data for some common materials.

Type of pipe or surface		Roughness ϵ [ft]			C-Factor			
		Low	High	Design	Low	High	Clean	Design
STEEL	welded and seamless	0.0001	0.008	0.0002	80	150	140	100
	corroded	0.0005	0.0133					
	riveted	0.003	0.03				139	100
	Stainless			0.00005				
	galvanized, plain	0.0002	0.0008	0.0005				
	corrugated						60	60
MINERAL	concrete	0.001	0.01	0.004	85	152	120	100
	cement	0.0013	0.004	0.003	140	160	150	140
	vitrified clays							110
	brick sewer							100
IRON	cast, plain	0.0004	0.0027	0.00085	80	150	130	100
	cast, tar (asphalt) coated	0.0002	0.0008	0.0004	50	145	130	100
	cast, cement lined	0.000008		0.000008			150	140
	cast, bituminous lined	0.000008		0.000008	130	160	148	140
	cast, centrifugally spun	0.00001		0.00001				
	wrought, plain	0.0001	0.0003	0.0002	80	150	130	100
MISC.	fiber						150	140
	copper and brass	0.000005	0.003	0.000005	120	150	140	130
	wood stave	0.0006	0.003	0.002	110	145	120	110
	transite	0.000008		0.000008				
	lead, tin, glass			0.000005	120	150	140	130
	plastic			0.000005	120	150	140	130

TABLE 11.10 Absolute roughness and C-factors of common pipe materials.

Laminar Flow, Turbulent Flow, and the Transition Zone The Reynolds Number is a dimensionless quantity that quantifies the localized flow patterns inside a flow stream. Laminar flow is the term used to describe flow patterns in which localized flow within the pipe occurs in layers. Turbulent flow is the term used to describe random three-dimensional flow. The gross motion through the pipe is from high-pressure to low-pressure, but the local flow is multi-directional.

There is a transition zone in which the fluid flow at the periphery of the pipe is laminar, and at the center is turbulent. The friction factor f is unknown in this transition zone.

For laminar flow, the friction factor is given not by the Moody Diagram, but instead by the simple formula

$$f = 64/Re \qquad \text{(Equation 11.11)}$$

The approximate values of Reynolds Numbers that describe each of these flow types are:

Laminar flow if $Re < 2300$

Transitional flow if $2300 < Re < 4000$

Turbulent flow if $4000 < Re$

Example 11.3
Given: Water at 95°F, flowing in a 6 in diameter Sch 40 steel pipe at 8 fps, 1000 ft equivalent length
Find: Moody Friction Factor and head loss due to friction

We begin by calculating the Reynolds Number

$$Re = \rho v D_h/\mu$$
$$\rho = 62.057 \text{ lbm/ft}^3 \text{ from Table 11.9}$$
$$v = 8 \text{ ft/sec}$$
$$D_h = 6.065 \text{ in}$$
$$\mu = 0.000484 \text{ lbm/ft sec from Table 11.8}$$
$$Re = (62.057 \text{ lbm/ft}^3)(8 \text{ ft/sec})(6.065 \text{ in})(\text{ft/12 in}) / (0.000484 \text{ lbm/ft sec})$$
$$= 5.2 \times 10^5$$

The relative roughness of new steel pipe is calculated to be

$$\epsilon/D = (0.0002 \text{ ft}/6.065 \text{ in})(12 \text{ in/ft}) = 0.0004$$

with values of ϵ coming from Table 11.10

We enter the Moody Diagram at $Re = 5.2 \times 10^5$ and find $\epsilon/D = 0.0004$ to arrive at a friction factor f of 0.017.

The head loss due to friction is given by Equation 11.3 as

$$h_f = f\frac{L_e}{D}\frac{V^2}{2g_c}$$
$$= (0.017)(1000 \text{ ft} / 6.065 \text{ in})(12 \text{ in/ft})(8 \text{ ft/sec})^2 / 2(32.2 \text{ ft/sec}^2)$$
$$= 33.4 \text{ ft}$$

Colebrook Equation

The Colebrook Equation (also known as the Colebrook-White Equation) is

$$1/\sqrt{f} = -0.86 \ln\left(\frac{\frac{\epsilon}{3.7D} + 2.51}{Re\sqrt{f}}\right) \quad \text{(Equation 11.12)}$$

This equation must be solved iteratively, and that is never viewed as a good thing by engineers. This difficulty led to the development of the Moody Diagram. But in addition to the Moody Diagram, approximations to the Colebrook Equation have been developed.

Haaland Equation

An approximation of the Colebrook Equation was developed in 1983 by S.E. Haaland.

$$1/\sqrt{f} = -1.8 \log_{10}\left[\left(\frac{\epsilon/D}{3.7}\right)^{1.11} + \frac{6.9}{Re}\right] \quad \text{(Equation 11.13)}$$

This equation does not require iteration to solve.

Swamee-Jain Equation

The Swamee-Jain Equation is another approximation of the Colebrook Equation. It provides an accuracy within 1 percent of the Colebrook equation for relative roughnesses between 0.000001 and 0.01, and Reynolds Numbers between 5000 and 10^8.

$$f = \frac{0.25}{\left[\log_{10}\left(\frac{\epsilon/D}{3.7} + \frac{5.74}{Re^{0.9}}\right)\right]^2} \quad \text{(Equation 11.14)}$$

Serghide's Solution

An even closer approximation (within 0.003 percent) to the Colebrook Equation is given by Serghide's Solution.

$$f = \left(A - \frac{(B - A)^2}{C - 2B + A}\right)^{-2} \quad \text{(Equation 11.15)}$$

where

$$A = -2\log_{10}\left(\frac{\epsilon/D}{3.7} + \frac{12}{Re}\right) \quad \text{(Equation 11.16)}$$

$$B = -2\log_{10}\left(\frac{\frac{\epsilon}{D}}{3.7} + \frac{2.51\,A}{Re}\right) \quad \text{(Equation 11.17)}$$

$$C = -2\log_{10}\left(\frac{\epsilon/D}{3.7} + \frac{2.51\,B}{Re}\right) \quad \text{(Equation 11.18)}$$

Hazen-Williams Formula

The Hazen-Williams Formula for calculating head loss due to friction applies to the following conditions:

- Fluids with Kinematic Viscosity approximately 1.2×10^{-5} ft^2/sec. This matches the viscosity of water at 60°F (15.6°C)
- Turbulent flow conditions
- "Moderate" temperatures

The Hazen-Williams Equation for head loss is

$$h_f = \frac{C_f\, L\, Q^{1.852}}{C^{1.852}\, D^{4.87}} \quad \text{(Equation 11.19)}$$

where

C_f = Unit Conversion Factor [4.72 for Imperial units; 10.67 for SI units]
L = Equivalent Length of pipe [ft or m]
Q = Flow rate [ft^3/sec or m^3/sec]
C = Hazen-Williams C-Factor
D = Inside diameter [ft or m]

C-factors represent the carrying capacity of the pipe (independent of the diameter), with high C-factors indicating smoother pipe (see Table 11.10).

Example 11.4
Given: The same conditions as in Example 11.3
Find: Head loss due to friction using the Hazen-Williams Equation

C = 100 for steel pipe, according to Table 11.3. This is a "design" value that is conservative and allows for future scaling of the internal pipe surfaces. (The American Iron and Steel Institute's Committee of Steel Pipe Producers has recommended a value of C = 140).

$C_f = 4.72$

$L = 1000$ ft

$Q = (8 \text{ ft/sec})(0.20063 \text{ ft}^2) = 1.605 \text{ ft}^3\text{/sec}$

$D = (6.065 \text{ in})(\text{ft}/12 \text{ in}) = 0.505 \text{ ft}$

$$h_f = \frac{C_f \, L \, Q^{1.852}}{C^{1.852} \, D^{4.87}}$$

$$= (4.72)(1000)(1.605)^{1.852} / (100)^{1.852} (0.505)^{4.87}$$

$$= 11337 / 181.6$$

$$= 62.4 \text{ ft}$$

Note that if we use the "clean" or "new pipe" C-factor of 140, we get

$$h_f = 11337 / (140)^{1.852} (0.505)^{4.87}$$
$$= 33.5 \text{ ft}$$

We calculated 33.4 ft using the Moody Diagram. Conversely, if we use the values of roughness for corroded steel pipe in Example 11.3, we get

$$\epsilon/D = 0.013 \text{ ft}/(6.065 \text{ in})(\text{ft}/12 \text{ in}) = 0.03$$

and from the Moody Diagram, f = 0.0265
which increases the h_f by a factor of 0.0265/0.017 which yields

$$h_f = 33.4 \text{ ft } (0.0265/0.017) = 52.1 \text{ ft}$$

The results of these calculations are summarized below.

Pipe Condition	Darcy-Weisbach	Hazen-Williams
New Steel Pipe	33.4 ft	33.5 ft
Corroded Steel Pipe	52.1 ft	62.4 ft

This shows that

- There is good agreement between the Hazen Williams Equation and the Darcy-Weisbach Equation for clean pipe.
- The results are sensitive to the ϵ roughness and C-factor values for old or corroded pipe. This is likely due to the interpretation of "corroded" and the roughness of the corroded pipe that was used to experimentally derive the values.

Fanning Friction Factor

Occasionally one will encounter the Fanning Friction Factor, although its use is far less common than the Moody Friction Factor.

The Fanning Equation for friction uses the "Hydraulic Radius," defined as the cross sectional area of flow divided by the wetted perimeter. Thus,

$$R_h = A/U \qquad \text{(Equation 11.20)}$$

where

R_h = Hydraulic Radius [ft or m]

A = Cross sectional area [ft^2 or m^2]

U = Wetted Perimeter [ft or m]

The Fanning head loss is given as

$$h_f = f_f\left(\frac{L}{R_h}\right)\left(\frac{V^2}{2g_c}\right) \qquad \text{(Equation 11.21)}$$

where

f_f = Fanning Friction Factor [dimensionless]

This is similar to the friction head loss calculation used by the more common Darcy-Weisbach Equation, and by cancelling out terms

$$h_f = f\left(\frac{L_e}{D}\right)\left(\frac{v^2}{2g_c}\right) = f_f\left(\frac{L}{R_h}\right)\left(\frac{v^2}{2g_c}\right) \text{ we see that}$$

$$f/D = f_f/R_h \text{ and } D_h = D \text{ for full-pipe flow}$$

$$f = f_f\,(D_h/R_h) = f_f\,(4A/U)/(A/U) = 4f_f$$

Therefore, the Darcy Friction Factor is four times the Fanning Friction Factor.

Tabulated or Graphic Solutions

By now it is clear that there are multiple routes to our final destination of solving pressure drops. We indicated earlier that there are some shortcuts available for common problems, and that is indeed fortunate since the methods described above, while satisfactory, are mathematically complicated.

Tabulated values are available for water, air, and low-pressure natural gas. Once the equivalent length is determined, the friction loss is read from the tables and multiplied by the appropriate factor to give the friction loss. Linear interpolation must be used to determine losses at flow rates between tabulated values. This interpolation may be eliminated through the use of graphs that chart the relationships between flow rate and head loss. Some practical graphs are provided in the Appendix. These graphs are for asphalt-coated cast iron and various weights of steel pipe.

Note that manufacturers of other pipe materials can supply similar pressure drop data.

Minor Losses

Minor losses are those friction losses attributed to fittings, valves, and entrance and exit losses. Minor losses may be determined with any of three methods:

- Resistance Coefficient K
- Equivalent Length Method
- Flow Coefficient C_v

All other things being equal, the easiest method is the equivalent length method. However, equivalent length data are often not available for the components you need and conversions are necessary.

Resistance Coefficient K

Flow resistance data are available for a number of valves and fittings, although the user should be aware that significant variation may exist between data published by different sources. For critical applications, the valve or fitting manufacturer should be contacted for data specific to the item.

The friction head loss through fittings for which the resistance coefficient K is known is

$$h_f = K\,(V^2/2g_c) \qquad \text{(Equation 11.22)}$$

with $V^2/2g_c$ being the velocity head through the fitting. See Table 11.11 for Resistance Coefficient data for common valves and fittings.

There is little difference in the K values of different schedules of pipe fittings. Therefore, it is sufficient to use the data for Schedule 40 pipe. Sizes larger than 24-in diameter can be estimated using the K values for 24 in.

Reducers

The pressure drop literature is full of time-consuming equations that relate head loss to the angle of the reducer walls. While no one can be faulted for performing these equations and including them in the total loss calculations, the head loss through a single reducer is not significant.

Unless your system contains many, many reducers, a faster and more practical approach is simply to include the length of the reducer among the equivalent lengths of the smaller of the two diameters.

Equivalent Length Method

Since friction losses distill down to units of feet or meters, conversions have been developed to describe minor losses simply as an equivalent length of straight pipe. The lengths for a given diameter are summed.

If tables exist for friction losses (as they do for water and air), the total equivalent length is multiplied by the factor in the table to yield the friction loss for that particular diameter. The equivalent length is given by

$$L = KD/f \qquad \text{(Equation 11.23)}$$

where

K = Resistance Coefficient

D = Inside pipe diameter [ft]

f = Friction factor in the turbulent flow region

Otherwise, the equivalent length is used in the Darcy-Weisbach Equation with the friction factor f taken from the Moody Diagram.

This is a very convenient method, as it requires no extra step in the calculation. Data exists for many common fittings and valves. See Table 11.12.

Nominal Pipe Size		in	1/2	3/4	1	1 1/2	2	2 1/2	3	4
d (based on Sch 40)		in	0.622	0.824	1.049	1.61	2.067	2.469	3.068	4.026
D (based on Sch 40)		ft	0.052	0.069	0.087	0.134	0.172	0.206	0.256	0.336
f			0.027	0.025	0.023	0.021	0.019	0.018	0.018	0.017
Fitting										
Valves		L/D	K = f (L/D)							
	Ball Valve	3	0.08	0.08	0.07	0.06	0.06	0.05	0.05	0.05
	Gate Valve	8	0.22	0.20	0.18	0.17	0.15	0.14	0.14	0.14
	Butterfly Valve	35	—	—	—	—	0.67	0.63	0.63	0.60
	Swing Check Valve up to 6"	100	2.70	2.50	2.30	2.10	1.90	1.80	1.80	1.70
	Swing Check Valve 24" to 48"	50	—	—	—	—	—	—	—	—
	Angle Valve - 90 degree	150	4.05	3.75	3.45	3.15	2.85	2.70	2.70	2.55
	Angle Valve - Y Pattern	55	1.49	1.38	1.27	1.16	1.05	0.99	0.99	0.94
	Globe Valve	340	9.18	8.50	7.82	7.14	6.46	6.12	6.12	5.78
	Lift Check Valve	600	16.20	15.00	13.80	12.60	11.40	10.80	10.80	10.20
Tees	Tee - Through (Run)	20	0.54	0.50	0.46	0.42	0.38	0.36	0.36	0.34
	Tee - Branch	60	1.62	1.50	1.38	1.26	1.14	1.08	1.08	1.02
Bends	Elbow - 90 Short Radius	30	0.81	0.75	0.69	0.63	0.57	0.54	0.54	0.51
	Elbow - 90 Long Radius	16	0.43	0.40	0.37	0.34	0.30	0.29	0.29	0.27
	Miter Bend - 45 deg	15	—	—	—	—	0.29	0.27	0.27	0.26
	Miter Bend - 90 deg	60	—	—	—	—	1.14	1.08	1.08	1.02
Exits - All Sizes	Projecting		1.0							
	Sharp-edged		1.0							
	Rounded		1.0							
Entrances - All Sizes	Extending inward		0.78							
	Flush, Sharp-edged		0.5							
	Flush, Sharp-edged		1.5							

TABLE 11.11 Resistance Coefficient K for a variety of valves and fittings. Reprinted with permission from "Flow of Fluids Through Valves, Fittings and Pipe, Technical Paper 410" 1988 Crane Co. All Rights Reserved.

Flow Coefficient C_v

Suppose a water system has a need to throttle pressure or control flow, so you insert a control valve with a control loop to modulate the valve position. Naturally, the pressure drop through the valve is important to the overall pressure drop. The flow through the valve may be described as gallons per minute of 60°F water at a pressure drop of one psi across the valve. Control valve manufacturers use this as a reference point to compare flow characteristics of their valve designs. For other process liquids the flow rate is mathematically converted to an equivalent flow rate of water. Gaseous flows are similarly converted using air at standard temperature and pressure as the reference fluid.

6	8	10	12	14	16	18	20	24	30	36	42	48
6.065	7.981	10.02	11.94	13.12	15	16.88	18.81	22.63	28	34	40	46
0.505	0.665	0.835	0.995	1.094	1.250	1.406	1.568	1.886	2.333	2.833	3.333	3.833
0.015	0.014	0.014	0.013	0.013	0.013	0.012	0.012	0.012	0.011	0.011	0.01	0.01
K = f (L/D)												
0.05	0.04	0.04	0.04	0.04	0.04	0.04	0.04	0.04	0.03	0.03	0.03	0.03
0.12	0.11	0.11	0.10	0.10	0.10	0.10	0.10	0.10	0.09	0.09	0.08	0.08
0.53	0.49	0.49	0.46	0.46	0.46	0.42	0.42	0.42	0.39	0.39	0.35	0.35
1.50	—	—	—	—	—	—	—	—	—	—	—	—
—	0.70	0.70	0.65	0.65	0.65	0.60	0.60	0.60	0.55	0.55	0.50	0.50
2.25	2.10	2.10	1.95	1.95	1.95	1.80	1.80	1.80	1.65	1.65	1.50	1.50
0.83	0.77	0.77	0.72	0.72	0.72	0.66	0.66	0.66	0.61	0.61	0.55	0.55
5.10	4.76	4.76	4.42	4.42	4.42	4.08	4.08	4.08	3.74	3.74	3.40	3.40
9.00	8.40	8.40	7.80	7.80	7.80	7.20	7.20	7.20	6.60	6.60	6.00	6.00
0.30	0.28	0.28	0.26	0.26	0.26	0.24	0.24	0.24	0.22	0.22	0.20	0.20
0.90	0.84	0.84	0.78	0.78	0.78	0.72	0.72	0.72	0.66	0.66	0.60	0.60
0.45	0.42	0.42	0.39	0.39	0.39	0.36	0.36	0.36	0.33	0.33	0.30	0.30
0.24	0.22	0.22	0.21	0.21	0.21	0.19	0.19	0.19	0.18	0.18	0.16	0.16
0.23	0.21	0.21	0.20	0.20	0.20	0.18	0.18	0.18	0.17	0.17	0.15	0.15
0.90	0.84	0.84	0.78	0.78	0.78	0.72	0.72	0.72	0.66	0.66	0.60	0.60

The term "Standard Temperature and Pressure" (STP) is an oxymoron, since the terms described are anything but standard. Different organizations apply different conditions to it, with the variables being not only temperature and pressure, but also relative humidity. Further, some organizations have adopted more than one set of conditions to describe STP. See Table 11.13.

It therefore is imperative that when we speak of air at STP, we define the conditions. One good definition of STP is air at 0.075 lb/ft^3 and 0 percent RH.

C_v and K are related according to the following formula

$$C_v = \frac{29.9\, d^2}{\sqrt{K}} \qquad \text{(Equation 11.24)}$$

where

d = pipe inside diameter [in]

Nominal Pipe Size		in	1/2	3/4	1	1 1/2	2	2 1/2	3	4
d (based on Sch 40)		in	0.622	0.824	1.049	1.61	2.067	2.469	3.068	4.026
D (based on Sch 40)		ft	0.052	0.069	0.087	0.134	0.172	0.206	0.256	0.336
	Fitting	L/D	L = Equivalent Length (ft)							
Valves	Ball Valve	3	0.2	0.2	0.3	0.4	0.5	0.6	0.8	1.0
	Gate Valve	8	0.4	0.5	0.7	1.1	1.4	1.6	2.0	2.7
	Butterfly Valve	35	—	—	—	—	6.0	7.2	8.9	11.7
	Swing Check Valve up to 6"	100	5.2	6.9	8.7	13.4	17.2	20.6	25.6	33.6
	Swing Check Valve 24" to 48"	50	—	—	—	—	—	—	—	—
	Angle Valve - 90 degree	150	7.8	10.3	13.1	20.1	25.8	30.9	38.4	50.3
	Angle Valve - Y-Pattern	55	2.9	3.8	4.8	7.4	9.5	11.3	14.1	18.5
	Globe Valve	340	17.6	23.3	29.7	45.6	58.6	70.0	86.9	114
	Lift Check Valve	600	31.1	41.2	52.5	80.5	103	123	153	201
Tees	Tee - Through (Run)	20	1.0	1.4	1.7	2.7	3.4	4.1	5.1	6.7
	Tee - Branch	60	3.1	4.1	5.2	8.1	10.3	12.3	15.3	20.1
Bends	Elbow - 90 Short Radius	30	1.6	2.1	2.6	4.0	5.2	6.2	7.7	10.1
	Elbow - 90 Long Radius	16	0.8	1.1	1.4	2.1	2.8	3.3	4.1	5.4
	Miter Bend - 45 deg	15	—	—	—	—	2.6	3.1	3.8	5.0
	Miter Bend - 90 deg	60	—	—	—	—	10.3	12.3	15.3	20.1

TABLE 11.12 Equivalent Lengths for various valves and fittings. Reprinted with permission from "Flow of Fluids Through Valves, Fittings and Pipe, Technical Paper 410" 1988 Crane Co. All Rights Reserved.

Organization	Temperature	Absolute pressure	Relative Humidity
International Union of Pure and Applied Chemistry	0°C	100 kPa	
International Standards Organization, European Environment Agency	15°C	101.325 kPa	0%
US Environmental Protection Agency, National Institute of Standards and Technology	20°C	101.325 kPa	
US Environmental Protection Agency	25°C	101.325 kPa	
Compressed Air and Gas Association	20°C	100 kPa	0%
Society of Petroleum Engineers	15°C	100 kPa	
Occupational Safety and Health Administration	60°F	14.696 psia	
Organization of Petroleum Exporting Countries	60°F	14.73 psia	
International Standards Organization	59°F	14.696 psia	60%
Air Movement and Control Association	70°F	29.92 in Hg	0%

TABLE 11.13 Various definitions of STP for air.

6	8	10	12	14	16	18	20	24	30	36	42	48
6.065	7.981	10.02	11.94	13.12	15	16.88	18.81	22.63	28	34	40	46
0.505	0.665	0.835	0.995	1.094	1.250	1.406	1.568	1.886	2.333	2.833	3.333	3.833
L = Equivalent Length (ft)												
1.5	2.0	2.5	3.0	—	—	—	—	—	—	—	—	—
4.0	5.3	6.7	8.0	8.7	10.0	11.3	12.5	15.1	18.7	22.7	26.7	30.7
17.7	23.3	29.2	34.8	38.3	43.8	49.2	54.9	66.0	81.7	99.2	117	134
50.5	—	—	—	—	—	—	—	—	—	—	—	—
—	33.3	41.8	49.7	54.7	62.5	70.3	78.4	94.3	117	142	167	192
75.8	99.8	125	149	164	188	211	235	283	350	425	500	575
27.8	36.6	45.9	54.7	60.2	68.8	77.3	86.2	104	128	156	183	211
172	226	284	338	372	425	478	533	641	793	963	1133	1303
303	399	501	597	656	750	844	941	1131	—	—	—	—
10.1	13.3	16.7	19.9	21.9	25.0	28.1	31.4	37.7	46.7	56.7	66.7	76.7
30.3	39.9	50.1	59.7	65.6	75.0	84.4	94.1	113	140	170	200	230
15.2	20.0	25.1	29.8	32.8	37.5	42.2	47.0	56.6	70.0	85.0	100	115
8.1	10.6	13.4	15.9	17.5	20.0	22.5	25.1	30.2	37.3	45.3	53.3	61.3
7.6	10.0	12.5	14.9	16.4	18.8	21.1	23.5	28.3	35.0	42.5	50.0	57.5
30.3	39.9	50.1	59.7	65.6	75.0	84.4	94.1	113	140	170	200	230

Pump Head Terminology

Total Suction Head

The Total Suction Head H_s is defined as

$$H_s = (z_1 - z_{pump}) - h_{f\,suction} + (p_1/\rho) \qquad \text{(Equation 11.25)}$$

where

z_1 = Elevation of the level of the liquid to be pumped [ft]

z_{pump} = Elevation of the pump impeller centerline [ft]

$h_{f\,suction}$ = Head loss due to friction (including entrance losses)

p_1 = Pressure at the level of the liquid to be pumped [lb/ft^2]

ρ = Liquid density [lb/ft^3]

Note that p_1 may be negative if a vacuum exists above the liquid.

Suction Head

Suction Head h_s occurs when the liquid level is above the impeller centerline. "Static Suction Head" h_{st} is the vertical distance in feet from the impeller centerline to the free level of the liquid being pumped.

Suction Lift

Suction Lift occurs when the level of the liquid lies below the impeller centerline. This is a static lift condition. An equivalent suction lift condition occurs when there is no static lift (i.e. the liquid level is above the impeller centerline) but there is a vacuum above the liquid. In that case, the lift is equal to the static head minus the vacuum in feet. "Static Suction Lift" is the vertical distance in feet from the impeller centerline to the free level of the liquid being pumped.

> When p_1 starts out as atmospheric but changes to a vacuum, the results can be dramatic. Catastrophic tank failures have occurred due to improper venting of tanks during draw-down or pumping. This is often the result of a sheet of plastic having been placed over the vent during maintenance such as painting.
>
> The vent is there to permit atmospheric air to enter the tank and displace the liquid that is removed from the tank. If the air is prevented from entering, a vacuum is produced over the liquid and in a short period of time atmospheric pressure acts on a large external tank area. This results in a tremendous force being applied to the external surface of the tank. Tanks are not designed for this condition and the shells buckle and often rupture. It takes only a thin sheet of polyethylene to create this very hazardous condition over a vent. Vents must be inspected after maintenance and during start-ups.

Static Discharge Head

Static Discharge Head is the vertical distance in feet from the pump centerline to the discharge point or the surface of the liquid in a discharge tank whose free level is above the discharge point.

Total Static Head

Total Static Head is the vertical distance in feet between the free level of the liquid being pumped and the discharge point or the surface of the liquid in a discharge tank whose free level is above the discharge point.

Total Discharge Head

Not to be confused with Total Dynamic Head (which is commonly abbreviated TDH), the Total Discharge Head is defined as

$$H_d = (\text{Static Discharge Head}) + (h_{f\,discharge}) + (p_2/\rho), \text{ or}$$

$$H_d = (z_2 - z_{pump}) + (h_{f\,discharge}) + (p_2/\rho) \qquad \text{(Equation 11.26)}$$

where

$h_{f\,discharge}$ = Friction head losses in the discharge piping, including exit losses

p_2 = Pressure at the discharge point [lb/ft²]

ρ = Liquid density [lb/ft³]

Total Dynamic Head

Total Dynamic Head (TDH) is sometimes also known as the "Total System Head" or "Total Head". The term TDH is more commonly used. It is the sum of the Total Discharge Head and the Total Suction Head. If a common datum is used (for example, the impeller centerline) and attention is paid to the signs of the elevations, then the TDH can be summed algebraically[4]. Thus

$$\text{TDH} = H_d + H_s \quad \text{(Equation 11.27)}$$
$$= [(z_2 - z_{pump}) + h_{f\,discharge} + p_2/\rho] - [(z_1 - z_{pump}) + h_{f\,suction} + p_1/\rho]$$
$$= (z_2 - z_1) + (h_{f\,discharge} + h_{f\,suction}) + (p_2 - p_1)/\rho \quad \text{(Equation 11.28)}$$

We recognize the similarities to Bernoulli's Equation (Equation 11.4) and note that when:

- No head is extracted via a turbine
- The velocity heads may be ignored

Then

$$\text{TDH} = h_A \quad \text{(Equation 11.29)}$$

And in fact this is usually the case. For this reason, pump heads are normally specified according to the TDH.

Power Requirements

Once the Total Dynamic Head is found, the power required to operate the pump can be determined. The "Theoretical Horsepower", also called the "Hydraulic Horsepower" is defined as

$$\text{Hydraulic HP} = (\text{lb of liquid per minute})(\text{head in feet})/33{,}000$$

The actual horsepower required will be greater than the hydraulic horsepower, since the pump will be less than 100 percent efficient. In fact, the "Best Efficiency Point" or BEP of a centrifugal pump does not often exceed 85 percent. The BEP of a centrifugal pump is usually about 80 to 85 percent of the maximum head, or shut-off head.

The Brake Horsepower is defined as

$$\text{BHP} = (\text{Hydraulic HP})/\text{Pump Efficiency}$$

or

$$\text{BHP} = \frac{(\text{GPM})(\text{TDH})(\text{SG})}{(3960)(\eta)} \quad \text{(Equation 11.30)}$$

where

GPM = Flow rate in GPM

TDH = Total Dynamic Head in ft

SG = Specific Gravity (dimensionless)

η = Pump Efficiency (dimensionless). A reasonable estimate for pump efficiency is 70 percent

[4] Some references distinguish between a Suction Lift or a Suction Head condition in calculating the TDH, i.e., if there is a Suction Head condition (the liquid level is above the impeller centerline), then $\text{TDH} = H_d - H_s$ and if there is a Suction Lift condition (the liquid level is below the impeller centerline), then $\text{TDH} = H_d + H_s$. This seems unnecessarily confusing. If the elevations are treated as positive or negative with respect to the datum, then this confusion can be avoided.

Example 11.5

Given: A preliminary piping system layout as shown in Figure 11.2. The system takes contaminated water from a tank with a free surface elevation that is 10 ft above the pump suction, and delivers it across the plant to a treatment tank through a nozzle that is 40 ft higher than the pump suction. The developed length of the discharge piping is 1100 ft. After running through a building for 440 ft, the pipe exits onto a rack to a waste water treatment tank. The suction piping has a 20 ft developed length.

Required flow rate is 750 GPM.

Find: Pipe size, required pump head (TDH) and pump size.

Solution: At this point, we do not yet know what the pressure of the system will be (that is really the problem) so we start by assuming that the pipe schedule will be standard weight. If the pressure was calculated to be very high, then we might need to choose a heavier schedule of pipe.

Consulting Table 11.4, we see that for 6 in Schedule 40 pipe, the velocity falls between 8 and 9 FPS, which is within the reasonable range of velocities. This is a rather aggressive velocity though, so we look at 8 in diameter and see that the velocity will be close to 5 FPS, which leans to the conservative side. At 6 in diameter, the head loss per 100 ft is still less than 4 ft. See Figure A9.24. (This matches the ASHRAE 90.1 criteria, which sets established guidelines for energy conservation, and even though this is not a hydronics problem, it reinforces our belief that this is a reasonable pipe size.) So we select 6 in diameter.

We do not know how many fittings the system will finally have, so we have to estimate. Depending on the type of installation, we may be faced with more or less elbows to clear obstructions. In the basement of a paper machine, the direction may need to change every 10 or 20 ft. In a long mill building, perhaps every 50 ft.

Let's assume that this installation is dense with obstacles inside the building. This leads us to assume that for the 440 ft inside the building there will be perhaps 22 elbows. Outside on the pipe rack there may be 15.

Because we have a duplex pump arrangement, we know we will have isolation valves (gate valves in this case) and check valves at each pump. We can't be sure at this point how the piping will be arranged, so we will take the conservative approach and assume that the flow at either end of the pump will be through the branch outlet of a

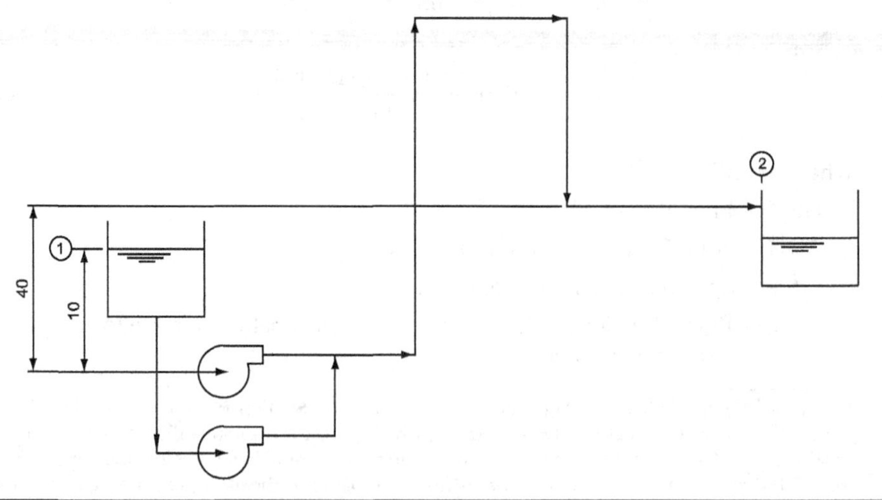

Figure 11.2 Schematic of piping for Example 11.5.

tee. The head loss through a branch outlet is always higher than through the run of a tee. In any case, we recognize that for an N+1 pump arrangement, all of the pumps will be identical.

We organize a table to account for the discharge losses, letting "L_e" stand for "Equivalent Length".

6 in Diameter (Discharge)

Item	Quantity	Unit L_e	L_e ft
Straight Pipe	1100 LF	1	1100
Elbows, 90° Long Radius	37	8.1	299.7
Gate Valve	1	4.04	4.04
Check Valve	1	50.5	50.5
Tee, Branch Flow	1	30.3	30.3
TOTAL			1485 ft

At this point we can determine the friction loss for the discharge piping. We get the friction loss for water in a 6 in Schedule 40 CS pipe from Figure A9.24 in the Appendix. At 750 GPM, the friction loss is 3.55 ft/100 ft. We can apply this value to the total equivalent length, having equated the friction loss of the system with the valves and fittings to an equivalent length of straight pipe.

$$h_{f\,discharge} = 1485 \text{ ft} \times 3.55 \text{ ft}/100 \text{ ft} = 52.7 \text{ ft}$$

We also need to find the friction loss in the suction portion of the pipe. We select a velocity of 2 to 5 fps, and select a pipe diameter from Table 11.4. We see that 8 in diameter will give us a velocity in the desired range.

Each time we move to a different pipe size or a different flow rate (as flows peel off from a header for instance) we organize another table. The reason for this is because the friction losses change when diameters or flows change. Separate tables must be created for each different diameter or flow rate.

8 in Diameter (Suction)

Item	Quantity	Unit L_e	L_e ft
Straight Pipe	20 LF	1	20
Elbows, 90° Long Radius	2	10.6	21.2
Gate Valve	1	5.32	5.32
Tee, Branch Flow	1	39.9	39.9
TOTAL			86.4 ft

The friction loss in the suction line is solved by applying the friction loss taken from Figure A9.25 in the Appendix. We see that the friction drop for 750 GPM through an 8 in Schedule 40 pipe is 0.9 ft/100 ft. Therefore the total friction loss on the suction side is

$$h_{f\,suction} = 86.4 \text{ ft} \times 0.9 \text{ ft}/100 \text{ ft} = 0.78 \text{ ft}$$

Therefore the total head loss due to friction is 52.7 ft + 0.78 ft = 53.5 ft. Examining Bernoulli's Equation

$$h_A = \left(\frac{p_2 - p_1}{\rho}\right) + \left(\frac{v_2^2 + v_1^2}{2g_c}\right) + (z_2 - z_1) + h_E + h_f$$

we see that there is no difference in pressure heads since $p_1 = p_2$ and

$$v = 0 \text{ fps}$$

$$v_2 = (750 \text{ gal/min})(0.1337 \text{ ft}^3/\text{gal})(\text{min}/60 \text{ sec}) / 0.2006 \text{ ft}^2 = 8.33 \text{ fps}$$

Had the discharge point been below the liquid level in the tank, we would have defined point two as the free surface of the liquid level, and v_2 would have been essentially zero, since the velocity of the large surface area would not rise very quickly. As it is, we discharge above the liquid level, and

$$v_2^2 / 2 g_c = (8.33 \text{ ft/sec})^2 / 2(32.2 \text{ ft/sec}^2) = 1.08 \text{ ft}$$

The velocity head components are usually very small.

$$z_2 - z_1 = 40 \text{ ft} - 10 \text{ ft} = 30 \text{ ft}$$

$h_E = 0$ ft since no head has been extracted from the system.

$$h_A = 0 \text{ ft} + 1.08 \text{ ft} + 30 \text{ ft} + 0 \text{ ft} + 53.5 \text{ ft} = 85 \text{ ft}$$

In practice, we take a 15 percent factor of safety on top of this, and round up to the nearest 5 ft. The 15 percent safety factor takes into account aging of the pipe (scaling or tuberculation) as well as a contingency for unknown factors. The rounding up to the nearest 5 ft increment is a convenience. An additional 5 ft water column is only 2.2 psi.

Note that the 15 percent factor applies to all of the terms in the Bernoulli Equation (the velocity head, the elevation head, and the friction head).[5]

Note also that in high head problems, the velocity head may be ignored. In this case, it contributed only 1.3 percent of the head required.

85 ft × 1.15 = 98 ft, so we call it 100 ft, and say we need a pump with a capacity of 750 GPM at 100 ft TDH.

We next consult a manufacturer's "hydraulic coverage curve" such as the one shown in Figure 11.3. This curve directs us to a set of pumps that will satisfy our requirements. Now that we know the TDH, we enter the curve at the flow and head requirements and see that there are possibly two pumps that overlap the operating point. Both pumps have 6 in diameter flanges on the suction end and 4 in diameter flanges on the discharge end. This implies that we will need reducers.

As we discussed earlier, the calculations for reducers, while not complex, take more time than they are worth. But we recognize that we will need an 8 in × 6 in reducer on the suction end and a 6 in × 4 in reducer on the discharge end. We want to place these reducers as close as possible to the pump so that we do not increase the friction losses with smaller pipe. Especially on the suction side, high friction losses are detrimental to pump operation as they can create cavitation. Attaching the valves directly to the pump would be the wrong thing to do. The valves must be sized for the flows we decided on, namely 8 in suction and 6 in discharge. Saving money by using smaller valves and attaching them directly to the pump would be unwise. At least two diameters of straight pipe should lead into the pump suction to reduce the possibility of cavitation.

The suction reducer must be an eccentric reducer installed with the "flat-on-top" to prevent an air pocket from forming and getting sucked into the pump suction and causing erosion and imbalanced forces on the impeller.

[5] The Factor of Safety may be adjusted to suit unusual conditions or where losses are expected to vary significantly over time.

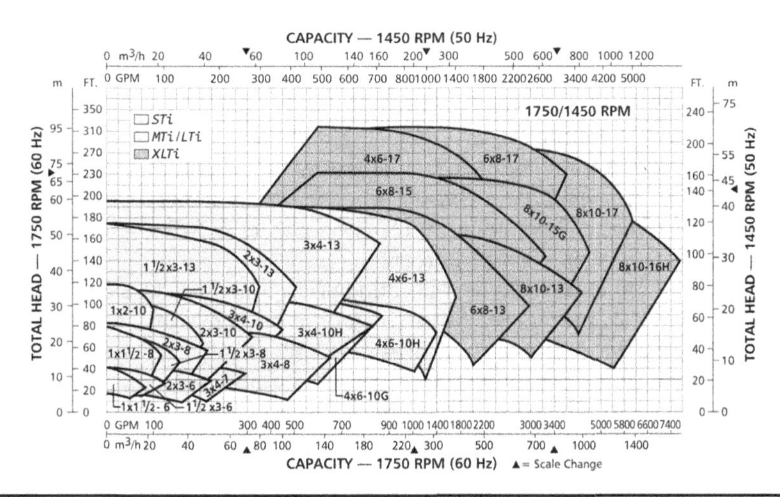

FIGURE 11.3 Hydraulic coverage curves for a typical industrial pump. *ITT Goulds Pumps. All rights reserved.*

Let's examine the losses due to entrance and exit effects to determine their contribution. Let's assume a sharp-edged, flush entrance from the tank to the pump suction line. The entrance loss is given by

$$h_{f\,entrance} = K_{entrance}\,(V^2/2g_c)$$

where

$K_{entrance}$ = 0.5 from Table 11.11

V^2 = The velocity of the fluid as it enters the 8 in pipe.

$v_1^2/2\,g_c$ = [(750 gal/min)(0.1337 ft³/gal)(min/60 sec)/0.3474 ft²]² / 2(32.2 ft/sec²) = 0.36 ft

Therefore,

$$h_{f\,entrance} = (0.5)\,(0.36\text{ ft}) = 0.18\text{ ft}$$

The exit loss at point 2 is given by the same equation, with K = 1.0 for all exits and all sizes.

V_2 = (750 gal/min)(min/60 sec)(0.1337 ft³/gal)/(0.20063 ft²), since the internal area of a 6 in Sch 40 line is 0.20063 ft² according to Table A.1 in the Appendix.

Therefore

V_2 = 8.3 ft/sec and

$$h_{f\,exit} = K_{exit}\,(V^2/2g_c) = (1.0)\,(8.3\text{ ft/sec})^2/\,(2)(32.2\text{ ft/sec}^2) = 1.1\text{ ft.}$$

So the total entrance and exit losses are 0.36 ft + 1.1 ft = 1.5 ft. These losses are typically very low, and once again, while it is never incorrect to include entrance and exit losses in head loss calculations, the effort required is generally not worth it for most practical calculations.

The exit losses depend only on the velocity, since they are based on K and this is constant regardless of type or pipe size. The entrance losses are generally going to be for flush, sharp-edged entrances since commercial fittings are not available for smooth entrances, and no one is going to spend money to fabricate an especially smooth, rounded entrance to save less than a foot of head loss. Certainly there may be special cases that warrant this kind of attention to detail, but most engineers will never experience these in practice.

Because the hydraulic coverage chart shows a close overlap at the operating conditions, we examine both curves to determine which gives us the better fit. At 750 GPM and 100 ft TDH, Figure 11.4 shows us that we will need a 4 × 6 pump with a 10.25 in impeller. This will require a 30 HP motor at 1775 RPM, and will provide a pump efficiency of approximately 73 percent.

Figure 11.5 shows that at the operating point we will need a 4 × 6 pump with a 25 HP motor and a 10.25 in impeller. Since this pump has a maximum impeller diameter of 13 in, we select this one, since we can use a smaller motor and we still have some future head capacity (larger impeller) if operating conditions should change. This operating point occurs at an efficiency of 75 percent. The pump selected is shown in Figure 11.6.

> Sometimes the hydraulic coverage in an 1800 rpm pump is just not sufficient to meet the TDH demand, and we must move to a 3600 rpm pump. Some owners and engineers prefer to stay out of that speed range since higher speeds mean more wear and more maintenance. This is a justified concern, since the bearings will be subjected to twice the wear. Even so, there are plenty of 3600 rpm pumps and motors that give satisfactory service. Life-cycle costs should always be considered when making a selection.

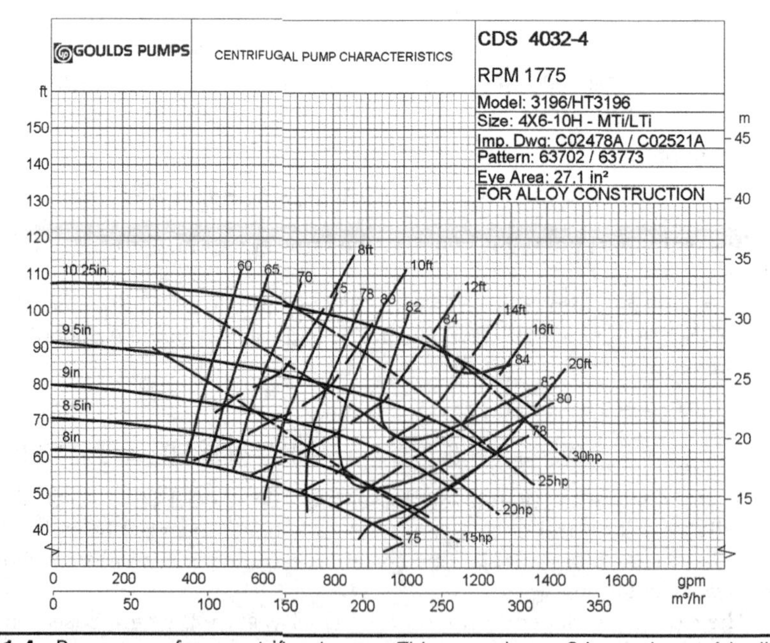

FIGURE 11.4 Pump curve for a centrifugal pump. This pump has a 6 in suction, a 4 in discharge, and a 10 in maximum impeller diameter. It is driven by a motor at 1775 rpm. *ITT Goulds Pumps. All rights reserved.*

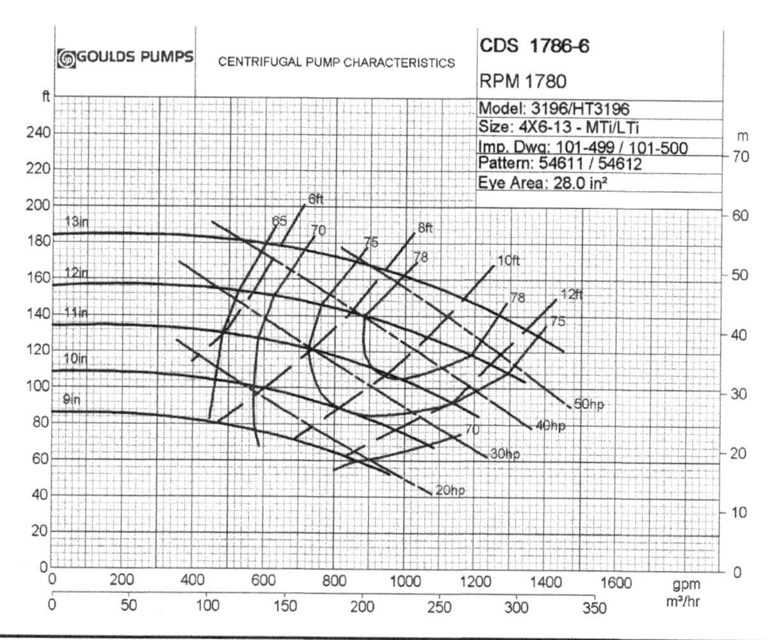

FIGURE 11.5 Pump curve for a centrifugal pump with a 6 in suction, 4 in discharge, and a 13 in maximum impeller diameter. It is driven by a motor at 1780 rpm. *ITT Goulds Pumps. All rights reserved.*

Suction Piping and Cavitation

The reason for trying to maintain low suction velocities at pumps is that fluids at low pressures may vaporize. Naturally, low pressures occur at pump suctions. If the liquid temperature is sufficiently high and the fluid pressure falls below the fluid vapor pressure at that temperature, then small bubbles will form at the pump entrance and collapse inside the pump volute as the pump increases pressure. This phenomenon is known as "cavitation," and the action of the fluid changing phases within the pump can lead to erosion of the wetted surfaces.

Keeping the velocities reasonably low in pump suctions reduces the friction losses and subsequently aids in maintaining the pressures above the fluid vapor pressure.

FIGURE 11.6 A typical end suction centrifugal pump with motor and coupling mounted on a cast iron baseplate. Note the grout hole to assist in placing grout under the baseplate. Cast iron is often used for baseplates due to its excellent vibration dampening properties. *ITT Goulds Pumps. All rights reserved.*

One application for pumps in which cavitation is very difficult to prevent is in the evaporator plant of a pulp mill. The weak black liquor is pumped from one evaporator tower to another. The towers are under negative pressure to purposely reduce the vapor pressure so that less heat must be added to drive off the water. The combination of low pressure and high temperature leads to the characteristic sound of cavitation. It really does sound like there are marbles inside the pump.

The method for determining whether or not a pump will be prone to cavitation relies on comparing the Net Positive Suction Head Required (NPSHR) to the Net Positive Suction Head Available (NPSHA). If NPSHA exceeds NPSHR, then cavitation will not occur. The NPSHR is a characteristic of the pump, and is read from the pump curves supplied by the manufacturer. The NPSHA is a characteristic of the pump system and the fluid conditions therein. It is defined as

$$\text{NPSHA} = h_a - h_{vpa} - h_{st} - h_{f\,suction} \text{ for Suction Lift} \quad \text{(Equation 11.31)}$$

where the liquid supply level is below the impeller centerline, or

$$\text{NPSHA} = h_a - h_{vpa} + h_{st} - h_{f\,suction} \text{ for flooded suction} \quad \text{(Equation 11.32)}$$

where the liquid supply level is above the impeller centerline, and where

h_a = Absolute pressure (in feet) acting on the surface of the liquid supply. In an open tank, this will be atmospheric pressure.

h_{vpa} = Vapor pressure (in feet) of the liquid being pumped. This is temperature dependent.

h_{st} = Static Suction Head is the vertical distance (in feet) between the liquid supply level and the impeller centerline.

$h_{f\,suction}$ = Head loss (in feet) due to friction in the suction line.

Example 11.6

Given: The system described in Example 11.5. Assume that the water being pumped is at 100°F (38°C)

Find: The NPSHA

h_a = 33.9 ft

h_{vpa} = 2.2 ft from Table 11.9

h_{st} = 10 ft

$h_{f\,suction}$ = 0.78 ft

$\text{NPSHA} = h_a - h_{vpa} + h_{st} - h_{f\,suction}$ since the suction is flooded

$\text{NPSHA} = 33.9 \text{ ft} - 2.2 \text{ ft} + 10 \text{ ft} - 0.78 \text{ ft} = 40.9 \text{ ft}$

Because this value exceeds the NPSHR shown on the pump curve in Figure 11.5 (NPSHR = 9 ft), we do not expect cavitation to occur.

Example 11.7

Compare the calculated BHP to the HP shown on the pump curve. Use the pump efficiency from the curve of 75 percent.

From Equation 11.30, $BHP = \frac{(GPM)(TDH)(SG)}{(3960)(\eta)}$

$BHP = (750\ GPM)(100\ ft)(1.0)/[(3960)(0.75)] = 25\ HP$. This agrees with the curve.

Example 11.8

Suppose we learn from the client in Example 11.5 that there is an 80 ft tall obstacle that has to be overcome inside the building immediately before reaching the pipe rack. See Figure 11.7. Would the pump selected be able to overcome this height?

We designate the point at which we exit the building as 1A. We have 440 ft of piping, 22 elbows, a tee, a gate valve, and a check valve before we land on the rack:

6" Diameter (Discharge)

Item	Quantity	Unit L_e	L_e ft
Straight Pipe	440 LF	1	440
Elbows, 90° Long Radius	22	8.1	178.2
Gate Valve	1	4.04	4.04
Check Valve	1	50.5	50.5
Tee, Branch Flow	1	30.3	30.3
TOTAL			703 ft

$$h_{f1A} = 703\ ft \times 3.55\ ft/100\ ft = 25.0\ ft$$

We still have the suction losses to account for, so $h_{f\,suction} = 0.78\ ft$

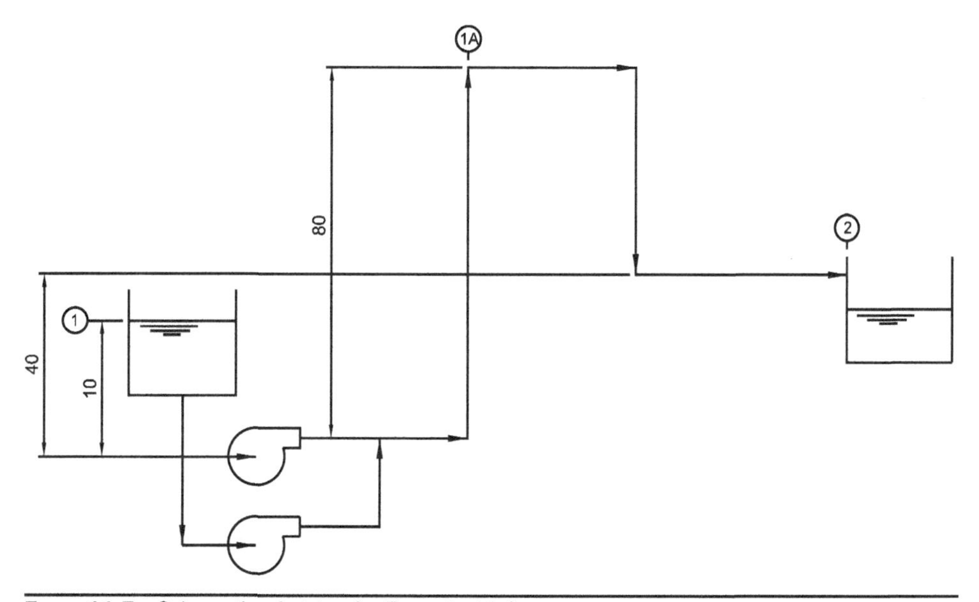

FIGURE 11.7 Schematic of piping for Example 11.8.

The head required to move the water to point 1A is given by:

$$h_A = \left(\frac{p_{1A} - p_1}{\rho}\right) + \left(\frac{v_{1A}^2 + v_1^2}{2g_c}\right) + (z_{1A} - z_1) + h_E + (h_{f1A} + h_{f\,suction})$$

p_{1A} = is still atmospheric pressure, and is equal to p_1, so this term will cancel

$$v_{1A} = 8.33 \text{ ft/sec}$$

$$h_A = (0 \text{ ft}) + [(8.33 \text{ ft/sec})^2 - 0]/[2(32.2 \text{ ft/sec}^2)] + (80 \text{ ft} - 10 \text{ ft}) + (0 \text{ ft}) + (25 \text{ ft} + 0.78 \text{ ft}) = 96.8 \text{ ft}$$

Therefore we can see that the pump with TDH = 100 ft would theoretically be sufficient to overcome this obstacle. The remaining head required to flow from point 1A to the tank at point 2 would be overcome by the potential energy of the fluid at point 1A. But with only a 3.1 ft surplus of head, we decide that the factor of safety here is too small, and we select a pump rated for 115 TDH at 750 GPM. This choice allows us to retain the same model of pump, but with a larger impeller and motor.

This example illustrates that the elevation of the entire route must be examined to ensure that the pump would be able to overcome an intermediate elevation. It may not be sufficient to look at only the beginning and ending elevations.

CHAPTER 12

Piping Project Anatomy

Depending on the size of a project, more or less preparation may be required. This preparation takes the form of a variety of engineering design tools that the piping engineer employs to build a project from a blank sheet of paper.

Simple projects may not require any of these tools. Large complex projects may require all of them. This chapter will explain how these design documents are used and developed.

A good place to start is by asking the question: "Why is engineering required at all?" One obvious answer is that for high pressure, dangerous, or critical systems, it seems like a good idea to have engineering oversight to ensure that systems are operated in a safe manner. But what about less hazardous systems? What if pipefitters were left to their own devices to route a 24 in cooling water line? Why do they need an engineer to be involved?

There are several answers to this. One is that the engineer serves as the information clearinghouse and coordinates and plans for the various uses of the cooling water. The engineer must size the lines and account for the pressure drops, energy consumption, and economic routing of the pipe to each of the devices that require cooling water.

A second reason is that contractors can reduce their installation costs by fabricating as much piping as possible in their shops rather than in the field. This is intuitively clear. It is easier to cut and weld at ground level, rather than at 20 ft in the air. Piping drawings can therefore be used by the contractor to shop fabricate (pre-assemble) as much of the piping as possible. The field work then becomes an exercise in connecting the pieces and making them all fit-up to the equipment connections, realizing that there will be construction tolerances that prevent spools from being fabricated to the theoretical dimensions in the shop. This means that there will be field cuts required.

Still a third reason relates to knowledge of the codes and safety considerations as they pertain to both the public and the employees who will work inside the facility. It is true that not all engineers are familiar with the various codes, just as it is certainly true that many tradesmen are familiar with codes. But the responsibility for the safe design of a system should always be left to an engineering professional.

An Archetypical Project

Let's assume that we have been awarded a contract for the design of piping in a large steel mill facility that sprawls across many acres. The major equipment vendors have provided a proposed layout for the client, and have provided preliminary utility consumption data. All of the utilities must be routed from their source to the appropriate equipment connection points. And even though we are responsible for the piping, we understand that we will have to coordinate with the electrical engineers not only to prevent interferences, but also to provide an intelligent design.

The equipment vendors must be relied upon to provide much of the incipient data. They will describe their requirements for both the quantity and the quality of the fluids demanded by their equipment. Each of the various vendors will have requirements for the various fluids needed to make their machines perform as they have promised in their sales documents.

Utility Consumption Table

An excellent place to start such a project is by designing a utility consumption spreadsheet. This will total the required flow rates from each piece of equipment. A good format for such a spreadsheet will incorporate the following items:

- An identification number
- The description of the connection purpose
- The flow rate
- The reference revision from which the information was obtained

This information will be prepared for each consumer of utilities throughout the plant, with like utilities (e.g. Natural Gas) grouped together so that a total load can then be summed. It is best to provide a separate sheet for each utility.

The identification number will be used later to match the use point in plan on a "Flow Map."

The description of the connection point is used to identify the user, e.g., "width gage."

The flow rate is the heart of the matter, and obviously consistent flow rates must be used throughout the project. Again, spreadsheets are an excellent way to enter the data, especially if the consumption data are provided in a combination of metric and Imperial units.

Including a reference source is also extremely useful, since in any project the data may be changing rapidly. Engineering projects are often quite dynamic, with the vendors and Owners frequently revising the requirements. An accurate means of accounting for such rapid revisions is essential to maintain the integrity of the document, and including the reference source will allow you to pick up a cold trail. See Table 12.1 for a sample Utility Consumption Table.

Diversity Factors

Not all users demand the maximum flow simultaneously. For process uses, one can assume that the demand will be 100 percent all the time. Even if there is not likely to be simultaneous flow to various process users, it is wise to design as though there will be. The reason is that while unlikely, there is a chance that the process sequencing may require simultaneous flow at some point. Another reason is that future production improvements may require additional flow rates, and it is easier to accommodate that increase in capacity during the construction phase rather than through a retrofit later. The incremental cost in utilizing one larger pipe size is small compared to having to add another line later. Most of the labor cost would be absorbed by installing the smaller line size (it does not cost much more to weld an 8 in pipe than a 6 in pipe), and the difference in material cost is not usually great.

Description		Flow (GPM)		Press (psig)		
		Avg.	Max.	Min.	Max.	Reference
CGL	entry hyd cooler	50	70	60	85	Utility Quality & Consumption Data Rev D 4/15/97
	pay-off reel	5	5	60	85	Utility Quality & Consumption Data Rev D 4/15/97
	cleaning section	0	0			
	cleaning dryer	0	0			
	cleaning fume exhaust	0	0			
	pot pre-heater	0	0			
	pot	72	72	60	85	Utility Quality & Consumption Data Rev D 4/15/97
	wiper	150	150	60	85	Utility Quality & Consumption Data Rev D 4/15/97
	edge heater	0	0			
	GA furnace	950	950	60	85	Utility Quality & Consumption Data Rev D 4/15/97
	quench tank	1260	1800	60	85	Utility Quality & Consumption Data Rev D 4/15/97
	skinpass mill	0	0			
	tension leveler	150	150	60	85	Utility Quality & Consumption Data Rev D 4/15/97
	roll force hyd cooler	45	55	60	85	Utility Quality & Consumption Data Rev D 4/15/97
	chromating section	0	0			
	chromating fume exhaust	0	0			
	chromate dryer	0	0			
	phosphating section	20	20	60	85	Utility Quality & Consumption Data Rev D 4/15/97
	phosphate dryer	0	0			
	width gauge	1	1	60	85	Utility Quality & Consumption Data Rev D 4/15/97
	tension reel EPC	5	5	60	85	Utility Quality & Consumption Data Rev D 4/15/97
	exit hyd cooler	50	70	60	85	Utility Quality & Consumption Data Rev D 4/15/97
	annealing furnace	3150	3400	60	85	Utility Quality & Consumption Data Rev D 4/15/97
	APC	0	0			Letter to MIC dated 6/10/97 DC-L-MHIA-186
	induction equipment	800	800	60	85	Utility Quality & Consumption Data Rev D 4/15/97
	CGL SUBTOTAL	6708	7548			

Table 12.1 Galvanizing line cooling water requirements.

On the other hand, utilities like potable water may take advantage of a diversity factor if the facility is large enough. For a two-unit apartment, all of the tenants might be showering, running the dishwasher and doing laundry at once. For a four-unit apartment, that may be unlikely.

One of the site Project Managers called me to complain about the water pressure in the maintenance building at a very large site. I had designed the utility piping for the site, and routed the potable water from the city water service through a maze of mill buildings and over to the maintenance building that was serving as a construction office. The maintenance building was tucked into a corner of the site, diagonally opposite the service entrance; they could not have been further apart.

"The sinks lose pressure when you flush the toilet!" he complained.

"What do you want me to do? Route a new line 2000 ft all the way over to the building?"

The next time I was on site, I turned on all the faucets, including the janitor's slop sink. I turned on a shower, and commenced flushing the toilets.

There was no discernible drop in pressure at any of the faucets. I concluded that the line size I chose was adequate. It is conceivable that when the PM tested it, another major user elsewhere in the plant was robbing pressure. It was impossible to tell.

There is not much guidance in the literature regarding appropriate diversity factors. The IPC states that water distribution systems shall be designed such that under conditions of peak demand the capacities at the fixtures will meet specified minimum flow rates[1]. But it provides no instructions on how to determine the peak demand.

Utility Quality Spreadsheets

The vendors will not merely state that they need water to cool a piece of equipment. They will specify that the water must be a minimum of 85°F, for instance, and they may even place limits on chlorides that could corrode their metallurgy, or calcium that could plate out on heat exchange surfaces and thus degrade the efficiency of the heat transfer that may be required. Table 12.2 shows an example of a Utility Quality Table for Non-Contact Cooling Water. This may also be used to compare vendors' proposals, as well as to determine what level of treatment may be necessary to bring the utility service into compliance with the vendor's requirements.

All of these requirements for the individual utility services must be compared among the various process users, and usually the more stringent requirement will be applied throughout the utility service distribution system. There may be a break point where economics will dictate that it is too expensive to "clean" the utilities for all of the users. It may be feasible to clean a sidestream and provide that flow to the specific piece of equipment, but these are details that have to be decided on an individual basis.

[1] 2003 International Plumbing Code, Paragraph 604.3.

Description Process Line	Vendor 1 CPL–APL	Vendor 2 CPL–APL	Vendor 1 H2 Anneal	Vendor 2 H2 Anneal	River Water Untreated	River Water Filtered
Pressure	60 psig.	60 psig	40–60 psig	58 psig		
Temperature	90 °F max.	90 °F	80 °F	68 °F	35–90 °F	35–90 °F
pH - Value	6.5–7.5	6.5–7.5		8.5	8.5–9.5	8.5–9.5
Total Hardness	120 PPM $CaCO_3$	120 PPM $CaCO_3$			140 PPM $CaCO_3$	140 PPM $CaCO_3$
Suspended Solids	< 10 PPM	< 20 PPM	10 mg/l	10 mg/l	64.09 mg/l	5–10 mg/l
Dissolved Solids	< 100 PPM				280 mg/l	280 mg/l
Chloride Content	< 100 PPM	< 100 PPM	70 mg/l	70 mg/l	21.32 mg/l	21.32 mg/l
Silica Content	< 25 PPM	< 25 PPM			4.23 mg/l	4.23 mg/l
Iron Content	< 2 PPM	< 2 PPM		.2 mg/l	2.14 mg/l	0.23 mg/l
Turbidity	< 10 deg.	< 10 deg.				
Sulfate	< 10 PPM	< 10 PPM	70 mg/l		93.67 mg/l	93.67 mg/l
Conductivity	< 500 Micro S/cm	<500 Micro S/cm	500 Micro S/cm	100 Micro S/cm		
Alkalinity	< 55	< 55			70 mg/l	70 mg/l
Anticorrosive Conc.	30–40 PPM	40 PPM				
Filtration Level			1/32″ mesh	.012″ mesh		

TABLE 12.2 Utility quality table for non-contact water.

Block Flow Diagrams

Block Flow Diagrams (Figure 12.1) are the next step in preparing the system design. These are basic flow diagrams that show the major users in block form, as well as the source from where the utility will be drawn. Because the Block Flow Diagram is diagrammatic in nature, there is no sense of scale at this point, although there may be some advantage in locating the blocks in relative positions based on their geography.

Flow rates are applied to the lines connecting the utility source to the users, and so this is now an appropriate time to consider performing a gross mass balance.

P&IDs

P&ID stands for "Piping & Instrumentation Diagram", or in some circles, "Process & Instrumentation Diagram." These documents are relied upon by piping designers to specify line sizes and in-line components such as shut off valves, control valves, flow meters, pressure transmitters, and so on.

The P&ID effort usually is coordinated with the process engineers and the instrumentation engineers, although for small projects or those employing well-rounded piping engineers, these additional responsibilities may be assumed by the piping engineer himself.

The International Society of Automation (formerly the Instrument Society of America) has published standards for how the information shown on a P&ID should be represented. In practice, different companies may have their own special symbology that should be shown in a legend, as discussed in Chapter 10.

Line identifiers are shown at every pipeline to indicate the diameter, service, and line identifier. Some line numbering schemes also include a pipe and or insulation spec.

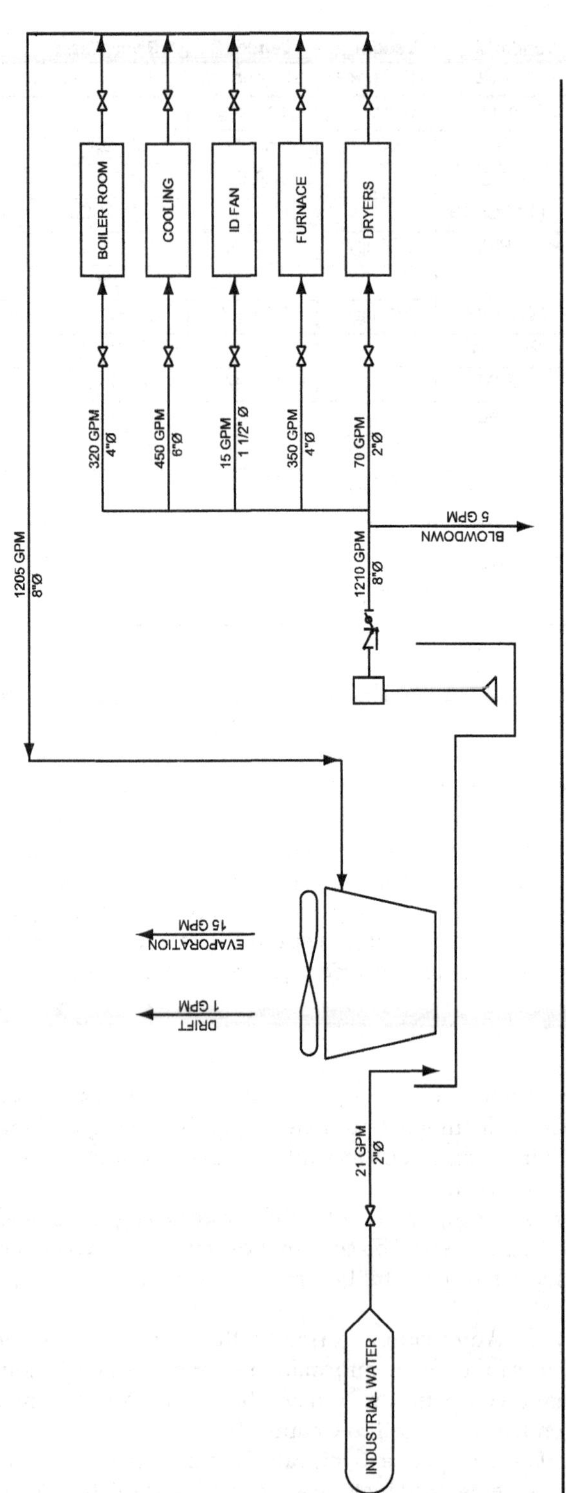

FIGURE 12.1 Block flow diagram for a simple cooling water system.

Some projects require that each individual line have a unique line number. This is commonly the case in the chemical, petrochemical, and pulp & paper industries. Other industries such as metals are content to only identify the diameter of the line and the service. The line number designations for a chemical plant are often very detailed, and regardless of what information is presented, it must remain consistent throughout the project. An example line number might look like:

D – SERVICE – ID – AREA – PIPE SPEC – INSULATION

Where

D is the nominal pipe size.

SERVICE is the fluid conveyed.

ID is a unique line number.

AREA is an area designation inside the plant which may relate to a specific processing area.

PIPE SPEC is an alpha-numeric code which identifies the piping material specification.

INSULATION may be an insulation spec, thickness and/or an indication of whether the line is heat traced.

A common precept of piping engineering is that a proper P&ID, a piping specification, and a general arrangement are all that is theoretically necessary to design the piping for a project. In practice of course there are many other influences on a project such as vendor cut sheets of the equipment, structural steel supports, and so on. The concept here is that the P&ID is the main document from which physical piping drawings are developed.

Engineers are not in the business of creating documents for their own sake. While the P&ID is a valuable reference tool that can be used in later years to troubleshoot systems, it is also a valuable design tool that allows the designer to examine and develop the behavior of the system in relation to the equipment and controls. Feedback and control schemes are developed during this stage of design, and the act of preparing the drawings forces the designer to consider how the system is to be controlled.

It is generally not sufficient to power up a pump and walk away from it. A control scheme is required. What will cause the pump to start? What will cause the pump to stop? What could go wrong during the operation of the pump? What process variables must be measured during operation? These are the types of questions that must be answered during the incipient design phase of a piping system. P&ID development forces one to consider these issues.

Example 12.1

Consider a cooling water system that must deliver from a coldwell to a piece of equipment. See Figure 12.2. A vertical turbine pump must start automatically prior to operation of the equipment in order to prevent damage to the equipment internals. The pump must stop when the level in the coldwell reaches a low level in order to prevent damage to the pump. The operators desire to know when the pump is running so a pressure transmitter is provided to advise that the system is operating. We can also use this transmitter to alarm certain conditions such as low pressure and high pressure. If the cooling capacity of the piece of equipment is critical, we can adjust it by modulating the flow of water to it. We can do this with a control valve or by applying a Variable Frequency Drive to the pump's AC motor. If we choose the less expensive and lower-tech control valve, we need to consider what happens when the control valve throttles. Will the pressure of the system rise more than we want it to? Should we shut off the pump? Is turning off the pump even an acceptable alternative?

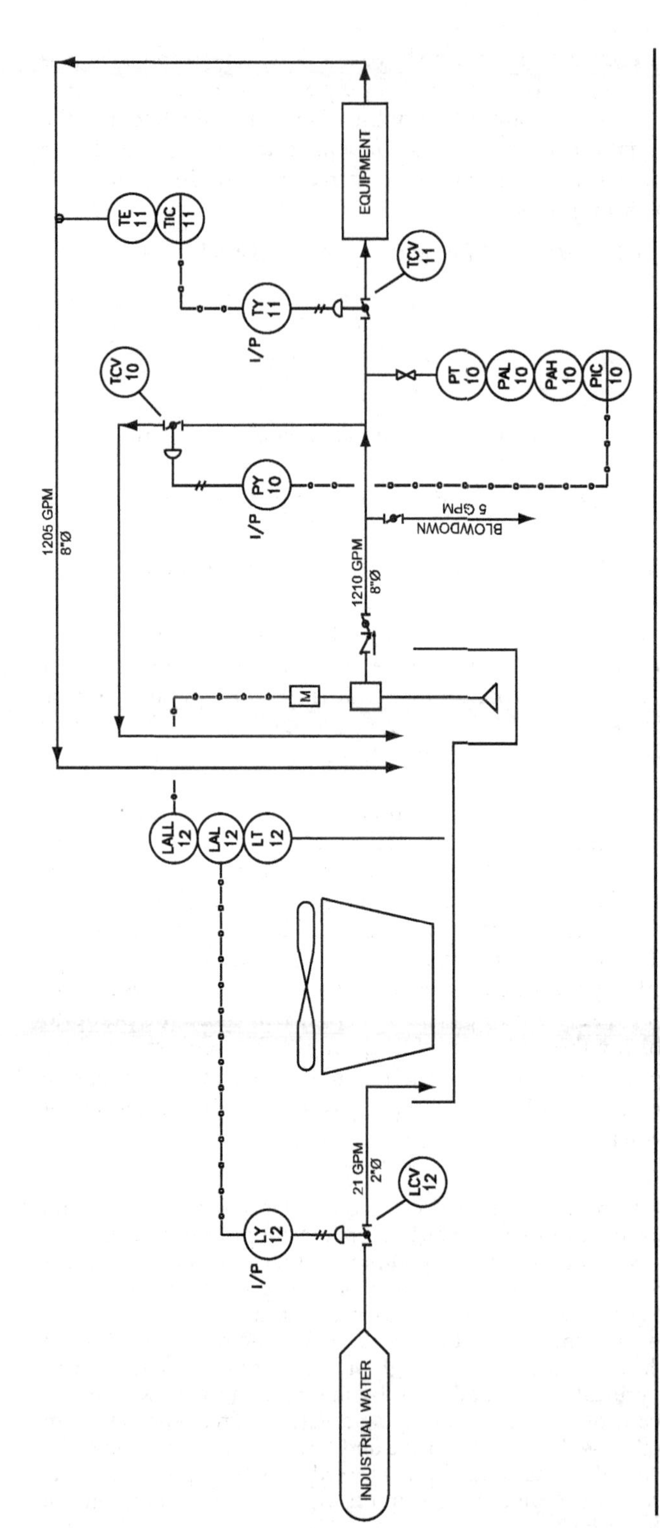

FIGURE 12.2 Example 12.1. Control of a cooling tower is represented with a P&ID.

Shutting the pump down completely is probably not the best option since:

- Frequent pump cycling is poor operating practice. The reason for this is due to wear on the motor starter, as well as energy surge in overcoming the inertia of a stagnant system (the rotational inertia of the pump impeller, as well as the inertia of the water in the system). This energy surge is expressed as an inrush current which may be 8 to 10 times higher than the motor's rated current, and for large motors, there is often a penalty applied by the utility for frequent starting.
- There may be an advantage in keeping a small amount of cooling water flowing to prevent thermal cycles, or perhaps even boiling of the water if the item cooled is sufficiently hot.

While we could "turn down" the pump speed with a VFD, we choose for this exampple to install a pressure bleed valve that will allow the pump to remain on at low flow to the equipment. Each of the different control loops is assigned a unique loop number. In this example, the pressure control loop is loop number 10, the temperature control loop is loop number 11, and a level control loop is loop number 12. The level control loop ensures that the water level is sufficiently high in the cooling tower basin to keep from starving the pump. The pump operation would become unstable if the water level became so low that vortices formed. Vortices at vertical pumps are always unwelcome since they can draw air into the pump suction.

General Notes

On small projects there may not be a need to include a complete piping specification. A General Notes sheet will inform the contractor of the project's general requirements. The General Notes provide an opportunity to instruct the contractor to:

- Verify dimensions. This is sometimes viewed as a shirking of the responsibilities of the engineer, but one must realize that there are good reasons for verifying dimensions. Consider first the case of new construction. The tolerance on setting structural steel columns has been observed in the field to be as much as 2 in. Since piping and equipment is often dimensioned from a column line, unless the contractor has verified the dimensions he cannot expect that a spool will fit. This is also a reason why it is important for the pipe fabricator to allow sufficient field trim on spools. Next consider a project in which piping must be designed into an existing installation. It would be the rare engineer who has the capacity to measure exactly the dimensions from tie point to tie point. Engineers simply are not as well equipped as contractors are for reaching high areas and drawing accurate dimensions in existing installations. This does not excuse bad dimensions on a drawing, but the engineer must recognize his limitations and make the contractor aware of them as well.
- Provide high point vents and low point drains. These are especially important during start-up and hydrotests, since it is impossible to secure a good hydro with air in the line. Also, it is good practice to provide vents and drains for normal maintenance.
- Provide sufficient field trim for spools and end connections at equipment.

- Provide adequate pipe supports. Engineered support details must still be provided by the engineer.
- Allow for gasket thickness between flanges, since it is never accounted for in normal piping layout drawings.
- Supply, install, and remove any start-up strainers required for flushing.
- Coordinate the work of other crafts prior to fabrication or erection.

Design Basis

A Design Basis should be included which states what codes are to be applied, as well as what assumptions had to be made in order to progress with the design. Some consulting firms provide a separate Design Basis, and these are often tomes that have evolved into lengthy design detail documents. These are almost always the product of an ambitious Project Manager who felt that "more is better."

In the case of a Design Basis, less is more. The idea is to convey the general requirements of the project and permit the design professionals to exercise their engineering judgment in satisfying the project requirements, within the broad criteria established by the Design Basis.

A good Design Basis must address the concerns of the other disciplines, since engineering is a combined effort. Items to include in a Design Basis:

Recommended Data

- Geographical data so that environmental loads may be determined from the appropriate codes, and so that the proper Authority Having Jurisdiction (AHJ) may be identified.
- Applicable codes, that will certainly include the piping code, but may also include building codes. The building codes will provide information regarding the environmental loads, for example ASCE 7 "Minimum Design Loads for Buildings and Other Structures."
- The source of piping and valve specifications. Will it be the Owner's existing specification?
- The source of pipe support details, and the criteria used to determine whether a line should be stress analyzed.
- Environmental load data for seismic, wind, snow, and rain. Cite the source of such information.
- Temperature data, again citing the source and the criterion used, e.g., 2 percent dry-bulb temperature. Temperature data may not seem to have relevance to the piping engineer, but consider the task of specifying cooling towers. These data are available in the ASHRAE Fundamentals Handbook, Chapter 28, which now comes with a database on CD-ROM.

Optional Data

- Identity and contact information for the Authority Having Jurisdiction (AHJ) if the project will require permitting, or if the AHJ's opinion will be sought during the design phase.
- CAD standards may be included if the Owner has specific requirements for layer or level structures, or for the CAD platform to be used.
- Hazardous Area Classifications should be included if the project will include such areas. These will impact electrical equipment and cable and conduit installation.

Inappropriate Data

The following information should be represented elsewhere in the design or contract documents. They have no place in a thoughtful Design Basis:

- Schedules and milestones belong in a schedule.
- Performance guarantees belong in a contract.
- Utility consumption data, unless it is trivial. For a complex project, utility consumption data should be in its own spreadsheet.
- Safety and health requirements. While always integral to every project, mention of safety and health in a design basis will force the document to become unnecessarily burdened with issues relating to liability.

General Arrangement

The General Arrangement (GA) drawings show the equipment and structures. There may be more or less detail represented on such drawings, but generally all that is required of a GA is a dimensioned layout.

As with any plan drawing, the following elements must be represented:

- A north arrow for orientation. This is usually shown pointing to the top or left of the sheet. It is NEVER shown pointing to the bottom of the sheet. The actual north direction shown on all plans is "Plant North" which may deviate from "True North" by 45°. Sometimes both Plant North and True North are shown, but Plant North must always be shown.
- Column lines and identifying column bubbles should be shown. It is best to locate column bubbles along the top and right hand side of the drawings, and to remain consistent with the scheme. The reason for placing bubbles on the right hand side instead of the left is because they are easier to locate when bound in a stick file.
- On geographically large projects in which the design area of the project does not fit entirely on one sheet, it is useful to make a key plan to determine where the items represented on one sheet fit into the scheme of the plant grid. In that case, match lines are also required.

GAs traditionally mean plan views, but they are by no means restricted to plan views. It is often necessary to cut sections for GAs to show the relationships between equipment, to show lines-of-sight, to determine access to valves or equipment, or for any other reason.

Design and Construction Schedules

Most schedules for large projects are prepared using canned software that is capable of producing reams of paper. These are most often printed on 8 1/2 in × 11 in paper, but even on 11 in × 17 in sheets they are incapable of showing the interrelationships of the various tasks at this scale. They are often prepared by Project Managers or Schedulers who delve too far into the minutia, leaving the project team with a bad case of data overload.

The best schedules will highlight the main milestones and project commitments, but will also clearly show what tasks are on the Critical Path. The Critical Path is the path through the schedule that determines how long the project will ultimately take to complete. It should show the earliest and the latest that each activity can start and finish without making the project any longer. "Slack Time" or "Float" is the amount of available time in a given activity that exceeds the time required to perform that activity.

A good Critical Path schedule can actually become a design tool for the engineer, since it forces one to consider constructability issues. Examples of this are:

- When must vendor certified data be available in order to complete piping tie-ins or equipment sizing?
- Will structural steel be available to hang piping?
- When will concrete foundations be strong enough to set equipment?

Owners and Project Managers love aggressive schedules, but the clever PM will allow time for the unforeseen. This is more critical on small projects than on large projects, since anomalies may be absorbed without much notice on a very large project with many participants and lots of inertia. But on smaller projects, the absence of the IT specialist when the CAD system decides to crash may be significant. If everyone's cars, families, and health were always reliable, there would be no need for insurance. Think of added slack time in a schedule as insurance.

A self-evident truth for design engineers is that on any project you can select any two of the following three:

- Schedule
- Budget
- Quality

One of these is going to suffer on a project, but Owners and Project Managers are always reluctant to acknowledge this. Quality better always be there. No one will be happy if the system does not work, and poor designs will lead to problems with budget or schedule. So a reasonable question at the beginning of a project is, "What matters more? Maintaining the schedule or meeting the budget?"

Flow Maps or Utility Distribution Diagrams

In order to determine the sizes and general routings of headers, it is convenient to label the consumers of a single utility on a GA. The headers can be routed and the velocities, and hence the diameters, can be computed. This is a valuable tool during the incipient design phase. The consumption data is collected from the utility consumption charts.

Equipment Lists

All but the smallest of owners keep track of their assets with equipment numbers. This facilitates maintenance, and ensures that the proper pump, for instance, is installed in the proper location. Equipment Lists are the tools used to keep track of the assets during the design phase. They will include data such as:

- Equipment number
- Name (e.g., "Pump, CWS")
- Technical data (e.g., 450 GPM at 120 ft TDH)
- Electrical data such as HP, Volts, Phase

Piping Plans, Sections, and Details

These drawings are really at the heart of piping projects. If the project is sufficiently large, the piping contractor will use these drawings to develop his spool drawings. Contractors (and hence Owners) can save money by having as much piping pre-fabricated in a pipe shop as possible, rather than out in the field. Shop-fabbed piping is less costly because:

- The piping is fabricated under more controlled conditions (more consistent temperature, out of the weather) and so quality issues may be mitigated.
- No work at height is required, so no manlifts or scaffolding are needed.
- There is less time spent by crews having to reach the location where the pipe is to be installed.
- Field erectors make higher wages than shop fabricators, since the field work is more demanding.

The fabricator will have likely seen these drawings at least once before, during the bidding phase. In order to develop his bid price, the contractor needs to perform a Material Take-Off, or MTO. The MTO is a list of all of the pipe, fittings, and specialties that he must purchase and install. A sample MTO is shown in Table 12.3. Quantities from the MTO are then entered into an estimating spreadsheet to develop the installed cost estimate.

Pipe Support Plans and Instrument Location Plans

Past practice was to provide separate drawings to show the locations of supports or instrumentation. The piping plan was screened[2] and the support of instrument was drawn darker so that it stood out on the drawing.

Because engineering has been forced into a competitive marketplace, these features are now most often shown on the regular piping plan, without the need for a separate drawing.

Isometrics

Piping isometrics are most often used for conceptual layouts and stress analysis. The use of isometrics for routing small bore piping is very useful for the field engineer who must develop a bill of materials (a "take-off") for procurement.

[2] In the not-too-distant past, a darker-than-normal sepia could be made of the piping plan and the background erased to highlight the support or instrument.

IDEA, LLC Piping Material Take Off							
Project No: 166				By BKS		Date 4/18/2007	
Project:	Emergency Vent System						
System:	Relief						
Dwg No:	166-01 thru 166-05						
Pipe Spec:	SS-01						
Notes:	304 SS, Sch 40						
Item	Part	Sizes					
No.	Name	1	2	6	8	10	12
1	Pipe	20	20	1	71	25	102
2	90°	3	6		1	1	3
3	45°				2	1	
4	Tee	1	1				
5	Cap						
6	Reducer						
7	Coupling	1	1				
8	Union	1	1				
9	Flange Incl. stub end			1	6	4	9
10	Bolt and gasket kit			4	8	6	14
11	Threadolet						
12	Cuts	2	2		1		
13	Nozzle weld					2	1
14	B/W	2	2				
15	Joint M.-up						
16	Valve, ball						
17	Vavle, BF			2			
18	Blind flange			2			
19							
20							

Table 12.3 Typical Material Take-Off Worksheet. *Courtesy of Innovative Design Engineering of America, LLC.*

While a stress isometric provides an excellent means of identifying nodes and may offer some insight into problem stress or deflection areas, the development of the iso may be waived, with the nodes being neatly labeled directly on the plans and elevations. This saves much time.

Checklists

Being highly organized, engineers are drawn to checklists in the hope of preventing errors of the past from being repeated:

- Was at least 6′-8″ of headroom provided?
- Are fire hoses spaced sufficiently close together?
- Can strainer elements be removed?

The problem is that checklists are often too vague ("Was the piping code observed?") or so detailed that they are not applicable to every job ("Have stainless steel tees been used for oxygen service?"). Adding to the difficulty is the fact that most of these checklists are rolled out during the checking phase when the overnight mail truck is idling outside the front door, waiting to whisk the drawings to the construction site.

Checklists are fine, especially if your designs often follow the same routine, job after job. That is not always the case though, and the tendency is to pay little attention to a lengthy list during checking. It may sometimes be better to rely on a solid base of experience and technical knowledge than to try to distill every detail down into a recipe.

Document Control

Drawings and specifications are assigned a revision number whenever they are issued outside of the engineering organization. The common convention is to assign letters and numbers for the various issues:

- A – APPROVAL. This is issued to the Owner to ensure that the engineer is representing the Owner's requirements for the project. Vendors also issue drawings to the owner or engineer for approval. These drawings are marked up as required, and sent back to the drawing originator for consideration.
- B – BID. Bid issues are contractually sensitive, since these drawings are what the contractor bases his price on. Changes made after the bid issue should be clouded so that the extent of the revision is clear to all parties. The contractor will likely adjust his price or the schedule to account for any changes that occur after the Bid Issue, and this often has significant commercial consequences.
- 0 – CONSTRUCTION. Drawings are Issued For Construction once the engineer is satisfied that the current scope of the work is shown correctly on the drawings. Some engineering houses have a policy of removing all clouds from IFC drawings. This may be to the contractor's disadvantage if any item that could affect cost or schedule has changed between the Bid Issue and the Construction Issue.
- 1, 2, etc. Revisions are issued as necessary to correct any problems or changes that have occurred since the preceding issue. These changes are clouded and identified with a revision triangle. Previous clouds are removed so that the drawing does not become cluttered, and so the latest change is immediately apparent to the reader of the drawing. It is a good idea to provide as detailed an explanation of the revision as space will permit, since the revision history may be important if a dispute should arise.
- CERTIFIED. Engineers always seem to be working with preliminary information, and against schedules that demand final data in order to complete the design. In order to establish tie-ins, it is imperative that the engineer knows the exact location of a nozzle, the flange rating and size, and the dimensions of the piece of equipment. The vendor provides this final information as a "certified" drawing. This certifies that the equipment will arrive on site exactly as shown on the drawings.

The terms "Issued for..." and "Released for..." are synonymous.

Holds are sometimes placed on drawings if certain information is missing. Rather than delay the issue of an entire drawing, a cloud is placed around the missing information with the word "HOLD." It is a good idea to describe why the hold is there, for example "HOLD FOR VENDOR DIMENSION."

Document issues are often accompanied by a formal transmittal, which is a list that describes what was issued. These transmittals are tiresome to prepare, but often form the only basis for confirming whether or not a particular drawing was sent. The use of a transmittal is encouraged for every project.

The transmittal will list the drawing number, the title, the revision, and the date of the issue. Sophisticated databases are sometimes used to enhance this effort.

After IFC

Once the drawings are Issued For Construction, there are often still questions or revisions. Questions or clarifications may be formally requested by the contractor through a "Request For Information" format, which will list the particular drawing, the revision, the date of the request, and the question. Lists of these RFIs are maintained, with all of the accompanying correspondence, and these should be retained by the engineer at the end of the project for two reasons:

1. They may form the basis for a defense against claims relating to scope, cost, or schedule.
2. They are an excellent review of information that may be better described in the next project.

Field Engineering

The benefits of placing qualified engineers in the field during the construction phase cannot be underestimated. Very often these are the very engineers that designed the project, and in that case, so much the better. Aside from providing the field with dedicated technical resources, field work also provides the engineer with the opportunity to improve future designs by observing construction techniques and materials. It also affords an opportunity to build rapport with the owner and contractor, as well as defend the design.

The need to have the engineering firm's interests represented should be obvious. The qualities possessed by field engineers should include:

- Loyalty to the engineering firm
- Technical competence
- The ability to function effectively in a field environment
- The ability to deal calmly with antagonism
- The fortitude to stand up to conflicting opinions regarding the design, when the design can be defended

The reader is advised to avoid being "too eager to please," and to avoid being manipulated by the field personnel. After spending careful hours designing something that will work, one is often approached by a foreman or superintendent who offers a different solution on the basis of cost or time savings. I know what I designed originally will work, but I don't know that the suggestion will work. Being put on the spot and making a snap decision to appease someone in the field may only benefit the field.

On one construction project, I was being badgered by the contractor in a design review meeting to specify a pipe sealant for threads on a hot oil system. Admittedly, until then, I had not given it any thought. The pipe dope was not indicated in the pipe spec. The contractor was adamant, and acted as if the lack of this information would set the project back by a month. It was a big deal, so after I returned to the office I tracked down the information and submitted a product and manufacturer to the field. On a subsequent visit, I learned that they had ignored my spec, and substituted another product.

This brings attention to the phrase "or approved equal," which although it allows the engineer the opportunity to reject a substitute, it may also require much additional time to compare the properties of every suggested substitute. "Or equal" may be good enough, without requiring additional oversight. "Or equal" can also save time if delivery of the specified product becomes a problem.

CHAPTER 13

Specifications

Engineers hate to write specifications. It is difficult and often dull work, with a lot of potential dire consequences if something is incorrect. The best specifications are well-organized, consistent, and easy to read.

A lot of trouble can be saved by requesting that the contractor review the specifications ahead of time, if possible. This is viewed by many engineers as opening the door to unwanted criticism, but having the contractor review the specs prior to issue is preferable to having them request substitutions during the construction phase. It also obviates the risk of having criticism leveled at the engineer. However, many times it is simply not possible to solicit input from the contractor.

Engineering is in many ways the science of finding shortcuts. We use CAD (and before CAD, sepias) to reproduce similar details and even entire layouts quickly, without having to start from a blank sheet of paper. Specs are no different. Old specs are recycled from one project to the next, with bits added or deleted as the engineer experiences new problems and remembers to file them for future reference for use in subsequent projects. The result is often a tangle of miscellaneous requirements listed in no particular order, and often in conflict within the same document. What is needed in such cases is a complete overhaul.

Types of Specifications

When we think of specifications, the first that comes to mind is the piping spec. But we recognize that because of our unique position in projects, we may also be responsible for certain equipment specifications, as well as for construction specifications.

The requirements of the specifications are to define the expectations of the owner. The specifications will form part of the bid documents and will be relied upon to determine the cost of the project as well as the quality of the finished product.

Specification Formats

Depending on the particular specification, its format may be:

- An outlined narrative, which is more appropriate for equipment and construction specifications, but which has also been used for many years to prepare piping specifications.

- A table, which is more appropriate for some equipment and for most piping specifications. Tabulated datasheets for mechanical process equipment can be particularly useful since they condense the salient points in an easy to read format.

Equipment Specifications

The piping engineer will often be required to prepare specifications for process equipment such as pumps, compressors, cooling towers, heat exchangers, and so on. One must recognize that the specifying engineer is not the equipment designer however, and the distinction between those roles should be respected in order to prevent the specification from being written in such a way that only one manufacturer meets the spec. Generally the whole idea behind writing specifications is to prepare a document from which competitive bids may be fairly solicited. Writing a pump spec that requires a proprietary feature stacks the deck in favor of the manufacturer who has the patent on that feature.

Performance based specifications are the best at arriving at competitive bids. Specific details should be limited to those absolutely necessary for the requirements of the application.

One capital project for a new steel-making process had an intensive equipment procurement component. The company I was working for had the Engineering and Procurement contract (E&P) and was writing many specifications for the ancillary equipment like conveyors, blowers, stoves, etc.

We needed a very large shell-and-tube heat exchanger. It turns out that there are not many suppliers of such equipment, but even so, a specification had to be written. I wrote the technical specification and the technical data sheet that described the performance required of the exchanger. Once the purchasing manager added the engineering company's standard terms and conditions, various onerous appendices, and the owner's addenda, the specification was approximately 3 in thick.

We waited for replies to our RFQ, and heard nothing. After a couple of weeks passed, I called one of the vendors (the one who I thought had the best chance of designing such an exchanger). He told me that they did not intend to bid on the project because they felt that the sheer size of the specification indicated that it was full of liability for them. We were back where we started.

Sample Outline

1. Definitions. This is an opportunity to identify terms to be used within the specification. For instance, "Owner" may be synonymous with "Purchaser." For an Engineering, Procurement, and Construction Management contract, the "Engineer" will be the same entity as the "Purchaser."
2. Project location. It is important for bidders to understand where the equipment is to be delivered in order to account for freight costs, scheduling, and any environmental features that might need to be considered, such as freeze protection.

3. Standards. A list of the applicable standards that are expected to be applied to the design and fabrication of the equipment should be included. But care should be taken, as this is the most abused portion of any equipment specification. A specification that lists "ASME, ANSI, UL, …" is meaningless. Does the specifier really intend that a manufacturer retains access to the hundreds of individual ASME codes? This shotgun approach is often viewed by the engineer as offering protection[1] against the unknown. In reality, it is unenforceable, as it is too vague. On the other hand, if one indicates that flanges are to comply with ANSI B16.5, then that has meaning.

   A common statement that is appended to referenced standards is "latest edition," indicating that the specifier expects the manufacturer to design to a different standard than the one the engineer used. Many standards change rarely, and are issued triennially as a matter of course (the B16.5 dimensions are probably going to be the same 10 years from now as they are now). But certain codes may change, and in order to avoid applying a double standard, consider indicating the revision date used for the design.

   Conflicts between specified standards may also arise. See the sidebar discussion in Chapter 7 regarding the API flanges on a basket strainer.

   Another pitfall is to specify an obsolete standard.

4. Scope of supply. This is likely to be the longest section of the spec, and may include references to drawings that clearly outline the scope of supply. On very large projects, the Owner may desire to purchase all of the motors in bulk from a motor manufacturer. This would be important to a pump bidder who is accustomed to providing drivers with the pumps.

A process engineer was managing an installation at a paper mill in Georgia. The project required a rectangular tank, a vessel supported within a structural steel framework, and lots of piping. Upon delivery of the tank, the process engineer discovered that a square manway cover was not provided by the vendor.

The cover plate was not called out on the tank drawing, nor in the scope of supply in the spec. The vendor had assumed that something else was going to bolt up to the square flange. The process engineer had assumed that the vendor would know to supply the cover. After all, what else would bolt up to a square manway?

5. Functional requirements. These should almost always be delineated in the form of a data sheet. This provides a convenient and organized means of assembling and reviewing the data. Figure 13.1 shows a sample data sheet for a centrifugal pump.

6. Schedule. As important as the date the equipment is needed on site for installation is the date that vendor data is required by the engineer to complete the design. Milestones for Approval Issues as well as Certified Drawings must be stated in the specification. Clever Purchasing Managers will tie the delivery of these milestones to the payment schedule.

[1] The common term in engineering offices is "CYA," enveloping a philosophy that is intended to protect the engineer from criticism. This philosophy never assisted a project in meeting a schedule or coming in under budget.

ITEM -	IDEA	IDEA PROJECT -
		CUSTOMER -
DATA SHEET - PMP-1		LOCATION -
ITEM #		SPEC. NO. -
# REQ'D. -	HORIZONTAL CENTRIFUGAL	SHEET # - 1 OF 2

	DEPARTMENT - MECHANICAL	
1	LIQUID PUMPED (SERVICE)	
2	VISCOSITY (cp) / VAPOR PRESSURE (PSIA)	/
3	TEMPERATURE (°F)- MIN/ MAX/ S.G.@ °F	/
4	FLOW RATING MIN/ MAX/ DES (GPM)	/ /
5	SUCTION PRESSURE @ FLANGE (PSIG)	
6	DISCHARGE PRESSURE @ FLANGE (PSIG)	
7	TDH (w/s.f.) / SHUTOFF PRESSURE (FEET)	/ *
8	NPSH - AVAIL / REQ'D (FEET)	/ *
9	TYPE OF PUMP / MODEL / # OF STAGES	HORIZONTAL CENTIFUGAL / * / ONE
10	RPM / ROTATION (VIEWED FROM MOTOR	* / *
11	FACING PUMP)	
12	EFFICIENCY (%) / BHP @ RATING /	* / * / *
13	MAX BHP @ RUNOUT	
14	IMPELLER DIA (IN)-BID / MIN / MAX	* / * / *
15	IMPELLER -EYE AREA (SQ IN) /	* / *
16	PERIPHERAL VELOCITY	
17	WORKING PRESSURE (PSIG) / MAX /	* / * / *
18	HYDROTEST PRESSURE	
19	CLEARANCE (IN) WEAR RING / BEARING /	* / * / *
20	IMPELLER	
21	HYDRAULIC THRUST (FT-LBS) -	* / * / *
22	RATING / MAX / UPWARD	
23	SUCTION CONNECTION - SIZE (IN) /	* / CL150 / END
24	RATING (LBS) / POSITION	
25	DISCHARGE CONNECTION - SIZE (IN) /	* / CL150 / TOP
26	RATING (LBS) / POSITION	
27	BASEPLATE - MODEL # OR SIZE (LxW-IN)	*
28	COUPLING - TYPE / MFR / FURNISHED BY	* / * / SELLER
29	BEARING LUBRICATION - TYPE / METHOD	* / *
30	MECHANICAL SEAL - TYPE / MFR / MODEL # /	SINGLE BALANCED / * / * / *
31	MATERIAL OF FACES	
32	**MATERIAL OF CONSTRUCTION**	
33	COUPLING GUARD MATERIAL / FURNISHED BY	* / SELLER
34	CASING	
35	SHAFT - MAT'L / DIA (INCHES)	/ *
36	SHAFT SLEEVES	
37	WEAR RINGS - CASING / IMPELLER	
38	IMPELLER(S)	
39	BEARING HOUSING	
40	BASEPLATE	
41	**DRIVER:**	
42	TYPE - MOTOR OR TURBINE / RPM / HP	ELECTRIC MOTOR / * / *
43	FURNISHED BY / WEIGHT (LBS) /	SELLER / * / * / *
44	DRAWING REF. # / MANUFACTURER	
45	DRIVE END BEARING - DESCRIPTION /	* / *
46	THRUST RATING (UNITS)	
47	OPPOSITE END BEARING - DESCRIPTION /	* / *
48	THRUST END BEARING (UNITS)	
49		
50		
51		
52		
53		

Figure 13.1 Sample data sheet for a centrifugal pump. *Courtesy of Innovative Design Engineering of America, LLC*

ITEM -	IDEA	EEI PROJECT -
		CUSTOMER -
DATA SHEET - PMP-1		LOCATION -
ITEM #		SPEC. NO. -
# REQ'D. -	HORIZONTAL CENTRIFUGAL	SHEET # - 2 OF 2

DEPARTMENT - MECHANICAL						
54	DRIVER: (CONTINUED)					
55	DRIVER LUBRICATION - TYPE /	*	/ *			
56	RECOMMENDED LUBRICANT					
57	FRAME DESIGNATION / VOLTS / PHASE	*	/	/		
58	MANUFACTURER / INSULATION CLASS /	*	/ *	/ TEFC		
59	ENCLOSURE TYPE					
60	PUMP PERFORMANCE CURVE DRAWING NUMBER	*				
61	NET WEIGHTS (LBS):					
62	COMPLETE ASS'Y / BASEPLATE	*	/ *			
63	BARE PUMP / GEARBOX	*	/			
64	GEARBOX INFORMATION - MANUFACTURER /	*	/	/		
65	MODEL / RATIO					
66	INSPECTION REQUIRED - YES OR NO / TYPE	YES	/ MFR. STD. HYDRAULIC TEST			
67						
68	WITNESS OF TEST REQUIRED - YES OR NO	NO				
69	/ BY WHO					
70	QUALITY ASSURANCE - MFR'S STD / OR	MFR STD				
71	HOW DOCUMENTED					
72	APPLICABLE STANDARDS - API, ANSI,	ANSI				
73	NUCLEAR, ETC. AND SECTION.					
74						
75	PAINT	MFR STD				
76						
77						
78						
79						
80						
81						
82						
83						
84						
85						
86						
87						
88						
89						
90						
91						
92						
93						
94						
95						
96						
97						
98						
99						
100	MISCELLANEOUS					
101	* = SUPPLIED BY VENDOR					
102						
103						

104	REVISION #	A	0	1	2	3	
105	DATE						
106	INITIALS						

FIGURE 13.1 *(continued).*

7. Details of supply. Suppose you are writing a specification for a large piece of equipment; one too large to be handled in a single shipment. Usually the individual components arrive at different times, and rarely are they installed immediately. So once it gets on site, it must be unloaded, stored, moved to the installation location, and assembled. A procedure must be in place to ensure that nothing gets misplaced, and that the assembly progresses smoothly. Some suggestions to facilitate this are:
   A. Start-up troubles can be detected in the manufacturer's shop if the equipment lends itself to a shop test there. The vendor has the added incentive of correcting the problem because the equipment is taking up space on the shop floor. (Not that a vendor would not give your equipment sitting in the field top priority.)
   B. Match marking prior to knocking the equipment down for shipment is a simple procedure that can greatly reduce erection time.
   C. Designing a system to be skid-mounted removes many of the variables that add to project duration and cost. Pre-piping and pre-wiring to terminal blocks should also be specified.
   D. Insist that every crate, box, or component is labeled with the vendor's name, vendor's telephone number, purchase order number, and equipment number. This is useful to the field engineers in any case, but is imperative where a vendor is supplying multiple pieces of similar equipment. When something is missing or damaged in the field, there is nothing like immediate access to the vendor without having to track down the paperwork. The first thing the vendor asks is, "What is the PO number?" Having it labeled on the equipment is a definite advantage.
   E. Every piece of equipment should be labeled indelibly with its equipment number. Nameplates must be legible and remain unpainted.
8. Delivery details. If the project has special requirements for unloading equipment at the site, they must be stated in the bid documents. This may be addressed in an appendix or in the body of the specification.
9. Technical data. It is easy to get carried away in requesting technical data, and one must take care to not request lubrication data for a heat exchanger. Some bureaucratic document control departments may not approve progress payments if they have not received all of the documentation they are supposed to have. (This is more likely if the receipt of technical data has been tied to the schedule of payments.)

   A matrix of required data is a convenient way to display what data is required, and how many copies of each will be needed to distribute among the design engineers, erectors, and owner's maintenance files.
10. Intent to bid. An intent to bid form should be included with instructions to the bidders to return a signed copy a short time after the bid packages have been issued. The bidders should indicate that they will or will not submit a bid. If competitive bids are required, this may give the engineer time to locate and solicit bids from alternate vendors.

Provisions should be made for questions from bidders to be answered. The fairest way to deal with questions is to list all of the questions and answers from all of the bidders, and to distribute the Q&A to all of the bidders. This ensures that everyone has access to the same information. Questions may be submitted in person or over the phone, at the engineer's and purchasing agent's discretion, but answers should be in writing to all interested parties.

Bid Tabulation

A clever way to prepare bid tabs is to provide the spreadsheet format to the vendors, and require them to complete the information electronically (price, terms, delivery, technical data). These completed columns may then be cut and pasted into the master bid tab to provide an accurate comparison of the important details that will be considered in the purchase decision.

In this way, the engineer does not have to sift through varied bid formats to extract the information requested. Nor does he have to re-enter information (risking an incorrect entry or typographical error). Make the vendor use your format.

Bid tabulations are required to determine which vendor will be awarded the purchase order. In tabulated form the bids are easily compared to each other, and gaps in the information are more easily identified.

Bid tabs should include all of the contact information as well as the technical and commercial details.

Pipe Specifications

CSI Format

Editable outlines are available from organizations such as the Construction Specifications Institute[2] to provide engineers and architects with an organized format for the preparation of specifications. The advantages of using such a format are:

- It provides a standard format that contractors, owners, engineers and architects recognize.
- It permits every project spec to look like every other project spec.
- It saves time in the preparation of the specs.

While these specifications excel in HVAC piping or plumbing projects (as well as the architectural projects for which they are tailored), they are poorly adapted to industrial projects, which may have dozens of piping services, many of which are not part of the pre-written specs.

Prior to 2004, the entire format consisted of 16 "Divisions," with the Mechanical Specifications listed under Division 15. Each specification was assigned a five-digit number, with the first two digits indicating the division. The format was expanded in 2004 to 50 divisions, with each specification consisting of a six-digit number. The piping-related specifications were moved to Division 22. Again, the first two digits identify the division number.

Each specification contains three parts:

1. General. This part identifies related sections in the specification, reference standards, required submittals, quality assurance requirements, the need for pre-installation meetings, delivery and storage requirements, warranty information, and whether any spare materials may be required.

[2] CSI, 99 Canal Center Plaza, Suite 300, Alexandria, VA 22314, http://www.csinet.org.

2. Products. This section details materials to be used.
3. Installation. This section describes installation requirements that may or may not be shown on the detail drawings.

That there may be discrepancies between the specifications and the drawings may be surprising, but it is common enough that the contract documents often contain language specifying the "Order of Precedence of Documents," which may look like:

- Contract
- Special Provisions
- General Provisions
- Specifications
- Details on Drawings
- Plan Drawings

This indicates that the contract language reigns supreme and that specifications would take precedence over drawings.

Outlined Narrative

This is the most common format of industrial piping specification. Each specification describes the requirements for each of the various piping services. Thus, there may be separate specifications for cooling water, natural gas, compressed air, etc.

The problem with writing specifications for each individual service is that it becomes difficult to compare multiple specifications. There may be only slight differences between two or more specifications, and these differences could go undetected, losing an opportunity to combine specs that are only slightly different.

There are advantages to combining similar specifications:

- The number of specifications are fewer and therefore are easier to manage.
- The specifications are simpler and there may be fewer different fittings or valves. From the contractor's viewpoint, the fewer different fittings and valves, the better.
- The specs may be combined into one of several service classifications, like standard weight carbon steel with 150 lb flanges. A single service classification may be able to include a number of fluid services.

There will likely be a sub-specification for lines 2 in and smaller. The reason for this is twofold: for welded lines, it is far easier to use socket-weld fittings instead of butt-welds, and for other systems, it will be more economical to use threaded and coupled, or screwed connections. (Even on a system such as compressed air, in which threaded connections lead to excessive leakage, it may be easier to install T&C piping and seal weld the connections.) So there will usually be a break within the spec to describe the differences between the smaller bore piping and the larger bore piping. That break most often falls between 2 and 2 1/2 in.

There also may be entirely separate specifications for a single fluid service that runs both above and below ground. Flanges may not be permitted below ground, or the engineer may desire to use a material like HDPE below ground and steel above ground. The HDPE is a good choice below ground, but would be extremely difficult to support above ground.

Single pipelines sometimes have a "spec break" along their route. This often happens if a carbon steel system terminates in a strainer or filter, and continues out of the strainer as stainless in order to maintain the cleanliness of the system. Such transitions need to be identified on the drawings.

Sample outlined narratives are shown in Figures 13.2 and 13.3. Note that while suitable CSI specs may not be available for industrial services, many industrial specs

SECTION 15408 - COMPRESSED AIR AND FUEL OIL (CARBON STEEL)

PART 1 - GENERAL

1.1 WORK INCLUDED

A. This Section specifies furnishing and installing the following piping system:

System Abbreviation	System Name	Design Conditions	Test Conditions
CA	Compressed Air	150 psig, 80°F	225 psig(hydrostatic)
FO	Fuel Oil	100 psig, 70°F	150 psig(hydrostatic)

PART 2 - PRODUCTS

2.1 MATERIALS

A. Pipe:

1. 2-1/2" and Larger: ASTM A53 Gr B, ERW, standard weight, bevelled ends.
2. 2" and Smaller: ASTM A53 Gr F, ERW, Schedule 40, T&C.

B. Fittings:

1. 2-1/2" and Larger: ASTM A234, Gr B, standard weight, butt weld.
2. 2" and Smaller:

   a. Aboveground: ASTM A197, 150#, malleable iron, screwed.

   b. Belowground: ASTM A105, 3000#, forged steel, socket weld.

C. Unions: ASTM A197, 150#, screwed, brass seats.

D. Joint Compound: TFE tape 1/2" wide x 3 mils thick.

E. Flanges:

1. 2-1/2" and Larger: ASTM A105, 150#, RFWN.
2. 2" and Smaller: ASTM A105, 150#, RFSW.
3. Provide matching flat faced flanges at joints mating against flat face flanged valves, specialties, or equipment.

F. Gaskets: 1/8" Garlock Blue-Gard 3000, 150#. Provide full faced gaskets at all flat face flanged joints. Butterfly valves with raised sealing rings on seat faces require no gaskets.

G. Bolts: ASTM A307 Gr B with ASTM A307 Gr B heavy hex nuts.

H. Valves:

1. Gate: G-1010
2. Globe: L-1010
3. Check: C-0010
4. Ball: B-1010
5. Butterfly: F-1480

FIGURE 13.2 Sample outlined narrative. *(continued on next page)*

I. Strainers:

1. 2" and Smaller: 250 lb., bronze (ASTM B62), Y-type, threaded ends, brass or stainless screen.
2. 2-1/2" and Larger: 125#, cast iron (ASTM A126, Class B), flanged, flat face, Y-type, 1/8" perforated brass or stainless screen, provide with gate or ball valve for blowdown.

J. Hoses:

1. All hose material is to be compatible with the fluid used in the system.
2. All hose lengths, diameters and end connections shall be as shown on the Drawings.
3. 1/8" thru 2" diameter hoses must be rated for 150 psi and 175°F and be a stainless steel tube hose or high pressure rubber hose. For hoses exposed to heat the outer surface must be protected by a glass wool and wire braid. Provide swivel fitting on one end of hose.
4. For hoses over 2" diameter, material shall be stainless steel tube hose or high pressure rubber hose. Hoses must be rated for 150 psi and 175°F. For hoses exposed to heat, the outer surface must be protected with glass wool and wire braiding. Provide swivel fitting on one end of hose.

PART 3 - EXECUTION 3.1

INSTALLATION

A. Install and clean per these Specification and ANSI B31.9. Coat and wrap underground pipe per these Specifications.

B. System to be tested hydrostatically.

C. Underground Pipe: Coat and wrap underground steel pipe. Provide dielectric insulator kit at flange wherever pipe rises aboveground.

D. Install low point drains and high point vents in accessible locations.

End of Section

FIGURE 13.2 *(continued).*

adapt the CSI format. Note also the unlikely pairing of Fuel Oil and Compressed Air in the same spec in Figure 13.2. This is a common problem with narrative specs. Services that seem unrelated are grouped within a common spec.

Upon comparing Figure 13.2 with Figure 13.3, we see that an opportunity exists to perhaps combine the portions of the services into a single carbon steel spec for above ground locations. Such opportunities become much clearer if the specifications are tabulated.

Tabulated Piping Specifications

If a table is prepared to itemize the various requirements of the individual services, it becomes easier to group similar specifications. In some cases the engineer may be induced to eliminate a certain type of fitting if it occurs infrequently and may be substituted for one that is more common to the project. This has the advantage of saving purchasing costs as well as eliminating a potential source of confusion for the contractor. See Figure 13.4 for a tabulated specification that includes the previous spec examples, as well as some other services.

Another advantage of this type of specification is the ability to reproduce the entire pipe service specification in several drawings rather than in many pages.

SECTION 15410 - COOLING WATER PIPING SYSTEM (150# ANSI CARBON STEEL)

PART 1 - GENERAL

1.1 WORK INCLUDED

A. This Section specifies furnishing and installing various water piping systems. Water piping systems inside the caster spray chamber shall be stainless steel.

System Abbreviation	System Name	Design Conditions	Test Conditions
BD	Blowdown	100 psig, 70°F	150 psig
CCW	Caster Contact Cooling Water	235 psig, 120°F	353 psig
CNCW	Caster Non-contact Cooling Water	100 psig, 140°F	150 psig
ECW	Emergency Cooling Water Pumped	100 psig, 140°F	150 psig
TCW	Tower Cooling Water	50 psig, 140°F	75 psig

B. Where Contractor elects to use polyethylene pipe for underground lines, refer to Section 15414.

PART 2 - PRODUCTS

2.1 MATERIALS

A. Pipe:

1. 10" and Larger: ASTM A53 Gr B, ERW, Schedule 20, may be used if available.
2. 2-1/2" and Larger: ASTM A53 Gr B, ERW, standard weight, bevelled ends. Rigid grooved couplings for aboveground piping in sizes up to 14" may be used, except in the Melt Shop.
3. 2" and Smaller: ASTM A53 Gr F, ERW, Schedule 40, threaded and coupled.
4. Nipples: ASTM A 53 Gr B, Seamless, Sch 80, TBE (no all-threads).

B. Fittings:

1. 10" and Larger: ASTM A234 Gr B, Schedule 20, butt weld, if available.
2. 2-1/2" to 8": ASTM A234, Gr B, standard weight, butt weld.
3. 2" and Smaller: ASTM A197, 150# screwed.

C. Unions: ASTM A197, 150#, brass seats, screwed.

D. Joint Compound: TFE tape 1/2" wide x 3 mils thick.

E. Flanges:

1. 2-1/2" and Larger: ASTM A105, 150#, RFWN; RFSO for 12" and larger.
2. 2" and Smaller: ASTM A105, 150#, RF screwed.
3. Provide matching flat faced flanges at joints mating against flat face flanged valves, specialties, or equipment.

Figure 13.3 Sample outlined narrative. *(continued on next page)*

F. Gaskets: 1/8" Garlock Blue-Gard 3000, 150#. Provide full faced gaskets at all flat face flanged joints.

G. Bolts: ASTM A307 Gr B w/ASTM A307 Gr B heavy hex nuts.

H. Valves:

1. Gate: G-1010
2. Globe: L-1010
3. Check: C-0010
4. Ball: B-1010
5. Butterfly: F-1480

J. Strainers:

1. 2" and Smaller: 250 lb., bronze (ASTM B62), threaded ends, brass or stainless screen.
2. 2-1/2" and Larger: 125#, cast iron (ASTM A 126, Class B), flanged, flat face, Y-type, 1/8" perforated brass or stainless screen, provide with gate valve for blowdown.

K. Branch Connections: Reinforcement required for branch connection shall meet ANSI Code for Pressure Piping. Commercial weld-end tees shall meet ASTM A234 and ANSI B16.9. Weldolets, threadolets, and sockolets (Bonney Forge or approved equal) shall be full strength to match that of pipe.

L. Hoses:

1. All hose material is to be compatible with the fluid used in the system.
2. All hose lengths, diameters and end connections shall be as shown on the Drawings.
3. 1/8" thru 2" diameter hoses must be rated for 150 psi and 175°F and be a stainless steel tube hose or high pressure rubber hose. For hoses exposed to heat the outer surface must be protector by a glass wool and wire braid. Provide swivel fitting on one of hose.
4. For hoses over 2" diameter, material shall be stainless steel tube hose or high pressure rubber hose. Hoses must be rated for 150 psi and 175°F. For hoses exposed to heat, the outer surface must be protected with glass wool and wire braiding. Provide swivel fitting on one of hose.

PART 3 - EXECUTION 3.1

INSTALLATION

A. Install, clean and hydrotest in accordance with these Specifications and ANSI B31.9.

B. Install high point vents and low point drains in accessible locations.

C. Underground Pipe: Coat and wrap underground steel pipe. Provide dielectric insulator kit at flange wherever pipe rises aboveground.

D. Install start-up screens on the suction side of all horizontal pumps and any sensitive instruments not protected by permanent strainers. Remove start-up screens during commissioning and button-up piping.

End of Section

FIGURE 13.3 *(continued).*

SERVICE LEGEND AND PIPING MATERIAL SPECIFICATION:

SYMBOL	SERVICE	SERVICE RANGES	HYDRO PRESS AND MEDIUM	CODE	PIPING MATERIAL		FITTINGS
					2" & SMALLER	2 ½ " & LARGER	2" & SMALLER
CA FO TCW	COMPRESSED AIR (ABOVE GROUND) FUEL OIL TOWER COOLING WATER	150 PSIG 80 DEG F 100 PSIG 70 DEG F 50 PSIG 140 DEG F	225 PSIG WATER 150 PSIG WATER (BLOW DRY WITH NITROGEN AFTER DRAINING) 75 PSIG WATER	ASME B31.3	SCH 40 ASTM A53, GRADE F, ERW, THREADED ENDS	STD WT ASTM A53, GRADE B, ERW, BEVELLED ENDS SCH 20 MAY BE USED FOR SIZES LARGER THAN 10" IF AVAILABLE	150# ASTM A197 MALLEABLE IRON, THREADED, ANSI B16.3
CND	CONDENSATE	115 PSIG 350 DEG F	173 PSIG WW	ASME B31.1	SCH 80 ASTM A53, GRADE B, SEAMLESS, THREADED ENDS	STD WT ASTM A53, GRADE B, SEAMLESS, BEVELLED ENDS	150# ASTM A197 MALLEABLE IRON, THREADED, ANSI B16.3
CW	CITY WATER	115 PSIG 70 DEG F	90 PSIG CW	ASME B31.3	0.049" WALL ASTM A269 TP 304LSS, PLAIN ENDS, VICTAULIC PRESSFIT SYSTEM, GRADE "E" EPDM SEALS	SCH 40 ASTM A53, GRADE A, SEAMLESS OR WELDED, LEAD-FREE GALVANIZED, WITH VICTAULIC GROOVED ENDS	VICTAULIC PRESSFIT
FP	FIRE PROTECTION	115 PSIG 70 DEG F	90 PSIG CW	NFPA	SCH 40 ASTM A53, GRADE A OR B, SEAMLESS, THREADED ENDS	STD WT, ASTM A53, GRADE A OR B, SEAMLESS, BEVELLED ENDS OR GROOVED FOR VICTAULIC	150# ASTM A197 MALLEABLE IRON, THREADED, ANSI B16.3

Figure 13.4 Tabulated specification. *(continued on next page)*

A disadvantage is that much more information may be used in a narrative style specification. This may be handy for providing specific instructions.

Once grouped according to common materials, the entire specification for the various fluid services may be designated with a descriptor that identifies the pipe material group. Pipe Group 01 for instance might refer to a specification which uses standard weight carbon steel and 150 lb flanges. The number of such groups may be limitless but the general idea is to limit the number of individual service specs by combining those with similar characteristics, if technically possible and economically feasible.

FITTINGS	JOINTS		BOLTING	GASKETS	INSULATION	VALVES
2 ½" & LARGER	2" & SMALLER	2 ½" & LARGER				
SCH 40 ASTM A234 GRADE B, SEAMLESS OR WELDED, ANSI B16.9 BUTT WELD SCH 20 BUTT WELD MAY BE USED IF AVAILABLE	T & C OR 150# MI UNION, ASTM A197 SCREWED ENDS, BRASS SEATS	150# ASTM A105, ANSI B16.5 RFWN FLANGES. MATCH FLAT FACE MATING FLANGES	ASTM A307 GR B, FIT BOLT STUDS W/ HEAT TREATED SEMI-FINISHED HVY HEX NUTS ASTM A307 GR B	1/8" THICK GARLOCK STYLE 3000 USE FULL FACE GASKETS WITH FLAT FACE FLGS	N/A	GATE: G-1010 GLOBE: L-1010 CHECK: C-0010 BALL: B-1010 BUTTERFLY: F-1480
SCH 40 ASTM A234 GRADE B, SEAMLESS OR WELDED, ANSI B16.9 BUTT WELD	T & C OR 3000# FORGED CS UNION, ASTM A105, STEEL SEATS, GROUND JOINT, SW ENDS	150# ASTM A105, ANSI B16.5 RFSO OR RFWN FLANGES	ASTM A193 GR B7, CL 2A FIT MACHINE BOLT W/ HEAT TREATED SEMI-FINISHED HVY HEX NUTS ASTM A194, GR 2H, CL 2B	1/16" THICK GORETEX STYLE G2F	2" CELLULAR GLASS WITH 0.020" THICK ASTM E-84 PVC COVER	GATE: G-1010 GLOBE: L-1010 CHECK: C-0010
VICTAULIC GROOVED END		DI BODY, AWWA C-110 FOR USE WITH VICTAULIC STYLE 741	ASTM A307, GR B HVY HEX MACHINE BOLT, COARSE THD, W/ ASTM A307 HVY HEX NUTS	VICTAULIC SYSTEM	1" FOAM RUBBER FOR ANTI-SWEAT PROTECTION	GATE: G-1010 GLOBE: L-1010 CHECK: C-1010 BALL: B-1010 BUTTERFLY: F-1020
SCH 40 ASTM A234 GRADE B, SEAMLESS OR WELDED ANSI B16.9 BUTT WELD OR VICTAULIC	250# MI UNION, ASTM A197 SCREWED ENDS, BRONZE TO IRON SEATS	150# ASTM A105 ANSI B16.5 FFSO OR FFWN FLANGES	ASTM A307, GR B HVY HEX MACHINE BOLT, COARSE THD, W/ ASTM A307 HVY HEX NUTS	1/16" THICK GARLOCK STYLE 3400	N/A	GATE: G-1010 GLOBE: L-1010 CHECK: C-0010 BALL: B-1010 BUTTERFLY: F-1480

FIGURE 13.4 *(continued).*

CHAPTER 14

Field Work and Start-up

An electrical engineer once told me that he can tell if someone doesn't know what they are doing in an electrical room by watching how they act around the equipment. "If you see someone leaning against an MCC enclosure, right away you know he doesn't know anything about electrical gear." He explained that some electricians in a paper mill he worked at exposed a bus bar to perform maintenance. They found that a mound of wood dust had accumulated on one of the buses, reaching almost to the top of the enclosure. Had the dust touched it, the entire enclosure could have been energized.

NEVER LEAN AGAINST ELECTRICAL GEAR.

When opening or closing an MCC switch, stand to the side of the bucket (the term for the switch compartment). Press the palm of your non-dominant hand against the switch handle (don't grasp it), look away, and throw the switch. This procedure is recommended to minimize your exposure in the event of an arc-fault in the equipment. Better yet, get an electrician to energize the gear.

There are countless ways to get injured in an industrial environment. It is not sufficient to keep your wits about you; you must also be aware of what others are doing so that their careless actions do not injure you. An injury can be a life-changing event.

Many safety rules are common sense and some are just onerous. I worked in a steel mill where hard hats and safety glasses were mandatory even while driving a car through the site. At some point, common sense really does have to play a role.

Safety

Safety should be the top priority always, but engineers who perform field work would be wise to take a moment to consider the particular hazards of their environment. This is particularly vital since the field may be a bit foreign after having been confined for awhile to the office.

Acute safety hazards may be obvious, but some of the latent hazards are particularly insidious. These generally relate to environmental conditions such as what is present in air (asbestos, lead fumes, mold, chemicals, and so on).

Another consideration is that start-ups are essentially "boundary conditions" in the parlance of differential equations, and all of the interesting things seem to happen during those times. Certainly there may be equipment or instrumentation failures during steady-state operations, but start-ups are notorious for damage to equipment, property, and even personnel. The reasons may include:

- Lack of familiarity with operation of the system
- Dynamic instability of the system
- Poor housekeeping as a result of construction activities
- Equipment or instrumentation malfunction
- A perceived sense of urgency

A start-up plan should also contain a safety plan, but as a minimum the following rules should apply any time you are in a field environment:

1. Never walk under a load. Loads shift, hoists fail, and crane operators have enough to worry about without being distracted by someone wandering under a load.
2. Never lean against electrical equipment.
3. Never look at a welding arc. This leads to a "flash" which is a painful eye condition. It can also seriously damage your eyesight.
4. Do not stand in front of the heads of reciprocating pumps or compressors. The heads do not often blow, but why take the chance?
5. If using a ladder behind another person, wait until the other person clears it before ascending or descending, in case of slipping, falling, or dirt falling off of shoes from above.
6. Make sure pressurized systems are safely vented before attempting repairs or opening flanges.
7. Be always mindful of where a tape measure could fall if it buckles when extended. Beware of open buses, as at crane hot rails.
8. Be aware that many aging installations still contain asbestos pipe insulation, and many of these installations are in poor condition.
9. Check with the site manager to determine risks of exposure to hazards like asbestos, ionizing radiation, lead fumes from abatement programs, etc.
10. Cultivate the habit of never taking a step backward without first looking behind you.
11. Pass through personnel doors instead of overhead doors.
12. Always let the operators know you are in the area. It's both a courtesy and an opportunity for them to alert you to any potential hazards.
13. Establish eye contact with forklift operators or other equipment operators when sharing the same work area.

I was taking photos to document the progress on a black liquor recovery boiler rebuild. The boiler had suffered a steam explosion inside the firebox and I was assigned to follow progress for a client. The boilermakers and welders were taking their morning break, and I crawled inside a polyethylene enclosure that had been erected to minimize drafts that could interfere with the critical weld puddles at the boiler tubes. I noticed a strong garlic odor and wondered what the boilermakers were having for breakfast. After I snapped a few flash photos of the tube membranes, I heard a hissing. I traced the hissing to a cutting torch lying on the scaffold planking. As I turned the torch in my hand, a blast of acetylene gas streamed into my face, and I understood where the garlic odor originated. I also realized that the poly tent was rapidly filling with highly combustible gas, and I quickly exited, notified the foreman, and realized how lucky I was that my flash had not ignited the gas.

Walkdowns

In an operating plant, or on a start-up, you often need to operate valves and energize equipment. A walkdown of the system is required to verify that the line is ready to be pressurized. This will include checking that all of the valving is in the correct position, that equipment is connected (or not connected), that blinds are in the correct orientation, and finally, that the system is actually complete. Many construction projects have fairly elaborate schemes for alerting personnel to the readiness for operation. These often take the form of color coded tags to indicate that the fitters have signed off on the system, that the line has been pressure tested, and that pumps have been lubricated and aligned. Another walkdown can't hurt.

Check that test gags have been removed from Pressure Safety Valves. Check that limit stops have been removed from spring hangers, and that the hangers have been adjusted for the cold (or once heated, the hot) position.

Pumping systems need to be checked for the ability of the pump to maintain a prime. But prior to filling a well or vessel with liquid, it is imperative to examine the vessel for cleanliness. A QA program during construction should also minimize the amount of debris left inside piping, which can be considerable. The lines should be inspected wherever possible, and during construction of potable water lines this is especially important. Water mains under construction must be sealed at the end of the workday to prevent wildlife from entering.

A cooling water system was being started up at a steel mill. The fitters had been spitting sunflower shells into the coldwell throughout construction, and no one thought it was important to sweep them out before water was added. Unfortunately, the strainers had been purchased without automatic blowdowns in an effort to save money. These were large basket strainers, and when the vertical turbine pumps were started, the bottoms of the baskets were blown out of the strainers since the sunflower shells blinded the screens.

The tendency to develop an all-encompassing list of things to check has limited value. The reason is that certainly something important will be omitted from the list, and a too-cumbersome list will quickly lose the interest of the individual doing the checking. A better strategy is to develop an ongoing, project-specific checklist as the design progresses.

Pipe Cleanliness

Pipe cleanliness is an issue that is often ignored. Cleanliness may not be crucial to many systems, but it is often a problem due to sanitary reasons, abrasion of delicate components such as instrumentation or rotary joints, or contamination of the fluid transported.

Some of the problems associated with pipe cleanliness may be obviated during the design phase by specifying materials that are easier to clean, or even less likely to get dirty to begin with. One example is the use of stainless steel piping, which does not contain mill scale or introduce slag during welding the joints. Another example is the use of pre-cleaned hydraulic piping. Some proprietary hydraulic piping uses mechanical joints to eliminate welding, and the number of fittings is further reduced through bending the pipe rather than having to include elbows. The pipes arrived at the site pre-cleaned for hydraulic service. Hydraulic service is one of the most demanding for cleanliness due to tight clearances in control valves.

In many cases, the MOC will be carbon steel, and it will need to be cleaned. Some procedures such as banging on the pipes with hammers to knock off loose scale have dubious merit. Other procedures suggest that hot fluids be circulated in the system, the theory being that the expansion and contraction of the metal pipe will loosen mill scale and dirt. Flushing the lines relies on rather high velocities, on the order of 12 to 15 ft per second in order to entrain particulates and carry them to a collection point. The flushing fluid is selected in most cases to be compatible with the operating fluid.

Specifying a TIG weld for the root weld can help minimize the introduction of welding slag inside the pipe. The welds can then be completed with the more common stick welding.

Good practice requires that fabricated reservoirs are cleaned in the fab shop and sealed against the entry of contaminants (which may include humidity) prior to shipping to the site. Other precautions for critical systems will include magnets to trap ferrous particles in reservoirs, non-bypass filters, baffles to drop out particulates, and low-point drains.

Some very elaborate flushing procedures have been developed by contractors who specialize in cleaning pipes. These can include acid cleaning at high temperatures, filtration, draining, and passivation of the internal surfaces. One should be alert to the limitations of even elaborate procedures however.

During the start-up of a hot oil calender system, the cleaning solution was circulated for several hours at a temperature of approximately 140°F. After passivation, the system was blown out with dry air, but a small amount of water from the passivation solution remained. As the system was started up with hot oil, the water flashed to steam resulting in tremendous water hammer. Worse, the operating temperature of 400°F was sufficient to melt the varnish-like coating inside the pipes. The acid etch at 140°F was not sufficient to remove this coating, and it peeled off in sheets, collecting in the strainers and quickly blinding them.

Start-up strainers are often used to catch miscellaneous debris that collects in lines during start-up. These so-called "witches' hats" are installed in spool pieces just before the pump suction to protect the pump internals. After the fluid has been circulated for several hours the system is drained and the strainers are removed. The orientation of the strainers is always with the point of the hat upstream. This reduces the chance of damage to the strainer due to collection of debris at the center.

Vigorous construction management should enforce pipe cleanliness. Frequent inspections should ensure that fitters are not using pipes as garbage containers. Hard hats, cigarette butts, soda cans, wooden blocking, and plastic bags must never enter a pipe.

Sample Bearing Lube Oil System Cleaning Procedure

1. Clean the inside of the reservoir with clean rags.
2. Bypass the bearings with tubing.
3. Secure a filter bag (or nylon stocking) on the return line.
4. Add 55 gallons of low quality straight mineral oil (no additives) of 100 SUS at 100°F to the reservoir.
5. Circulate the oil for 1/2 hour, change filter bag, and repeat.
6. Drain flushing oil and fill with working fluid.
7. Circulate lube oil.

Sample Hydraulic Oil System Cleaning Procedure

1. Clean the inside of the reservoir with clean lint-free rags.
2. Bypass restricting valves, pumps, cylinders, and equipment.
3. For older systems that have been in use, flush with a solvent that can attain 100 SUS at the circulating temperature. For new systems, fill and flush with the working fluid.
4. Secure a filter bag (or nylon stocking) inside the return line to prevent contaminants from entering the reservoir.
5. Flush for four hours.

Additional information on pipe cleanliness may be found within PFI-ES5 "Cleaning of Fabricated Piping."

Pumps

Pumps can be damaged in a number of ways during start-ups. Inadequate lubrication, improper coupling alignment, poor suction conditions, debris entering the pump, and incorrect direction of rotation are all hazards that await the start-up engineer. The direction of pump rotation should always be checked prior to installing the coupling[1]. This is generally done by "bumping" the motor. This involves turning the motor on and off very quickly. Once rotation is verified, the coupling may be aligned. Incorrect rotation on a centrifugal pump can loosen the impeller in some pump designs, so the best method is to bump the motor BEFORE connecting it to the pump shaft. Even in this

[1] This coupling connects the driver shaft to the pump shaft.

case, the motor should only be energized for a short time since the mounting bolts are not always properly torqued.

Extreme care and caution must be exercised when energizing a motor. The area should be cleared of unnecessary personnel and good housekeeping practices should be enforced prior to start-up. In some cases, the pump may already be connected to the motor, and it may be impractical to have them uncoupled. In this case, ensure that someone has signed off on the correct rotation prior to energizing the motor. And then only bump it to verify the rotation.

Most of the poor suction conditions that may occur should have been eliminated through the application of proper design. Some, however, cannot be designed out of the system. Consider a plastic bag that has been drawn against the suction strainer of a foot valve. Short of emptying the reservoir or entering it, probably the best technique is to start and stop the pump in the hope of allowing the falling column of water to push the bag far enough away from the foot valve that it can be snagged with a hook.

Where large horsepower pumps are concerned, it is worthwhile to purchase the assistance of the manufacturer's field service technician. They will examine the pump for proper installation, in some cases even verifying impeller clearances. The ability to have access to an equipment expert on-site is not to be underestimated during start-up.

It is worth mentioning that thc contractor should be made aware of the necessity of assigning qualified personnel to the task of installing pumps. Some union-staffed projects will make arguments that the job of installing the pumps ought to belong to the pipefitters since the pumps are connected to piping. Some will insist that because the pumps have motors the setting of pumps ought to belong to the electricians. Probably the best choice is to have the pumps set by millwrights, but anyone qualified can set a pump. The danger in having unqualified crafts set pumps is the risk of damage to the shafts and keyways, the lack of a proper coupling alignment, or improper shimming.

Venting

High point vents should be used to rid air from liquid systems. This is imperative in hydraulics systems.

The operation of vents must always be checked prior to filling or draining tankage. This is particularly important after construction turnover or after maintenance has occurred on the tank.

If a vent is blocked, even by a thin sheet of polyethylene, and air is not admitted through the breather vent when the tank is pumped out, the tank wall can collapse with spectacular results due to the negative pressure resulting from the liquid leaving the tank. Pressure equalization during pump-out is paramount to tank operation.

Steam Systems

Travel stops on spring hangers must be removed prior to placing the steam system in service. Spring hanger manufacturers generally paint these red for visibility and band them into place for shipping. They also take the additional precaution of tagging them with a caution to remove them prior to placing the steam system into service. But the start-up engineer will want to verify that they have indeed been removed. Prior to start-up, spring hangers need to be checked that they are set at the cold position.

Steam systems produce a considerable amount of condensate on start-up due to the cool mass of the piping. Large systems should be designed with supervised start-up, bypassing the steam traps and wasting the condensate. Smaller systems may be designed with an unsupervised start-up. Supervised start-up means that the steam traps are manually bypassed until the piping has heated up sufficiently to produce a normal amount of condensate that can be handled efficiently by the steam traps. Such large systems are usually not cycled on and off frequently, so the manual or supervised start-up is not onerous. Smaller systems may be shut down more frequently and so the unsupervised start-up is more convenient.

On large steam systems, warm-up jumpers across large valves are used to prevent large temperature swings which could shock equipment like turbine blades or create large quantities of condensate which may lead to water hammer. Small diameter globe valves are used to bypass the header gate valves, and some large gate valves are supplied with bosses cast into the bodies to accommodate the taps for warm-up bypass valves.

Steam lines should not be started up unless the insulation has been installed. Aside from the obvious safety hazard, the loss of heat leads to poor economy and excessive condensate.

Once cleared for start-up, the steam system should be blown down to a safe location outside the building. The steam system needs to be brought to saturation temperature to loosen mill scale and dirt, and to melt or soften any lacquer or oils that may be present inside the piping. The steam traps and heat transfer equipment must be valved off at this time, entirely bypassing the condensate return system so that the traps are not damaged by the sludge inside the steam lines. Once the steam system is clean enough to place in service, the traps and condensate return system can be placed in service, but the condensate should be sewered until it is clean enough for re-use.

Once the steam system is up to temperature, any variable spring hangers should be checked to ensure that the indicators are pointing to the "hot" position. If they are not, the hanger should be adjusted until the spring is loaded to the hot position. Once this adjustment is made, no further adjustment to the hanger is required.

Steam systems should be blown down to eliminate air (noncondensibles) from the system. Be aware that blowing down a steam line creates a considerable amount of noise. The design should include provisions to blow down the steam line to a safe location outside, preferably terminating in a silencer.

Compressed Air

As mentioned elsewhere, rental compressed air is often required during start-up. This will often be on the critical path for starting up the other piping systems. Once a source of air is ready and the system is ready for pressure testing, the lines are blown down to remove as much particulate matter and oil and water as feasible. Because this will create an intense amount of noise, hearing protection is required of all personnel in the area.

CHAPTER 15

What Goes Wrong

Fires

Fires would not normally have much effect on a carbon or stainless steel water piping system. Consider the metallurgy. Steel pipe is formed at extremely high temperatures, so a pipe subject to a structural fire would not necessarily be exposed to a temperature higher than that at which it was manufactured. A visual examination of the piping (absent structural damage to the supporting system or physical damage due to falling debris) may be enough to merit performing a hydrotest to determine fitness for service. Materials used for valve seats and packings, and gaskets need to be investigated to determine whether these need to be replaced prior to the pressure test, since many of these materials will have melted or distorted in a fire.

For critical piping systems handling steam, compressed gases, or dangerous chemicals, it is advisable to remove sample coupons for metallurgical testing.

Floods

Piping is required to be pressure-tight, so the chances of floodwaters entering a closed system are minimal, especially when one considers that the external pressure induced by a flood is perhaps a few feet of water. An external pressure of 5 psi would be quite a flood, and the floodwaters would be unable to penetrate a system rated at 150 psi.

Damage to piping systems by flooding is usually confined to:

- Physical damage due to shifting debris
- Wetting of insulation
- Salt water contamination of the exterior of pipes, leading to higher-than-normal corrosion rates
- Empty pipes having floated off of supports. Empty plastic underground piping can lift out of soil in non-flood conditions if it was installed without anchors and the water table rises.

Wet insulation has lost its insulating properties and must be replaced. This can be a significant project, and one that could have been avoided through the use of closed-cell insulation, which will not become wet. Cellular glass insulation is an example of this.

Saltwater contamination may simply be washed off of uninsulated piping and equipment. Very often after a flood, mud and silt accumulate around equipment and low piping. The sooner this is removed the better, as it will continue to hold the moisture

Figure 15.1 This structural steel storage rack and gas tanks for a fire protection system were exposed to a saltwater flood, but were left in the mud for a period of several weeks. The portions embedded in the wet mud began to corrode.

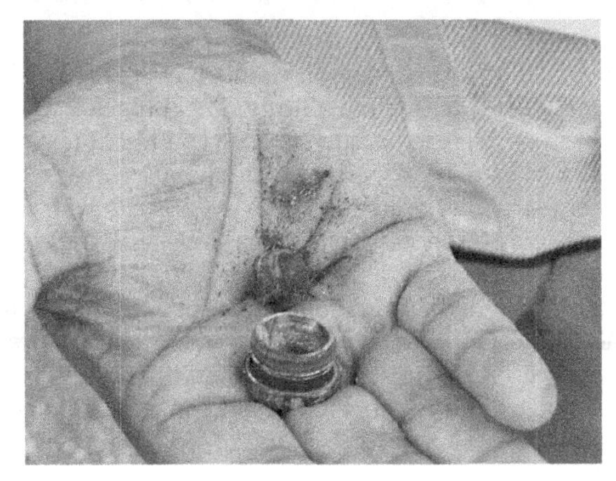

Figure 15.2 Check valve internals corroded due to exposure to sea water flooding. These valves were on gas utility drops in an industrial facility.

and accelerate corrosion. See Figure 15.1. In other cases, it may be impractical to sufficiently clean pipe or valve internals soon enough after a flood to prevent corrosion. See Figure 15.2.

Piping insulated with closed cell insulation exposed to saltwater may require to have the insulation removed and the piping and insulation washed, although it is doubtful if this is necessary, since once an insulated pipe heats up, the moisture necessary to promote corrosion would evaporate. In such cases it may be prudent to save the time and expense of removing and washing piping and insulation, and to instead monitor the corrosion rate over time to determine if it is necessary to provide other remedies.

Earthquakes

Any piping that has been subject to seismic loads must be thoroughly examined to determine whether the supports are intact and properly located. Damage due to an earthquake can be reduced during the design phase by making the system less rigid and more pliable. This is often accomplished through the use of mechanical couplings that rotate about the axis of the pipe, by forming loops or by including flex hoses specially designed for seismic events. Fire protection systems located in seismically-active regions require flexibility to be designed into the piping.

Another design approach is to perform a dynamic stress analysis on critical piping. These analyses can predict the behavior of the piping system under the influence of

Figure 15.3 This natural gas service entry was gouged when a stump grinder struck it. The "One-Call" center had not been notified prior to excavation so the landscaper was unaware of the location of the line until the grinder struck it.

seismic loads. The analysis may be performed by applying a fraction of the "G" load in the horizontal directions or a dynamic analysis with seismic forcing functions may be modeled. Refer to Chapter 9 for more on seismic analysis.

Critical piping[1] subject to seismic loads should be taken out of service and hydrotested.

Unanticipated Thermal Growth

Sometimes a pipe may move more than expected during operation. If it moves far enough, the shoe may slip off the supporting steel (see Figure 9.8). This can place enormous stresses on the pipe when it cools back down, since the pipe will be unable to return to its ambient position.

Condensation

Lines that may operate below the dew point must often be insulated to prevent the formation of condensation. This is especially true in commercial and residential construction (the liquid refrigerant line from split system air conditioning compressor units must always be insulated). Aside from aesthetic issues like ceiling tiles or finishes being damaged, there is a concern that materials that contain moisture make biological growth much easier. Mold remediation is very costly.

Another area of concern is where cold lines might pass above electrical gear in an industrial environment. These could be re-routed perhaps, but drips on floors may also be an annoyance. Further, there must be an associated loss of energy resulting from exposure of cold lines to the ambient atmosphere.

Damage to Underground Utilities

One Call and 811

In the U.S., we now have a national program to mark the location of underground utilities before excavation commences. The Common Ground Alliance sponsors the "Call Before You Dig" program, and offers a single phone number (811) to call to notify utilities of a pending excavation.

[1] "Critical" in this sense means any piping service that could cause damage to human health or pose an unacceptable risk to property (including the environment) if a leak were to occur.

Even small excavations such as planting a tree or installing a fence require a call to be placed to the One Call center or 811. Some states have legislated penalties for damages incurred after failing to call 811 prior to excavation. It is the responsibility of the person performing the excavation to place the call. It generally takes several days for the local utilities to mark the locations at the site.

When using power equipment to excavate at any depth, a call is required, even for do-it-yourself projects. A stump grinder can wreak havoc on a natural gas line. See Figure 15.3.

Of course, this program extends to electrical utilities as well as piping.

Underground Markers

Where piping passes below roads and rail beds, line markers must be used to designate the right-of-way. See Figure 15.4. To prevent having to shut down a roadway for a line replacement, underground utilities are usually sleeved under the roadway. Not only does this facilitate replacement but it isolates traffic loads from the pipe that contains the fluid. The ends of the casing sleeve are sealed to the carrier pipe and the sleeve is vented above grade to provide indication of leaks.

Plastic marking tape is often specified to be buried 6 to 12 in below grade to alert future excavators of the presence of underground lines. Some of these contain an aluminum foil core to facilitate location with a metal detector. This is especially important for plastic piping, which will not be easily detected otherwise. RFID tags are now available for identifying underground pipes with asset numbers.

FIGURE 15.4 A densely-populated underground pipeline right-of-way is identified with above-ground markers.

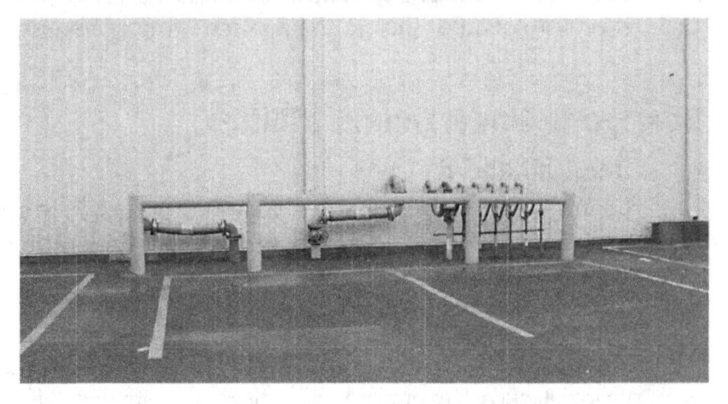

FIGURE 15.5 Pipe bollards and rails protect a service entrance near a parking area.

Legionella

A mysterious respiratory disease killed 34 people attending an American Legion conference in July 1976. The cause was identified as a previously unknown bacterium, *Legionella pneumophila*, that thrives in water in the temperature range of 77°F to 108°F (25°C to 42°C). The bacteria do not begin to die rapidly until temperatures exceed 131°F (55°C), and above 140°F (60°C), they die almost immediately[2]. This temperature range places cooling towers and even domestic hot water supplies at risk of sustaining legionella.

Infection by legionella may lead to Legionnaire's Disease, an acute respiratory infection that produces pneumonia, or a milder infection known as Pontiac Fever, which is similar to acute influenza. The atomization of water as it passes through a cooling tower or exits a bath shower nozzle provides a convenient pathway for inhalation of the bacteria. The bacteria are not spread from person to person.

Legionella is thought to enter the water supply through the municipal reservoir. Being chlorine tolerant, it is unaffected by most municipal water treatments. Once it enters a hospitable environment, the bacteria multiply and take up residence in the biofilms that line most pipes.

Cooling Towers

Recommendations to prevent infection by legionella at cooling towers systems include:

1. Locate the tower so that the tower drift is away from building air intakes.
2. Regularly apply biocide.
3. Maintain the tower in accordance with the manufacturer's instructions, and record maintenance on the tower.

While the first two recommendations are obvious, the third will work only if there is a systematic inspection of the maintenance records.

Hot Water Systems

Legionella may be present in any hot water system, but a closed system such as a radiant heating hydronic system would pose little threat to human health due to the lack of a convenient pathway for inhalation. Therefore, we are most concerned with hot water systems that distribute water that comes into direct contact with humans. These may be defined as potable hot water or domestic hot water systems.

Hyperchlorination

The Centers for Disease Control recommends shock hyperchlorination. This consists of a flush of at least 5 minutes with a concentration of 10 mg chlorine per liter of water. Continuous hyperchlorination can corrode piping, so its use is not recommended.

Copper-Silver Ionization

The Vikings used copper fibers on their ship hulls to discourage the growth of shells and algae, and nomads are known to have used silver coins to improve well water

[2] "Report of the Maryland Scientific Working Group to Study Legionella in Water Systems in Healthcare Institutions," State of Maryland, Department of Health and Mental Hygiene, June 14, 2000.

quality. Since the 1950s, copper-silver ionization techniques have been used for water disinfection. This technique has been applied in hospitals to reduce the growth of Legionella.

The ionization of the metals occurs through electrolysis and forms positively charged copper (Cu^{+} and Cu^{2+}) and silver (Ag^{+}) ions. Because chlorine is not required, this technique was used to disinfect water for NASA's space program.

Ultraviolet Radiation

UV radiation has been applied successfully to municipal water systems to eliminate legionella. UV disinfection of drinking water was first used in the U.S. in 1916. The disinfection occurs because the UV radiation penetrates the cell walls of bacteria and disrupts its ability to reproduce.

Compared to chemical treatments, UV radiation offers no residual disinfection capabilities. There are no chemical residues carried downstream to continue to act on the bacteria. This may be viewed either as a positive or negative aspect of UV treatment.

Monochloramine

Monochloramine has been used for some time as a disinfectant in public water supplies. Studies prepared by the CDC[3] indicate that monochloramine performs better at eliminating legionella than traditional chlorine doses.

Temperature

Another way of eliminating and preventing legionella from inhabiting a domestic or potable hot water system is by elevating the temperature above 140°F (60°C). Chemical treatments such as chlorine shocks are deemed to be ineffective against existing colonies of legionella due to their inability to penetrate the biofilms that line most pipes. The risk of scalding has discouraged facilities from operating domestic hot water systems above 120°F (49°C). But this lies within the range of temperatures at which legionella thrive. So if the temperatures must be elevated to 140°F (60°C), ASSE anti-scald devices must be installed at the points-of-use.

Operator Error

Petrochemical plants have a higher level of technical training than other facilities in general industry. The facilities that rely on the common knowledge of their operations and maintenance staff should realize that such knowledge is rarely common. Training should be provided to teach personnel how to recognize when a valve is open or closed.

Emergency shut-off valves should be represented diagrammatically on sketches throughout facilities and the valves should be labeled. Curb valve wrenches need to be stored conveniently and a maintenance schedule should be developed to exercise these valves.

It is also important to inspect fire protection valves. Technicians that perform backflow preventer inspections will sometimes accidentally leave a main valve in the closed position. A facility manager that simply accepts the inspector's report, and

[3] B. Flannery, L.B. Gelling, D.J. Vugia, J.M. Weintraub, J.J. Salerno, M.J. Conroy, et al. *Reducing Legionella colonization of water systems with monochloramine.* Available from http://www.cdc.gov/ncidod/EID/vol12no04/05-1101.htm.

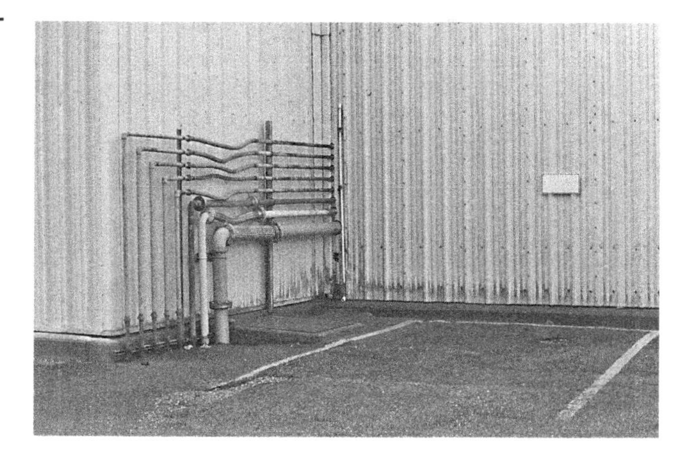

Figure 15.6 No protection at a parking area from physical damage of utility lines. Some of these lines contained oxygen and flammable gases.

maintains a hands-off attitude with regard to the hardware may have lost his fire protection until someone discovers that the valve that serves the sprinkler system is actually closed.

Signage

Product identifiers and arrows to indicate direction of flow are available from a variety of suppliers. The proper labeling of lines should be viewed as an opportunity to reduce product loss, mitigate environmental contamination, and reduce the chances of accidents.

ANSI/ASME A13.1 "Scheme for the Identification of Piping Systems" is the international standard, but individual companies often adopt their own standards. A plant in which every pipe is painted blue may fit someone's aesthetic view, but engineers should apply the maxim "form follows function."

Protection from Physical Damage

In areas where damage may occur due to fork trucks or vehicular traffic, pipe bollards are embedded in the ground to physically protect piping. See Figures 15.5 and 15.6.

Freezing

Freeze protection is required on liquid lines exposed to outdoor air temperatures that will stay below freezing. Freeze protection may be provided by ensuring that the fluid remains moving, or by heat tracing and insulating the pipes.

The maximum density of liquid water occurs at 39.16°F (3.98°C). But upon freezing it becomes less dense, expanding to 109 percent of its volume. During that time enough force can be placed on the inside of the pipe to cause it to burst, as in Figure 15.7 This property actually can be advantageous to temporarily block the flow in a line long enough to perform line maintenance such as installing a new in-line valve.

Freezing of lines is a serious design issue, and much time may be spent determining the best strategy to apply to prevent freezing. Consider a water tank that provides emergency water to a continuous caster. In the event of a cooling water failure, emergency water must quickly be provided to prevent the copper mold from melting due to the molten steel that it contains. Once a mold fails, the molten steel could come

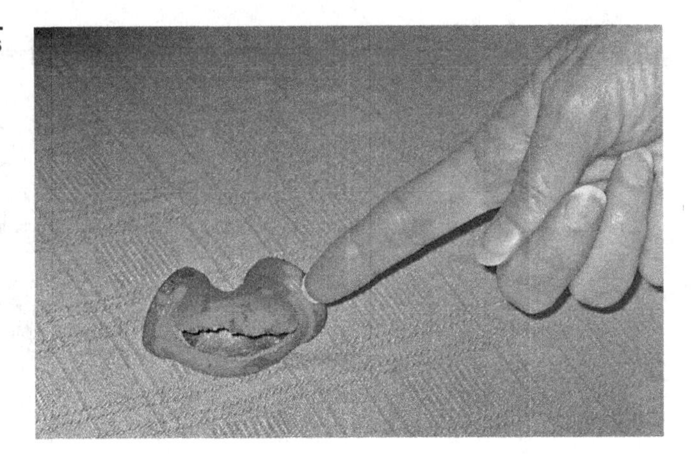

Figure 15.7 This cast brass elbow ruptured during shipping. After tests at the factory in north Georgia the skid was trucked to Indiana. The line had not been blown dry prior to shipping and during the transit through cold weather this small line froze, rupturing the elbow. The owner refused receipt and the unit had to be re-tested at the factory.

into contact with water resulting in a flash steam explosion that could shower the area with globules of red-hot liquid steel.

Water tanks that supply potable water for municipalities rely on the constant movement of the water inside the standpipe and into the tank to prevent freezing of the water. But an emergency water tank would not normally have flow. Other strategies are required, such as insulation and heat tracing, inducing a flow through the use of a side stream line that pumps the water back into the tank, or bubbling air through the standpipe to stir the water.

Where water is not wasted, drain lines can be opened to prevent freezing of headers. Aside from concerns over wasting the water, the creation of nuisance freezing on road or walking surfaces must also be considered.

Burying lines sufficiently deep (below the frost line) is an effective means of preventing freezing. But eventually the lines must return to the surface, and sometimes that must occur outside of a heated structure, for instance at a valve stand. Insulated enclosures such as shown in Figure 15.8 are available for such applications. These are electrically heated to protect the components. Some manufacturers even disguise these enclosures as boulders for aesthetic reasons.

Figure 15.8 An insulated cover protecting a valve stand against freezing (and tampering) in New York.

Design and Construction Errors

One observation I have made throughout my career is that whenever a project has big problems, it is never caused by one single issue. Some may point to an inept colleague, a too-demanding owner, an incompetent contractor, the weather, or the alignment of the planets. But when a project turns sour, it is always from a combination of multiple problems. There is never a shortage of things that can go wrong, so the engineer must always be on the alert for what can go wrong in the design, construction, or start-up phases.

Drawing Issues

Though it sounds like it ought to be a simple thing to keep track of, it is in fact too common that different drawings appear in the field that bear the same revision number. A strict document control policy will go a long way toward ameliorating this problem, but it does happen.

The dangers are obvious. Conflicting information appears in the construction documents, resulting in rework or acceptance of an inferior design (perhaps a dimension is changed to allow better maintenance access to a pump, but the correct revision is discovered in the field after the equipment is set).

The document control policy must make a distinction between drawings in progress and drawings that have already been issued. Drawing databases are difficult to maintain but they are very useful in preventing such errors. A note on the CAD drawing that clearly indicates that the drawing is "Revision-in-Progress" is also worthwhile, to prevent a premature drawing issue.

Interferences

Probably the most common design errors are interferences, in which one feature occupies the same space where another feature is located. A common problem involves the structural engineer's penchant for stick figure representations of structural steel. These sticks have no scaled width and so it is not obvious visually that an interference is present.

Other interferences involve cable trays or lighting fixtures. The order of precedence in a project may list structural steel first, except on a project like a pulp mill, in which the structure must be built around the large pulp lines. These are large diameter lines made of light gauge stainless steel, and are polished on the inside so that the pulp fibers will not attach themselves to the surface of the pipe (this is known as the "cotton ball test"). If a cotton ball is swiped along the surface of the weld joint, its fibers must glide over all of the surfaces without snagging. A line with such requirements will not easily be relocated.

So the order of precedence is:

1. Large process lines like pulp lines in a paper mill, or black liquor lines in an evaporator plant
2. Structural steel
3. Other process or utility lines along established pipe routes such as racks

4. Electrical cable trays
5. Piping drops to equipment
6. Electrical drops to equipment

Very often the real estate will be carved out at the beginning of a project, establishing at what elevations the piping and electrical routings will occur. This saves a lot of trouble since it eliminates the need to cross-check between disciplines within that elevation envelope.

Contractor Errors

I once provided some preliminary, undimensioned mechanical room concept layouts to my contractor client who wanted to review progress and comment on the layout. During a site visit later in the week, I saw the foreman hovering over the 11 × 17s in the mechanical room, holding a ruler to the drawings. Chalk outlines appeared on the floor.

"What are you doing?" I asked nervously.

"Laying out the pumps and chillers," came his matter-of-fact reply.

"Those drawings aren't issued for construction," I explained.

He had been scaling a reduced-sized drawing with a 6-in flexible ruler with markings in sixteenths of an inch. An unchecked drawing. It was an opportunity for error piled on top of error, so I waved them off until I could provide checked, dimensioned drawings that could physically accommodate the space constraints and the piping configuration.

Contractors are always eager to get started with construction and they may be too eager. On occasion they need to be reined in.

The desire to reduce costs can manifest itself in too little field trim being left on prefabbed pipe spools. Another problem that can occur is excessive out-of-tolerance on setting structural steel. Columns have been set as much as 2 in off from where they were designed. Since piping is referenced to column lines in plans, any deviation from where the structural members were intended can accumulate and cause problems with piping alignment.

Coordination with other disciplines (as described in Chapter 10) is an ongoing process. But even the diligent review of other discipline's drawings may not catch errors like conduit being field-routed across intended access points. Early detection by knowledgeable field engineers and construction managers is required, because once a cable tray or conduit is full of wires, few possess the temerity to demand relocating it.

CHAPTER 16
Special Services

Natural Gas

Natural gas is essentially methane, which has a heat of combustion of approximately 1000 BTU/ft^3. The supplying utility can provide detailed information regarding the exact constituents in the supply gas, as well as an accurate heat of combustion. Methane is an odorless, colorless gas, so the utilities add an odorant, often tert-butyl mercaptan to the stream to provide ready indication of leaks.

> To provide natural gas to a continuous casting operation in a steel mill, the vendor recommended passing the gas stream through a water seal as a measure to prevent backflashing. One of the engineers for the company I worked for discounted this as an unsafe practice, claiming that water would absorb the odorant. Additional research indicated that the mercaptan is not soluble in water, and therefore would be unaffected by passing the gas stream through a water seal.
>
> We decided not to use the water seal anyway, since we judged the chance of oxygen entering the natural gas piping to be acceptably low. Unless oxygen is mixed inside the gas pipe, there is no chance of explosion. This may not be the case with burning torches however, which should always be equipped with backflash arrestors.

The natural gas infrastructure is routed cross-country using steel or plastic piping. Plastic piping requires a copper tracer to be buried with the pipe. The pressure is often stepped down at the user through a pressure reducing valve that is equipped with a vent to atmosphere. While these reducing valves are usually located outside, it may be convenient at times to locate them inside a facility, in which case the vent must be piped outside[1] unless the regulator is equipped with and labeled for utilization with an approved vent-limiting device. Attention should be paid at the beginning of a project to determine whether it is worthwhile to provide a lower pressure header system or have many penetrations through the building roof or walls for each of the vents.

Piping to individual appliances must be supplied with a shut-off valve in the same room, and below that, a dirt leg to catch particles and prevent them from fouling the burner orifices. The dirt leg utilizes a tee to make the bend from vertical to horizontal and has a nipple with a cap extending below the tee.

The gas piping must be protected from physical damage. Except for devices that must be piped up and then moved into place (for example, a gas clothes dryer or gas range), the piping should be rigid, although it has become common to see flex connectors

[1] 2003 International Fuel Gas Code, Paragraph 410.3, "Venting of regulators."

used for appliance hook-ups in stationary installations (water heaters, gas-fired unit heaters), since it alleviates the contractor from having to make an exacting rigid fit-up.

Plastic gas piping must never be used above ground[2] unless it is:

- Terminated outside in a riser specially made for that application (an "anodeless riser")
- Terminated with a wall head adapter inside a building[3]

Capacity of Natural Gas Pipelines

An equation to determine the capacity of a gas line is provided by The Institute of Gas Technology. It is applicable for pressures between 5 and 60 psig and pipe sizes between 2 and 20 in.

$$Q = 3501\left(\frac{p_2^2 - p_1^2}{L}\right)^{0.556} D^{2.667} \qquad \text{(Equation 16.1)}$$

where

Q = Flow [ft³/hour]

p = Pressure [psia]

L = Length [ft]

D = Inside diameter [in]

Sealing Natural Gas Threaded Connections

During a trip to one of the big box stores I noticed a variety of PTFE (Teflon®) tape available in different colors. I asked someone what the difference was and was told that the different colors are intended for different services. A comparison of the labels indicated no difference in materials, so I was at a loss as to the applications. I checked a few plumbing forums online, and learned that the knowledge base regarding Teflon® tape is minimal.

It seems that some of the colors like yellow (which is intended for fuel gas) are thicker, but the literature incorrectly identifies this as "denser." So if the only difference is that the tape is thicker, it seems like one could apply more wraps to the joint.

There are hundreds of thousands of low-pressure natural gas joints performing satisfactorily with only the application of white PTFE tape on the joint. There is not much technique to applying tape to a joint, other than:

- The tape should be wrapped in the direction of the threads so that it does not loosen as the joint is started.
- The tape should not be permitted to enter the inside of the pipe.

Therefore, whether the installer takes 2 or 10 wraps, the various codes and inspectors do not care. All that matters is that a leak-tight joint is made. Manufacturers of the yellow tape claim it is shred-resistant. Pipe dope may also be used on gas pipe threads.

---

[2] Ibid, Paragraph 404.14.1.

[3] The wall head adapter is a transition (from plastic to steel) fitting that permits new plastic tubing to be fed through a foundation wall. The fitting attaches to the steel pipe and terminates the plastic tubing within the fitting on the interior side of the wall. Since many gas utilities now prefer the transition to steel be made outside the building with an anodeless riser, they move their meters outside for easy access. This makes the use of these wall head adapters infrequent.

Nominal Pipe Size		Length of Piping Requiring Purging	
(in)	DN	(ft)	(m)
2 1/2	65	> 50	> 15
3	80	> 30	> 9
4	100	> 15	> 4.5
6	150	> 10	> 4
8 or larger	200	Any length	Any length

Table 16.1 Purge requirements for fuel gas

Purging

Fuel gas lines must be purged before placing into service to prevent a combustible mixture from forming inside the piping. The safest way to do this is to provide a nitrogen purge. Indeed, a nitrogen purge is required to prevent dangerous gas pockets from detonating in a boiler. Two inch diameter lines and smaller may be purged with the fuel gas if the flow is adequate to prevent air from entering the line, and if the gas can be vented outside. Lines 2 1/2 in in diameter to 6 in must be purged with inert gas if their length exceeds the values shown in Table 16.1.

Compressed Air

Compressed air may be categorized as plant air or instrument air. Plant air is ordinary compressed air. It is usually provided at a pressure of approximately 100 psig (6.9 bar) and at a dew point of 40°F (4°C). This dew point is achievable by passing the gas stream through aftercoolers that are supplied with the compressor and cooled with plant cooling water. An inlet filter is always used upstream of the compressor to protect the compressor internals.

The quality of the air downstream of the compressor is generally poor. The airstream contains particulate matter, oil from the compressor, and often moisture. These can be cleaned from the air through the use of local filters which are often fitted with regulators and drip lubricators as well. This type of compressed air is suitable for use in pneumatic tools.

Some portion of the particulates will be a result of the material of construction of the compressed air piping. This is usually carbon steel, but there are also aluminum and stainless steel systems that are designed specifically for compressed air. Leakage is often estimated to be as much as 10 percent of a carbon steel system (threaded connections not seal-welded), and since the cost of producing compressed air is considerable, all means should be employed to minimize leaks.

Branches off of headers should always be from the top of the header in any compressed air system in order to decrease the amount of moisture and particulates delivered to the user.

Even though the most benefit is derived with systems that have high delivery pressures and lower user pressures, there is also an advantage to installing air receivers near high capacity users. Air receivers act as reservoirs to decrease the effects of sudden losses in system pressure. This is especially useful for large systems with centralized air compressors. If there is economic justification to use a very high delivery pressure out of the compressor, then the use of receivers allows the system pressure to drop without the need for the compressor to run every time a user demands flow.

Instrument Air

Instrument-quality air is required for "cleaner" applications such as pneumatic controls and instruments that are used in HVAC controls and to operate diaphragms in control valve operators. The International Society of Automation provides a standard in conjunction with ANSI for defining the quality of an airstream to be used for instrument air. This standard is ANSI/ISA 7.0.01 "Quality Standard for Instrument Air." This standard limits moisture content, entrained particle size and oil content, standard air supply pressures, ranges of pneumatic transmission signals, and criteria for testing compliance with instrument-quality air standards.

The basic requirements for instrument air are:

1. The dew point for instrument air (if the system is completely indoors) is limited to 35°F (2°C) although it is usually specified to be –40°F (–40°C). In any case, the dew point must be 18°F (10°C) below the lowest temperature to which any part of the system might be exposed.
2. The particulate size is limited to 0.00012 in (3 micron).
3. The oil content is limited to 1 ppm, but shall be as close to zero as possible. Some compressors are oil-free, so this is not difficult to achieve.
4. There are to be no corrosive or hazardous gases present in the airstream.
5. The operating pressures are generally 15 to 50 psig (1.0 to 3.4 bar).

The trend in industry is to provide instrument quality air for both instruments and general compressed air. This prevents deterioration of the air piping due to the high moisture content, and prevents the need for two separate systems.

Compressed air systems should be provided with a tee on the main header and a valve to the outside of the building along a road, so that whenever supplemental air is needed it can easily be connected. Rental air compressors are extremely useful for this, and during the lifetime of the plant may be required at start-up, or during maintenance to one of the main compressors.

The compressors will most likely require cooling water, so they may not be available at start-up because the cooling water system may need air to regulate the control valves. To sidestep this chicken-and-egg problem, an air-cooled compressor may be rented and connected to the tee in the main header.

Compressed air systems will be full of bits of dirt, mill scale, and oil. During commissioning, the headers should be blown down to eliminate as much of this as possible. Some specifications call for a target to be made out of a board painted white, and for the air to be blown at the board. Through successive trials, the number of particles decreases to an acceptable level.

Like pump systems, compressors are often sized according to "N+1," where N is the number of compressors required to operate the system, and the extra compressor is down for maintenance, or available as an in-line spare.

Oxygen

A steel mill in Alabama wanted to replace some very special valves in a lime conveying system. The lime was pneumatically conveyed into Basic Oxygen Furnaces using oxygen, since the furnaces need both lime for slag production and oxygen for increasing the combustion of the fuel and to lower the carbon content. When I learned that the conveying medium was oxygen, I wondered if they had any devices in place to prevent stones, iron nuts or bolts, or whatever else might find its way into a bulk lime delivery and create a spark inside the steel pipe. They had nothing like that.

I explained to them that this was a risky operation, since the presence of a spark inside a high-oxygen atmosphere was likely to cause a fire. Fires fed by high-pressure oxygen are exceedingly dangerous. But this mill had had experience operating in this fashion for 30 or more years and they were comfortable with the situation.

Oxygen is one of the most hazardous fluid services to deal with. Oxygen supports combustion and a fire fed by high-pressure oxygen will be spectacular and deadly. Oxygen systems must be free of grease or oil, since the heat of compression alone may be sufficient to ignite the organic fuel. Particulates inside the pipe could cause a spark which could lead to ignition and burning of the metal pipe. Impingement is also a concern, as when the flow enters from the branch of a tee. The risks of creating a spark are reduced if the velocity is kept below 100 fps. Once ignited, the fire will burn even the carbon steel pipe that is usually used to contain oxygen systems.

The Compressed Gas Association standard CGA G-4.4 "Industrial Practices for Gaseous Oxygen Transmission and Distribution Piping Systems" details acceptable practices for the design of oxygen piping systems. The reader is encouraged to carefully review those recommendations, some of which follow.

1. When ordering valves, the purchaser must specify that the valves are for oxygen service. Such valves must be specially cleaned and kept free of oil or grease.
2. For carbon steel piping, the design velocity of oxygen must be kept below 200 ft per second (61 m/sec) for pressures below 200 psig (13.8 bar). The reason for this is to reduce the chance of ignition by particles impacting the inside of the pipe. Copper, stainless steel, and nickel alloys may be subjected to higher velocities. Aluminum and aluminum alloys should never be used.
3. Lengths of copper piping are sometimes installed in a carbon steel system as fire stops. The thinking is that if the carbon steel catches on fire, the fire will not spread upstream of the copper, since copper (as well as nickel, Monel, tin bronze, and Inconel) resists burning.
4. As an economical consideration, carbon steel is safe to use in oxygen systems, but it is best to use stainless steel tees where the flow is into the tee from the branch leg (a side entrance tee). This again is because of the desire to limit the chances of ignition due to impingement by foreign particles.
5. A system is considered clean for oxygen service once all internal organic, inorganic, and particulates have been removed. Internal welding slag and spatter should be mitigated through the use of TIG welds on root welds.

Oxy-Fuel Cutting

Where utility drops are provided for the convenience of maintenance in a facility, consideration must be given to the chances of connecting pneumatic equipment to an oxygen drop. A pneumatic hammer that contains oil could explode into flames if connected to an oxygen supply. Therefore, the inadvertent connection must be prevented by signage and the use of different hose connections for the different uses.

Sample end connections for a utility drop:

- Natural gas drops equipped with an on-off valve which has a ½ in NPT male inlet and a Type 'C' 7/8-14 LH male thread. This will connect to a natural gas regulator which has the mating female thread. The outlet of the regulator will be a Type 'B' 9/16-18 LH male thread, for direct connection to a torch hose. These are known as CGA 025 connections.
- Oxygen drops equipped with an on-off valve which has a 1/2 in NPT male inlet and a Type 'C' 7/8-14 RH male thread. This will connect to an oxygen regulator which has the mating female thread. The outlet of the regulator will be a Type 'B' 9/16-18 RH male thread, for direct connection to a torch hose. These are known as CGA 024 connections.
- Compressed air drops equipped with a ball valve. The end connection will be a "twist-to-connect" connection with a locking pin.

Burning torches have the appropriate connecting nuts for attachment to the natural gas and oxygen regulators. The oxygen hoses are green and the nuts are RH threaded and are smooth. The fuel gas hoses are red, the nuts are LH threaded and have dimples on the corners.

It should be noted that where on-off valves are threaded, these safety measures can be defeated by a determined individual with a wrench. A typical utility drop is shown in Figure 16.1.

Oxygen systems should be pressure tested only with clean dry "instrument-quality" air or an inert gas like nitrogen. Compressed air will leave an oil residue, and water will cause corrosion and may contain chemicals that could react with the oxygen.

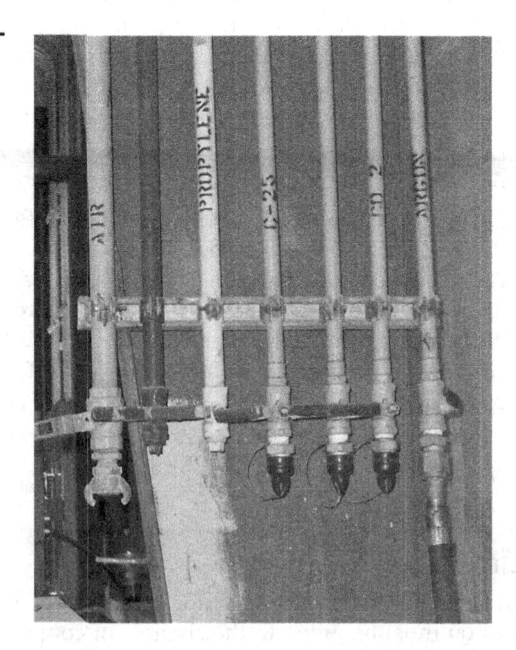

FIGURE 16.1 A typical utility drop with valves and identification of services.

Hydrogen

Hydrogen is the lightest of all gases. It is colorless, odorless, and tasteless, and burns with an almost invisible flame.

Hydrogen gas may be conveyed using carbon steel pipe, but it is advisable to use Schedule 80 pipe to further protect against physical damage. Because hydrogen gas can sometimes have a low operating temperature due to cryogenic production, thermal displacements must also be considered.

Hydraulics

Hydraulic systems are available in non-flammable water-glycol or petroleum based oils. The flammable oil systems must be fire-protected, so much effort can be saved by coordinating with suppliers of hydraulic systems to use non-flammable water glycol as an alternative. This is a discussion that must take place early in the bid phase of the equipment to permit the bidders to design their systems with the appropriate components and seals.

Hydraulic systems always operate at very high pressures, so one can imagine the problems associated with leaks; especially leaks of flammable fluids at high pressures near hot equipment. Hence the concerns over fire protection. Velocity fuse hoses are available that stop the flow of fluids should a hose fail. These have an internal valve that shuttles closed, stopping the flow when the fluid velocity exceeds the normal velocity, as would happen during a break in the hose. Pinhole leaks would not be sufficient to cause the internal valve to stop the flow however.

Cleanliness of hydraulic piping is paramount. Proprietary assembly systems are available to reduce the number of fittings by utilizing bends, and by clamping fittings like tees to the adjoining pipe to eliminate welds. Stick welding introduces slag to the inside of the pipe, so at a minimum, the root pass of any welded hydraulic pipe should be TIG welded.

In the past, contamination limits were set by NASA NAS 1638. This standard has been superseded by SAE AS4059 "Aerospace Fluid Power - Cleanliness Classification for Hydraulic Fluids."

Hydraulic piping can be supported with any support designed for high pressure piping, but some manufacturers have developed special supports that are configured for this application. These consist of plastic or aluminum blocks that grab the pipe around the circumference. The blocks are clamped to structural members by through bolts, and the blocks may also be stacked one atop another to permit the convenient routing of hydraulic piping without consuming a lot of space. These blocks are also useful for supporting high pressure descale lines in rolling mills. There are limits to the ability of these blocks to grip the piping axially, and the manufacturers have developed charts to describe the amount of force that can be withstood in the axial and radial directions. Sometimes dead legs may be welded to a long run to act as trunnions if the axial force is too great to be grabbed by a single block. In this way, the axial force is resolved into two radial forces that are more easily absorbed by the blocks. See Figure 16.2.

These blocks have the further advantage of having some, but not much, "give" to them. This permits some cushioning of dynamic loads, dampening vibrations.

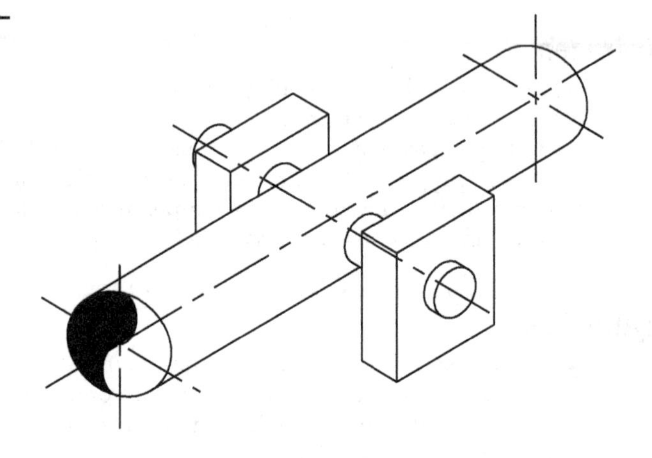
FIGURE 16.2 Trunnion support for a descale system.

Pigging

A pig (Figure 16.3) is a cylindrical or bullet-shaped plug that is inserted into a pipeline. Some pigs used for cleaning may be designed to scrape, brush, wipe, or dry the inside surface of the pipe. Other pigs are used to separate two different fluid slugs within the same pipe (batching). Still other pigs are used for inspecting the pipe and providing data about the wall thickness and corrosion of the line. Inspection techniques include magnetic flux leakage and ultrasonic inspection.

Pigs may be constructed of soft foam or a more rigid material. Soft pigs are able to negotiate a standard 1.5 D LR elbow, but rigid pigs may require 3D bends in order to pass. Sometimes transition bars will be welded inside a tee branch to ensure smooth passage of a pig through the run of a tee. Hard pigs may be cast as a plastic cylinder with a series of wiper blades. Other rigid pigs may be fabricated of steel, with brushes or plastic wipers. There are many styles depending on the application.

Spheres are also used as pigs. These are primarily used for meter testing in the oil and gas industry, but they may also be used for batching systems with tight radius bends.

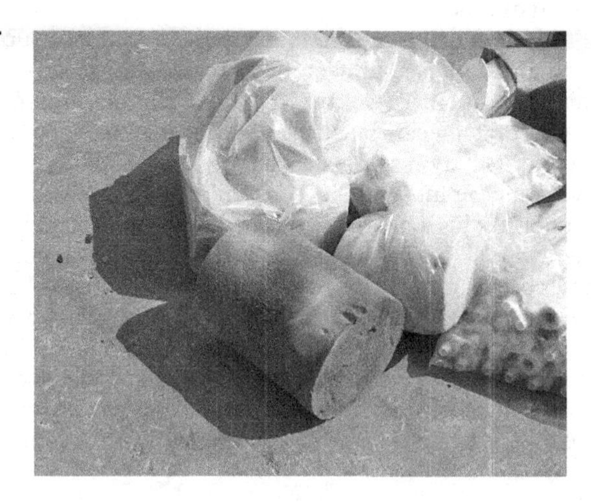
FIGURE 16.3 A cylindrical foam pig that has been used to wipe a chemical line.

CHAPTER 17

Infrastructure

Piping engineering seems like a mundane topic to the population-at-large. Laying out a piping system is certainly not as exciting as designing something like the Akashi-Kaikyo bridge in Japan, the Burj Dubai, or a subatomic particle accelerator. Public sentiment regarding piping design may be summed up as "out-of-sight, out-of-mind."

Yet the mechanical or chemical engineer who has driven around East Houston cannot help but be impressed with the maze of piping that controls the landscape. The sympathetic engineer's pulse quickens and the mind reels in appreciation of the sheer number of drawings that must have been required to produce such an array of piping.

The business of satisfying the world's thirst for energy is the business of extracting resources from underground, processing them into a usable product, and delivering them to the consumer.

But there is another more literal thirst that must be satisfied by every living creature: drinkable water. The demands placed on potable water resources have already led to legal actions, and some quarters feel that it does not stretch the imagination to think that future wars will be waged over water rights.

All the more reason to select materials and design systems to utilize the resources efficiently.

Infrastructure

Infrastructure generally refers to the network of engineering works that sustain activity in a modern society. This includes highways and the distribution of utilities. We will isolate our discussion to utilities other than electrical.

On a discrete level, this may be the engineering works required to provide utilities within the boundaries of a steel-making facility. On a more macroscopic level it may be the systems that provide water, steam, and sewage disposal to an entire region.

When one is starting with a clean sheet of paper, many desired features can be incorporated within the limits of schedule, budget, and the cleverness of the engineer or owner. A far greater challenge is dealing with existing infrastructure. Interferences with other utilities, settling of the supporting soil, and access for something as simple as inspection are some issues that the engineer must face when modifying or operating the aging infrastructure in most developed countries. Failure prediction is not an exact science as applied to utility piping, since so much of it is buried.

- A steam explosion in Manhattan on July 18, 2007 killed one person and injured dozens more. The crater near Grand Central Station swallowed a tow truck. The steam line was installed in 1924 and was determined to have failed due to a water hammer event triggered by a failed steam trap. At least 12 other steam pipe explosions have occurred in New York City since 1987.

- A 36 in cast iron water main (that was located directly under the recently re-paved Fort Duquesne Boulevard) failed in August 2005 in downtown Pittsburgh, Pennsylvania, flooding the basements of local buildings and parking garages. The line was 90 years old, and had been rehabilitated with a cement lining in 1986.

The rehabilitation of aging pipelines is obviously an issue that will only become increasingly important until these utility lines can be replaced in their entirety. It should be noted that the design life for modern industrial installations is usually set at 30 or 40 years. And even though many industrial installations are far older, few efficient systems are operating at 80 or 90 years old.

Diagnostic Tools

Aside from seeing a puddle forming on the road, or hearing or smelling a gas leak, there are some sophisticated diagnostic tools available to detect leaks.

Acoustical Leak Detectors

Sophisticated acoustical leak detectors are available at moderate prices to listen for the sound of fluids leaking from the pipe. Some are used as a sensitive wand that is walked over the route of the pipeline to listen for the leak. Others are connected to two end points of the pipeline. A correlator detects the sound of the leak at both ends of the pipe. The sound produced by the leak arrives at the nearer end first, and the time delay for the sound to arrive at the farther point is measured and entered into an algorithm to calculate the position of the leak.

Infrared Imaging

Infrared (thermal) scans can be made from the ground or air to check for leaks. High resolution images may be obtained through the use of low-level flights over a pipeline as it travels cross-country. This can even determine the relative amount of leakage so that repairs can be prioritized.

Where the exact position of a pipeline is not known, a higher-level flight can be flown. Depending on ground cover, the pipe bed may be visible on the thermal scan a little over half of the time.

Infrared imaging can be very sensitive, and can indicate leaks before they would otherwise be obvious as ground water. This technique is used for steam and water, and may be applied to other liquids as well.

Smoke and Dye Detection

Older municipalities often have combined storm and sewer systems. When it rains, it overwhelms the sewage treatment plants which causes flooding of raw sewage into rivers and streams. These municipalities are forced to eliminate storm water from the sanitary sewer systems.

Aside from large trunk lines that are interconnected below grade, another culprit is the connection of roof drains to the sanitary sewer system. Periodically the sewage authority will conduct smoke tests to determine if downspouts are connected to the sanitary sewer. They discharge a high-density smoke generator inside the sanitary sewer, and look around the neighborhood to observe smoke rising from downspouts. Offenders are warned to disconnect the downspouts from the sewer or face fines.

In some municipalities a dye test is required as part of a real estate transaction. A registered plumber is required to place a dye down any downspout that discharges below ground. The pipe is flooded with water, and a sanitary sewer manhole downstream is observed to see if there is any evidence of the dye. Observation of the dye fails the test, and the downspouts must be disconnected. If no dye is observed the plumber certifies that the downspouts are not connected to the sanitary sewer.

Disconnecting the downspouts is an easy task, but it should be noted that the downspouts should discharge onto splash blocks directed away from the foundation in order to prevent erosion and water penetration into the basement of structures.

Infiltration of ground water into the sanitary sewer is still another problem for sewage treatment plants. Older sewer lines are susceptible to leaks and infiltration, and trenchless technologies are sometimes employed to seal them.

Rehabilitation and Replacement of Pipelines

Aside from excavating and removing an existing line and then replacing it in total with a new pipe, there are technologies available to repair and replace lines without the expense of excavation. These are known as "trenchless technologies" and the lining of pipes in place is described as "in situ" relining.

Problems with Trenching

To begin with, any form of excavation requires knowing where other utilities are located. In the U.S., this requires a call to the national one call number 811. Canada also offers one-call response for utility identification.

Besides above ground markers for utility rights-of-way and buried plastic strips to warn excavators of the presence of underground utilities, RFID tags are now available to place on the piping to locate and uniquely identify the line with an asset number. These markers operate at a frequency between 65 and 135 kHz.

Trenchless technologies are suitable for replacement or rehabilitation of pipelines. But repairs generally require that a trench be opened for access. Many of these lines are far below grade and special precautions are required to ensure the safety of workers who enter these excavations. See Table 17.1.

Sloping of trenches is usually only available on greenfield sites. The trench slope is laid back at an angle shallower than the angle of repose of the soil. This often requires a wide right-of-way.

Cement Mortar Lining

Piping may be lined with cement to protect against corrosion, reduce leaks, increase flow, or extend the life of the pipe. A thin coating, between 1/8 inch and 3/16 inch (approximately 4 mm) is flung against the sides of the pipe and is sometimes even troweled smooth.

The result is a layer of mortar that creates a zone of alkalinity against the wall of the pipe, neutralizing the pH of slightly acidic fluids that would normally accelerate corrosion of ferrous metals.

Cement exudes calcium hydroxide in the presence of moisture, so that if shrinkage cracks occur during setting, the calcium hydroxide fills and seals the cracks. This phenomenon is known as "autogenous healing."

Cured-In-Place Pipe (CIPP) Relining

One method used to improve the interior surface of pipes and provide a corrosion-resistant interior is to apply an epoxy coating.[1] This method is used primarily for hot and cold water systems. The basics steps are:

[1] Refer to ASTM F1743 "Standard Practice for Rehabilitation of Existing Pipelines and Conduits by Pulled-in-Place Installation of Thermosetting Resin Pipe (CIPP)."

OSHA Standard 29 CFR	Description
1926.21(b)	Employee safety and health training
1926.651(a)	Protect employees with shoring or sloping
1926.651(b)	Locate the utilities; contact the utilities (one-call system)
1926.651(c)	Provide egress from the trench or excavation
1926.651(h)	Protect employees from water accumulation in trench
1926.651(j)	Spoil pile or excavated material must be placed at least two feet from edge of excavation
1926.651(k)	Inspect trench or excavation on a daily or as-needed basis

TABLE 17.1 Some OSHA standards for trenching.

1. Valves in the portion of the pipe to be relined are replaced with spool pieces.
2. A high velocity air stream with entrained abrasive particles is introduced into the pipe to clean corrosion and scale from the inside surfaces of the pipe. The removed corrosion and the abrasives are collected in a filter and additional air is circulated to remove as much of the dust fines as possible.
3. The pipe is heated and an epoxy coating is sprayed into the pipe and cured in the presence of conditioned air.
4. The removed valves are reinstalled after the lining cures. The epoxy cures within 24 hours.
5. A leak test is performed and water quality is tested.

This process has been applied to lines from 1/2 inch to 8 in, in lengths up to 1000 ft.

For potable, sanitary, storm, and process lines that require relining, other methods may be used. One involves the inversion inside the pipe of a felt tube that has been saturated with a thermosetting resin:

1. The tube is positioned at the end of the pipe.
2. Water pressure is applied to the outside of the tube, forcing it into the host pipe. This inverts the tube so that what was the outside surface of the tube is now in contact with the ID of the host pipe.
3. The tube continues to be unfurled inside the host pipe over the desired length.
4. The thermosetting resin is cured in place by circulating hot water throughout the length of the tube. The ends are cut and sealed and the pipe is returned to service.

Lateral branch connections are opened with robotically controlled cutters. The pipe may then be inspected with closed circuit television.

Pipe Bursting

Fracturable pipes made of clay, concrete, cast iron, and some plastics with diameters between 2 and 54 inches can be burst with a percussive tool that is pulled through the old pipe. A new flexible plastic pipe, usually HDPE, is dragged through the cavity that is left in the soil where the old pipe has been burst. Other methods use bladed cutting wheels to split the host pipe, and expanders to displace the original pipe fragments into the soil.

Sliplining

Sliplining is similar to pipe bursting in that a new pipe is dragged into the host pipe. The difference is that the host pipe remains intact. Slip lining can be used on nearly any type of host pipe, and the host pipe may even be left in service while the sliplining is taking place. The final result is a reduction in flow area, but the flow characteristics may actually be improved due to the improved flow coefficient of the liner. The void created between the liner and the host pipe may be filled with grout.

Sliplining is practical in sizes ranging from 8 to 96 in. Frequent elbows render the method impractical, and branch connections must of course be opened and sealed.

Other Trenchless Technologies

Pipe Ramming Pipe ramming is a useful installation method for placing steel pipes under rail beds and roadways. Distances up to 150 ft (60 m) and diameters as large as 60 in (1500 mm) can be accommodated, although longer and larger diameter pipes have been installed.

Pipe ramming requires a pit or other lateral access at the elevation at which the line is to be driven. Compressed air is used to drive the pipe percussively into the soil. Usually the leading edge of the pipe is left open, with the soil displaced by the pipe entering the inside of the pipe where it is later removed. The effort of driving the pipe through the soil may be reduced through the use of lubricants such as bentonite or polymers.

Impact Moles Impact moles are pneumatic devices that launch a tool below grade, forming a tunnel through which a new pipe may be inserted. Some moles are percussive and move at a slow rate, while others, especially those used by contractors to replace small bore utilities, are shot underground over a length of 50 ft or more from a single blast of air.

Energy Considerations

Centrifugal Machines

A common control strategy for pumping or compressing applications in industry is to relieve flow through a dump valve. The valve opens when the pressure rises above a preset limit and returns the fluid to a tank or reservoir. This prevents the centrifugal machine from operating below its recommended minimum flow. While this achieves the desired result, it wastes energy through circulating the fluid.

A better strategy is to control the centrifugal machine speed through the use of a Variable Frequency Drive. Because the power consumption of a centrifugal machine varies inversely as the cube root of the speed, tremendous energy savings can be realized with only slight variations in the speed. Another advantage is that the transition from low or no-flow to full flow and vice versa is smoother than simply opening a valve, and the effects of water hammer in pumping systems can be ameliorated.

The disadvantage is that another complex piece of equipment is required, but the cost of VFDs has come down since they first became commercially available. Attention must be paid to the location of the VFD in relation to drive motors operating at 480 V or more, since cable length can become a limiting factor due to the voltage rise in the cable that results from the pulses generated by the VFD.

The return on investment can be significant. In one project, a galvanizing facility reduced its operating cost by $1000 per week by installing a VFD on a single 250 HP blower.

Compressed Air Systems

Designers commonly oversize compressed air systems by 10 percent to take leakage rates into account. Leaks are notorious energy wasters and must be eliminated. Seal

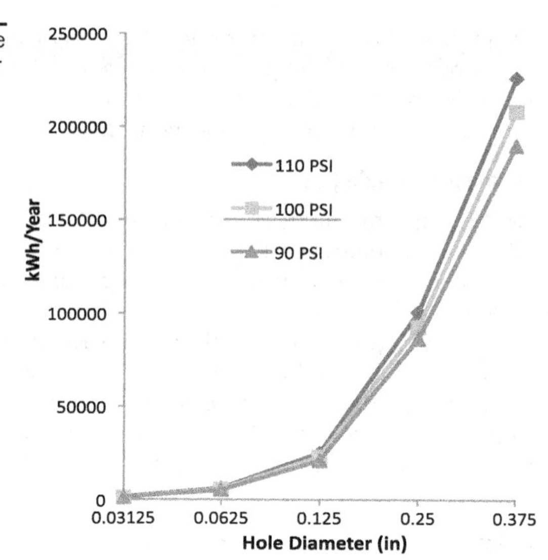

FIGURE 17.1 Energy lost due to leaks in a compressed air system. *Printed with permission "Handbook of Energy Management," published by Fairmont Press.*

welding threaded joints is a good practice, but so is the intelligent routing of air lines through accessible areas where they may be maintained. Lines routed underground or through trenches will never receive adequate maintenance.

The Association of Energy Engineers estimates that the energy lost through a hole that is 1/16 in diameter is 5800 kWh/year in a system th at operates at 100 psi. At a modest cost of $0.03/kWh, this equates to $174 per year. Refer to Figure 17.1 for losses at other conditions.

Water Conservation

Water shortages occur throughout the world due to lack of precipitation, overconsumption, misuse of resources, and pollution of fresh water reserves. In the United States, the traditional dry region was the southwest, which is fed by the relatively small Colorado River. In recent years, other areas of the country have suffered the effects of low water reservoirs. The phenomenon of urban sprawl throughout the Sunbelt has stretched the ability of the existing infrastructure to meet the demands of modern life in a developed country. Water restrictions are often implemented as a stop-gap measure of dealing with a diminished supply of fresh water.

Prior to a series of federal laws that were enacted to protect water resources, industries would routinely dump waste water into rivers and streams. Such pollution led to fires on the Cuyahoga River as early as 1936. The 1969 Cuyhoga River fire helped to spawn the Clean Water Act of 1977, which required industries to limit the amount of effluent discharged into bodies of water.

Waste water treatment is an expensive proposition, and industries realized that it was cheaper to conserve water than to treat it. As a result water efficiency in industry as a whole improved. While low-flow toilets and shower heads decrease water consumption on the household level, the greater effects on water consumption and surface water quality are realized on the industrial level.

Other opportunities exist for water conservation, and engineers designing new utility systems look to material less prone to leakage, while water utilities work to rehabilitate aging infrastructure that used leadite sealed joints on cast iron lines.

CHAPTER 18

Strategies for Remote Locations

During a visit to Saint Vincent College a few years ago, I ran into one of my calculus professors who I knew was involved in charity work in Haiti. He invited me into his office and I asked about his work there. Out came an engineering report that was prepared by a team of Cuban engineers to address a water distribution problem. The good professor optimistically said "Maybe you'd like to look at this." Then upon reflection, he lowered it and said cautiously, "But this is probably more civil engineering than mechanical."

I explained that I had lots of experience moving lots of water in lots of places, and I would be happy to review the report. The report was remarkable for the level of detail it contained. The conclusions were very good, but I did take exception to some of the recommendations.

First, they wanted to use a diesel motor to run an electric generator, to run an electric pump. This would occur near a water source, and my concern was that diesel spills would contaminate the source, the cost of fuel would get expensive, and there would in general be quite a lot of pollution. I decided to investigate solar powered pumps.

Second, they wanted to bury galvanized threaded pipe to route the piping from the source in the valley up to the top of a mountain where the road crested at a restaurant. While I knew that the galvanized pipe would assemble rapidly, I also knew that it would restrict flow, eventually corrode, and would likely contain trace lead from the galvanizing process. I suggested HDPE, with fewer fittings and field-applied fusion sleeves. The cost would also be low, it would still be easy to assemble, and the HDPE would last indefinitely.

Third, I also noticed that the lengths and elevations had been estimated by the Cubans. We would need accurate elevations and distances to size the equipment. It turned out that my brother had a hand-held GPS unit that provided elevation, so a few months after I submitted my review of the original report, my professor and I, along with my brother (who yearns for excitement), left with a small group of people bound for Port au Prince, and other parts unknown.

The result was a week of camaraderie, sight-seeing, minor repairs to the hostel in which we stayed, miserable digestive issues, errant late-night gunfire that was far too close for comfort, and the appreciation of having clean, potable water at the tap when you want it back home. We also performed the survey and I designed the system.

It was not installed as designed however. The Owner decided to purchase a diesel generator instead--which proves that you can lead a client to water, but you cannot make him use the best technology.

Remote or undeveloped regions do not contain the infrastructure to conveniently deliver potable water. Water may still be delivered in such regions from suitable sources such as springs or wells, however, alternate technologies are often required to provide the motive force for pumping applications.

Such motive force may derive its energy from carbon-based fuels and engines, solar power, kinetic energy, or even manpower. The quantities of water to be moved are generally small in such cases, and this is fortunate, since large volumes require large power sources[1]. The pressure required is often high though, and this requires low-flow, high-head pumps, which forces the design toward positive displacement pumps.

HDPE is a good choice for long runs of underground piping if it can be delivered to the site. HDPE is impervious to corrosion, has a low friction factor, and can be purchased in long lengths on spools. This minimizes the number of field joints, and these can be made using thermofusion sleeves that are normally used for repairs.

Field surveys can be made with hand-held GPS devices, which can provide data for elevation above sea level as well as northing and easting coordinates. This forms the basis of pressure drop calculations used to size the pumps and determine power requirements.

Motive Power Technologies

Solar Power

Use of solar panels is limited to areas that receive lots of sunshine for many hours. In the continental United States, the combined solar incidence and current panel efficiencies yield a power density of approximately 0.015 HP/ft^2 (120 W/m^2), so the application of solar panels is further limited to relatively low power requirements. Still, they are an economical and environmentally sound choice for many remote projects. See Figure 18.1.

Hydraulic Ram Pump

The hydraulic ram pump, also known as a Bélier hydraulic pump, was invented in 1796 by Joseph Michel Montgolfier[2]. The design and construction are simple, although their practical use is limited to remote areas that lack electricity. They also require a significant change in elevation of the water source.

The hydraulic ram pump consists of a pressure chamber filled with air and a spring or weight loaded check valve known as a "waste" or "clack" valve from the sound it makes during operation. A delivery check valve keeps the pumped water column from falling back through the device.

A source of water at a higher elevation is required, as at a waterfall or pond. The waste valve is initially open and as the water falls from a higher elevation, the kinetic energy forces the weighted waste valve closed. This causes a water hammer that forces some of the falling water through the delivery check valve and out of the pump to its point of use. The pressure chamber acts as a shock absorber. See Figure 18.2.

The hydraulic ram pump acts essentially as a positive displacement pump, with spurts of water delivered in cycles rather than a continuous flow. Experiments by the University of Warwick indicate that the pressure chamber should have a minimum

[1] Some clever sources of man-made power include merry-go-rounds and see-saws. As children play in a park, they are actually pumping water to a storage tank. This technology has been successfully applied in Gaviotas, Colombia.

[2] Of the flying Montgolfier brothers, inventors of the hot air balloon.

volume of 20 times the expected flow rate per cycle, with the flow per cycle depending on the diameter of the inlet pipe. The pressure chamber should include a bladder or else the entrapped air will dissolve into the water. An expansion tank may be used, but in remote applications, a partially inflated bicycle tire jammed into the tank (which is really nothing more than a larger diameter pipe) may be used.

The inlet pipe should be rigid, although good results have been reported with PVC pipe. This inlet pipe from the source to the pump is also known as a drive pipe, and the optimum length to diameter ratio appears to be between 150 and 100, based on studies performed by N. G. Calvert. The delivery of the pump depends on the fall from the source, the flow available from the source, and the height of the delivery point above the pump. A common equation for use in estimating the delivery is:

$$0.6 \times Q \times F/E = D \qquad \text{(Equation 18.1)}$$

where

Q = Source flow in gallons per minute

F = Fall in feet from the water source to the pump

E = Elevation from the pump to the water outlet

D = Capacity of the pump in gallons per minute.

The 0.6 is an efficiency factor and will differ among various ram pumps. Note that much water is "wasted" through the waste check valve, in the sense that it is not pumped. It often returns to the natural stream from which it came though, so in that sense the water is conserved.

Water Treatment

Once the water is delivered to a tank or cistern, some minimal degree of water treatment should be considered. These will be customized to the area. For example, Guinea Worm Disease may be avoided simply through water filtration. This disease, once prevalent throughout Africa and Asia, remains endemic only in five countries in sub-Saharan Africa.

Water-borne diseases are caused by:

- Protozoa
- Bacteria
- Viruses
- Parasites.

Boiling

Boiling is considered to be the best method of treating water to make it potable, but this is considered to be a "point-of-use" treatment, and would not be appropriate to a remote piping system. Some pathogens that cause botulism can survive at temperatures up to 244°F (118°C) so boiling at pressures higher than atmospheric would be required.

Filtration

Bacteria, parasites, and protozoa can be filtered successfully, but viruses are too small to be filtered. This requires chemical additives or ultraviolet disinfection after filtration. Chemicals may include chlorine, iodine, chlorine dioxide, or sodium hypochlorite

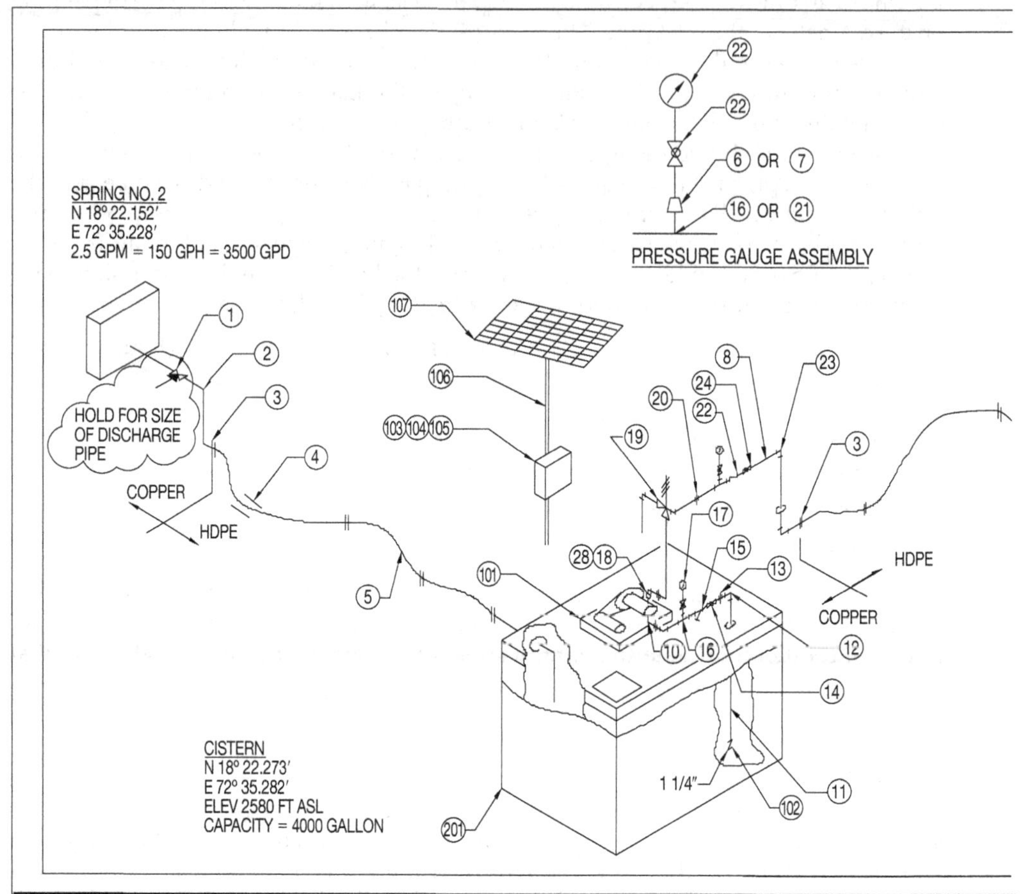

Figure 18.1 A remote solar-powered water pumping project for a village in Haiti. *Courtesy of Innovative Design Engineering of America, LLC.*

(bleach). Care must be taken with filters to prevent the filter media from becoming a colonization site for the microbes. Bonding silver particles to the media decreases the growth of these microbes in the filters. This is due to the germicidal properties of many metals.

Chemical Disinfection

Iodine kills many pathogens, and is available in liquid or solid form. Ascorbic acid is frequently added to water treated with iodine to help remove the iodine taste. Chlorine bleach can also be used as a disinfectant. Two drops of 5% sodium hypochlorite is added per quart (liter) of water and left covered for an hour. While iodine and chlorine are each effective against Giardia, neither is sufficient to kill Cryptosporidia.

Silver ion/chlorine dioxide is effective against Cryptosporidia and Giardia, but the disinfection time is up to four hours.

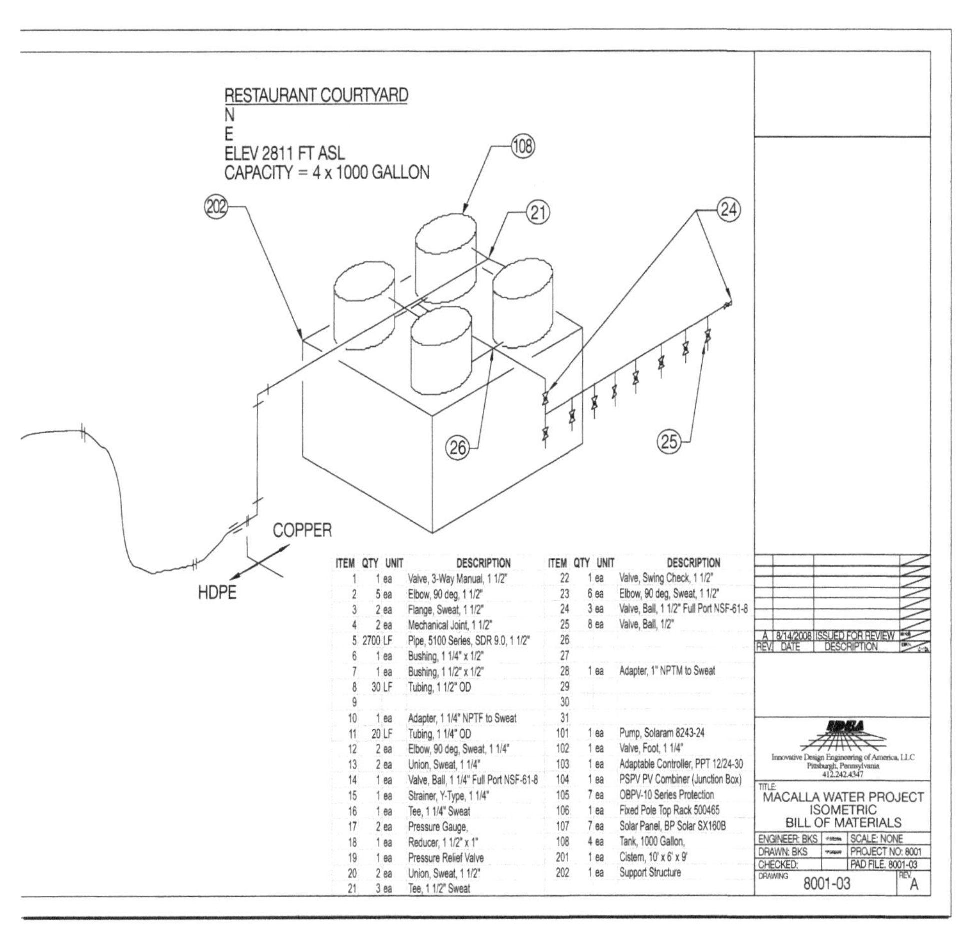

ITEM	QTY	UNIT	DESCRIPTION	ITEM	QTY	UNIT	DESCRIPTION
1	1	ea	Valve, 3-Way Manual, 1 1/2"	22	1	ea	Valve, Swing Check, 1 1/2"
2	5	ea	Elbow, 90 deg, 1 1/2"	23	6	ea	Elbow, 90 deg, Sweat, 1 1/2"
3	2	ea	Flange, Sweat, 1 1/2"	24	3	ea	Valve, Ball, 1 1/2" Full Port NSF-61-8
4	2	ea	Mechanical Joint, 1 1/2"	25	8	ea	Valve, Ball, 1/2"
5	2700	LF	Pipe, 5100 Series, SDR 9.0, 1 1/2"	26			
6	1	ea	Bushing, 1 1/4" x 1/2"	27			
7	1	ea	Bushing, 1 1/2" x 1/2"	28	1	ea	Adapter, 1" NPTM to Sweat
8	30	LF	Tubing, 1 1/2" OD	29			
9				30			
10	1	ea	Adapter, 1 1/4" NPTF to Sweat	31			
11	20	LF	Tubing, 1 1/4" OD	101	1	ea	Pump, Solaram 8243-24
12	2	ea	Elbow, 90 deg, Sweat, 1 1/4"	102	1	ea	Valve, Foot, 1 1/4"
13	2	ea	Union, Sweat, 1 1/4"	103	1	ea	Adaptable Controller, PPT 12/24-30
14	1	ea	Valve, Ball, 1 1/4" Full Port NSF-61-8	104	1	ea	PSPV PV Combiner (Junction Box)
15	1	ea	Strainer, Y-Type, 1 1/4"	105	7	ea	OBPV-10 Series Protection
16	1	ea	Tee, 1 1/4" Sweat	106	1	ea	Fixed Pole Top Rack 500465
17	2	ea	Pressure Gauge,	107	7	ea	Solar Panel, BP Solar SX160B
18	1	ea	Reducer, 1 1/2" x 1"	108	4	ea	Tank, 1000 Gallon,
19	1	ea	Pressure Relief Valve	201	1	ea	Cistern, 10' x 6' x 9'
20	2	ea	Union, Sweat, 1 1/2"	202	1	ea	Support Structure
21	3	ea	Tee, 1 1/2" Sweat				

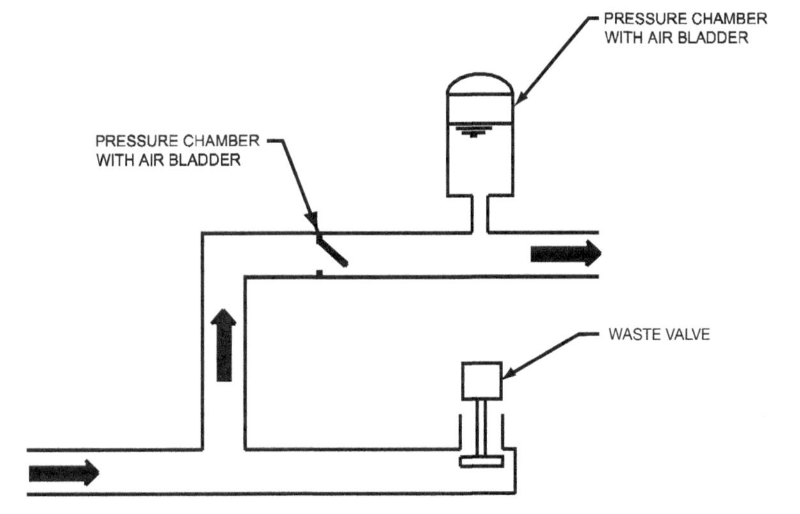

FIGURE 18.2 A hydraulic ram pump.

Ultraviolet Radiation

Ultraviolet radiation destroys DNA and this prevents microbes from reproducing. While this technology may seem inappropriate to remote locations, it is feasible to apply it as a batch method of treatment using a technique known as "solar disinfection." Water is shaken in a transparent bottle to oxygenate it, and is left in full sunlight for six hours. This raises the temperature of the water and exposes the microbes to UVA radiation.

Appendices

Appendix 1—Carbon Steel Pipe Schedule

Appeddix 2—PVC Pipe Schedule

Appendix 3—Copper Tubing Schedule

Appendix 4—Material Properties

Appendix 5—NEMA Enclosures

Appendix 6—IP Codes

Appendix 7—Steam Tables (English Units)

Appendix 8—Steam Tables (SI Units)

Appendix 9—Friction Losses

Nominal Pipe Size	DN	Outside Diameter	Schedule	Wall Thickness	Wall Thickness less Mill Tolerance	Inside Diameter	Inside Diameter	Flow Area	Flow Area	Metal Area
in	mm	in		in	in	in	ft	in²	ft²	in²
1/8	6	0.405	10S	0.049	0.043	0.307	0.026	0.074	0.00051	0.055
			40/Std	0.068	0.060	0.269	0.022	0.057	0.00039	0.072
			80/XS	0.095	0.083	0.215	0.018	0.036	0.00025	0.093
1/4	8	0.540	10S	0.065	0.057	0.410	0.034	0.132	0.00092	0.097
			40/Std	0.088	0.077	0.364	0.030	0.104	0.00072	0.125
			80/XS	0.119	0.104	0.302	0.025	0.072	0.00050	0.157
3/8	10	0.675	10S	0.065	0.057	0.545	0.045	0.233	0.00162	0.125
			40/Std	0.091	0.080	0.493	0.041	0.191	0.00133	0.167
			80/XS	0.126	0.110	0.423	0.035	0.141	0.00098	0.217
1/2	15	0.840	5S	0.065	0.057	0.710	0.059	0.396	0.00275	0.158
			10S	0.083	0.073	0.674	0.056	0.357	0.00248	0.197
			40/Std	0.109	0.095	0.622	0.052	0.304	0.00211	0.250
			80/XS	0.147	0.129	0.546	0.046	0.234	0.00163	0.320
			160	0.188	0.165	0.464	0.039	0.169	0.00117	0.385
			XXS	0.294	0.257	0.252	0.021	0.050	0.00035	0.504
3/4	20	1.050	5S	0.065	0.057	0.920	0.077	0.665	0.00462	0.201
			10S	0.083	0.073	0.884	0.074	0.614	0.00426	0.252
			40/Std	0.113	0.099	0.824	0.069	0.533	0.00370	0.333
			80/XS	0.154	0.135	0.742	0.062	0.432	0.00300	0.433
			160	0.219	0.192	0.612	0.051	0.294	0.00204	0.572
			XXS	0.308	0.270	0.434	0.036	0.148	0.00103	0.718
1	25	1.315	5S	0.065	0.057	1.185	0.099	1.103	0.00766	0.255
			10S	0.109	0.095	1.097	0.091	0.945	0.00656	0.413
			40/Std	0.133	0.116	1.049	0.087	0.864	0.00600	0.494
			80/XS	0.179	0.157	0.957	0.080	0.719	0.00500	0.639
			160	0.250	0.219	0.815	0.068	0.522	0.00362	0.836
			XXS	0.358	0.313	0.599	0.050	0.282	0.00196	1.076
1 1/4	32	1.660	5S	0.065	0.057	1.530	0.128	1.839	0.01277	0.326
			10S	0.109	0.095	1.442	0.120	1.633	0.01134	0.531
			40/Std	0.140	0.123	1.380	0.115	1.496	0.01039	0.669
			80/XS	0.191	0.167	1.278	0.107	1.283	0.00891	0.881
			160	0.250	0.219	1.160	0.097	1.057	0.00734	1.107
			XXS	0.382	0.334	0.896	0.075	0.631	0.00438	1.534
1 1/2	40	1.900	5S	0.065	0.057	1.770	0.148	2.461	0.01709	0.375
			10S	0.109	0.095	1.682	0.140	2.222	0.01543	0.613
			40/Std	0.145	0.127	1.610	0.134	2.036	0.01414	0.799
			80/XS	0.200	0.175	1.500	0.125	1.767	0.01227	1.068
			160	0.281	0.246	1.338	0.112	1.406	0.00976	1.429
			XXS	0.400	0.350	1.100	0.092	0.950	0.00660	1.885
2	50	2.375	5S	0.065	0.057	2.245	0.187	3.958	0.02749	0.472
			10S	0.109	0.095	2.157	0.180	3.654	0.02538	0.776
			40/Std	0.154	0.135	2.067	0.172	3.356	0.02330	1.075
			80/XS	0.218	0.191	1.939	0.162	2.953	0.02051	1.477
			160	0.344	0.301	1.687	0.141	2.235	0.01552	2.195

Nominal Pipe Size	DN	Outside Diameter	Schedule	Wall Thickness	Wall Thickness less Mill Tolerance	Inside Diameter	Inside Diameter	Flow Area	Flow Area	Metal Area
in	mm	in		in	in	in	ft	in²	ft²	in²
			XXS	0.436	0.382	1.503	0.125	1.774	0.01232	2.656
2 1/2	65	2.875	5S	0.083	0.073	2.709	0.226	5.764	0.04003	0.728
			10S	0.120	0.105	2.635	0.220	5.453	0.03787	1.039
			40/Std	0.203	0.178	2.469	0.206	4.788	0.03325	1.704
			80/XS	0.276	0.242	2.323	0.194	4.238	0.02943	2.254
			160	0.375	0.328	2.125	0.177	3.547	0.02463	2.945
			XXS	0.552	0.483	1.771	0.148	2.463	0.01711	4.028
3	80	3.500	5S	0.083	0.073	3.334	0.278	8.730	0.06063	0.891
			10S	0.120	0.105	3.260	0.272	8.347	0.05796	1.274
			40/Std	0.216	0.189	3.068	0.256	7.393	0.05134	2.228
			80/XS	0.300	0.263	2.900	0.242	6.605	0.04587	3.016
			160	0.438	0.383	2.624	0.219	5.408	0.03755	4.213
			XXS	0.600	0.525	2.300	0.192	4.155	0.02885	5.466
3 1/2	90	4.000	5S	0.083	0.073	3.834	0.320	11.545	0.08017	1.021
			10S	0.120	0.105	3.760	0.313	11.104	0.07711	1.463
			40/Std	0.226	0.198	3.548	0.296	9.887	0.06866	2.680
			80/XS	0.318	0.278	3.364	0.280	8.888	0.06172	3.678
4	100	4.500	5S	0.083	0.073	4.334	0.361	14.753	0.10245	1.152
			10S	0.120	0.105	4.260	0.355	14.253	0.09898	1.651
			40/Std	0.237	0.207	4.026	0.336	12.730	0.08840	3.174
			80/XS	0.337	0.295	3.826	0.319	11.497	0.07984	4.407
			120	0.438	0.383	3.624	0.302	10.315	0.07163	5.589
			160	0.531	0.465	3.438	0.287	9.283	0.06447	6.621
			XXS	0.674	0.590	3.152	0.263	7.803	0.05419	8.101
5	125	5.563	5S	0.109	0.095	5.345	0.445	22.438	0.15582	1.868
			10S	0.134	0.117	5.295	0.441	22.020	0.15292	2.285
			40/Std	0.258	0.226	5.047	0.421	20.006	0.13893	4.300
			80/XS	0.375	0.328	4.813	0.401	18.194	0.12635	6.112
			120	0.500	0.438	4.563	0.380	16.353	0.11356	7.953
			160	0.627	0.549	4.309	0.359	14.583	0.10127	9.723
			XXS	0.750	0.656	4.063	0.339	12.965	0.09004	11.340
6	150	6.625	5S	0.109	0.095	6.407	0.534	32.240	0.22389	2.231
			10S	0.134	0.117	6.357	0.530	31.739	0.22041	2.733
			40/Std	0.280	0.245	6.065	0.505	28.890	0.20063	5.581
			80/XS	0.432	0.378	5.761	0.480	26.067	0.18102	8.405
			120	0.562	0.492	5.501	0.458	23.767	0.16505	10.705
			160	0.719	0.629	5.187	0.432	21.131	0.14674	13.340
			XXS	0.864	0.756	4.897	0.408	18.834	0.13079	15.637
8	200	8.625	5S	0.109	0.095	8.407	0.701	55.510	0.38549	2.916
			10S	0.148	0.130	8.329	0.694	54.485	0.37837	3.941
			20	0.250	0.219	8.125	0.677	51.849	0.36006	6.578
			30	0.277	0.242	8.071	0.673	51.162	0.35529	7.265
			40/Std	0.322	0.282	7.981	0.665	50.027	0.34741	8.399
			60	0.406	0.355	7.813	0.651	47.943	0.33294	10.483

Nominal Pipe Size	DN	Outside Diameter	Schedule	Wall Thickness	Wall Thickness less Mill Tolerance	Inside Diameter	Inside Diameter	Flow Area	Flow Area	Metal Area
in	mm	in		in	in	in	ft	in²	ft²	in²
			80/XS	0.500	0.438	7.625	0.635	45.664	0.31711	12.763
			100	0.594	0.520	7.437	0.620	43.440	0.30166	14.987
			120	0.719	0.629	7.187	0.599	40.568	0.28172	17.858
			140	0.812	0.711	7.001	0.583	38.495	0.26733	19.931
			XXS	0.875	0.766	6.875	0.573	37.122	0.25779	21.304
			160	0.906	0.793	6.813	0.568	36.456	0.25317	21.970
10	250	10.750	5S	0.134	0.117	10.482	0.874	86.293	0.59926	4.469
			10S	0.165	0.144	10.420	0.868	85.276	0.59219	5.487
			20	0.250	0.219	10.250	0.854	82.516	0.57303	8.247
			30	0.307	0.269	10.136	0.845	80.691	0.56035	10.072
			40	0.365	0.319	10.020	0.835	78.854	0.54760	11.908
			Std	0.365	0.319	10.020	0.835	78.854	0.54760	11.908
			60/XS	0.500	0.438	9.750	0.813	74.662	0.51849	16.101
			80	0.594	0.520	9.562	0.797	71.810	0.49868	18.952
			100	0.719	0.629	9.312	0.776	68.104	0.47295	22.658
			120	0.844	0.739	9.062	0.755	64.497	0.44789	26.266
			140/XXS	1.000	0.875	8.750	0.729	60.132	0.41758	30.631
			160	1.125	0.984	8.500	0.708	56.745	0.39406	34.018
12	300	12.750	5S	0.156	0.137	12.438	1.037	121.504	0.84378	6.172
			10S	0.180	0.158	12.390	1.033	120.568	0.83728	7.108
			20	0.250	0.219	12.250	1.021	117.859	0.81846	9.817
			30	0.330	0.289	12.090	1.008	114.800	0.79722	12.876
			Std	0.375	0.328	12.000	1.000	113.097	0.78540	14.579
			40	0.406	0.355	11.938	0.995	111.932	0.77730	15.745
			XS	0.500	0.438	11.750	0.979	108.434	0.75301	19.242
			60	0.562	0.492	11.626	0.969	106.157	0.73720	21.519
			80	0.688	0.602	11.374	0.948	101.605	0.70559	26.071
			100	0.844	0.739	11.062	0.922	96.107	0.66741	31.569
			120/XXS	1.000	0.875	10.750	0.896	90.762	0.63030	36.914
			140	1.125	0.984	10.500	0.875	86.590	0.60132	41.086
			160	1.312	1.148	10.126	0.844	80.531	0.55925	47.145
14	350	14.000	5S	0.156	0.137	13.688	1.141	147.153	1.02190	6.785
			10S	0.250	0.219	13.500	1.125	143.139	0.99402	10.799
			20	0.312	0.273	13.376	1.115	140.521	0.97584	13.417
			30/Std	0.375	0.328	13.250	1.104	137.886	0.95754	16.052
			40	0.438	0.383	13.124	1.094	135.276	0.93942	18.662
			XS	0.500	0.438	13.000	1.083	132.732	0.92175	21.206
			60	0.594	0.520	12.812	1.068	128.921	0.89528	25.017
			80	0.750	0.656	12.500	1.042	122.718	0.85221	31.220
			100	0.938	0.821	12.124	1.010	115.447	0.80171	38.491
			120	1.094	0.957	11.812	0.984	109.581	0.76098	44.357
			140	1.250	1.094	11.500	0.958	103.869	0.72131	50.069
			160	1.406	1.230	11.188	0.932	98.309	0.68270	55.629
16	400	16.000	5S	0.165	0.144	15.670	1.306	192.853	1.33926	8.208

Nominal Pipe Size	DN	Outside Diameter	Schedule	Wall Thickness	Wall Thickness less Mill Tolerance	Inside Diameter	Inside Diameter	Flow Area	Flow Area	Metal Area
in	mm	in		in	in	in	ft	in²	ft²	in²
			10S	0.250	0.219	15.500	1.292	188.692	1.31036	12.370
			20	0.312	0.273	15.376	1.281	185.685	1.28948	15.377
			30/Std	0.375	0.328	15.250	1.271	182.654	1.26843	18.408
			40/XS	0.500	0.438	15.000	1.250	176.714	1.22718	24.347
			60	0.656	0.574	14.688	1.224	169.440	1.17666	31.622
			80	0.844	0.739	14.312	1.193	160.876	1.11719	40.186
			100	1.031	0.902	13.938	1.162	152.577	1.05957	48.484
			120	1.219	1.067	13.562	1.130	144.456	1.00317	56.605
			140	1.438	1.258	13.124	1.094	135.276	0.93942	65.785
			160	1.594	1.395	12.812	1.068	128.921	0.89528	72.141
18	450	18.000	5S	0.165	0.144	17.670	1.473	245.224	1.70294	9.245
			10S	0.250	0.219	17.500	1.458	240.528	1.67033	13.941
			20	0.312	0.273	17.376	1.448	237.131	1.64675	17.337
			Std	0.375	0.328	17.250	1.438	233.705	1.62295	20.764
			30	0.438	0.383	17.124	1.427	230.303	1.59933	24.166
			XS	0.500	0.438	17.000	1.417	226.980	1.57625	27.489
			40	0.562	0.492	16.876	1.406	223.681	1.55334	30.788
			60	0.750	0.656	16.500	1.375	213.824	1.48489	40.644
			80	0.938	0.821	16.124	1.344	204.190	1.41799	50.278
			100	1.156	1.012	15.688	1.307	193.297	1.34234	61.172
			120	1.375	1.203	15.250	1.271	182.654	1.26843	71.815
			140	1.562	1.367	14.876	1.240	173.805	1.20698	80.664
			160	1.781	1.558	14.438	1.203	163.721	1.13695	90.748
20	500	20.000	5S	0.188	0.165	19.624	1.635	302.458	2.10040	11.701
			10S	0.250	0.219	19.500	1.625	298.647	2.07394	15.512
			20/Std	0.375	0.328	19.250	1.604	291.039	2.02110	23.120
			30/XS	0.500	0.438	19.000	1.583	283.528	1.96895	30.631
			40	0.594	0.520	18.812	1.568	277.945	1.93018	36.214
			60	0.812	0.711	18.376	1.531	265.211	1.84174	48.948
			80	1.031	0.902	17.938	1.495	252.719	1.75499	61.440
			100	1.281	1.121	17.438	1.453	238.827	1.65852	75.332
			120	1.500	1.313	17.000	1.417	226.980	1.57625	87.179
			140	1.750	1.531	16.500	1.375	213.824	1.48489	100.335
			160	1.969	1.723	16.062	1.339	202.623	1.40710	111.536
22	550	22.000	5S	0.188	0.165	21.624	1.802	367.250	2.55035	12.883
			10S	0.250	0.219	21.500	1.792	363.050	2.52118	17.082
			20/Std	0.375	0.328	21.250	1.771	354.656	2.46289	25.476
			30/XS	0.500	0.438	21.000	1.750	346.360	2.40528	33.772
			60	0.875	0.766	20.250	1.688	322.062	2.23654	58.070
			80	1.125	0.984	19.750	1.646	306.354	2.12746	73.778
			100	1.375	1.203	19.250	1.604	291.039	2.02110	89.094
			120	1.625	1.422	18.750	1.563	276.116	1.91747	104.016
			140	1.875	1.641	18.250	1.521	261.586	1.81657	118.546
			160	2.125	1.859	17.750	1.479	247.449	1.71840	132.683

Nominal Pipe Size	DN	Outside Diameter	Schedule	Wall Thickness	Wall Thickness less Mill Tolerance	Inside Diameter	Inside Diameter	Flow Area	Flow Area	Metal Area
in	mm	in		in	in	in	ft	in²	ft²	in²
24	600	24.000	5S	0.218	0.191	23.564	1.964	436.101	3.02848	16.287
			10S	0.250	0.219	23.500	1.958	433.736	3.01205	18.653
			20/Std	0.375	0.328	23.250	1.938	424.556	2.94831	27.833
			XS	0.500	0.438	23.000	1.917	415.475	2.88524	36.914
			30	0.562	0.492	22.876	1.906	411.007	2.85422	41.382
			40	0.688	0.602	22.624	1.885	402.002	2.79168	50.387
			60	0.969	0.848	22.062	1.839	382.278	2.65471	70.111
			80	1.218	1.066	21.564	1.797	365.215	2.53621	87.174
			100	1.531	1.340	20.938	1.745	344.318	2.39110	108.071
			120	1.812	1.586	20.376	1.698	326.082	2.26446	126.307
			140	2.062	1.804	19.876	1.656	310.276	2.15469	142.113
			160	2.344	2.051	19.312	1.609	292.917	2.03414	159.472
26	650	26.000	10	0.312	0.273	25.376	2.115	505.750	3.51215	25.179
			Std	0.375	0.328	25.250	2.104	500.740	3.47736	30.189
			20/XS	0.500	0.438	25.000	2.083	490.873	3.40884	40.055
28	700	28.000	10	0.312	0.273	27.376	2.281	588.613	4.08759	27.139
			Std	0.375	0.328	27.250	2.271	583.207	4.05005	32.545
			20/XS	0.500	0.438	27.000	2.250	572.555	3.97607	43.197
			30	0.625	0.547	26.750	2.229	562.001	3.90278	53.751
30	750	30.000	5S	0.250	0.219	29.500	2.458	683.492	4.74647	23.366
			10S	0.312	0.273	29.376	2.448	677.758	4.70665	29.099
			Std	0.375	0.328	29.250	2.438	671.957	4.66637	34.901
			20/XS	0.500	0.438	29.000	2.417	660.519	4.58694	46.338
			30	0.625	0.547	28.750	2.396	649.180	4.50820	57.678
32	800	32.000	10S	0.312	0.273	31.376	2.615	773.187	5.36936	31.060
			Std	0.375	0.328	31.250	2.604	766.990	5.32632	37.257
			20/XS	0.500	0.438	31.000	2.583	754.767	5.24144	49.480
			30	0.625	0.547	30.750	2.563	742.642	5.15724	61.605
			40	0.688	0.602	30.624	2.552	736.569	5.11506	67.678
34	850	34.000	10	0.312	0.273	33.376	2.781	874.899	6.07569	33.020
			Std	0.375	0.328	33.250	2.771	868.306	6.02990	39.613
			20/XS	0.500	0.438	33.000	2.750	855.298	5.93957	52.622
			30	0.625	0.547	32.750	2.729	842.388	5.84992	65.532
			40	0.688	0.602	32.624	2.719	835.918	5.80499	72.001
36	900	36.000	20	0.312	0.273	35.376	2.948	982.895	6.82566	34.981
			Std	0.375	0.328	35.250	2.938	975.905	6.77712	41.970
			XS	0.500	0.438	35.000	2.917	962.112	6.68133	55.763
			30	0.625	0.547	34.750	2.896	948.417	6.58623	69.459
			40	0.750	0.656	34.500	2.875	934.819	6.49180	83.056
42	1100	42.000	20	0.375	0.328	41.250	3.438	1336.403	9.28058	49.038
			XS	0.500	0.438	41.000	3.417	1320.253	9.16842	65.188

Appendix 2

PVC Pipe Schedules

Nominal Pipe Size	DN	Outside Diameter	Schedule	Min Wall Thickness	Avg Inside Diameter	Flow Area	Flow Area	Nom. Wt./Ft.	Max. Working Pressure at 73°F
in	mm	in		in	in	in^2	ft^2	lb/ft	PSI
1/8	6	0.405	40	0.068	0.249	0.049	0.000338	0.051	810
		0.405	80	0.095	0.195	0.030	0.000207	0.063	1230
1/4	8	0.540	40	0.088	0.344	0.093	0.000645	0.086	780
		0.540	80	0.119	0.282	0.062	0.000434	0.105	1130
3/8	10	0.675	40	0.091	0.473	0.176	0.001220	0.115	620
		0.675	80	0.126	0.403	0.128	0.000886	0.146	920
1/2	15	0.840	40	0.109	0.602	0.285	0.001977	0.170	600
		0.840	80	0.147	0.526	0.217	0.001509	0.213	850
		0.840	120	0.170	0.480	0.181	0.001257	0.236	1010
3/4	20	1.050	40	0.113	0.804	0.508	0.003526	0.226	480
		1.050	80	0.154	0.722	0.409	0.002843	0.289	690
		1.050	120	0.170	0.690	0.374	0.002597	0.311	770
1	25	1.315	40	0.133	1.029	0.832	0.005775	0.333	450
		1.315	80	0.179	0.936	0.688	0.004778	0.424	630
		1.315	120	0.200	0.891	0.624	0.004330	0.464	720
1 1/4	32	1.660	40	0.140	1.360	1.453	0.010088	0.450	370
		1.660	80	0.191	1.255	1.237	0.008590	0.586	520
		1.660	120	0.215	1.204	1.139	0.007906	0.649	600
1 1/2	40	1.900	40	0.145	1.590	1.986	0.013789	0.537	330
		1.900	80	0.200	1.476	1.711	0.011882	0.711	470
		1.900	120	0.225	1.423	1.590	0.011044	0.787	540
2	50	2.375	40	0.154	2.047	3.291	0.022854	0.720	280
		2.375	80	0.218	1.913	2.874	0.019960	0.984	400
		2.375	120	0.250	1.845	2.674	0.018566	1.111	470
2 1/2	65	2.875	40	0.203	2.445	4.695	0.032605	1.136	300
		2.875	80	0.276	2.290	4.119	0.028602	1.500	420
		2.875	120	0.300	2.239	3.937	0.027342	1.615	470
3	80	3.500	40	0.216	3.042	7.268	0.050471	1.488	260
		3.500	80	0.300	2.864	6.442	0.044738	2.010	370

Nominal Pipe Size	DN	Outside Diameter	Schedule	Min Wall Thickness	Avg Inside Diameter	Flow Area	Flow Area	Nom. Wt./Ft.	Max. Working Pressure at 73°F
in	mm	in		in	in	in^2	ft^2	lb/ft	PSI
		3.500	120	0.350	2.758	5.974	0.041487	2.306	440
3 1/2	90	4.000	40	0.226	3.521	9.737	0.067617	1.789	240
		4.000	80	0.318	3.326	8.688	0.060335	2.452	350
4	100	4.500	40	0.237	3.998	12.554	0.087179	2.118	220
		4.500	80	0.337	3.786	11.258	0.078179	2.938	320
		4.500	120	0.437	3.574	10.032	0.069668	3.713	430
5	125	5.563	40	0.258	5.016	19.761	0.137228	2.874	190
		5.563	80	0.375	4.768	17.855	0.123994	4.078	290
6	150	6.625	40	0.280	6.031	28.567	0.198384	3.733	180
		6.625	80	0.432	5.709	25.598	0.177765	5.610	280
		6.625	120	0.562	5.434	23.191	0.161052	7.132	370
8	200	8.625	40	0.322	7.942	49.539	0.344022	5.619	160
		8.625	80	0.500	7.565	44.948	0.312137	8.522	250
		8.625	120	0.718	7.189	40.591	0.281880	11.277	380
10	250	10.750	40	0.365	9.976	78.163	0.542800	7.966	140
		10.750	80	0.593	9.493	70.778	0.491512	12.635	230
12	300	12.750	40	0.406	11.889	111.015	0.770935	10.534	130
		12.750	80	0.687	11.294	100.181	0.695701	17.384	230
14	350	14.000	40	0.437	13.073	134.227	0.932132	12.462	130
		14.000	80	0.750	12.410	120.958	0.839983	20.852	220
16	400	16.000	40	0.500	14.940	175.304	1.217386	16.286	130
		16.000	80	0.843	14.213	158.658	1.101789	26.810	220
18	450	18.000	40	0.562	16.809	221.908	1.541029	20.587	130
		18.000	80	0.937	16.014	201.414	1.398707	33.544	220
20	500	20.000	40	0.593	18.743	275.910	1.916043	24.183	120
		20.000	80	1.031	17.814	249.237	1.730812	41.047	220
24	600	24.000	40	0.687	22.544	399.164	2.771973	33.652	120
		24.000	80	1.218	21.418	360.286	2.501986	58.233	210

Appendix 3

Copper Tubing Schedules

Nominal Size	Outside Diameter	Schedule	Wall Thickness	Inside Diameter	Flow Area	Flow Area	Metal Area
in	in		in	in	in^2	ft^2	in^2
1/8	0.125	ACR - Annealed	0.030	0.065	0.003	0.00002	0.009
3/16	0.187	ACR - Annealed	0.030	0.127	0.013	0.00009	0.015
1/4	0.250	ACR - Annealed	0.030	0.190	0.028	0.00020	0.021
	0.375	K	0.035	0.305	0.073	0.00051	0.037
	0.375	L	0.030	0.315	0.078	0.00054	0.033
5/16	0.312	ACR - Annealed	0.032	0.248	0.048	0.00034	0.028
3/8	0.375	ACR - Annealed	0.032	0.311	0.076	0.00053	0.034
	0.375	ACR - Drawn	0.030	0.315	0.078	0.00054	0.033
	0.500	K	0.049	0.402	0.127	0.00088	0.069
	0.500	L	0.035	0.430	0.145	0.00101	0.051
	0.500	M	0.025	0.450	0.159	0.00110	0.037
1/2	0.500	ACR - Annealed	0.032	0.436	0.149	0.00104	0.047
	0.500	ACR - Drawn	0.035	0.430	0.145	0.00101	0.051
	0.625	K	0.049	0.527	0.218	0.00151	0.089
	0.625	L	0.040	0.545	0.233	0.00162	0.074
	0.625	M	0.028	0.569	0.254	0.00177	0.053
5/8	0.625	ACR - Annealed	0.035	0.555	0.242	0.00168	0.065
	0.625	ACR - Drawn	0.040	0.545	0.233	0.00162	0.074
	0.750	K	0.049	0.652	0.334	0.00232	0.108
	0.750	L	0.042	0.666	0.348	0.00242	0.093
3/4	0.750	ACR - Annealed	0.035	0.680	0.363	0.00252	0.079
	0.750	ACR - A or D	0.042	0.666	0.348	0.00242	0.093
	0.875	K	0.065	0.745	0.436	0.00303	0.165
	0.875	L	0.045	0.785	0.484	0.00336	0.117
	0.875	M	0.032	0.811	0.517	0.00359	0.085
7/8	0.875	ACR - A or D	0.045	0.785	0.484	0.00336	0.117
1	1.125	K	0.065	0.995	0.778	0.00540	0.216
	1.125	L	0.050	1.025	0.825	0.00573	0.169
	1.125	M	0.035	1.055	0.874	0.00607	0.120
1 1/8	1.125	ACR - A or D	0.050	1.025	0.825	0.00573	0.169
1 1/4	1.375	K	0.065	1.245	1.217	0.00845	0.268
	1.375	L	0.055	1.265	1.257	0.00873	0.228
	1.375	M	0.042	1.291	1.309	0.00909	0.176
	1.375	DWV	0.040	1.295	1.317	0.00915	0.168
1 3/8	1.375	ACR - A or D	0.055	1.265	1.257	0.00873	0.228
1 1/2	1.625	K	0.072	1.481	1.723	0.01196	0.351
	1.625	L	0.060	1.505	1.779	0.01235	0.295
	1.625	M	0.049	1.527	1.831	0.01272	0.243
-	1.625	DWV	0.042	1.541	1.865	0.01295	0.209
1 5/8	1.625	ACR - A or D	0.060	1.505	1.779	0.01235	0.295

Nominal Size	Outside Diameter	Schedule	Wall Thickness	Inside Diameter	Flow Area	Flow Area	Metal Area
in	in		in	in	in²	ft²	in²
2	2.125	K	0.083	1.959	3.014	0.02093	0.532
	2.125	L	0.070	1.985	3.095	0.02149	0.452
	2.125	M	0.058	2.009	3.170	0.02201	0.377
	2.125	DWV	0.042	2.041	3.272	0.02272	0.275
2 1/8	2.125	ACR - Drawn	0.070	1.985	3.095	0.02149	0.452
2 1/2	2.625	K	0.095	2.435	4.657	0.03234	0.755
	2.625	L	0.080	2.465	4.772	0.03314	0.640
	2.625	M	0.065	2.495	4.889	0.03395	0.523
2 5/8	2.625	ACR - Drawn	0.080	2.465	4.772	0.03314	0.640
3	3.125	K	0.109	2.907	6.637	0.04609	1.033
	3.125	L	0.090	2.945	6.812	0.04730	0.858
	3.125	M	0.072	2.981	6.979	0.04847	0.691
	3.125	DWV	0.045	3.035	7.234	0.05024	0.435
3 1/8	3.125	ACR - Drawn	0.090	2.945	6.812	0.04730	0.858
3 1/2	3.625	K	0.120	3.385	8.999	0.06249	1.321
	3.625	L	0.100	3.425	9.213	0.06398	1.107
	3.625	M	0.083	3.459	9.397	0.06526	0.924
3 5/8	3.625	ACR - Drawn	0.100	3.425	9.213	0.06398	1.107
4	4.125	K	0.134	3.857	11.684	0.08114	1.680
	4.125	L	0.110	3.905	11.977	0.08317	1.387
	4.125	M	0.095	3.935	12.161	0.08445	1.203
	4.125	DWV	0.058	4.009	12.623	0.08766	0.741
4 1/8	4.125	ACR - Drawn	0.110	3.905	11.977	0.08317	1.387
5	5.125	K	0.160	4.805	18.133	0.12593	2.496
	5.125	L	0.125	4.875	18.665	0.12962	1.963
	5.125	M	0.109	4.907	18.911	0.13133	1.718
	5.125	DWV	0.072	4.981	19.486	0.13532	1.143
6	6.125	K	0.192	5.741	25.886	0.17976	3.579
	6.125	L	0.140	5.845	26.832	0.18634	2.632
	6.125	M	0.122	5.881	27.164	0.18864	2.301
	6.125	DWV	0.083	5.959	27.889	0.19368	1.575
8	8.125	K	0.271	7.583	45.162	0.31362	6.687
	8.125	L	0.200	7.725	46.869	0.32548	4.979
	8.125	M	0.170	7.785	47.600	0.33056	4.249
	8.125	DWV	0.109	7.907	49.104	0.34100	2.745
10	10.125	K	0.338	9.449	70.123	0.48697	10.392
	10.125	L	0.250	9.625	72.760	0.50528	7.756
	10.125	M	0.212	9.701	73.913	0.51329	6.602
12	12.125	K	0.405	11.315	100.554	0.69829	14.912
	12.125	L	0.280	11.565	105.046	0.72949	10.419
	12.125	M	0.254	11.617	105.993	0.73606	9.473

Appendix 4

Material Properties of Some Common Piping Materials

Material	Density	Coefficient of Linear Expansion at 70°F	Young's Modulus at 70°F	Tensile Strength	Yield Strength	Specific Heat	Thermal Conductivity
	lb/in³	in/in/°F	psi	ksi	ksi	BTU/lb/°F	BTU in/ ft² hr °F
ABS	0.037	60×10^{-6}	0.340×10^{6}	7.0	—	0.34	1.35
Aluminum	0.093	12.3×10^{-6}	10.0×10^{6}	13 to 42	4.5 to 35	0.225	1140
Cast Iron	0.26	6.2×10^{-6}	13.4×10^{6}	20 to 60	—	0.10	312
Copper	0.323	9.3×10^{-6}	17.0×10^{6}	30 to 50	9 to 40	0.094	2700
Ductile Iron	0.26	5.7×10^{-6}	24×10^{6}	60 to 70	40 to 50	0.11	250
HDPE	0.035	100×10^{-6}	0.130×10^{6}	3.2	—	0.54	2.7
PEX	0.034	90×10^{-6}	0.123×10^{6}	2.8	—	0.55	3.2
PVC	0.053	30×10^{-6}	0.410×10^{6}	8.0	—	0.25	1.1
PVDF	0.064	70×10^{-6}	0.220×10^{6}	7.0	—	0.29	1.5
Reinforced Concrete	0.078	5.5×10^{-6}	4.4×10^{6}	—	—	0.21	—
Stainless	0.277	8.5×10^{-6}	28.3×10^{6}	75 to 80	30 to 35	0.109	120
Steel	0.2816	6.4×10^{-6}	29.5×10^{6}	48 to 70	30 to 40	0.102	360
Titanium	0.1278	4.6×10^{-6}	15.5×10^{6}	35 to 70	25 to 55	0.13	120

Appendix 5

NEMA Enclosures

Protection Against	Indoor or Outdoor	Hazardous Area	Solid Foreign Objects	Water	Rain, Sleet, Snow	Ice Formation	Corrosion	Oil and Non-Corrosive Coolants
Type 1	Indoor	No	Falling Dirt	No	No	No	No	No
Type 2	Indoor	No	Falling Dirt	Light Splashing	No	No	No	No
Type 3	Indoor or Outdoor	No	Falling Dirt Windblown Dust	Yes	Yes	Yes	No	No
Type 3R	Indoor or Outdoor	No	Falling Dirt	Yes	Yes	Yes	No	No
Type 3S	Indoor or Outdoor	No	Falling Dirt Windblown Dust	Yes	Yes	External Mechanisms Remain Operable	No	No
Type 3X	Indoor or Outdoor	No	Falling Dirt Windblown Dust	Yes	Yes	Yes	Yes	No
Type 3RX	Indoor or Outdoor	No	Falling Dirt	Yes	Yes	Yes	Yes	No
Type 3 SX	Indoor or Outdoor	No	Falling Dirt Windblown Dust	Yes	Yes	External Mechanisms Remain Operable	Yes	No
Type 4	Indoor or Outdoor	No	Falling Dirt Windblown Dust	Splashing or Hose-Directed	Yes	Yes	No	No
Type 4X	Indoor or Outdoor	No	Falling Dirt Windblown Dust	Splashing or Hose-Directed	Yes	Yes	Yes	No
Type 5	Indoor or Outdoor	No	Falling Dirt, Airborne Dust, Lint, Fibers, Flyings	Dripping and Light-Splashing	No	No	No	No
Type 6	Indoor or Outdoor	No	Falling Dirt	Hose-Directed Water, Occasional Submersion at Limited Depth	Yes	Yes	No	No

Protection Against	Indoor or Outdoor	Hazardous Area	Solid Foreign Objects	Water	Rain, Sleet, Snow	Ice Formation	Corrosion	Oil and Non-Corrosive Coolants
Type 6P	Indoor or Outdoor	No	Falling Dirt	Hose-Directed Water, Prolonged Submersion at Limited Depth	Yes	Yes	No	No
Type 7	Indoor	Class I, Division 1, Groups A, B, C, or D	Designed to contain an internal explosion					
Type 8	Indoor or Outdoor	Class I, Division 1, Groups A, B, C, and D	Designed to prevent combustion through use of oil-immersed equipment					
Type 9	Indoor	Class II, Division 1, Groups E, F, or G	Designed to prevent ignition of combustible dusts					
Type 10	Mines	Mine Safety and Health Administration, 30 CFR, Part 18	Designed to contain an internal explosion					
Type 12	Indoor (no knockouts)	No	Falling Dirt, Airborne Dust, Lint, Fibers, Flyings	Dripping and Light-Splashing	No	No	No	No
Type 12K	Indoor (with knockouts)	No	Falling Dirt, Airborne Dust, Lint, Fibers, Flyings	Dripping and Light-Splashing	No	No	No	No
Type 13	Indoor	No	Falling Dirt, Airborne Dust, Lint, Fibers, Flyings	Dripping and Light-Splashing	No	No	No	Yes

Appendix 6

IP Codes for Electrical Enclosures

IP Number	Solids (First Numeral)	Liquids (Second Numeral)
0	No protection against contact or entry of solids.	No protection.
1	Protection against accidental contact by hand, but not deliberate contact. Protection against large object (greater than 50mm)	Protection against drops of condensed water. Condensed water falling on housing shall have no effect.
2	Protection against contact by fingers. Protection against medium-size foreign objects (greater than 12mm)	Protection against drops of liquid. Drops of falling liquid shall have no effect when housing is tilted to 15 degrees from vertical.
3	Protection against contact by tools, wire, etc. Protection against small foreign objects (greater than 2.5 mm)	Protection against rain. No harmful effect from rain at angle less than 60 degrees from vertical.
4	Protection against contact by small tools and wires. Protection against small foreign objects (greater than 1mm)	Protection against splashing from any direction.
5	Complete protection against contact with live or moving parts. Protection against harmful deposits of dust.	Protection against water jets from any direction.
6	Complete protection of live or moving parts. Protection against penetration of dust.	Protection against conditions on ships' decks. Water from heavy seas will not enter.
7		Protection against immersion in water. Water will not enter under stated conditions of pressure and time.
8		Protection against indefinite immersion in water under a specified pressure.

IP (International Protection Rating) Codes are analogous to the NEMA ratings used in the U.S. The IP Code is defined in IEC 60529. It consists of the letters IP followed by two numerals. The first numeral describes the enclosure's level of protection against solid foreign objects and the second numeral describes the enclosure's level of protection against liquids.

Appendix 7

Steam Tables, English Units

TABLE A7.1 Steam table: saturation temperatures[1]

Temp, F t	Abs press., psi p	Specific volume, cu ft/lb Sat. liquid v_f	Specific volume Evap. v_{fg}	Specific volume Sat. vapor v_g	Enthalpy Btu/lb Sat. liquid h_f	Enthalpy Evap. h_{fg}	Enthalpy Sat. vapor h_g	Entropy Btu/(lb F) Sat. liquid s_f	Entropy Evap. s_{fg}	Entropy Sat. vapor s_g
32	0.08854	0.01602	3306	3306	0.00	1075.8	1075.8	0.0000	2.1877	2.1877
35	0.09995	0.01602	2947	2947	3.02	1074.1	1077.1	0.0061	2.1709	2.1770
40	0.12170	0.01602	2444	2444	8.05	1071.3	1079.3	0.0162	2.1435	2.1597
45	0.14752	0.01602	2036.4	2036.4	13.06	1068.4	1081.5	0.0262	2.1167	2.1429
50	0.17811	0.01603	1703.2	1703.2	18.07	1065.6	1083.7	0.0361	2.0903	2.1264
60	0.2563	0.01604	1206.6	1206.7	28.06	1059.9	1088.0	0.0555	2.0393	2.0948
70	0.3631	0.0606	867.8	867.9	38.04	1054.3	1092.3	0.0745	1.9902	2.0647
80	0.5069	0.01608	633.1	633.1	48.02	1048.6	1096.6	0.0932	1.9428	2.0360
90	0.6982	0.01610	468.0	468.0	57.99	1042.9	1100.9	0.1115	1.8972	2.0087
100	0.9492	0.01613	350.3	350.4	67.97	1037.2	1105.2	0.1295	1.8531	1.9826
110	1.2748	0.01617	265.3	265.4	77.94	1031.6	1109.5	0.1471	1.8106	1.9577
120	1.6924	0.01620	203.25	203.27	87.92	1025.8	1113.7	0.1645	1.7694	1.9339
130	2.2225	0.01625	157.32	157.34	97.90	1020.0	1117.9	0.1816	1.7296	1.9112
140	2.8886	0.01629	122.99	123.01	107.89	1014.1	1122.0	0.1984	1.6910	1.8894
150	3.718	0.01634	97.06	97.07	117.89	1008.2	1126.1	0.2149	1.6537	1.8685
160	4.741	0.01639	77.27	77.29	127.89	1002.3	1130.2	0.2311	1.6174	1.8485
170	5.992	0.01645	62.04	62.06	137.90	996.3	1134.2	0.2472	1.5822	1.8293
180	7.510	0.01651	50.21	50.23	147.92	990.2	1138.1	0.2630	1.5480	1.8109
190	9.339	0.01657	40.94	40.96	157.95	984.1	1142.0	0.2785	1.5147	1.7932
200	11.526	0.01663	33.62	33.64	167.99	977.9	1145.9	0.2938	1.4824	1.7762
210	14.123	0.01670	27.80	27.82	178.05	971.6	1149.7	0.3090	1.4508	1.7598
212	14.696	0.01672	26.78	36.80	180.07	970.3	1150.4	0.3120	1.4446	1.7566
220	17.186	0.01677	23.13	23.15	188.13	965.2	1153.4	0.3239	1.4201	1.7440
230	20.780	0.01684	19.365	19.382	198.23	958.8	1137.0	0.3387	1.3901	1.7288
240	24.969	0.01692	16.306	16.323	208.34	952.2	1160.5	0.3531	1.3609	1.7140
250	29.825	0.01700	13.804	13.821	216.48	945.5	1164.0	0.3675	1.3223	1.6998
260	35.429	0.01709	11.746	11.763	228.64	938.7	1167.3	0.3817	1.3043	1.6860
270	41.858	0.01717	10.044	10.061	238.84	931.8	1170.6	0.3958	1.2769	1.6727
280	49.203	0.01726	8.628	8.645	249.06	924.7	1173.8	0.4096	1.2501	1.6597
290	57.556	0.01735	7.444	7.461	259.31	917.5	1176.8	0.4234	1.2338	1.6472

[1] **Abridged from "Thermodynamic Properties of Steam" by Joseph H. Keenan and Frederick G. Keyes. Copyright, 1937, by Joseph H. Keenan and Frederick G. Keyes.** Reproduced with permission of John Wiley & Sons, Inc., New York.

(continued on next page)

TABLE A7.1 *(Continued)* Steam table: saturation temperatures[1]

Temp., F t	Abs press., psi p	Specific volume, cu ft/lb			Enthalpy Btu/lb			Entropy Btu/(lb F)		
		Sat. liquid v_f	Evap. v_{fg}	Sat. vapor v_g	Sat. liquid h_f	Evap. h_{fg}	Sat. vapor h_g	Sat. liquid s_f	Evap. s_{fg}	Sat. vapor s_g
300	67.013	0.01745	6.449	6.466	269.59	910.1	1179.7	0.4369	1.1980	1.6350
310	77.68	0.01755	5.609	5.626	279.92	902.6	1182.5	0.4504	1.1727	1.6231
320	89.66	0.01765	4.896	4.914	290.28	894.9	1185.2	0.4637	1.1478	1.6115
330	103.06	0.01776	4.289	4.307	300.68	887.0	1187.7	0.4769	1.1233	1.6002
340	118.01	0.01787	3.770	3.788	311.13	879.0	1190.1	0.4900	1.0992	1.5891
350	134.63	0.01799	3.324	3.342	321.63	870.7	1192.3	0.5029	1.0754	1.5783
360	153.04	0.01811	2.939	2.957	332.18	862.2	1194.4	0.5158	1.0519	1.5677
370	173.37	0.01823	2.606	2.625	342.79	853.5	1196.3	0.5286	1.0287	1.5573
380	195.77	0.01836	2.317	2.335	353.45	844.6	1198.1	0.5413	1.0059	1.5471
390	220.37	0.01850	2.0651	2.0836	364.17	835.4	1199.6	0.5539	0.9832	1.5371
400	247.31	0.01864	1.8447	1.8633	374.97	826.0	1201.0	0.5664	0.9608	1.5272
410	276.75	0.01878	1.6512	1.6700	385.83	816.3	1202.1	0.5788	0.9386	1.5174
420	308.83	0.01894	1.4811	1.5000	396.77	806.3	1203.1	0.5912	0.9166	1.5078
430	343.72	0.01910	1.3308	1.3499	407.79	796.0	1203.8	0.6035	0.8947	1.4982
440	381.59	0.01926	1.1979	1.2171	418.90	785.4	1204.3	0.6158	0.8730	1.4887
450	422.6	0.0194	1.0799	1.0993	430.1	774.5	1204.6	0.6280	0.8513	1.4793
460	466.9	0.0196	0.9748	0.9944	441.4	763.2	1204.6	0.6402	0.8298	1.4700
470	514.7	0.0198	0.8811	0.9009	452.8	751.5	1204.3	0.6523	0.8083	1.4606
480	566.1	0.0200	0.7972	0.8172	464.4	739.4	1203.7	0.6645	0.7868	1.4513
490	621.4	0.0202	0.7221	0.7423	476.0	726.8	1202.8	0.6766	0.7653	1.4419
500	680.8	0.0204	0.6545	0.6749	487.8	713.9	1201.7	0.6887	0.7438	1.4325
520	812.4	0.0209	0.5385	0.5594	511.9	686.4	1198.2	0.7130	0.7006	1.4136
540	962.5	0.0215	0.4434	0.4649	536.6	656.6	1193.2	0.7374	0.6568	1.3942
560	1133.1	0.0221	0.3647	0.3868	562.2	624.2	1186.4	0.7621	0.6121	1.3742
580	1325.8	0.0228	0.2989	0.3217	588.9	588.4	1177.3	0.7872	0.5659	1.3532
600	1542.9	0.0236	0.2432	0.2668	617.0	548.5	1165.5	0.8131	0.5176	1.3307
620	1786.6	0.0247	0.1955	0.2201	646.7	503.6	1150.3	0.8398	0.4664	1.3062
640	2059.7	0.0260	0.1538	0.1798	678.6	452.0	1130.5	0.8679	0.4110	1.2789
660	2365.4	0.0278	0.1165	0.1442	714.2	390.2	1104.4	0.8987	0.3485	1.2472
680	2708.1	0.0305	0.0810	0.1115	757.3	309.9	1067.2	0.9351	0.2719	1.2071
700	3093.7	0.0369	0.0392	0.0761	823.3	172.1	995.4	0.9905	1.1484	1.1389
705.4	3206.2	0.0503	0	0.0503	902.7	0	902.7	1.0580	0	1.0580

[1] Abridged from "Thermodynamic Properties of Steam" by Joseph H. Keenan and Frederick G. Keyes. Copyright, 1937, by Joseph H. Keenan and Frederick G. Keyes. Reproduced with permission of John Wiley & Sons, Inc., New York.

TABLE A7.2 Steam table: saturation pressures[1]

Abs press., psi p	Temp., F t	Specific volume cu ft/lb		Enthalpy, Btu/lb			Entropy, Btu/(lb F)			Btu/lb Internal energy,	
		Sat. liquid v_f	Sat. vapor v_g	Sat. liquid h_f	Evap. h_{fg}	Sat. vapor h_g	Sat. liquid s_f	Evap. s_{fg}	Sat. vapor s_g	Sat. liquid u_g	Sat. vapor u_g
0.491	79.03	0.01608	652.3	47.05	1049.2	1096.3	0.0914	1.9473	2.0387	47.05	1037.0
0.736	91.72	0.01611	444.9	59.71	1042.0	1101.7	0.1147	1.8894	2.0041	59.71	1041.1
0.982	101.14	0.01614	339.2	69.10	1036.6	1105.7	0.1316	1.8481	1.9797	69.10	1044.0
1.227	108.71	0.01616	274.9	76.65	1032.3	1108.9	0.1449	1.8160	1.9609	76.65	1046.4
1.473	115.06	0.01618	231.6	82.99	1028.6	1111.6	0.1560	1.7896	1.9456	82.99	1048.5
1.964	125.43	0.01622	176.7	93.34	1022.7	1116.0	0.1738	1.7476	1.9214	93.33	1051.8
2.455	133.76	0.01626	143.25	101.66	1017.7	1119.4	0.1879	1.7150	1.9028	101.65	1054.3
5	162.24	0.01640	73.52	130.13	1001.0	1131.1	0.2347	1.6094	1.8441	130.12	1063.1
10	193.21	0.01659	38.42	161.17	982.1	1143.3	0.2835	1.5041	1.7876	161.14	1072.2
14.696	212.0	0.01672	26.80	180.07	970.3	1150.4	0.3120	1.4446	1.7566	180.02	1077.5
15	213.03	0.01672	26.29	181.11	969.7	1150.8	0.3135	1.4415	1.7549	181.06	1077.8
16	216.32	0.01674	24.75	184.42	967.6	1152.0	0.3184	1.4313	1.7497	184.37	1078.7
18	222.41	0.01679	22.17	190.56	963.6	1154.2	0.3275	1.4128	1.7403	190.50	1080.4
20	227.96	0.01683	20.089	196.16	960.1	1156.3	0.3356	1.3962	1.7319	196.10	1081.9
25	240.07	0.01692	16.303	208.42	952.1	1160.6	0.3533	1.3606	1.7139	208.34	1085.1
30	250.33	0.01701	13.746	218.82	945.3	1164.1	0.3680	1.3313	1.6993	218.73	1087.8
35	259.28	0.1708	11.898	227.91	939.2	1167.1	0.3807	1.3063	1.6870	227.80	1090.1
40	267.25	0.01715	10.498	236.03	933.7	1169.7	0.3919	1.2844	1.6763	235.90	1092.0
45	274.44	0.01721	9.401	243.36	928.6	1172.0	0.4019	1.2650	1.6669	243.22	1093.7
50	281.01	0.01727	8.515	250.09	924.0	1174.1	0.4110	1.2474	1.6585	294.93	1095.3
55	287.07	0.01732	7.787	256.30	919.6	1175.9	0.4193	1.2316	1.6509	256.12	1095.7
60	292.71	0.01738	7.175	262.09	915.5	1177.6	0.4270	1.2168	1.6438	261.90	1097.9
65	297.97	0.01743	6.655	267.50	911.6	1179.1	0.4342	1.2032	1.6374	267.29	1099.1
70	302.92	0.01748	6.206	272.61	907.9	1180.6	0.4409	1.1906	1.6315	272.38	1100.2
75	307.60	0.01753	5.816	277.43	904.5	1181.9	0.4472	1.1787	1.6259	277.19	1101.2
80	312.03	0.01757	5.472	282.02	901.1	1183.1	0.4531	1.1676	1.6207	281.76	1102.1
85	316.25	0.01761	5.168	286.39	897.8	1184.2	0.4587	1.1571	1.6158	286.11	1102.9
90	320.27	0.01766	4.896	290.56	894.7	1185.3	0.4641	1.1471	1.6112	290.27	1103.7
100	327.81	0.01774	4.432	298.40	888.8	1187.2	0.4740	1.1286	1.6026	298.08	1105.2
110	334.77	0.01782	4.049	305.66	883.2	1188.9	0.4832	1.1117	1.5948	305.30	1106.5

[1] **Abridged from "Thermodynamic Properties of Steam" by Joseph H. Keenan and Frederick G. Keyes. Copyright, 1937, by Joseph H. Keenan and Frederick G. Keyes.** Reproduced with permission of John Wiley & Sons, Inc., New York.

(continued on next page)

Table A7.2 *(Continued)* Steam table: saturation pressures[1]

Abs press., psi p	Temp., F t	Specific volume cu ft/lb		Enthalpy, Btu/lb			Entropy, Btu/(lb F)			Btu/lb Internal energy,	
		Sat. liquid v_f	Sat. vapor v_g	Sat. liquid h_f	Evap. h_{fg}	Sat. vapor h_g	Sat. liquid s_f	Evap. s_{fg}	Sat. vapor s_g	Sat. liquid u_g	Sat. vapor u_g
120	341.25	0.01789	3.728	312.44	877.9	1190.4	0.4916	1.0962	1.5878	312.05	1107.6
130	347.32	0.01796	3.455	318.81	872.9	1191.7	0.4995	1.0817	1.5812	318.38	1108.6
140	353.02	0.01802	3.220	324.82	868.2	1193.0	0.5069	1.0682	1.5751	324.35	1109.6
150	358.42	0.01809	3.015	330.51	863.6	1194.1	0.5138	1.0556	1.5694	330.01	1110.5
160	363.53	0.01815	2.834	335.95	859.2	1195.1	0.5204	1.0436	1.5640	335.39	1111.2
170	368.41	0.01822	2.675	341.09	854.9	1196.0	0.5266	1.0324	1.5590	340.52	1111.9
180	373.06	0.01827	2.532	346.03	850.8	1196.9	0.5325	1.0217	1.5542	345.42	1112.5
190	377.51	0.01833	2.404	350.79	846.8	1197.6	0.5381	1.0116	1.5497	350.15	1113.1
200	381.79	0.01839	2.288	355.36	843.0	1198.4	0.5435	1.0018	1.5453	354.68	1113.7
250	400.95	0.01865	1.8438	376.00	825.1	1201.1	0.5675	0.9588	1.5263	375.14	1115.8
300	417.33	0.01890	1.5433	393.84	809.0	1202.8	0.5879	0.9225	1.5104	392.79	1117.1
350	431.72	0.01913	1.3260	409.69	794.2	1203.9	0.6056	0.8910	1.4966	408.45	1118.0
400	444.59	0.0193	1.1613	424.0	780.5	1204.5	0.6214	0.8630	1.4844	422.6	1118.5
450	456.28	0.0195	1.0320	437.2	767.4	1204.6	0.6356	0.8378	1.4734	435.5	1118.7
500	467.01	0.0197	0.9278	449.4	755.0	1204.4	0.6487	0.8147	1.4634	447.6	1118.6
550	476.94	0.0199	0.8424	460.8	743.1	1203.9	0.6608	0.7934	1.4542	458.8	1118.2
600	486.21	0.0201	0.7698	471.6	731.6	1203.2	0.6720	0.7734	1.4454	469.4	1117.7
650	494.90	0.0203	0.7083	481.8	720.5	1202.3	0.6826	0.7548	1.4374	479.4	1117.1
700	503.10	0.0205	0.6554	491.5	709.7	1201.2	0.6925	0.7371	1.4296	488.8	1116.3
750	510.86	0.0207	0.6092	500.8	699.2	1200.0	0.7019	0.7204	1.4223	598.0	1115.4
800	518.23	0.0209	0.5687	509.7	688.9	1198.6	0.7108	0.7045	1.4153	506.6	1114.4
850	525.26	0.0210	0.5327	518.3	678.8	1197.1	0.7194	0.6891	1.4085	515.0	1113.3
900	531.98	0.0212	0.5006	526.6	668.8	1195.4	0.7275	0.6744	1.4020	523.1	1112.1
950	538.43	0.0214	0.4717	534.6	659.1	1193.7	0.7355	0.6602	1.3957	530.9	1110.8
1,000	544.61	0.0216	0.4456	542.4	649.4	1191.8	0.7430	0.6467	1.3897	538.4	1109.4
1,100	556.31	0.0220	0.4001	557.4	630.4	1187.8	0.7575	0.6205	1.3780	552.9	1106.4
1,200	567.22	0.0223	0.3619	571.7	611.7	1183.4	0.7711	0.5956	1.3667	566.7	1103.0
1,300	577.46	0.0227	0.3293	585.4	593.2	1178.6	0.7840	0.5719	1.3559	580.0	1099.4
1,400	587.10	0.0231	0.3012	598.7	574.7	1173.4	0.7963	0.5491	1.3454	592.7	1095.4
1,500	596.23	0.0235	0.2765	611.6	556.3	1167.9	0.8082	0.5269	1.3351	605.1	1091.2
2,000	635.82	0.0257	0.1878	671.7	463.4	1135.1	0.8619	0.4230	1.2849	662.2	1065.6
2,500	668.13	0.0287	0.1307	730.6	360.5	1091.1	0.9126	0.3197	1.2322	717.3	1030.6
3,000	695.36	0.0346	0.0858	802.5	217.8	1020.3	0.9731	0.1885	1.1615	783.4	972.7
3,206.2	705.40	0.0503	0.0503	902.7	0	902.7	1.0580	0	1.0580	872.9	872.9

[1] Abridged from "Thermodynamic Properties of Steam" by **Joseph H. Keenan and Frederick G. Keyes. Copyright, 1937, by Joseph H. Keenan and Frederick G. Keyes.** Reproduced with permission of John Wiley & Sons, Inc., New York.

Table A7.3 Steam table: properties of superheated steam[1]

Abs Press., psi (sat. temp.)		Temperature												
		200	300	400	500	600	700	800	900	1000	1100	1200	1400	1600
1 (101.74)	v	392.6	452.3	512.0	571.6	631.2	690.8	750.4	809.9	869.5	869.5	988.7	1107.8	1227.0
	h	1150.4	1195.8	1241.7	1288.3	1335.7	1383.8	1432.8	1482.7	1533.5	1585.2	1637.7	1745.7	1857.5
	s	2.0512	2.1153	2.1720	2.2233	2.2702	2.2137	2.3542	2.3923	2.4283	2.4625	2.4952	2.5566	2.6137
5 (162.24)	v	78.16	90.25	102.26	114.22	126.16	138.10	150.03	161.95	173.87	185.79	197.71	221.6	245.4
	h	1148.8	1195.0	1241.0	1288.0	1335.4	1383.6	1432.7	1482.6	1533.4	1585.1	1637.7	1745.7	1857.4
	s	1.8718	1.9370	1.9942	2.0456	2.0927	2.1361	2.1767	2.2148	2.2509	2.2851	2.3178	2.3792	2.4363
10 (193.21)	v	38.85	45.00	51.04	57.05	63.03	69.01	74.98	80.95	86.92	92.88	98.84	110.77	122.69
	h	1146.6	1193.9	1240.6	1287.5	1335.1	1383.4	1432.5	1482.4	1533.2	1585.0	1637.6	1745.6	1857.3
	s	1.7927	1.8595	1.9172	1.9689	2.0160	2.0596	2.1003	2.1383	2.1744	2.2086	2.2413	2.3028	2.3598
14.696 (212.00)	v	....	30.53	34.68	38.78	42.86	46.94	51.00	55.07	59.13	63.19	67.25	75.37	83.48
	h	....	1192.8	1239.9	1287.1	1334.8	1383.2	1432.3	1482.3	1533.1	1584.8	1637.5	1745.5	1857.3
	s	....	1.8160	1.8743	1.9261	1.9734	2.0170	2.0576	2.0958	2.1319	2.1662	2.1989	2.2603	2.3174
20 (227.96)	v	....	22.36	25.43	28.46	31.47	34.47	37.46	40.45	43.44	46.42	49.41	55.37	61.34
	h	....	1191.6	1239.2	1286.6	1334.4	1382.9	1432.1	1482.1	1533.0	1584.7	1637.4	1745.4	1857.2
	s	....	1.7808	1.8396	1.8918	1.9392	1.9829	2.0235	2.0618	2.0978	2.1321	2.1648	2.2263	2.2834
40 (267.25)	v	....	11.040	12.628	14.168	15.688	17.198	18.702	20.20	21.70	23.20	24.69	27.68	30.66
	h	....	1186.8	1236.5	1284.8	1333.1	1381.9	1431.3	1481.4	1532.4	1584.3	1637.0	1745.1	1857.0
	s	....	1.6994	1.7608	1.8140	1.8619	1.9058	1.9467	1.9850	2.0212	2.0555	2.0883	2.1498	2.2069
60 (292.71)	v	....	7.259	8.357	9.403	10.427	11.441	12.449	13.452	14.454	15.453	16.451	18.446	20.44
	h	....	1181.6	1233.6	1283.0	1331.8	1380.9	1430.5	1480.8	1531.9	1583.8	1636.6	1744.8	1856.7
	s	....	1.6492	1.7135	1.7678	1.8162	1.8605	1.9015	1.9400	1.9762	2.0106	2.0434	2.1049	2.1621
80 (312.03)	v	....	....	6.220	7.020	7.797	8.562	9.322	10.077	10.830	11.582	12.332	13.830	15.523
	h	....	....	1230.7	1281.1	1337.5	1379.9	1429.7	1487.1	1531.3	1583.4	1636.2	1744.5	1856.5
	s	....	....	1.6791	1.7346	1.7836	1.8281	1.8694	1.9079	1.9442	1.9787	2.0115	2.0731	2.1303
100 (327.81)	v	....	....	4.937	5.589	6.218	6.835	7.446	8.052	8.656	9.259	9.860	11.060	12.258
	h	....	....	1227.6	1279.1	1329.1	1378.9	1428.9	1479.5	1530.8	1582.9	1635.7	1744.2	1856.2
	s	....	....	1.6518	1.7085	1.7581	1.8029	1.8443	1.8829	1.9193	1.9538	1.9867	2.0484	2.1056
120 (341.25	v	....	....	4.081	4.636	5.165	5.683	6.195	6.207	7.207	7.710	8.212	9.214	10.213
	h	....	....	1224.4	1277.2	1327.7	1377.8	1428.1	1478.8	1530.2	1582.4	1635.3	1743.9	1856.0
	s	....	....	1.6287	1.6869	1.7370	1.8722	1.8237	1.8625	1.8990	1.9335	1.9664	2.0281	2.0854

[1] **Abridged from "Thermodynamic Properties of Steam," by Joseph H. Keenan and Frederick G. Keyes. Copyright, 1937, by Joseph H. Keenan and Frederick G. Keyes.** Reproduced with permission of John Wiley & Sons, Inc., New York. **Temperature in deg F; specific volume v in cu ft/lb; enthalpy h in Btu/lb; entropy s in Btu/(lb F).**

(continued on next page)

TABLE A7.3 *(Continued)* Steam table: properties of superheated steam[1]

Abs Press., psi (sat. temp.)		Temperature												
		200	300	400	500	600	700	800	900	1000	1100	1200	1400	1600
140	v	....	....	3.468	3.954	4.413	4.861	5.301	5.738	6.172	6.604	7.035	7.895	8.752
(353.02)	h	....	....	1221.1	1275.2	1326.4	1376.8	1427.3	1478.2	1529.7	1581.9	1634.9	1743.5	1855.7
	s	....	....	1.6087	1.6683	1.7190	1.7645	1.8063	1.8451	1.8817	19.163	1.9493	2.0110	2.0683
160	v	....	....	3.008	3.443	3.849	4.244	4.631	5.015	5.396	5.775	6.152	6.906	7.656
(363.53)	h	....	....	1217.6	1273.1	1325.0	1375.7	1426.4	1477.5	1529.1	1581.4	1634.5	1743.2	1855.5
	s	....	....	1.5908	1.6519	1.7033	1.7491	1.7911	1.8301	1.8667	1.9014	1.9344	1.9962	2.0535
180	v	....	....	2.649	3.044	3.411	4.764	4.110	4.452	4.792	5.129	5.466	6.136	6.804
(373.06)	h	....	....	1214.0	1271.0	1323.5	1374.7	1425.6	1476.8	1528.6	1581.0	1634.1	1742.9	1855.2
	s	....	....	1.5745	1.6373	1.6894	1.7355	1.7776	1.8167	1.8534	1.8882	1.9212	1.9831	2.0404
200	v	....	....	2.361	2.726	3.060	3.380	3.693	4.002	4.309	4.613	4.917	5.521	6.123
(381.79)	h	....	....	1210.3	1268.9	1322.1	1373.6	1424.8	1476.2	1528.0	1580.5	1633.7	1742.6	1855.0
	s	....	....	1.5594	1.6240	1.6767	1.7232	1.7655	1.8048	1.8415	1.8763	1.9094	1.9713	2.0287
220	v	....	....	2.125	2.465	2.772	3.066	3.352	3.634	3.913	4.191	4.467	5.017	5.565
(389.86)	h	....	....	1206.5	1266.7	1320.7	1372.6	1424.0	1475.5	1527.5	1580.0	1633.3	1742.3	1854.7
	s	....	....	1.5453	1.6117	1.6652	1.7120	1.7545	1.7939	1.8308	1.8656	1.8987	1.9607	2.0181
240	v	....	....	1.9276	2.247	2.533	2.804	3.068	3.327	3.584	3.839	4.093	4.597	5.100
(397.37)	h	....	....	1202.5	1264.5	1319.2	1371.5	1423.2	1474.8	1526.9	1579.6	1632.9	1742.0	1854.5
	s	....	....	1.5319	1.6003	1.6546	1.7017	1.7444	1.7839	1.8209	1.8558	1.8889	1.9510	2.0084
260	v	....	....	....	2.063	2.330	2.582	2.827	3.067	3.305	3.541	3.776	4.242	4.707
(404.42)	h	....	....	....	1262.3	1317.7	1370.4	1422.4	1474.2	1526.3	1579.1	1632.5	1741.7	1854.2
	s	....	....	....	1.5897	1.6447	1.6922	1.7352	1.7748	1.8118	1.8467	1.8799	1.9420	1.9995
280	v	....	....	....	1.9047	2.156	2.392	2.621	2.845	3.066	3.286	3.504	3.938	4.370
(411.05)	h	....	....	....	1260.0	1316.2	1369.4	1421.5	1473.5	1525.8	1578.6	1632.1	1741.4	1854.0
	s	....	....	....	1.5796	1.6354	1.6834	1.7265	1.7662	1.8033	1.8383	1.8716	1.9337	1.9912
300	v	....	....	....	1.7675	2.005	2.227	2.442	2.652	2.859	3.065	3.269	3.674	4.078
(417.33)	h	....	....	....	1257.6	1314.7	1368.3	1420.6	1472.8	1525.2	1578.1	1631.7	1741.0	1853.7
	s	....	....	....	1.5701	1.6268	1.6751	1.7184	1.7582	1.7954	1.8305	1.8638	1.9260	1.9835
350	v	....	....	....	1.4923	1.7036	1.8980	2.084	2.266	2.445	2.622	2.798	3.147	3.493
(431.72)	h	....	....	....	1251.5	1310.9	1365.5	1418.5	1471.1	1523.8	1577.0	1630.7	1740.3	1853.1
	s	....	....	....	1.5481	1.6070	1.6563	1.7002	1.7403	1.7777	1.8130	1.8463	1.9086	1.9663
400	v	....	....	....	1.2851	1.4770	1.6508	1.8161	1.9767	2.134	2.290	2.445	2.751	3.055
(444.59)	h	....	....	....	1245.1	1306.9	1362.7	1416.4	1469.4	1522.4	1575.8	1629.6	1739.5	1852.5
	s	....	....	....	1.5281	1.5894	1.6398	1.6842	1.7247	1.7623	1.7977	1.8311	1.8936	1.9513

Abs Press., psi (sat. temp.)		Temperature													
		500	550	600	620	640	660	680	700	800	900	1000	1200	1400	1600
450 (456.28)	v	1.1231	1.2155	1.3005	1.3332	1.3652	1.3967	1.4278	1.4584	1.6074	1.7516	1.8928	2.170	2.443	2.714
	h	1238.4	1272.0	1302.8	1314.6	1326.2	1337.5	1348.8	1359.9	1414.3	1467.7	1521.0	1628.6	1738.7	1851.9
	s	1.5095	1.5437	1.5735	1.5845	1.5951	1.6054	1.6153	1.6250	1.6699	1.7108	1.7486	1.8177	1.8803	1.9381
500 (467.01)	v	0.9927	1.0800	1.1591	1.1893	1.2188	1.2478	1.2763	1.3044	1.4405	1.5715	1.6996	1.9504	2.197	2.442
	h	1231.3	1266.8	1298.6	1310.7	1322.6	1334.2	1345.7	1357.0	1412.1	1466.0	1519.6	1627.6	1737.9	1851.3
	s	1.4919	1.5280	1.5588	1.5701	1.5810	1.5915	1.6016	1.6115	1.6571	1.6982	1.7363	1.8056	1.8683	1.9262
550 (476.94)	v	0.8852	0.9686	1.0431	1.0714	1.0989	1.1259	1.1523	1.783	1.3038	1.4241	1.5414	1.7706	1.9957	2.219
	h	1223.7	1261.2	294.3	1306.8	1318.9	1330.8	1342.5	1354.0	1409.9	1464.3	1518.2	1626.6	1737.1	1850.6
	s	1.4751	1.5131	1.5451	1.5568	1.5680	1.5787	1.5890	1.5991	1.6452	1.6868	1.7250	1.7946	1.8675	1.9155
600 (486.21)	v	0.7947	0.8753	0.9463	0.9729	0.9988	1.0241	1.0489	1.0732	1.1899	1.3013	1.6208	1.6208	1.8279	2.033
	h	1215.7	1255.5	1289.9	1302.7	1315.2	1327.4	1339.3	1351.1	1407.7	1462.5	1516.7	1625.5	1736.3	1850.0
	s	1.4586	1.4990	1.5323	1.5443	1.5558	1.5667	1.5773	1.5875	1.6343	1.6762	1.7147	1.7846	1.8476	1.9056
700 (503.10)	v	....	0.7277	0.7934	0.8177	0.8411	0.8639	0.8860	0.9077	1.0108	1.1082	1.2024	1.3853	1.5641	1.7405
	h	....	1243.2	1280.6	1294.3	1307.5	1320.3	1332.8	1345.0	1403.2	1459.0	1513.9	1623.5	1734.8	1848.8
	s	....	1.4722	1.5084	1.5212	1.5333	1.5449	1.5559	1.5665	1.6147	1.6573	1.6963	1.7666	1.8299	1.8881
800 (518.23)	v	....	0.6154	0.6779	0.7006	0.7223	0.7433	0.7635	0.7833	0.8763	0.9633	1.0470	1.2088	1.3662	1.5214
	h	....	1229.8	1270.7	1285.4	1299.4	1312.9	1325.9	1338.6	1398.6	1455.4	1511.4	1621.4	1733.2	1847.5
	s	....	1.4467	1.4863	1.5000	1.5129	1,5250	1.5366	1.5476	1.5972	1.6407	1.6801	1.7510	1.8146	1.8729
900 (531.98)	v	....	0.5364	0.5873	0.6089	0.6294	0.6491	0.6680	0.6863	0.7716	0.8506	0.9262	1.0714	1.2124	1.3509
	h	....	1215.0	1260.1	1275.9	1290.9	1305.1	1318.8	1332.1	1393.9	1451.8	1508.1	1619.3	1731.6	1846.3
	s	....	1.4216	1.4653	1.4800	1.4938	1.5066	1.5187	1.5303	1.5814	1.6257	1.6656	1.7371	1.8009	1.8595
1000 (544.61)	v	....	0.4533	0.5140	0.5350	0.5546	0.5733	0.5912	0.6084	0.6878	0.7604	0.8294	0.9615	1.0893	1.2146
	h	....	1198.3	1248.8	1265.9	1297.0	1297.0	1311.4	1325.3	1389.2	1448.2	1505.1	1617.3	1730.0	1845.0
	s	....	1.3961	1.4450	1.4610	1.4757	1.4893	1.5021	1.5141	1.5670	1.6121	1.6525	1.7245	1.7885	1.8474
1100 (556.31)	v	....	....	0.4532	0.4738	0.4929	0.5100	0.5281	0.5445	0.6191	0.6866	0.7503	0.8716	0.9885	1.1031
	h	....	....	1236.7	1255.3	1272.4	1288.5	1303.7	1318.3	1384.3	1444.5	1502.2	1615.2	1728.4	1843.8
	s	....	....	1.4251	1.4425	1.4583	1.4728	1.4862	1.4989	1.5535	1.5995	1.6405	1.7130	1.7775	1.8363
1200 (567.22)	v	....	....	0.4016	0.4222	0.4410	0.4586	0.4752	0.4909	0.5617	0.6250	0.6843	0.7967	0.9046	1.0101
	h	....	....	1223.5	1243.9	1262.4	1279.6	1295.7	1311.0	1379.3	1440.7	1499.3	1613.1	1726.9	1842.5
	s	....	....	1.4052	1.4243	1.4413	1.4568	1.4710	1.4843	1.5409	1.5879	1.6293	1.7025	1.7672	1.8263

[1] **Abridged from "Thermodynamic Properties of Steam," by Joseph H. Keenan and Frederick G. Keyes. Copyright, 1937, by Joseph H. Keenan and Frederick G. Keyes.** Reproduced with permission of John Wiley & Sons, Inc., New York. **Temperature in deg F; specific volume v in cu ft/lb; enthalpy h in Btu/lb; entropy s in Btu/(lb F).**

(continued on next page)

Table A7.3 *(Continued)* Steam table: properties of superheated steam[1]

Abs Press., psi (sat. temp.)		Temperature													
		500	550	600	620	640	660	680	700	800	900	1000	1200	1400	1600
1400 (587.10)	v	....	....	0.3174	0.3390	0.3580	0.3753	0.3912	0.4062	0.4714	0.5281	0.5805	0.6789	0.7727	0.8640
	h	....	....	1193.0	1218.4	1240.4	1260.3	1278.5	1295.5	1369.1	1433.1	1493.2	1608.9	1723.7	1840.0
	s	....	....	1.3639	1.3877	1.4079	1.4258	1.4419	1.4567	1.5177	1.5666	1.6093	1.6836	1.7489	1.8083
1600 (604.90)	v	....	....	....	0.2733	0.2936	0.3112	0.3271	0.3417	0.4034	0.4553	0.5027	0.5904	0.6738	0.7545
	h	....	....	....	1187.8	1215.2	1238.7	1259.6	1278.7	1358.4	1425.3	1487.0	1604.6	1720.5	1837.5
	s	....	....	....	1.3489	1.3741	1.3952	1.4137	1.4303	1.4964	1.5476	1.5914	1.6669	1.7328	1.7926
1800 (621.03)	v	....	....	....	....	0.2407	0.2597	0.2760	0.2907	0.3502	0.3986	0.4421	0.5218	0.5968	0.6693
	h	....	....	....	....	1185.1	1214.0	1238.5	1260.3	1347.2	1417.4	1480.8	1600.4	1717.3	1835.0
	s	....	....	....	....	1.3377	1.3638	1.3855	1.4044	1.4765	1.5301	1.5752	1.6520	1.7185	1.7786
2000 (635.82)	v	....	....	....	....	0.1936	0.2161	0.2337	0.2489	0.3074	0.3532	0.3935	0.4668	0.5352	0.6011
	h	....	....	....	....	1145.6	1184.9	1214.8	1240.0	1335.5	1409.2	1474.5	1596.1	1714.1	1832.5
	s	....	....	....	....	1.2945	1.3300	1.3564	1.3783	1.4576	1.5139	1.5603	1.6384	1.7055	1.7660
2500 (668.13)	v	....	....	....	....	....	....	0.1484	0.1686	0.2294	0.2710	0.3061	0.3678	0.4244	0.4784
	h	....	....	....	....	....	....	1132.3	1176.6	1303.6	1387.8	1458.4	1585.3	1706.1	1826.2
	s	....	....	....	....	....	....	1.2687	1.3073	1.4127	1.4772	1.5273	1.6088	1.6775	1.7389
3000 (695.36)	v	....	....	....	....	....	....	....	0.0984	0.1760	0.2159	0.2476	0.3018	0.3505	0.3966
	h	....	....	....	....	....	....	....	1060.7	1267.2	1365.0	1441.8	1574.3	1698.0	1819.9
	s	....	....	....	....	....	....	....	1.1966	1.3690	1.4439	1.4984	1.5837	1.6540	1.7163
3206.2 (705.40)	v	....	....	....	....	....	....	....	....	0.1583	0.1981	0.2288	0.2806	0.3267	0.3703
	h	....	....	....	....	....	....	....	....	1250.5	1355.2	1434.7	1569.8	1694.6	1817.2
	s	....	....	....	....	....	....	....	....	1.3508	1.4309	1.4874	1.5742	1.6452	1.7080
3500	v	....	....	....	....	....	....	....	0.0306	0.1364	0.1762	0.2058	0.2546	0.2977	0.3381
	h	....	....	....	....	....	....	....	780.5	1224.9	1340.7	1424.5	1563.3	1689.8	1813.6
	s	....	....	....	....	....	....	....	0.9515	1.3241	1.4127	1.4723	1.5615	1.6336	1.6968
4000	v	....	....	....	....	....	....	....	0.0287	0.1052	0.1462	0.1743	0.2192	0.2581	0.2943
	h	....	....	....	....	....	....	....	763.8	1174.8	1314.4	1406.8	1552.1	1681.7	1807.2
	s	....	....	....	....	....	....	....	0.9347	1.2757	1.3827	1.4482	1.5417	1.6154	1.6795
4500	v	....	....	....	....	....	....	....	0.0276	0.0798	0.1226	0.1500	0.1917	0.2273	0.2602
	h	....	....	....	....	....	....	....	753.5	1113.9	1286.5	1388.4	1540.8	1673.5	1800.9
	s	....	....	....	....	....	....	....	0.9235	1.2204	1.3529	1.4253	1.5235	1.5990	1.6640
5000	v	....	....	....	....	....	....	....	0.0268	0.0593	0.1036	0.1303	0.1696	0.2027	0.2329
	h	....	....	....	....	....	....	....	746.4	1047.1	1256.5	1369.5	1529.5	1665.3	1794.5
	s	....	....	....	....	....	....	....	0.9152	1.1622	1.3231	1.4034	1.5066	1.5839	1.6499
5500	v	....	....	....	....	....	....	....	0.0262	0.0463	0.0880	0.1143	0.1516	0.1825	0.2106
	h	....	....	....	....	....	....	....	741.3	985.0	1224.1	1349.3	1518.2	1657.0	1788.1
	s	....	....	....	....	....	....	....	0.9090	1.1093	1.2930	1.3821	1.4908	1.5699	1.6369

[1] **Abridged from "Thermodynamic Properties of Steam," by Joseph H. Keenan and Frederick G. Keyes. Copyright, 1937, by Joseph H. Keenan and Frederick G. Keyes.** Reproduced with permission of John Wiley & Sons, Inc., New York. **Temperature in deg F; specific volume v in cu ft/lb; enthalpy h in Btu/lb; entropy s in Btu/(lb F).**

Appendix 8

Steam Tables, SI Units

Properties of Saturated Steam by Temperature
(SI units)

		specific volume (cm^3/g)		internal energy (kJ/kg)		enthalpy (kJ/kg)			entropy (kJ/kg·K)		
temp. (°C)	absolute pressure (bars)	sat. liquid v_f	sat. vapor v_g	sat. liquid u_f	sat. vapor u_g	sat. liquid h_f	evap. h_{fg}	sat. vapor h_g	sat. liquid s_f	sat. vapor s_g	temp. (°C)
0.01	0.006117	1.0002	205991	0.00	2374.9	0.0006	2500.9	2500.9	0.0000	9.1555	0.01
4	0.00814	1.0001	157116	16.81	2380.4	16.81	2491.4	2508.2	0.0611	9.0505	4
5	0.00873	1.0001	147011	21.02	2381.8	21.02	2489.1	2510.1	0.0763	9.0248	5
6	0.00935	1.0001	137633	25.22	2383.2	25.22	2486.7	2511.9	0.0913	8.9993	6
8	0.01073	1.0002	120829	33.63	2385.9	33.63	2482.0	2515.6	0.1213	8.9491	8
10	0.01228	1.0003	106303	42.02	2388.6	42.02	2477.2	2519.2	0.1511	8.8998	10
11	0.01313	1.0004	99787	46.22	2390.0	46.22	2474.8	2521.0	0.1659	8.8754	11
12	0.01403	1.0005	93719	50.41	2391.4	50.41	2472.5	2522.9	0.1806	8.8513	12
13	0.01498	1.0007	88064	54.60	2392.8	54.60	2470.1	2524.7	0.1953	8.8274	13
14	0.01599	1.0008	82793	58.79	2394.1	58.79	2467.7	2526.5	0.2099	8.8037	14
15	0.01706	1.0009	77875	62.98	2395.5	62.98	2465.3	2528.3	0.2245	8.7803	15
16	0.01819	1.0011	73286	67.17	2396.9	67.17	2463.0	2530.2	0.2390	8.7570	16
17	0.01938	1.0013	69001	71.36	2398.2	71.36	2460.6	2532.0	0.2534	8.7339	17
18	0.02065	1.0014	64998	75.54	2399.6	75.54	2458.3	2533.8	0.2678	8.7111	18
19	0.02198	1.0016	61256	79.73	2401.0	79.73	2455.9	2535.6	0.2822	8.6884	19
20	0.02339	1.0018	57757	83.91	2402.3	83.91	2453.5	2537.4	0.2965	8.6660	20
21	0.02488	1.0021	54483	88.10	2403.7	88.10	2451.2	2539.3	0.3107	8.6437	21
22	0.02645	1.0023	51418	92.28	2405.0	92.28	2448.8	2541.1	0.3249	8.6217	22
23	0.02811	1.0025	48548	96.46	2406.4	96.47	2446.4	2542.9	0.3391	8.5998	23
24	0.02986	1.0028	45858	100.64	2407.8	100.65	2444.1	2544.7	0.3532	8.5781	24
25	0.03170	1.0030	43337	104.83	2409.1	104.83	2441.7	2546.5	0.3672	8.5566	25
26	0.03364	1.0033	40973	109.01	2410.5	109.01	2439.3	2548.3	0.3812	8.5353	26
27	0.03568	1.0035	38754	113.19	2411.8	113.19	2436.9	2550.1	0.3952	8.5142	27
28	0.03783	1.0038	36672	117.37	2413.2	117.37	2434.5	2551.9	0.4091	8.4933	28
29	0.04009	1.0041	34716	121.55	2414.6	121.55	2432.2	2553.7	0.4229	8.4725	29
30	0.04247	1.0044	32878	125.73	2415.9	125.73	2429.8	2555.5	0.4368	8.4520	30
31	0.04497	1.0047	31151	129.91	2417.3	129.91	2427.4	2557.3	0.4505	8.4316	31
32	0.04760	1.0050	29526	134.09	2418.6	134.09	2425.1	2559.2	0.4642	8.4113	32
33	0.05035	1.0054	27998	138.27	2420.0	138.27	2422.7	2561.0	0.4779	8.3913	33
34	0.05325	1.0057	26560	142.45	2421.3	142.45	2420.4	2562.8	0.4916	8.3714	34
35	0.05629	1.0060	25205	146.63	2422.7	146.63	2417.9	2564.5	0.5051	8.3517	35
36	0.05948	1.0064	23929	150.81	2424.0	150.81	2415.5	2566.3	0.5187	8.3321	36
38	0.06633	1.0071	21593	159.17	2426.7	159.17	2410.7	2569.9	0.5456	8.2935	38
40	0.07385	1.0079	19515	167.53	2429.4	167.53	2406.0	2573.5	0.5724	8.2555	40
45	0.09595	1.0099	15252	188.43	2436.1	188.43	2394.0	2582.4	0.6386	8.1633	45
50	0.1235	1.0121	12027	209.33	2442.7	209.34	2382.0	2591.3	0.7038	8.0748	50
55	0.1576	1.0146	9564	230.24	2449.3	230.26	2369.8	2600.1	0.7680	7.9898	55
60	0.1995	1.0171	7667	251.16	2455.9	251.18	2357.6	2608.8	0.8313	7.9081	60
65	0.2504	1.0199	6194	272.09	2462.4	272.12	2345.4	2617.5	0.8937	7.8296	65
70	0.3120	1.0228	5040	293.03	2468.9	293.07	2333.0	2626.1	0.9551	7.7540	70
75	0.3860	1.0258	4129	313.99	2475.2	314.03	2320.6	2634.6	1.0158	7.6812	75
80	0.4741	1.0291	3405	334.96	2481.6	335.01	2308.0	2643.0	1.0756	7.6111	80
85	0.5787	1.0324	2826	355.95	2487.8	356.01	2295.3	2651.3	1.1346	7.5434	85
90	0.7018	1.0360	2359	376.97	2494.0	377.04	2282.5	2659.5	1.1929	7.4781	90
95	0.8461	1.0396	1981	398.00	2500.0	398.09	2269.5	2667.6	1.2504	7.4151	95
100	1.014	1.0435	1672	419.06	2506.0	419.17	2256.4	2675.6	1.3072	7.3541	100
105	1.209	1.0474	1418	440.15	2511.9	440.27	2243.1	2683.4	1.3633	7.2952	105
110	1.434	1.0516	1209	461.26	2517.7	461.42	2229.6	2691.1	1.4188	7.2381	110
115	1.692	1.0559	1036	482.42	2523.3	482.59	2216.0	2698.6	1.4737	7.1828	115
120	1.987	1.0603	891.2	503.60	2528.9	503.81	2202.1	2705.9	1.5279	7.1291	120

Properties of Saturated Steam by Temperature
(SI units)

		specific volume (cm^3/g)		internal energy (kJ/kg)		enthalpy (kJ/kg)			entropy (kJ/kg·K)		
temp. (°C)	absolute pressure (bars)	sat. liquid v_f	sat. vapor v_g	sat. liquid u_f	sat. vapor u_g	sat. liquid h_f	evap. h_{fg}	sat. vapor h_g	sat. liquid s_f	sat. vapor s_g	temp. (°C)
125	2.322	1.0649	770.0	524.83	2534.3	525.07	2188.0	2713.1	1.5816	7.0770	125
130	2.703	1.0697	668.0	546.10	2539.5	546.38	2173.7	2720.1	1.6346	7.0264	130
135	3.132	1.0746	581.7	567.41	2544.7	567.74	2159.1	2726.9	1.6872	6.9772	135
140	3.615	1.0798	508.5	588.77	2549.6	589.16	2144.3	2733.4	1.7392	6.9293	140
145	4.157	1.0850	446.0	610.19	2554.4	610.64	2129.2	2739.8	1.7907	6.8826	145
150	4.762	1.0905	392.5	631.66	2559.1	632.18	2113.8	2745.9	1.8418	6.8371	150
155	5.435	1.0962	346.5	653.19	2563.5	653.79	2098.0	2751.8	1.8924	6.7926	155
160	6.182	1.1020	306.8	674.79	2567.8	675.47	2082.0	2757.4	1.9426	6.7491	160
165	7.009	1.1080	272.4	696.46	2571.9	697.24	2065.6	2762.8	1.9923	6.7066	165
170	7.922	1.1143	242.6	718.20	2575.7	719.08	2048.8	2767.9	2.0417	6.6650	170
175	8.926	1.1207	216.6	740.02	2579.4	741.02	2031.7	2772.7	2.0906	6.6241	175
180	10.03	1.1274	193.8	761.92	2582.8	763.05	2014.2	2777.2	2.1392	6.5840	180
185	11.23	1.1343	173.9	783.91	2586.0	785.19	1996.2	2781.4	2.1875	6.5447	185
190	12.55	1.1415	156.4	806.00	2589.0	807.43	1977.9	2785.3	2.2355	6.5059	190
195	13.99	1.1489	140.9	828.18	2591.7	829.79	1959.0	2788.8	2.2832	6.4678	195
200	15.55	1.1565	127.2	850.47	2594.2	852.27	1939.7	2792.0	2.3305	6.4302	200
205	17.24	1.1645	115.1	872.87	2596.4	874.88	1920.0	2794.8	2.3777	6.3930	205
210	19.08	1.1727	104.3	895.39	2598.3	897.63	1899.6	2797.3	2.4245	6.3563	210
215	21.06	1.1813	94.68	918.04	2599.9	920.53	1878.8	2799.3	2.4712	6.3200	215
220	23.20	1.1902	86.09	940.82	2601.3	943.58	1857.4	2801.0	2.5177	6.2840	220
225	25.50	1.1994	78.40	963.74	2602.2	966.80	1835.4	2802.2	2.5640	6.2483	225
230	27.97	1.2090	71.50	986.81	2602.9	990.19	1812.7	2802.9	2.6101	6.2128	230
235	30.63	1.2190	65.30	1010.0	2603.2	1013.8	1789.4	2803.2	2.6561	6.1775	235
240	33.47	1.2295	59.71	1033.4	2603.1	1037.6	1765.4	2803.0	2.7020	6.1423	240
245	36.51	1.2403	54.65	1057.0	2602.7	1061.6	1740.7	2802.2	2.7478	6.1072	245
250	39.76	1.2517	50.08	1080.8	2601.8	1085.8	1715.2	2800.9	2.7935	6.0721	250
255	43.23	1.2636	45.94	1104.8	2600.5	1110.2	1688.8	2799.1	2.8392	6.0369	255
260	46.92	1.2761	42.17	1129.0	2598.7	1135.0	1661.6	2796.6	2.8849	6.0016	260
265	50.85	1.2892	38.75	1153.4	2596.5	1160.0	1633.5	2793.5	2.9307	5.9661	265
270	55.03	1.3030	35.62	1178.1	2593.7	1185.3	1604.4	2789.7	2.9765	5.9304	270
275	59.46	1.3175	32.77	1203.1	2590.3	1210.9	1574.3	2785.2	3.0224	5.8944	275
280	64.17	1.3328	30.15	1228.3	2586.4	1236.9	1543.0	2779.9	3.0685	5.8579	280
285	69.15	1.3491	27.76	1253.9	2581.8	1263.3	1510.5	2773.7	3.1147	5.8209	285
290	74.42	1.3663	25.56	1279.9	2576.5	1290.0	1476.7	2766.7	3.1612	5.7834	290
295	79.99	1.3846	23.53	1306.2	2570.5	1317.3	1441.4	2758.7	3.2080	5.7451	295
300	85.88	1.4042	21.66	1332.9	2563.6	1345.0	1404.6	2749.6	3.2552	5.7059	300
305	92.09	1.4252	19.93	1360.2	2555.9	1373.3	1366.1	2739.4	3.3028	5.6657	305
310	98.65	1.4479	18.34	1387.9	2547.1	1402.2	1325.7	2728.0	3.3510	5.6244	310
315	105.6	1.4724	16.85	1416.3	2537.2	1431.8	1283.2	2715.1	3.3998	5.5816	315
320	112.8	1.4990	15.47	1445.3	2526.0	1462.2	1238.4	2700.6	3.4494	5.5372	320
325	120.5	1.5283	14.18	1475.1	2513.4	1493.5	1190.8	2684.3	3.5000	5.4908	325
330	128.6	1.5606	12.98	1505.8	2499.2	1525.9	1140.2	2666.0	3.5518	5.4422	330
335	137.1	1.5967	11.85	1537.6	2483.0	1559.5	1085.9	2645.4	3.6050	5.3906	335
340	146.0	1.6376	10.78	1570.6	2464.4	1594.5	1027.3	2621.9	3.6601	5.3356	340
345	155.4	1.6846	9.769	1605.3	2443.1	1631.5	963.4	2594.9	3.7176	5.2762	345
350	165.3	1.7400	8.802	1642.1	2418.1	1670.9	892.8	2563.6	3.7784	5.2110	350
355	175.7	1.8079	7.868	1682.0	2388.4	1713.7	812.9	2526.7	3.8439	5.1380	355
360	186.7	1.8954	6.949	1726.3	2351.8	1761.7	719.8	2481.5	3.9167	5.0536	360
365	198.2	2.0172	6.012	1777.8	2303.8	1817.8	605.2	2423.0	4.0014	4.9497	365
370	210.4	2.2152	4.954	1844.1	2230.3	1890.7	443.8	2334.5	4.1112	4.8012	370
374	220.64000	3.1056	3.1056	2015.70	2015.70	2084.30	0	2084.3	4.4070	4.4070	373.9

Values in this table were calculated from *NIST Standard Reference Database 10*, "NIST/ASME Steam Properties," Ver. 2.11, National Institute of Standards and Technology, U.S. Department of Commerce, Gaithersburg, MD, 1997, which has been licensed to Professional Publications, Inc.

Properties of Saturated Steam by Pressure
(SI units)

		specific volume (cm^3/g)		internal energy (kJ/kg)		enthalpy (kJ/kg)			entropy (kJ/kg·K)		
absolute press. (bars)	temp. (°C)	sat. liquid v_f	sat. vapor v_g	sat. liquid u_f	sat. vapor u_g	sat. liquid h_f	evap. h_{fg}	sat. vapor h_g	sat. liquid s_f	sat. vapor s_g	absolute press. (bars)
0.04	28.96	1.0041	34791	121.38	2414.5	121.39	2432.3	2553.7	0.4224	8.4734	0.04
0.06	36.16	1.0065	23733	151.47	2424.2	151.48	2415.2	2566.6	0.5208	8.3290	0.06
0.08	41.51	1.0085	18099	173.83	2431.4	173.84	2402.4	2576.2	0.5925	8.2273	0.08
0.10	45.81	1.0103	14670	191.80	2437.2	191.81	2392.1	2583.9	0.6492	8.1488	0.10
0.20	60.06	1.0172	7648	251.40	2456.0	251.42	2357.5	2608.9	0.8320	7.9072	0.20
0.30	69.10	1.0222	5228	289.24	2467.7	289.27	2335.3	2624.5	0.9441	7.7675	0.30
0.40	75.86	1.0264	3993	317.58	2476.3	317.62	2318.4	2636.1	1.0261	7.6690	0.40
0.50	81.32	1.0299	3240	340.49	2483.2	340.54	2304.7	2645.2	1.0912	7.5930	0.50
0.60	85.93	1.0331	2732	359.84	2489.0	359.91	2292.9	2652.9	1.1454	7.5311	0.60
0.70	89.93	1.0359	2365	376.68	2493.9	376.75	2282.7	2659.4	1.1921	7.4790	0.70
0.80	93.49	1.0385	2087	391.63	2498.2	391.71	2273.5	2665.2	1.2330	7.4339	0.80
0.90	96.69	1.0409	1869	405.10	2502.1	405.20	2265.1	2670.3	1.2696	7.3943	0.90
1.00	99.61	1.0432	1694	417.40	2505.6	417.50	2257.4	2674.9	1.3028	7.3588	1.00
1.01325	99.97	1.0434	1673	418.95	2506.0	419.06	2256.5	2675.5	1.3069	7.3544	1.01325
1.50	111.3	1.0527	1159	466.97	2519.2	467.13	2226.0	2693.1	1.4337	7.2230	1.50
2.00	120.2	1.0605	885.7	504.49	2529.1	504.70	2201.5	2706.2	1.5302	7.1269	2.00
2.50	127.4	1.0672	718.7	535.08	2536.8	535.35	2181.1	2716.5	1.6072	7.0524	2.50
3.00	133.5	1.0732	605.8	561.10	2543.2	561.43	2163.5	2724.9	1.6717	6.9916	3.00
3.50	138.9	1.0786	524.2	583.88	2548.5	584.26	2147.7	2732.0	1.7274	6.9401	3.50
4.00	143.6	1.0836	462.4	604.22	2553.1	604.65	2133.4	2738.1	1.7765	6.8955	4.00
4.50	147.9	1.0882	413.9	622.65	2557.1	623.14	2120.2	2743.4	1.8205	6.8560	4.50
5.00	151.8	1.0926	374.8	639.54	2560.7	640.09	2108.0	2748.1	1.8604	6.8207	5.00
6.00	158.8	1.1006	315.6	669.72	2566.8	670.38	2085.8	2756.1	1.9308	6.7592	6.00
7.00	164.9	1.1080	272.8	696.23	2571.8	697.00	2065.8	2762.8	1.9918	6.7071	7.00
8.00	170.4	1.1148	240.3	719.97	2576.0	720.86	2047.4	2768.3	2.0457	6.6616	8.00
9.00	175.4	1.1212	214.9	741.55	2579.6	742.56	2030.5	2773.0	2.0940	6.6213	9.00
10.0	179.9	1.1272	194.4	761.39	2582.7	762.52	2014.6	2777.1	2.1381	6.5850	10.0
15.0	198.3	1.1539	131.7	842.83	2593.4	844.56	1946.4	2791.0	2.3143	6.4430	15.0
20.0	212.4	1.1767	99.59	906.14	2599.1	908.50	1889.8	2798.3	2.4468	6.3390	20.0
25.0	224.0	1.1974	79.95	958.91	2602.1	961.91	1840.0	2801.9	2.5543	6.2558	25.0
30.0	233.9	1.2167	66.66	1004.7	2603.2	1008.3	1794.9	2803.2	2.6455	6.1856	30.0
35.0	242.6	1.2350	57.06	1045.5	2602.9	1049.8	1752.8	2802.6	2.7254	6.1243	35.0
40.0	250.4	1.2526	49.78	1082.5	2601.7	1087.5	1713.3	2800.8	2.7968	6.0696	40.0
45.0	257.4	1.2696	44.06	1116.5	2599.7	1122.3	1675.7	2797.9	2.8615	6.0197	45.0
50.0	263.9	1.2864	39.45	1148.2	2597.0	1154.6	1639.6	2794.2	2.9210	5.9737	50.0
55.0	270.0	1.3029	35.64	1177.9	2593.7	1185.1	1604.6	2789.7	2.9762	5.9307	55.0
60.0	275.6	1.3193	32.45	1206.0	2589.9	1213.9	1570.7	2784.6	3.0278	5.8901	60.0
65.0	280.9	1.3356	29.73	1232.7	2585.7	1241.4	1537.5	2778.9	3.0764	5.8516	65.0
70.0	285.8	1.3519	27.38	1258.2	2581.0	1267.7	1504.9	2772.6	3.1224	5.8148	70.0
75.0	290.5	1.3682	25.33	1282.7	2575.9	1292.9	1473.0	2765.9	3.1662	5.7793	75.0
80.0	295.0	1.3847	23.53	1306.2	2570.5	1317.3	1441.4	2758.7	3.2081	5.7450	80.0
85.0	299.3	1.4013	21.92	1329.0	2564.7	1340.9	1410.1	2751.0	3.2483	5.7117	85.0
90.0	303.3	1.4181	20.49	1351.1	2558.5	1363.9	1379.0	2742.9	3.2870	5.6791	90.0
95.0	307.2	1.4352	19.20	1372.6	2552.0	1386.2	1348.2	2734.4	3.3244	5.6473	95.0
100	311.0	1.4526	18.03	1393.5	2545.2	1408.1	1317.4	2725.5	3.3606	5.6160	100

Properties of Saturated Steam by Pressure
(SI units)

		specific volume (cm^3/g)		internal energy (kJ/kg)		enthalpy (kJ/kg)			entropy (kJ/kg·K)		
absolute press. (bars)	temp. (°C)	sat. liquid v_f	sat. vapor v_g	sat. liquid u_f	sat. vapor u_g	sat. liquid h_f	evap. h_{fg}	sat. vapor h_g	sat. liquid s_f	sat. vapor s_g	absolute press. (bars)
100	311.0	1.4526	18.03	1393.5	2545.2	1408.1	1317.4	2725.5	3.3606	5.6160	100
110	318.1	1.4885	15.99	1434.1	2530.5	1450.4	1255.9	2706.3	3.4303	5.5545	110
120	324.7	1.5263	14.26	1473.1	2514.3	1491.5	1193.9	2685.4	3.4967	5.4939	120
130	330.9	1.5665	12.78	1511.1	2496.5	1531.5	1131.2	2662.7	3.5608	5.4336	130
140	336.7	1.6097	11.49	1548.4	2477.1	1571.0	1066.9	2637.9	3.6232	5.3727	140
150	342.2	1.6570	10.34	1585.3	2455.6	1610.2	1000.5	2610.7	3.6846	5.3106	150
160	347.4	1.7094	9.309	1622.3	2431.8	1649.7	931.1	2580.8	3.7457	5.2463	160
170	352.3	1.7693	8.371	1659.9	2405.2	1690.0	857.5	2547.5	3.8077	5.1787	170
180	357.0	1.8398	7.502	1699.0	2374.8	1732.1	777.7	2509.8	3.8718	5.1061	180
190	361.5	1.9268	6.677	1740.5	2339.1	1777.2	688.8	2466.0	3.9401	5.0256	190
200	365.7	2.0400	5.865	1786.4	2295.0	1827.2	585.1	2412.3	4.0156	4.9314	200
210	369.8	2.2055	4.996	1841.2	2233.7	1887.6	451.0	2338.6	4.1064	4.8079	210
220.64	373.95	3.1056	3.1056	2015.7	2015.7	2084.3	0	2084.3	4.4070	4.4070	220.64

Values in this table were calculated from *NIST Standard Reference Database 10*, "NIST/ASME Steam Properties," Ver. 2.11, National Institute of Standards and Technology, U.S. Department of Commerce, Gaithersburg, MD, 1997, which has been licensed to Professional Publications, Inc.

Properties of Superheated Steam
(SI units)

specific volume (v) in m^3/kg; enthalpy (h) in kJ/kg; entropy (s) in kJ/kg·K

absolute pressure (kPa) (sat. temp. °C)		temperature (°C)							
		100	150	200	250	300	360	420	500
10	v	17.196	19.513	21.826	24.136	26.446	29.216	31.986	35.680
(45.81)	h	2687.5	2783.0	2879.6	2977.4	3076.7	3197.9	3321.4	3489.7
	s	8.4489	8.6892	8.9049	9.1015	9.2827	9.4837	9.6700	9.8998
50	v	3.419	3.890	4.356	4.821	5.284	5.839	6.394	7.134
(81.32)	h	2682.4	2780.2	2877.8	2976.1	3075.8	3197.2	3320.8	3489.3
	s	7.6953	7.9413	8.1592	8.3568	8.5386	8.7401	8.9266	9.1566
75	v	2.270	2.588	2.900	3.211	3.521	3.891	4.262	4.755
(91.76)	h	2679.2	2778.4	2876.6	2975.3	3075.1	3196.7	3320.4	3489.0
	s	7.5011	7.7509	7.9702	8.1685	8.3507	8.5524	8.7391	8.9692
100	v	1.6959	1.9367	2.172	2.406	2.639	2.917	3.195	3.566
(99.61)	h	2675.8	2776.6	2875.5	2974.5	3074.5	3196.3	3320.1	3488.7
	s	7.3610	7.6148	7.8356	8.0346	8.2172	8.4191	8.6059	8.8361
150	v	....	1.2855	1.4445	1.6013	1.7571	1.9433	2.129	2.376
(111.35)	h	....	2772.9	2873.1	2972.9	3073.3	3195.3	3319.4	3488.2
	s	....	7.4208	7.6447	7.8451	8.0284	8.2309	8.4180	8.6485
400	v	....	0.4709	0.5343	0.5952	0.6549	0.7257	0.7961	0.8894
(143.61)	h	....	2752.8	2860.9	2964.5	3067.1	3190.7	3315.8	3485.5
	s	....	6.9306	7.1723	7.3804	7.5677	7.7728	7.9615	8.1933
700	v	....	....	0.3000	0.3364	0.3714	0.4126	0.4533	0.5070
(164.95)	h	....	....	2845.3	2954.0	3059.4	3185.1	3311.5	3482.3
	s	....	....	6.8884	7.1070	7.2995	7.5080	7.6986	7.9319
1000	v	....	....	0.2060	0.2328	0.2580	0.2874	0.3162	0.3541
(179.88)	h	....	....	2828.3	2943.1	3051.6	3179.4	3307.1	3479.1
	s	....	....	6.6955	6.9265	7.1246	7.3367	7.5294	7.7641
1500	v	....	....	0.13245	0.15201	0.16971	0.18990	0.2095	0.2352
(198.29)	h	....	....	2796.0	2923.9	3038.2	3169.8	3299.8	3473.7
	s	....	....	6.4536	6.7111	6.9198	7.1382	7.3343	7.5718
2000	v	....	....	....	0.11150	0.12551	0.14115	0.15617	0.17568
(212.38)	h	....	....	....	2903.2	3024.2	3159.9	3292.3	3468.2
	s	....	....	....	6.5475	6.7684	6.9937	7.1935	7.4337
2500	v	....	....	....	0.08705	0.09894	0.11188	0.12416	0.13999
(223.95)	h	....	....	....	2880.9	3009.6	3149.8	3284.8	3462.7
	s	....	....	....	6.4107	6.6459	6.8788	7.0824	7.3254
3000	v	....	....	....	0.07063	0.08118	0.09236	0.10281	0.11620
(233.85)	h	....	....	....	2856.5	2994.3	3139.5	3277.1	3457.2
	s	....	....	....	6.2893	6.5412	6.7823	6.9900	7.2359

Values in this table were calculated from *NIST Standard Reference Database 10*, "NIST/ASME Steam Properties," Ver. 2.11, National Institute of Standards and Technology, U.S. Department of Commerce, Gaithersburg, MD, 1997, which has been licensed to Professional Publications, Inc.

Appendix 9

Friction Losses

Note: Friction losses are given in ft per 100 ft ("cft" = 100 ft)

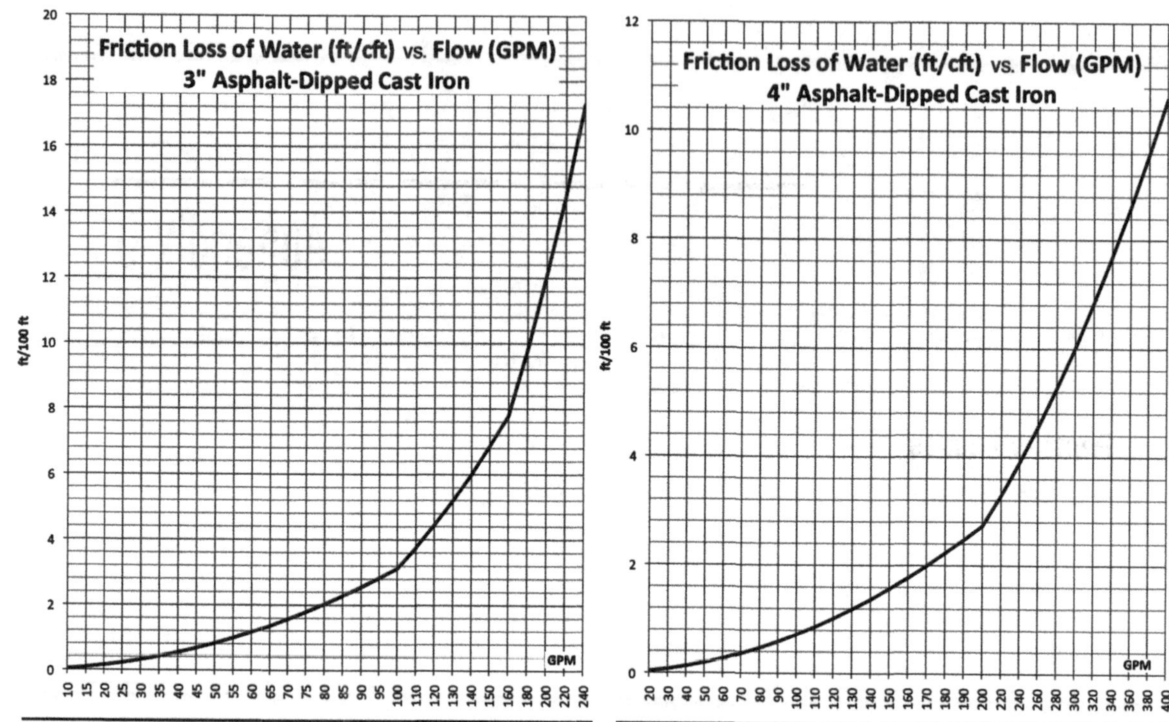

Figure A9.1 Cast Iron 3" diameter. Discontinuities occur at X-scale changes.

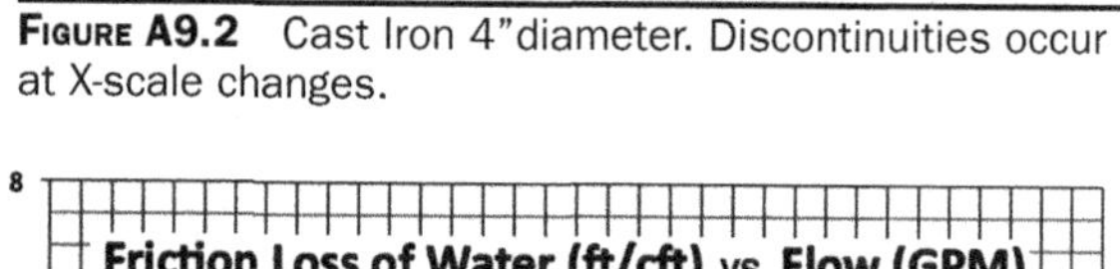

Figure A9.2 Cast Iron 4" diameter. Discontinuities occur at X-scale changes.

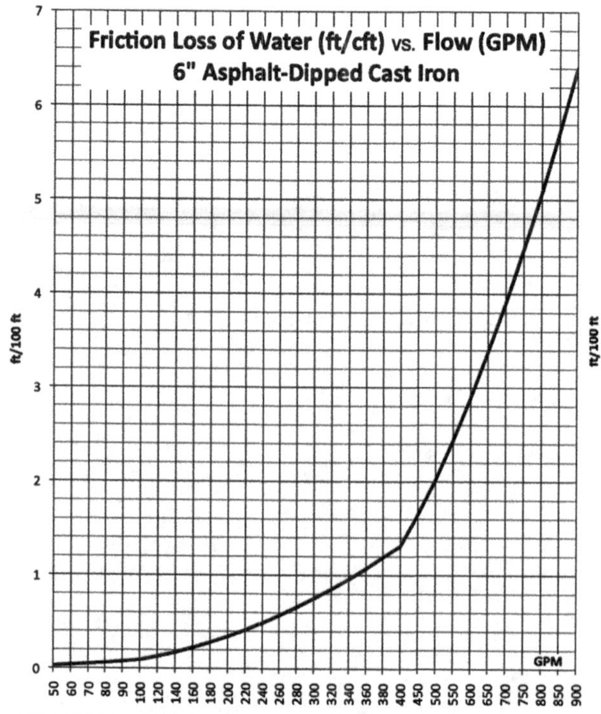

Figure A9.3 Cast Iron 6" diameter. Discontinuities occur at X-scale changes.

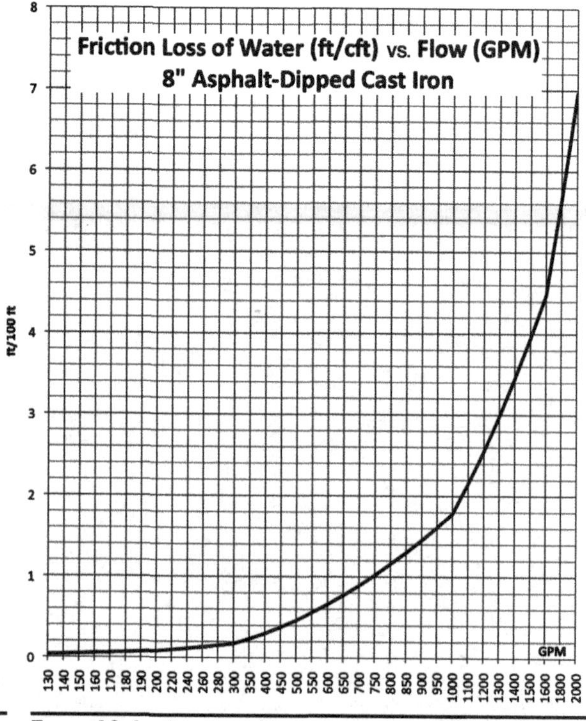

Figure A9.4 Cast Iron 8" diameter. Discontinuities occur at X-scale changes.

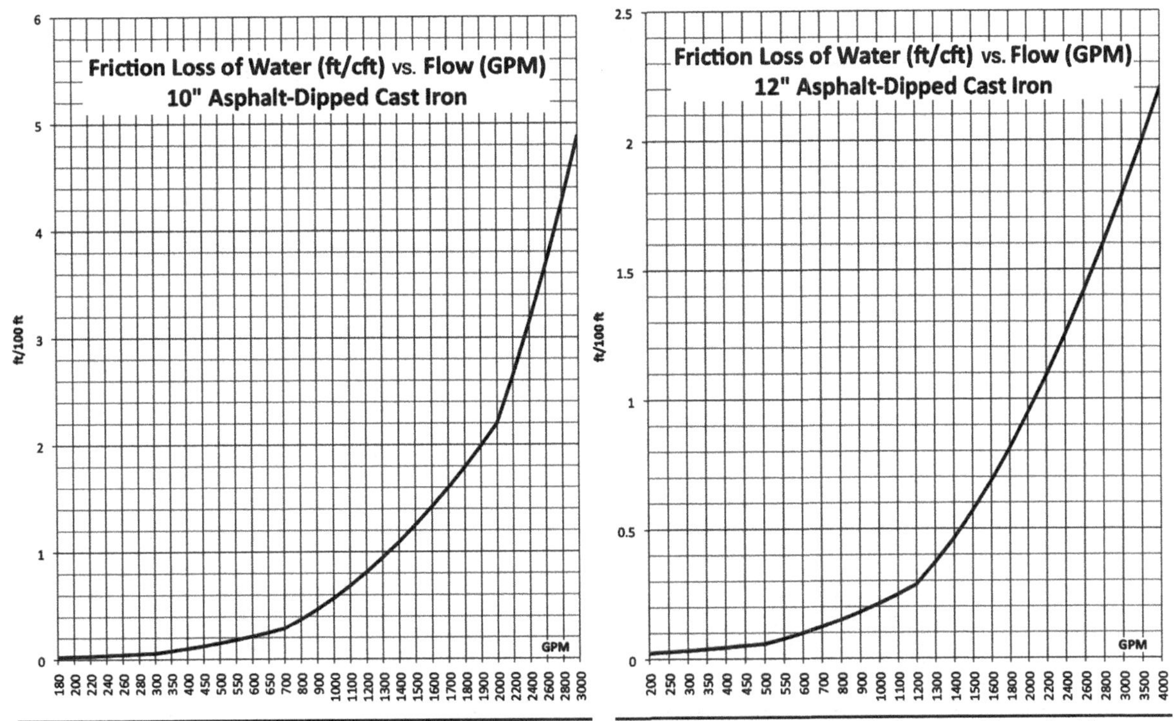

Figure A9.5 Cast Iron 10" diameter. Discontinuities occur at X-scale changes.

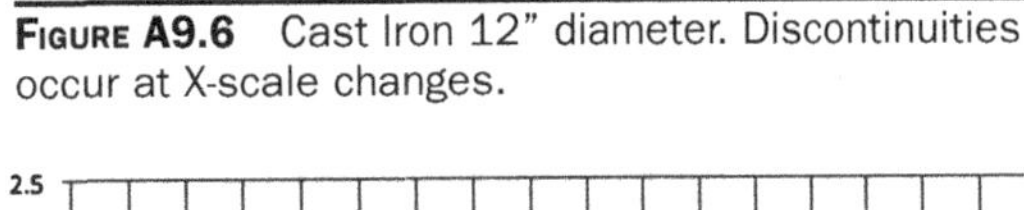

Figure A9.6 Cast Iron 12" diameter. Discontinuities occur at X-scale changes.

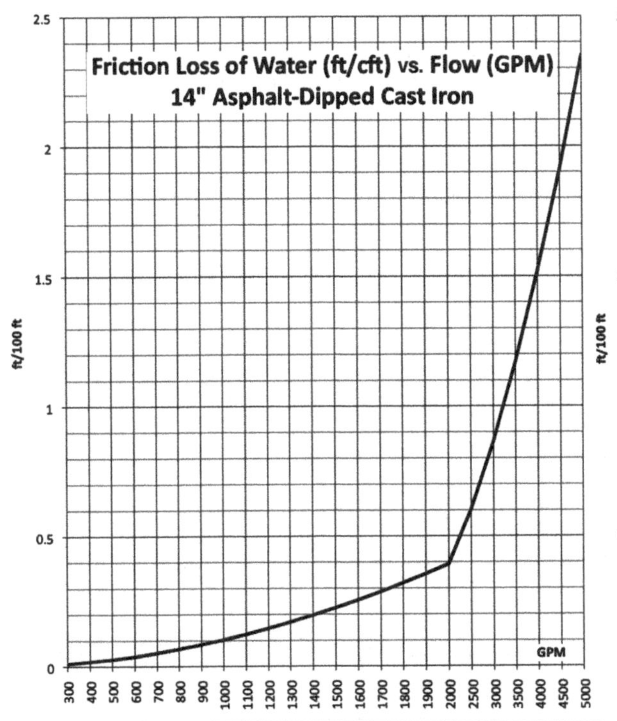

Figure A9.7 Cast Iron 14" diameter. Discontinuities occur at X-scale changes.

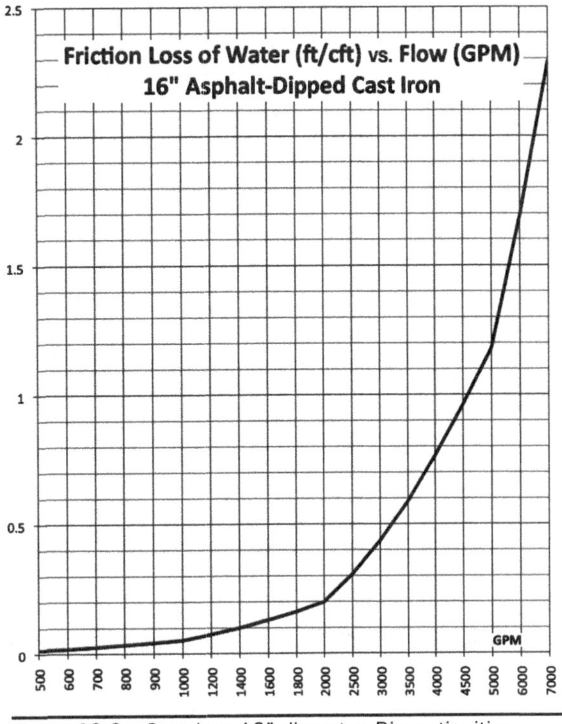

Figure A9.8 Cast Iron 16" diameter. Discontinuities occur at X-scale changes.

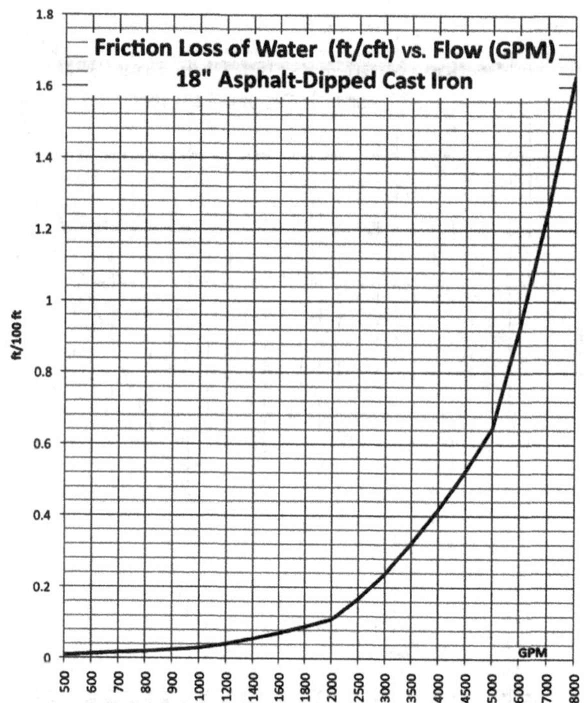

Figure A9.9 Cast Iron 18" diameter. Discontinuities occur at X-scale changes.

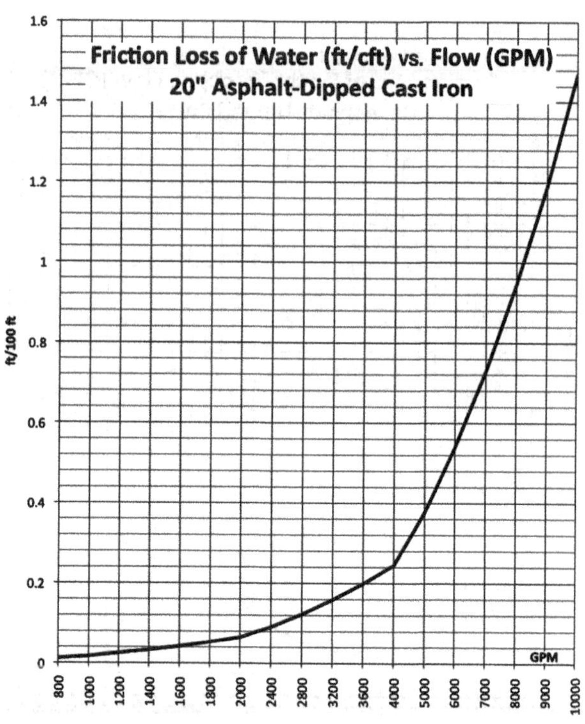

Figure A9.10 Cast Iron 20" diameter. Discontinuities occur at X-scale changes.

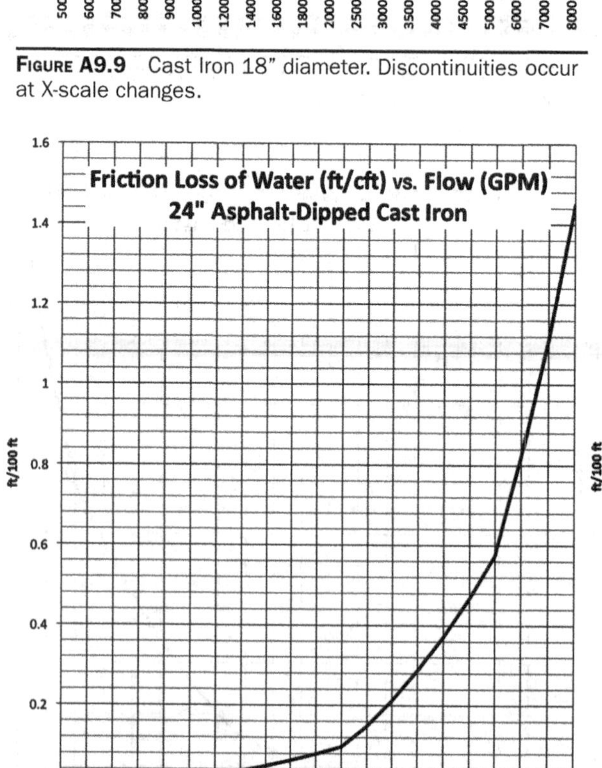

Figure A9.11 Cast Iron 24" diameter. Discontinuities occur at X-scale changes.

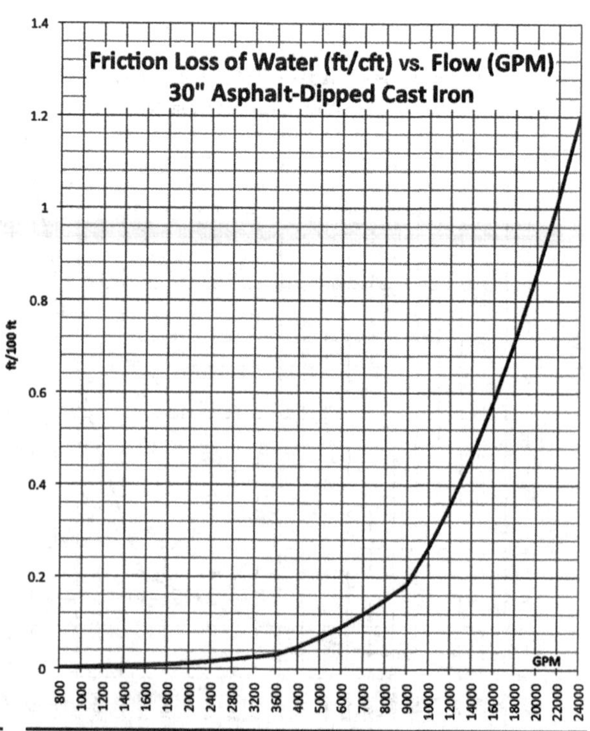

Figure A9.12 Cast Iron 30" diameter. Discontinuities occur at X-scale changes.

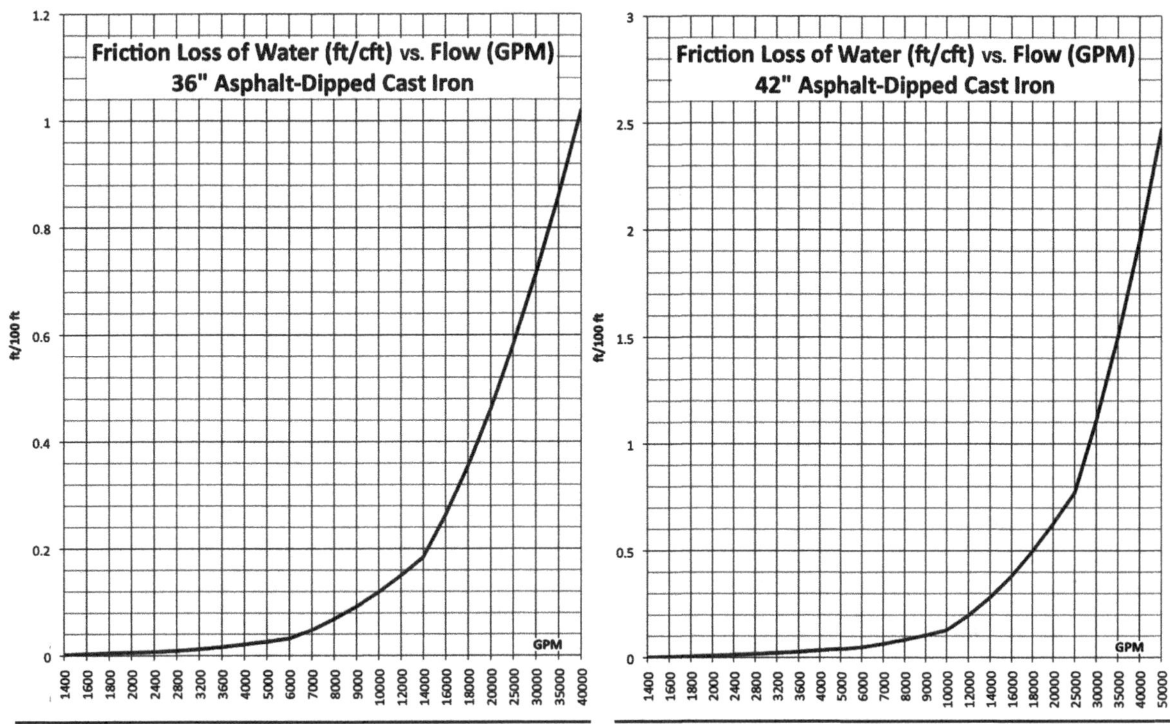

Figure A9.13 Cast Iron 36" diameter. Discontinuities occur at X-scale changes.

Figure A9.14 Cast Iron 42" diameter. Discontinuities occur at X-scale changes.

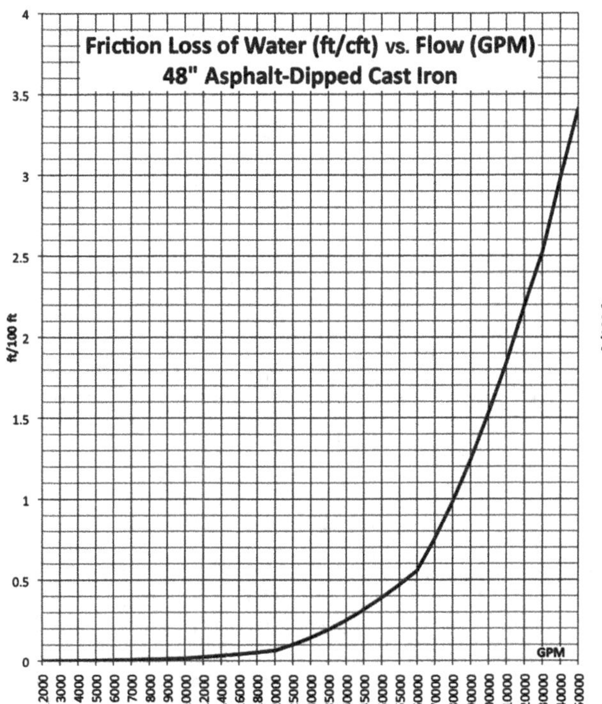

Figure A9.15 Cast Iron 48" diameter. Discontinuities occur at X-scale changes.

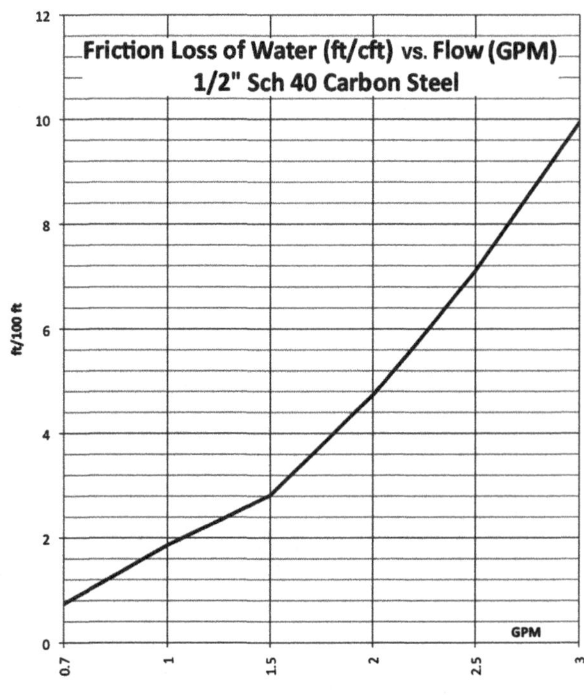

Figure A9.16 Sch 40 CS ½" diameter. Discontinuities occur at X-scale changes.

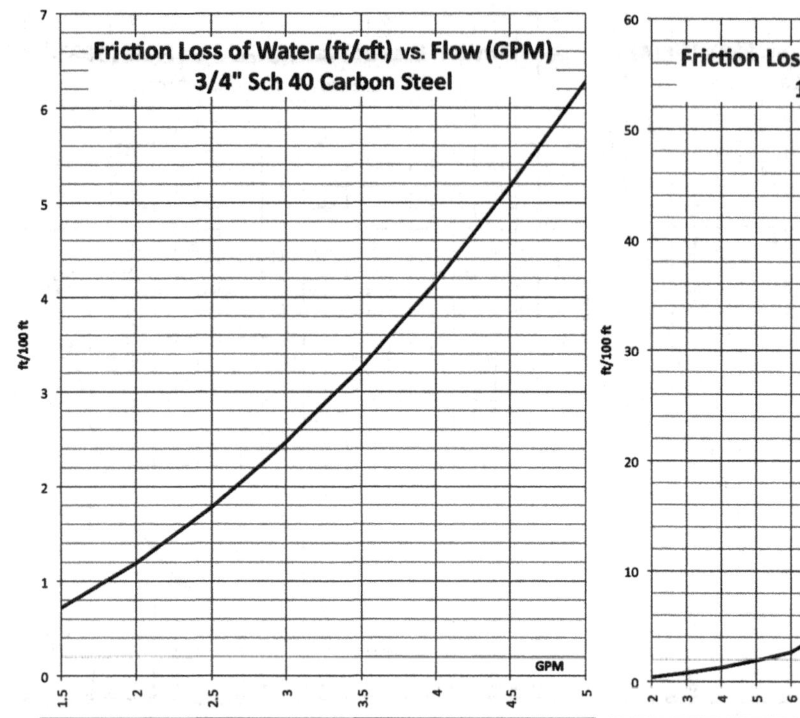

Figure A9.17 Sch 40 CS ¾" diameter. Discontinuities occur at X-scale changes.

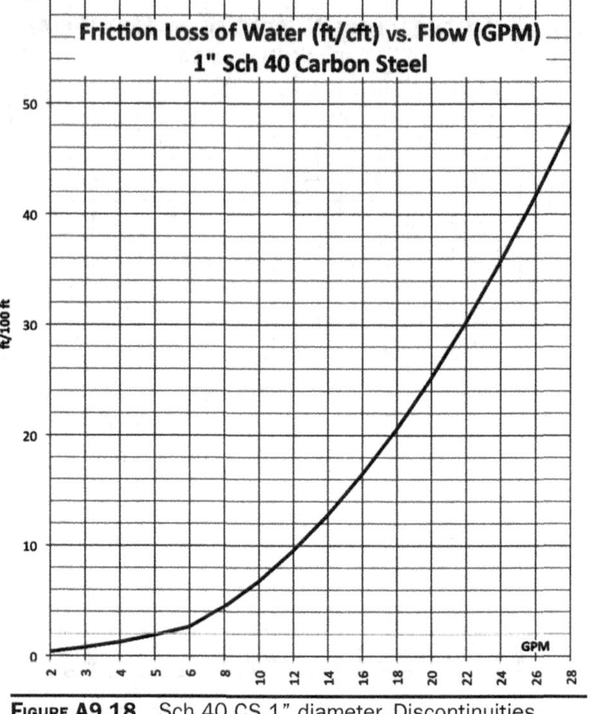

Figure A9.18 Sch 40 CS 1" diameter. Discontinuities occur at X-scale changes.

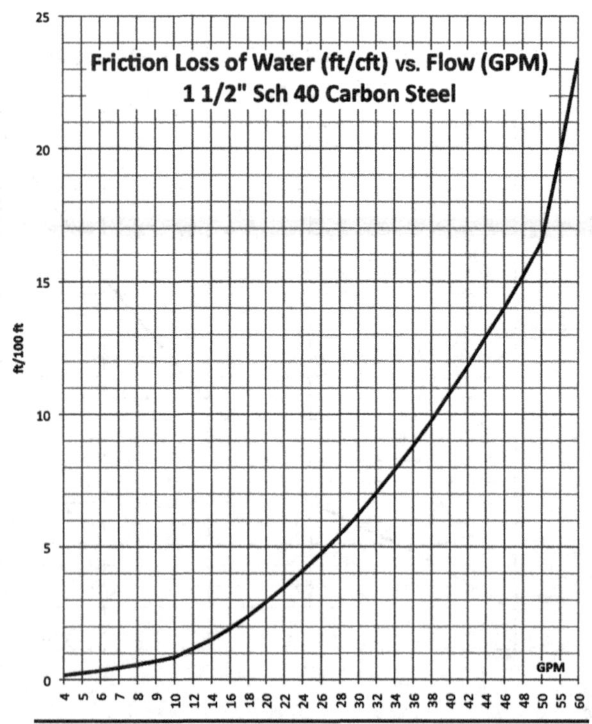

Figure A9.19 Sch 40 CS 1 ½" diameter. Discontinuities occur at X-scale changes.

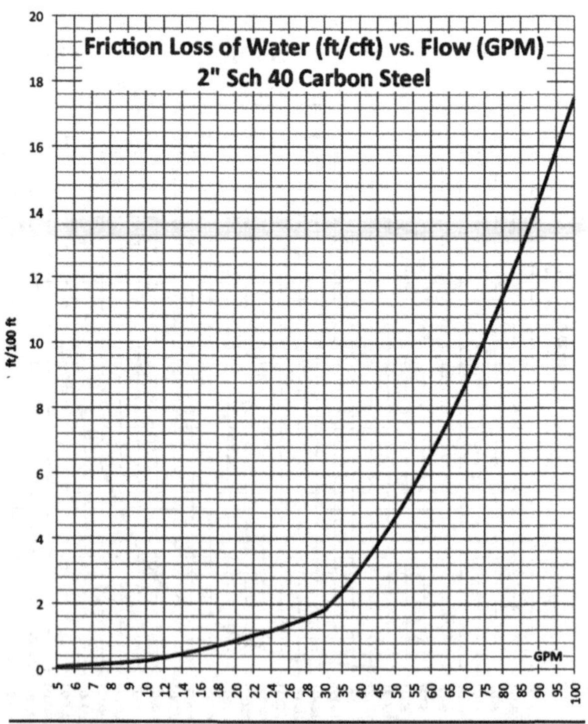

Figure A9.20 Sch 40 CS 2" diameter. Discontinuities occur at X-scale changes.

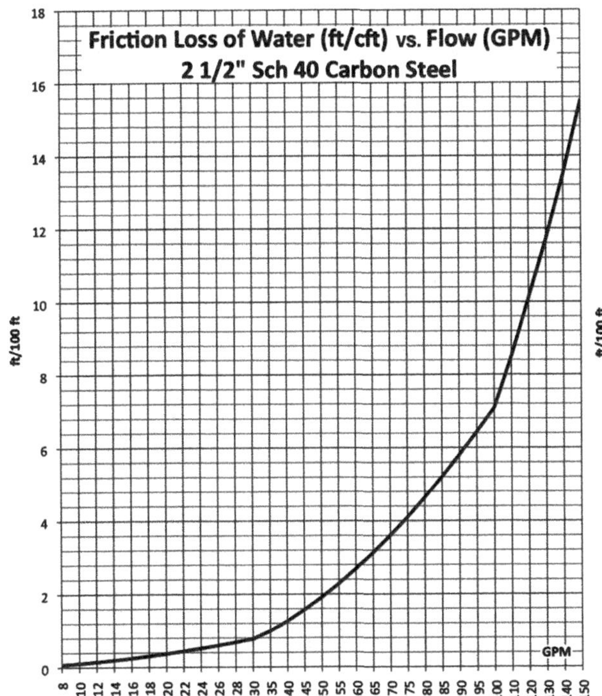

Figure A9.21 Sch 40 CS 2 ½" diameter. Discontinuities occur at X-scale changes.

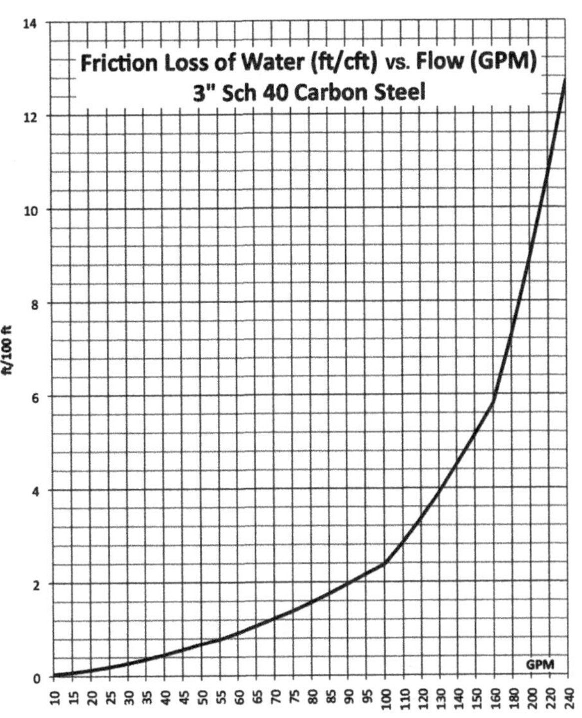

Figure A9.22 Sch 40 CS 3" diameter. Discontinuities occur at X-scale changes.

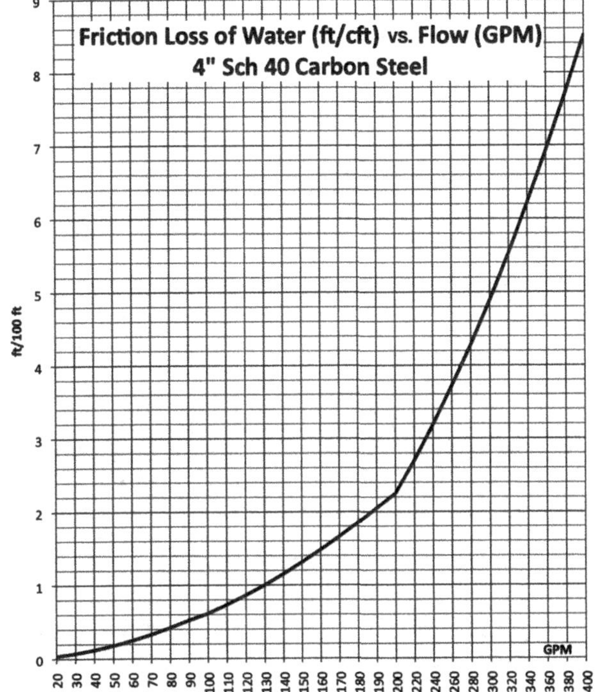

Figure A9.23 Sch 40 CS 4" diameter. Discontinuities occur at X-scale changes.

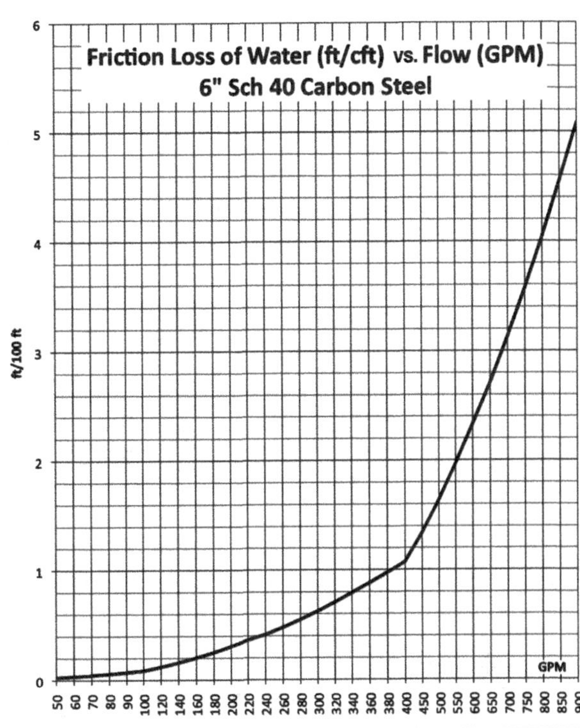

Figure A9.24 Sch 40 CS 6" diameter. Discontinuities occur at X-scale changes.

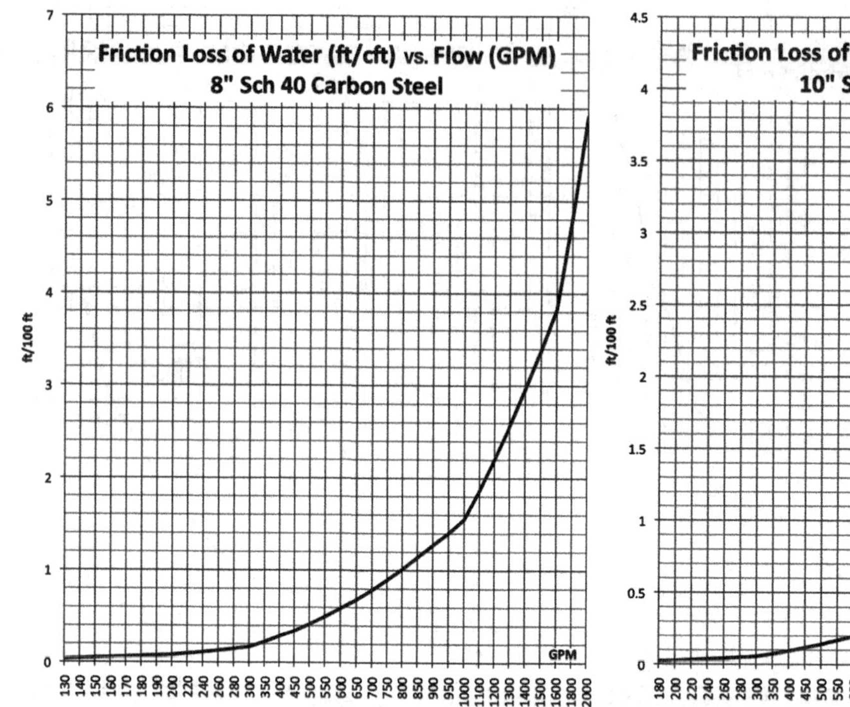

FIGURE A9.25 Sch 40 CS 8" diameter. Discontinuities occur at X-scale changes.

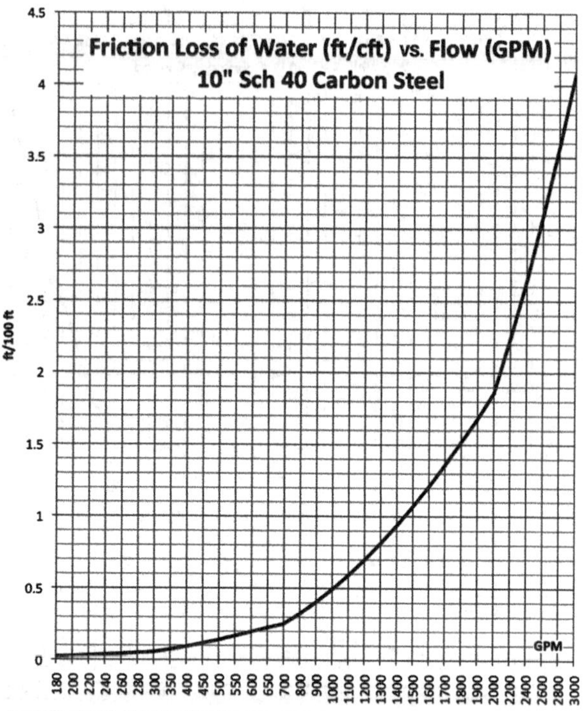

FIGURE A9.26 Sch 40 CS 10" diameter. Discontinuities occur at X-scale changes.

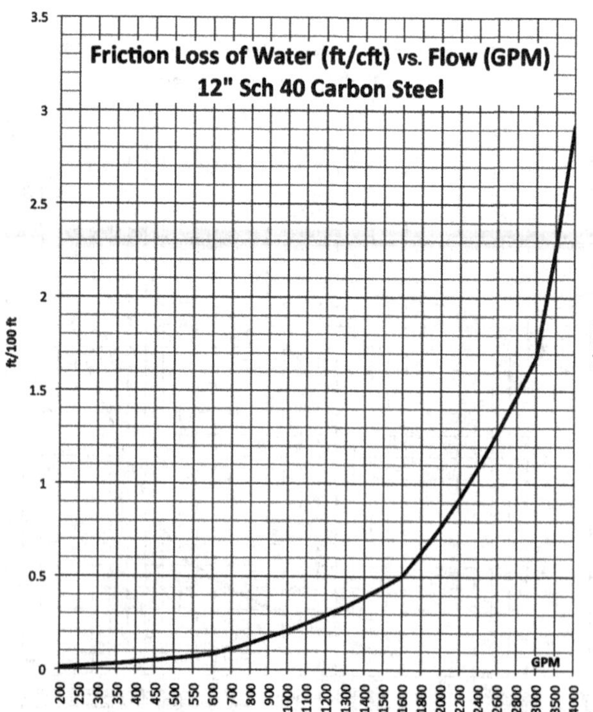

FIGURE A9.27 Sch 40 CS 12" diameter. Discontinuities occur at X-scale changes.

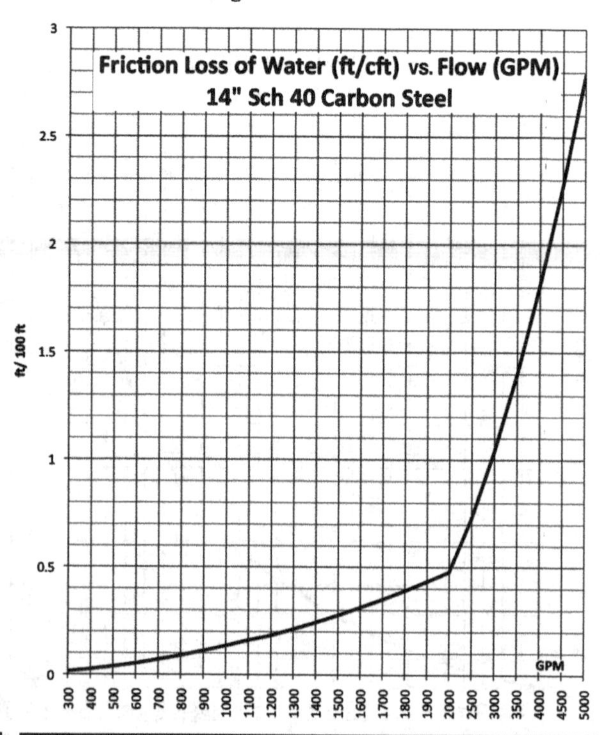

FIGURE A9.28 Sch 40 CS 14" diameter. Discontinuities occur at X-scale changes.

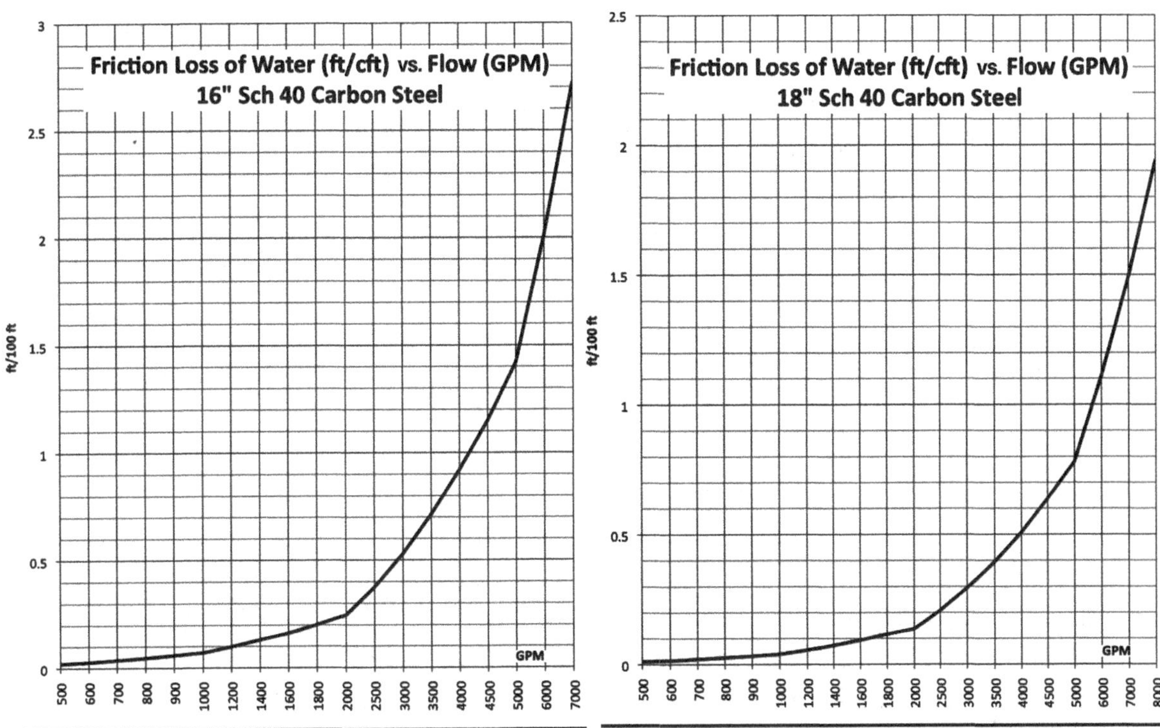

Figure A9.29 Sch 40 CS 16" diameter. Discontinuities occur at X-scale changes.

Figure A9.30 Sch 40 CS 18" diameter. Discontinuities occur at X-scale changes.

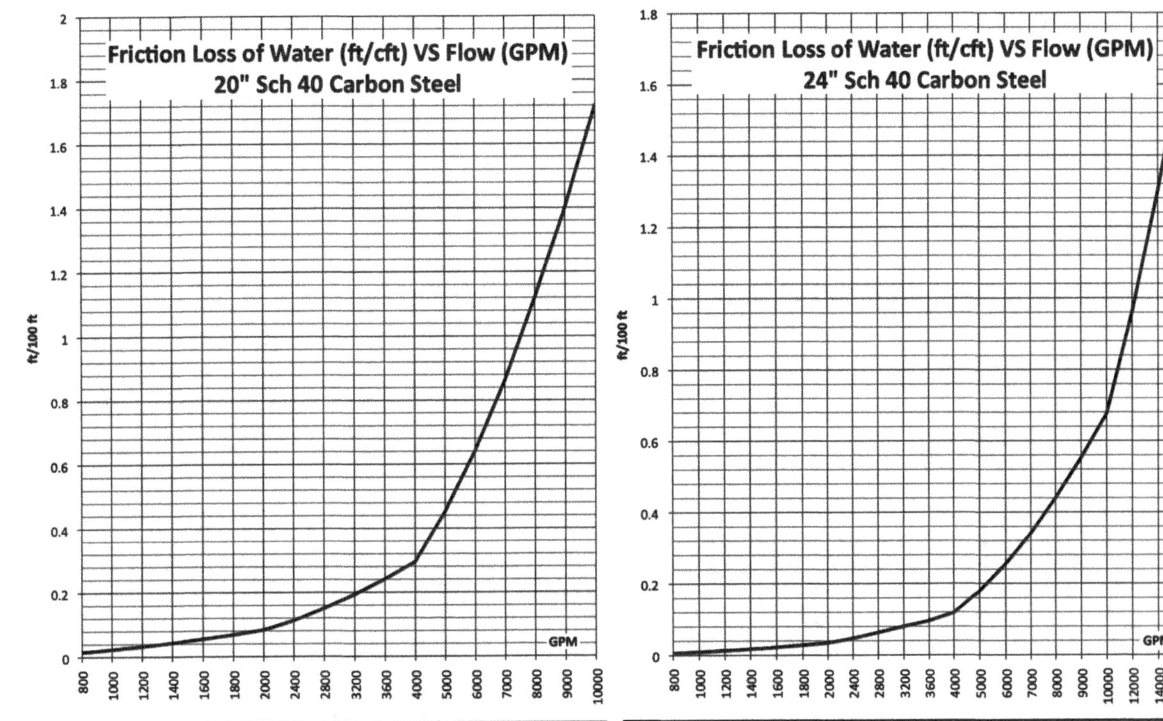

Figure A9.31 Sch 40 CS 20" diameter. Discontinuities occur at X-scale changes.

Figure A9.32 Sch 40 CS 24" diameter. Discontinuities occur at X-scale changes.

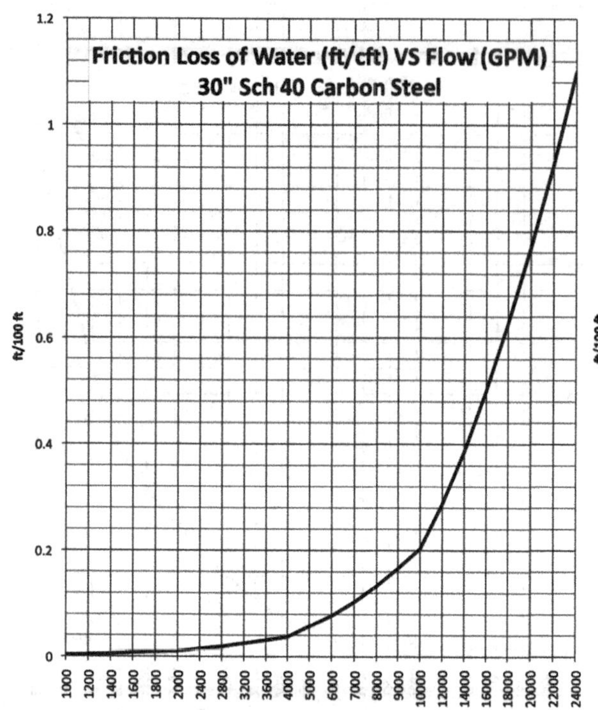

FIGURE A9.33 Sch 40 CS 30" diameter. Discontinuities occur at X-scale changes.

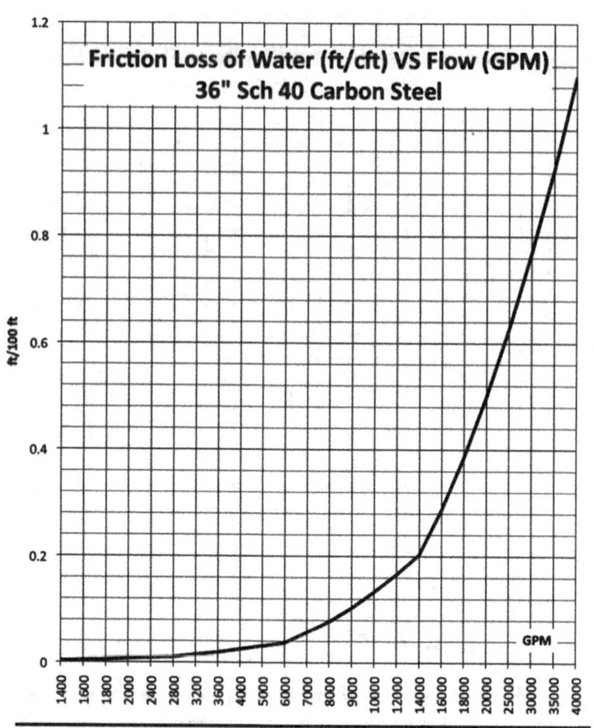

FIGURE A9.34 Sch 40 CS 36" diameter. Discontinuities occur at X-scale changes.

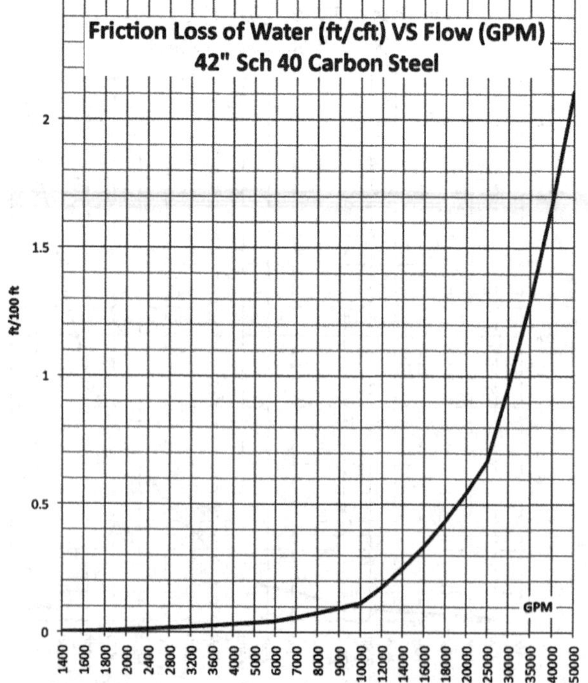

FIGURE A9.35 Sch 40 CS 42" diameter. Discontinuities occur at X-scale changes.

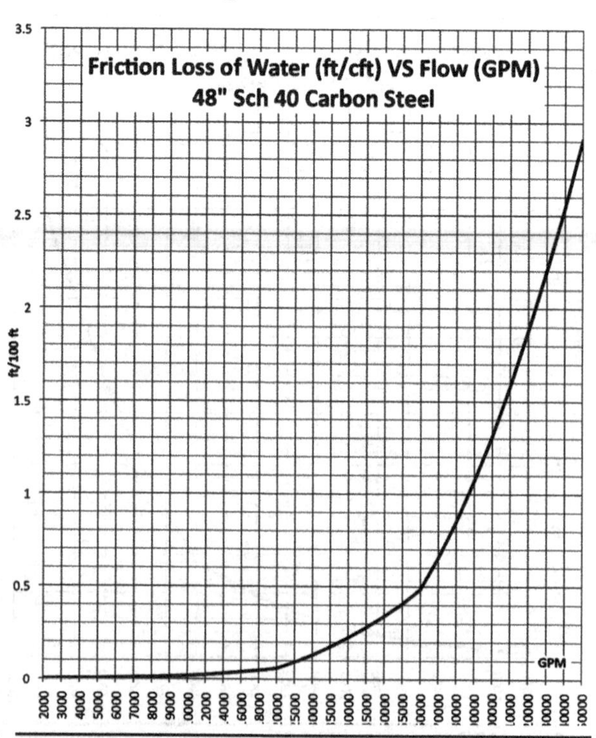

FIGURE A9.36 Sch 40 CS 48" diameter. Discontinuities occur at X-scale changes.

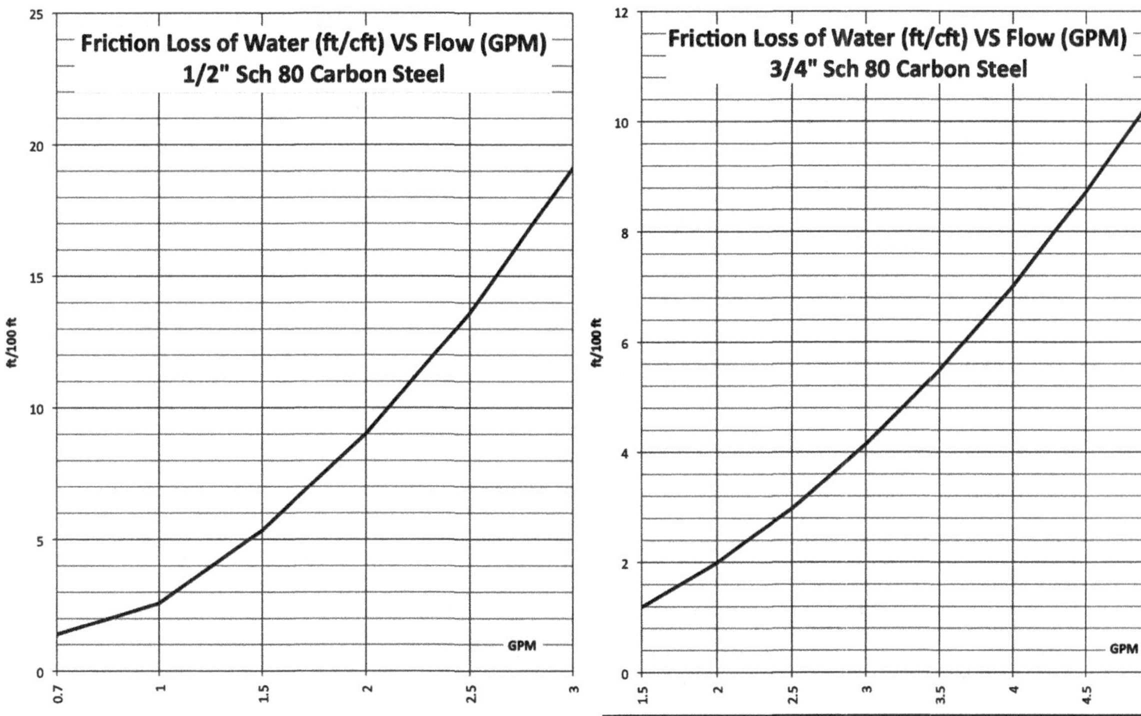

Figure A9.37 Sch 80 CS ½" diameter. Discontinuities occur at X-scale changes.

Figure A9.38 Sch 80 CS ¾" diameter. Discontinuities occur at X-scale changes.

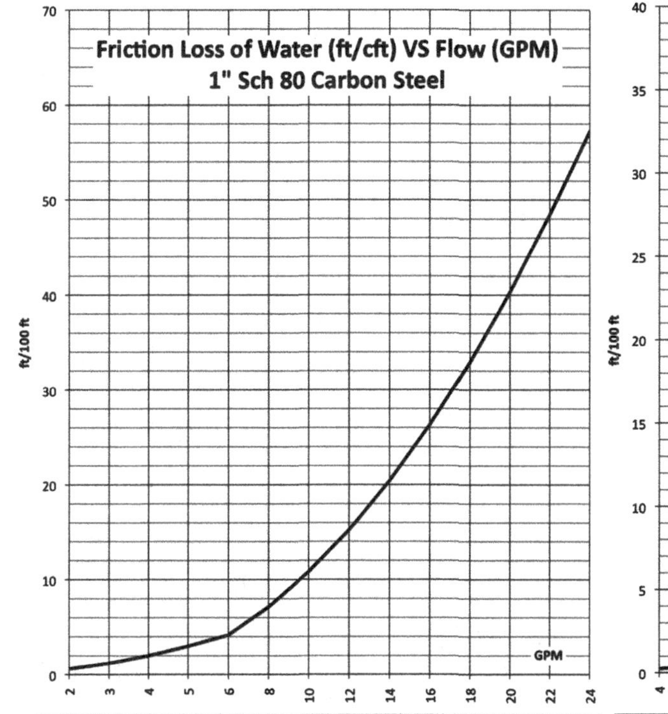

Figure A9.39 Sch 80 CS 1" diameter. Discontinuities occur at X-scale changes.

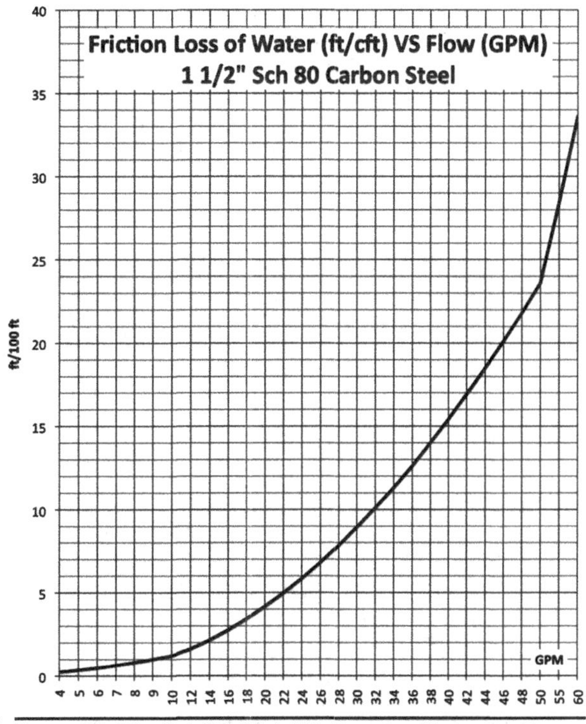

Figure A9.40 Sch 80 CS 1 ½" diameter. Discontinuities occur at X-scale changes.

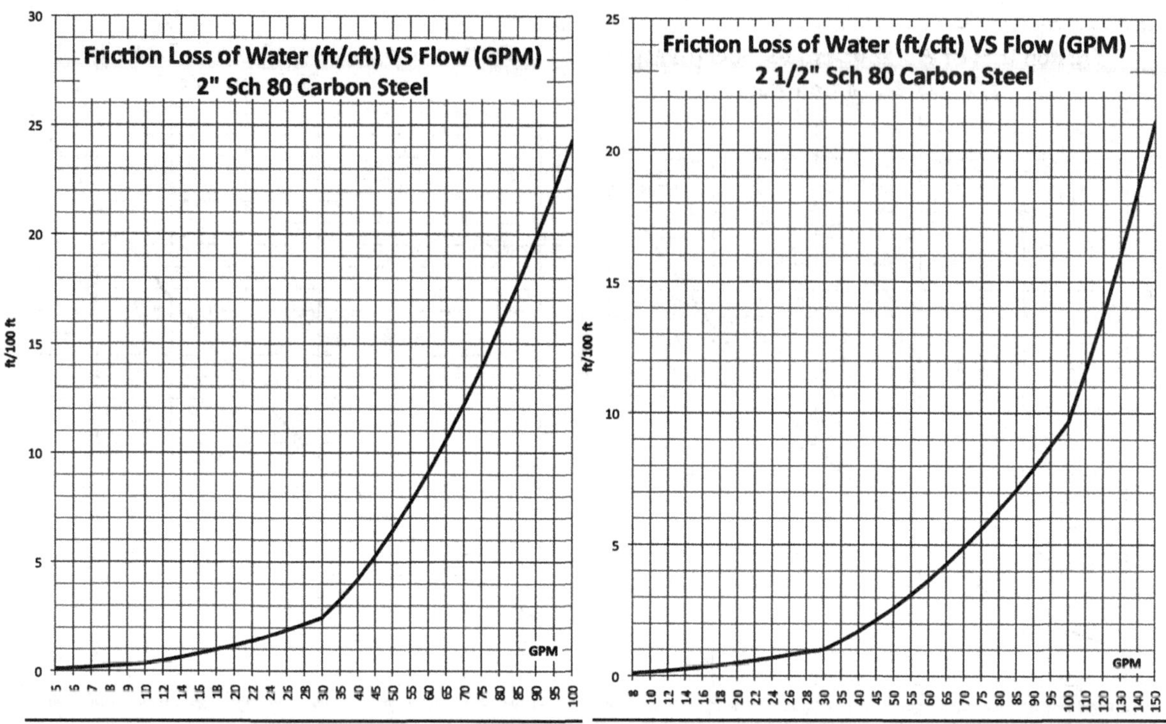

Figure A9.41 Sch 80 CS 2" diameter. Discontinuities occur at X-scale changes.

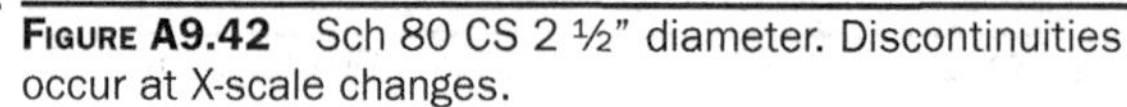

Figure A9.42 Sch 80 CS 2 ½" diameter. Discontinuities occur at X-scale changes.

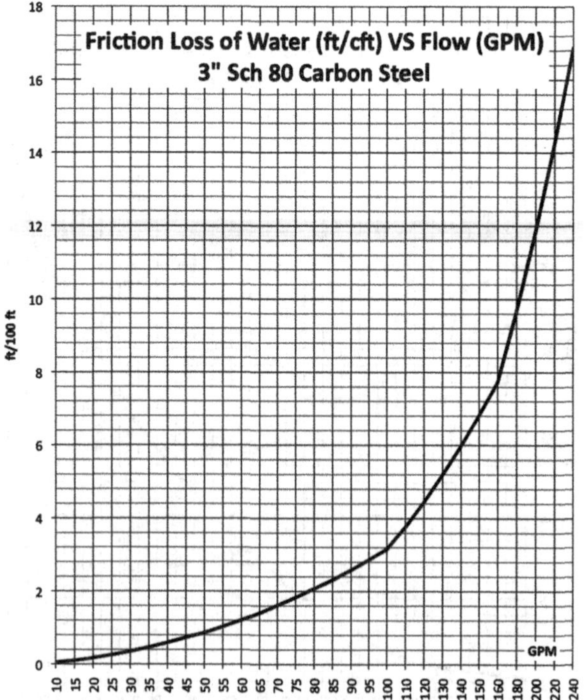

Figure A9.43 Sch 80 CS 3" diameter. Discontinuities occur at X-scale changes.

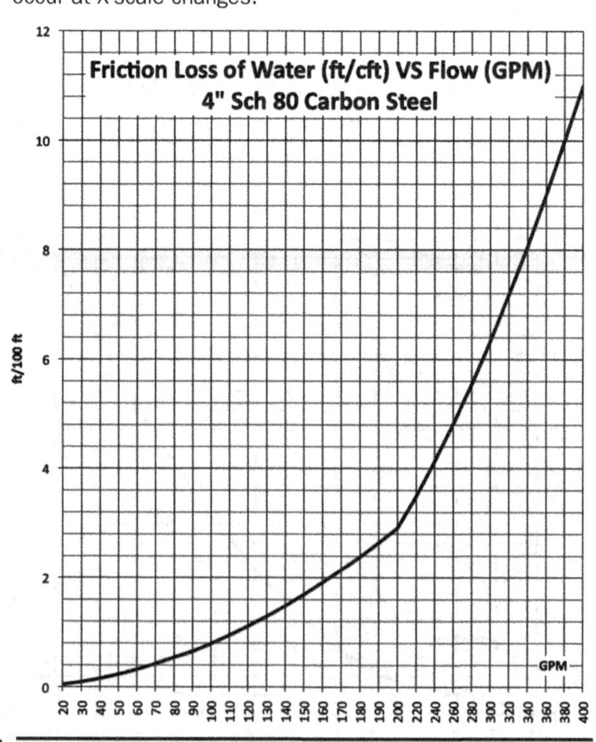

Figure A9.44 Sch 80 CS 4" diameter. Discontinuities occur at X-scale changes.

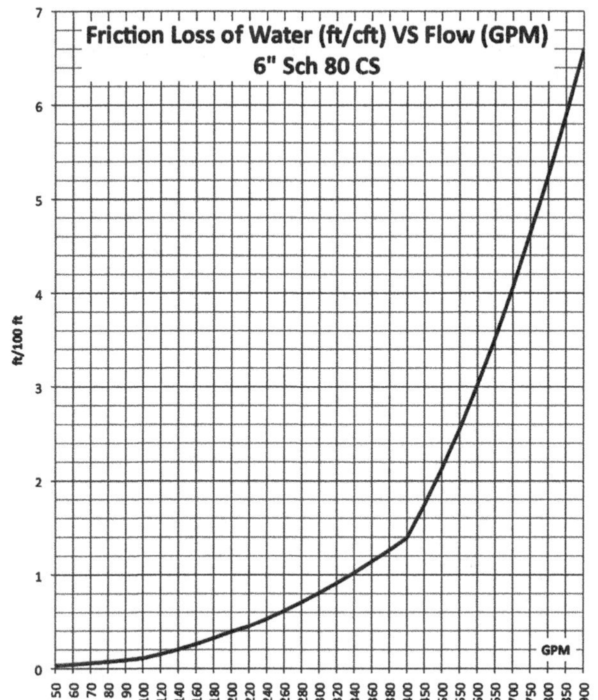

Figure A9.45 Sch 80 CS 6" diameter. Discontinuities occur at X-scale changes.

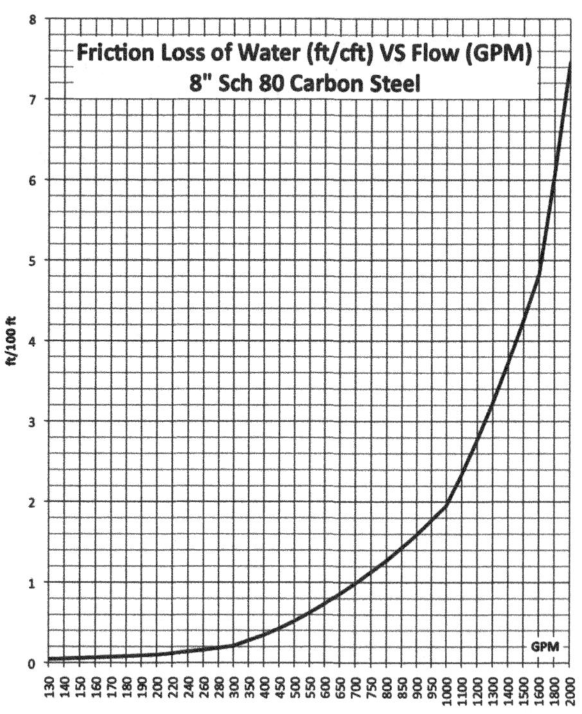

Figure A9.46 Sch 80 CS 8" diameter. Discontinuities occur at X-scale changes.

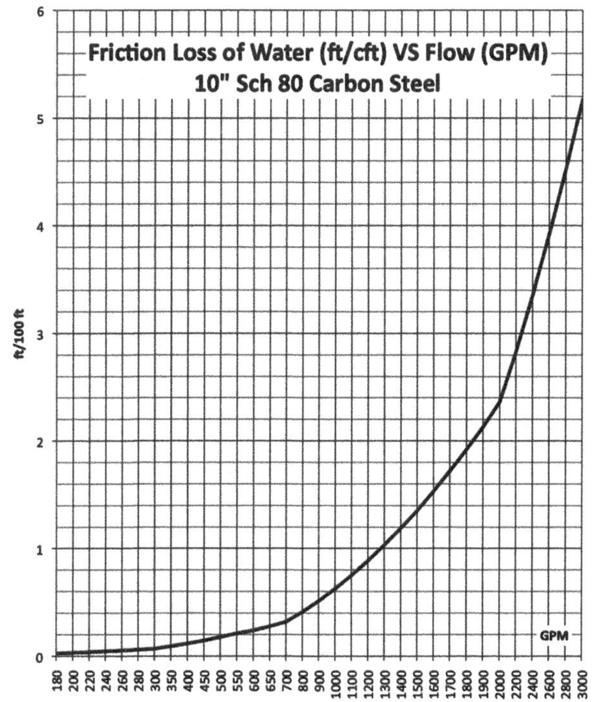

Figure A9.47 Sch 80 CS 10" diameter. Discontinuities occur at X-scale changes.

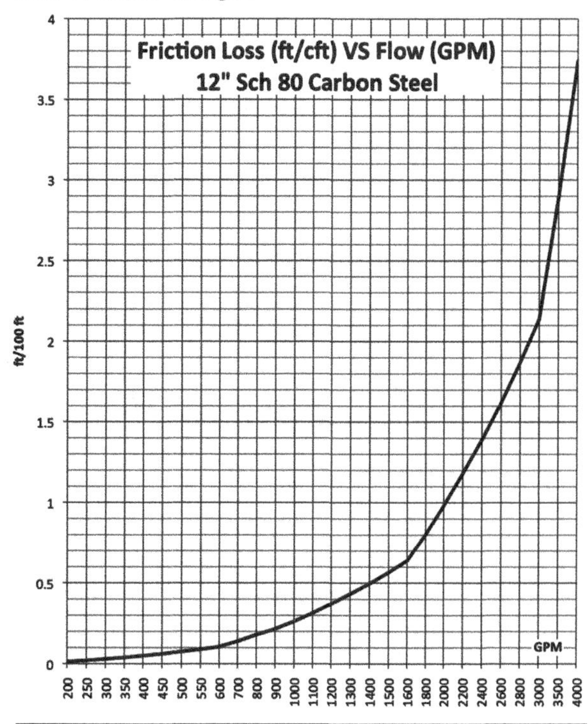

Figure A9.48 Sch 80 CS 12" diameter. Discontinuities occur at X-scale changes.

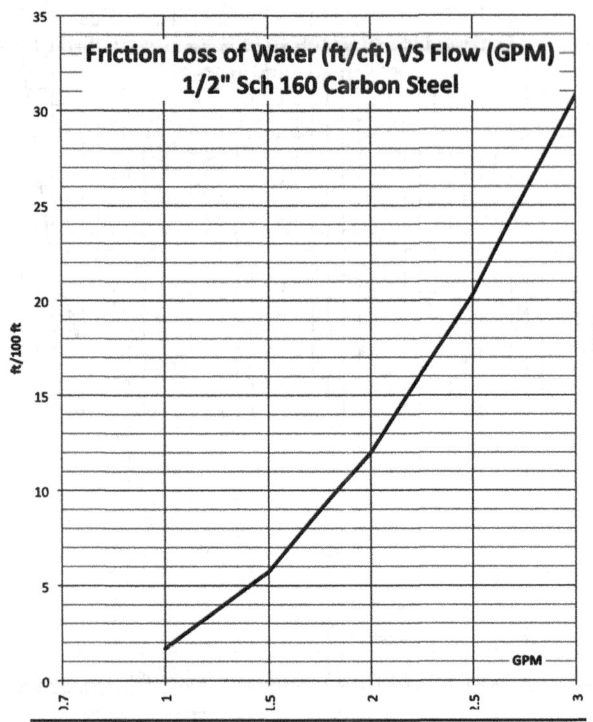

Figure A9.49 Sch 160 CS ½" diameter. Discontinuities occur at X-scale changes.

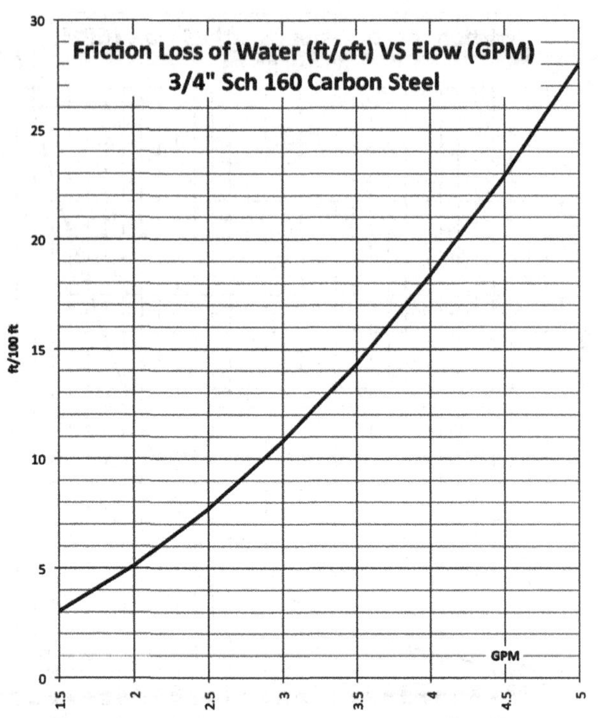

Figure A9.50 Sch 160 CS ¾" diameter. Discontinuities occur at X-scale changes.

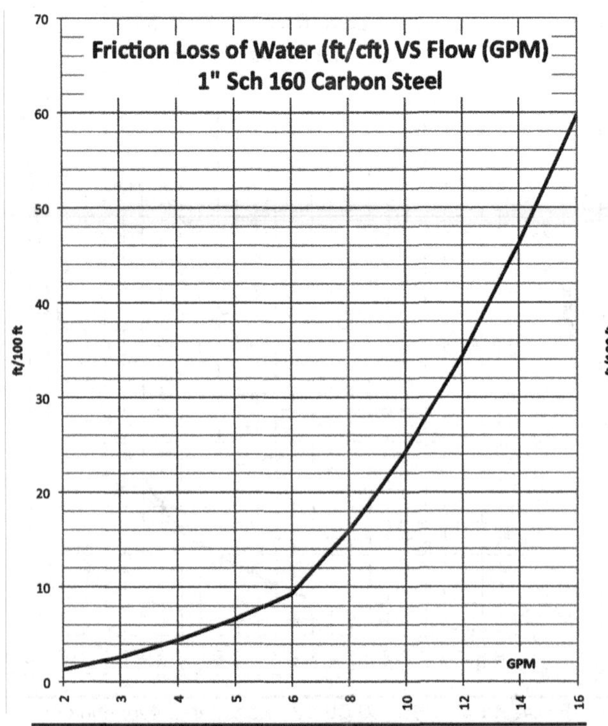

Figure A9.51 Sch 160 CS 1" diameter. Discontinuities occur at X-scale changes.

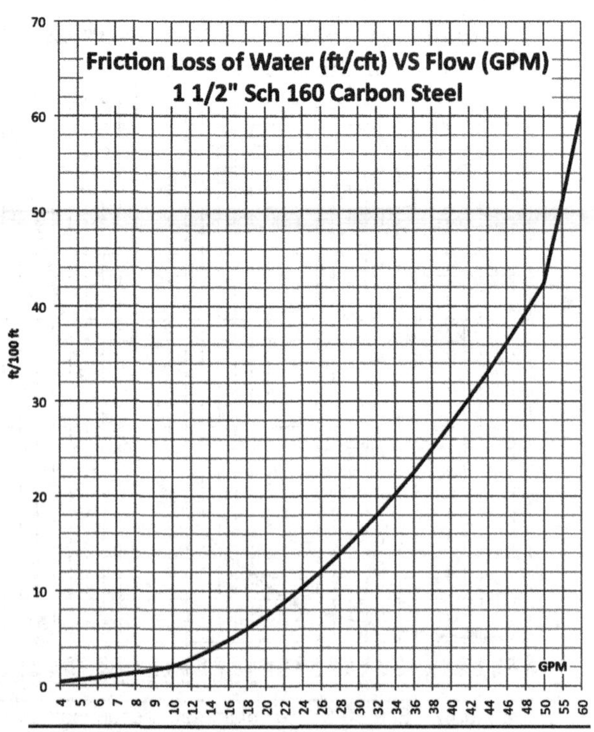

Figure A9.52 Sch 160 CS 1 ½" diameter. Discontinuities occur at X-scale changes.

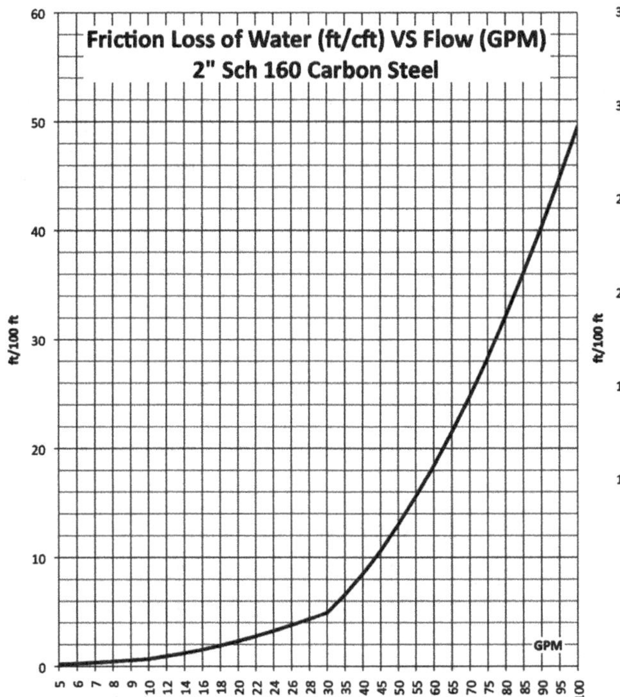

Figure A9.53 Sch 160 CS 2" diameter. Discontinuities occur at X-scale changes.

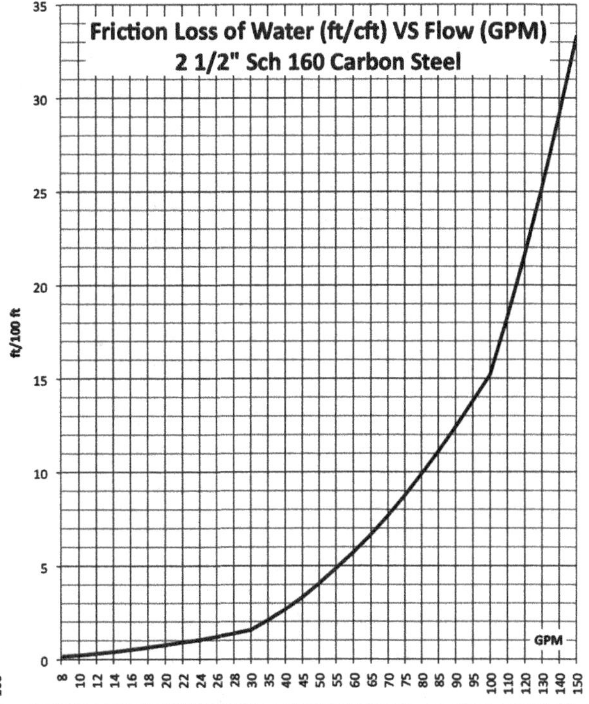

Figure A9.54 Sch 160 CS 2 ½" diameter. Discontinuities occur at X-scale changes.

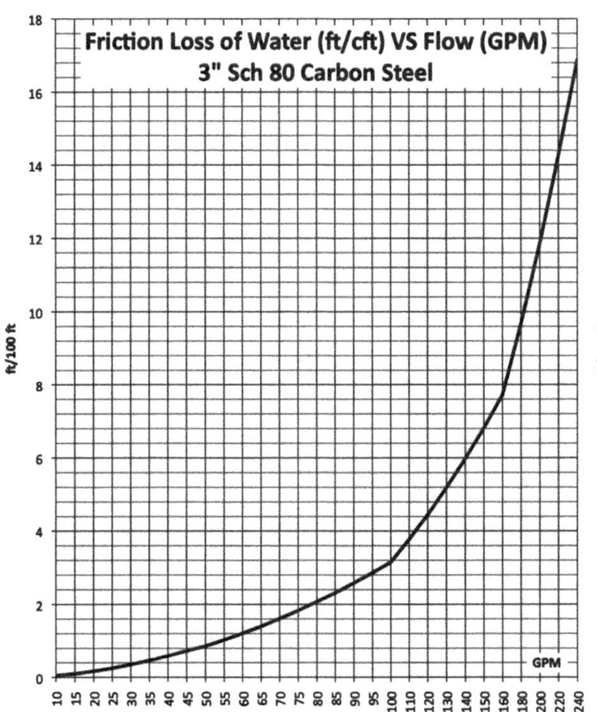

Figure A9.55 Sch 160 CS 3" diameter. Discontinuities occur at X-scale changes.

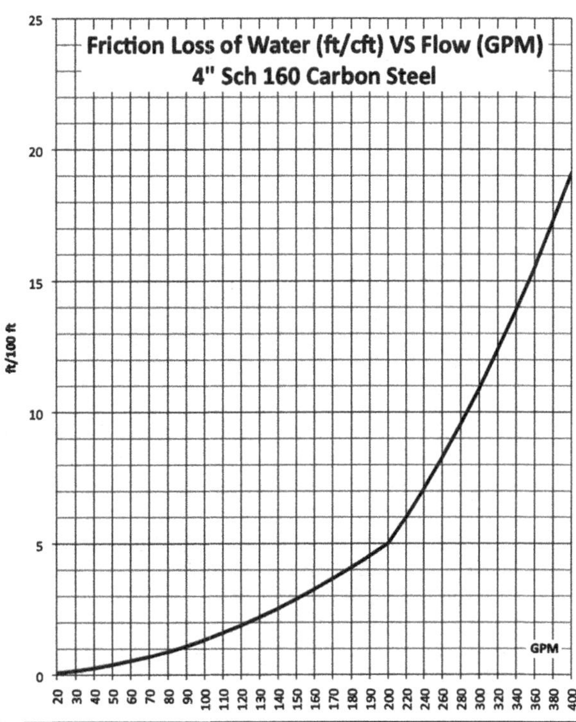

Figure A9.56 Sch 160 CS 4" diameter. Discontinuities occur at X-scale changes.

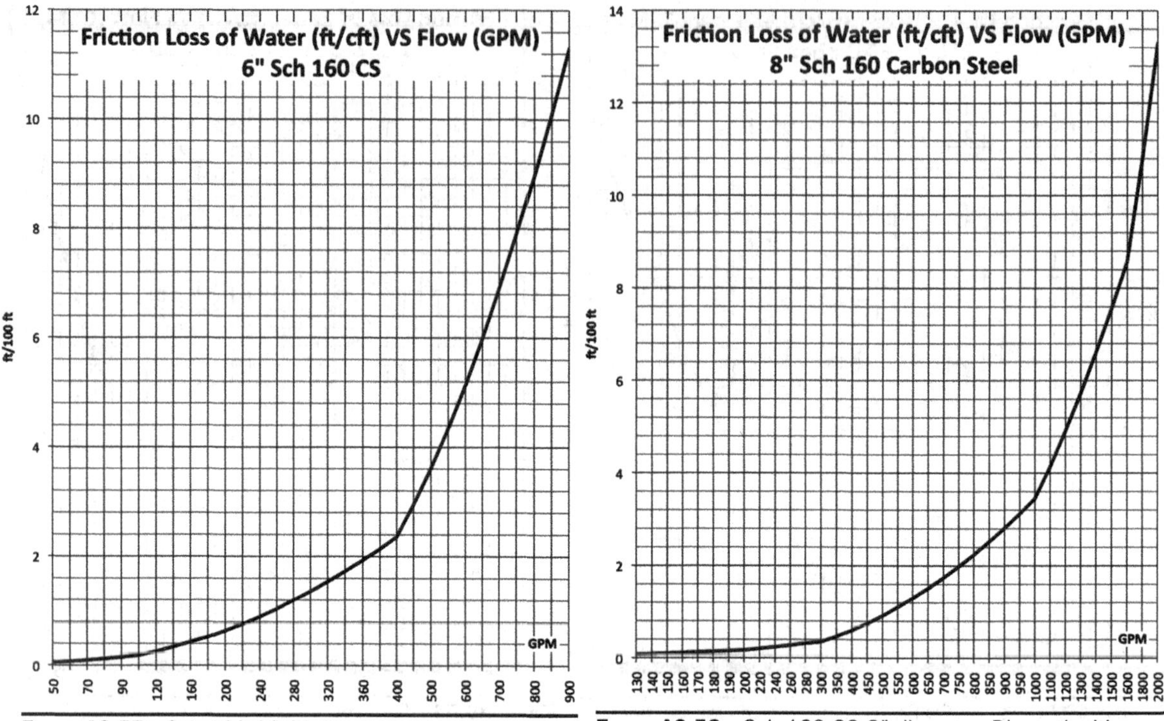

Figure A9.57 Sch 160 CS 6" diameter. Discontinuities occur at X-scale changes.

Figure A9.58 Sch 160 CS 8" diameter. Discontinuities occur at X-scale changes.

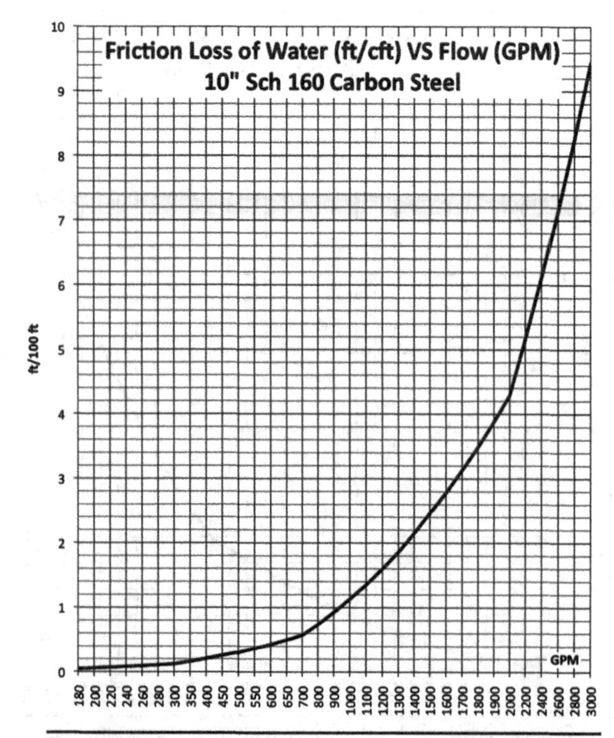

Figure A9.59 Sch 160 CS 10" diameter. Discontinuities occur at X-scale changes.

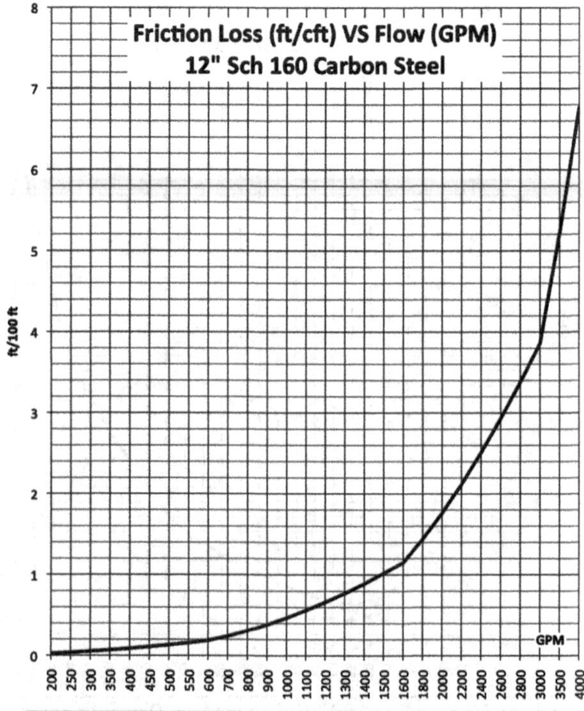

Figure A9.60 Sch 160 CS 12" diameter. Discontinuities occur at X-scale changes.

Index